MW00424816

Kinematic Design
of Machines
and Mechanisms

Kinematic Design
of Machines
and Mechanisms

Homer D. Eckhardt

McGraw-Hill

New York   San Francisco   Washington, D.C.   Auckland   Bogotá
Caracas   Lisbon   London   Madrid   Mexico City   Milan
Montreal   New Delhi   San Juan   Singapore
Sydney   Tokyo   Toronto

Library of Congress Cataloging-in-Publication Data

Eckhardt, Homer D.
    Kinematic design of machines and mechanisms / Homer D. Eckhardt.
        p.   cm.
    Includes index.
    ISBN 0-07-018953-6 (alk. paper)
    1. Machinery, Kinematics of.   2. Machine design.   I. Title.
TJ175.E32   1998
621.8'11--dc21   97-29215
                                                           CIP

McGraw-Hill

A Division of The **McGraw·Hill** Companies

    1 2 3 4 5 6 7 8 9 0   DOC / DOC   9 0 2 1 0 9 8 7

ISBN 0-07-018953-6

The sponsoring editor for this book was Harold B. Crawford, the editing supervisor was Frank Kotowski, Jr., and the production supervisor was Sherri Souffrance.

It was set in Century Schoolbook by Graphic World, Inc.

Printed and bound by R. R. Donnelly & Sons Company.

McGraw-Hill books are available at special quantity discounts to use as premiums and sales promotions, or for use in corporate training programs. For more information, please write to the Director of Special Sales, McGraw-Hill, 11 West 19th Street, New York, NY 10011. Or contact your local bookstore.

 This book is printed on recycled, acid-free paper containing a minimum of 50% recycled, de-inked fiber.

Contents

List of Procedures

Preface

The objective of this book is to provide practicing engineers and students with a practical understanding of the principles of kinematics, with an understanding of the connections between these principles and the behavior of actual machines, and with tools for the kinematic design of those machines.

The design of a machine or mechanism or any moving mechanical system always starts with a consideration of kinematics because kinematics is the study of the geometry of motion. That is, kinematics deals with (1) the functional relationships between the parts of a machine or other mechanical system, (2) how those parts are interconnected, and (3) how those parts move relative to each other. Only after choices have been made regarding those three factors can matters such as strengths, materials, fabrication techniques, and costs be seriously addressed. Failure to devote the proper attention to kinematics "up front" can, and often does, result in the design of a system with substandard or nonoptimum performance and/or with unsatisfactory reliability.

Fortunately, today, the ready availability of very powerful personal computers and the associated software allows kinematic synthesis and analysis, which were formerly laborious, to be performed quickly and cheaply. There is no longer an excuse for avoiding doing careful kinematic design up front. Because of the availability of these computer aids and the consequent incentives to apply kinematic principles in design, it is becoming *increasingly important* for the practicing engineer to have a good understanding of those kinematic principles.

Even before engineers can start to use a computer for synthesis or analysis of a machine, they must develop some initial concept of how the machine will operate. To assist engineers in generating such concepts, the first part of each of Chaps. 3 through 7 and all of Chap. 8 describe the operating capabilities of mechanisms which can be used as components of machines. Then step-by-step procedures (principally graphical) for synthesizing such mechanisms to provide the desired functions are given. If these initially synthesized configurations require further refinement, experience gained in this initial manual synthesis provides understanding and input for any further synthesis and analysis on a computer.

The graphical synthesis and analysis methods described require only the use of compass, ruler, protractor, and simple arithmetic. The CAD (computer-aided drafting or design) systems available today can also be very easily used to perform these graphical procedures and to obtain very accurate answers. Currently available spreadsheet software and equation-solving software can be used with the equations given in this book to obtain numerical values for the kinematic variables involved. Many of the equations have been left in a form in which the unknown variable has *not* been isolated on the left-hand side. When in this form:

1. the equations are closer to the derivation from the source phenomena involved and thus the significance of individual terms is more easily seen;

2. the equations are already in a form which many equation-solving software packages such as TKSolver®, MathCAD®, and others can evaluate iteratively; and

3. the equations are easily converted to a form suitable for programming into a spreadsheet by transposing the single term containing the unknown to the left-hand side and performing a simple operation such as dividing, taking a square root, taking an inverse tangent, etc. The equations can generally be evaluated in the order in which they are presented.

Very often, the answers which are provided by computerized aids are not unique and depend on details of the formulation of the problem as fed to the computer. If the computer-generated answer is not fully satisfactory to the engineer, a decision must be made as to what could and should be changed in that input formulation, how it should be changed, and what resulting change would be expected in the answer. Then new inputs must be fed to the computer and the process must be repeated, in the hope that the next answer will be more satisfactory. If the kinematic principles are not understood, this iteration process becomes one of wasteful trial and error, and valuable solutions can be missed. It is a primary aim of this book to provide help in more efficiently guiding such trial-and-error processes by providing engineers with the understanding which will allow them to visualize the connections between the relatively simple mathematics and the physical phenomena involved. Toward that aim, the later portions of each of Chaps. 3 through 7 and all of Chap. 9 describe analytical as well as graphical relationships between mechanism motion variables.

Several mathematical derivations are presented in this book. It is not necessary for readers to be familiar with, to read, or to understand these derivations in order to use the procedures and relationships presented, although derivations are usually helpful in providing understanding of the power and limitations of those procedures and relationships. The derivations are presented largely for readers who may wish to enlarge or build upon them in order to generate more elegant and/or more powerful procedures or relationships.

The only mathematics prerequisite assumed for use of the analytical portions of the book are high-school algebra, including complex numbers, high-school trigonometry, and a knowledge of differential calculus including the ability to differentiate e^x and the trigonometic functions. It is also assumed that the reader is aware that $F = ma$, and is able to add, subtract, and resolve two-dimensional vectors.

The first three chapters lay the foundations for the synthesis and analysis techniques and procedures which are presented in the remainder of the book.

Chapter 1 presents the definitions and basic concepts. Although the book deals essentially only with planar kinematics, Secs. 1.5 through 1.9 discuss three-dimensional or spatial phenomena. The remainder of the book does not depend upon the material in Secs. 1.6 through 1.9, so they could be skipped or lightly skimmed if time demands. However, because real machines are built and operated in three-dimensional space, it is important that practicing engineers have an appreciation for the phenomena covered in those sections.

Chapter 2 presents methods for analyzing the motions of rigid bodies in planar motion. The chapter covers displacements, velocities, and accelerations of isolated rigid bodies, and discusses the significant relationships between the mathematics and the physical phenomena. It provides analysis techniques which are used in subsequent chapters, where rigid bodies are connected together to form machines.

The crank-slider mechanism is a very useful mechanism, and Chap. 3 presents synthesis and analysis methods for use in its design. Because the geometry and motion of this mechanism are relatively simple and easily visualized, analysis of this mechanism is used as a basis from which the synthesis and analysis of the mechanisms in the subsequent chapters are treated as perturbations and extensions.

Chapters 4 through 7 describe the salient features and capabilities of increasingly complex mechanisms and present procedures for their synthesis. These chapters also briefly illustrate the adaptation of the previously described analysis procedures for use on these more complex mechanisms.

Chapters 8 and 9 present principles and procedures which are useful in preparation of concepts for machines which will be subjects for computer-assisted synthesis and analysis.

Chapters 10 and 11 give the principles of cam systems, gear systems, and timing-belt systems and give procedures for synthesizing such systems.

Although Chaps. 12 and 13 involve forces and inertial reactions and therefore are subjects in kinetics, the intimate involvement of geometry and thus of kinematics in the phenomena covered makes them important subjects for a kinematics text.

Chapter 12 describes the relationships between the static forces which occur in a mechanism, and presents procedures for static balancing of such mechanisms. Examples of such balancing are given. These examples involve the use of balance springs as well as the use of balance weights.

Chapter 13 extends the discussion of Chap. 12 to the relationships between the dynamic forces which occur in a mechanism, and presents procedures for dynamic balancing of mechanisms. Examples of such balancing are given.

Homer D. Eckhardt

How to Use This Book

It is intended that the step-by-step procedures presented in this book be usable without the need to refer to other portions of the text. However, to the extent that time permits, readers should become aware of the existence and nature of the background material in the remainder of this book and in other books on this subject.

Therefore, read this list and then:

1. Read the preface. (It takes only a few minutes!)

2. Read the table of contents and note the titles of sections.

3. Read the Introduction section of each chapter.

4. From the table of contents, select a section or sections of particular interest and read the introductory paragraphs of those sections. Also skim those sections for Procedures of interest (see List of Procedures also).

5. When you find a particular Procedure which is of immediate interest, read (or at least skim) the section containing that procedure, including Introduction and any Comments and suggestions.

6. Skim the index for terms which are of interest and skim the text where those terms appear. The index should be used together with the table of contents because information appears in different forms in these two places.

7. Naturally all authors want readers to read, understand, and treasure every word in their books. This particular author realizes that practicing engineers are often pressed for time. Nonetheless, although I, too, have often been pressed for time, I have found the background material in this book to be of great use over the course of many years.

Acknowledgments

I wish to express my appreciation to my many colleagues at Honeywell, Link Aviation, the Massachusetts Institute of Technology, RCA, Draper Loom, Polaroid, the Worcester Polytechnic Institute, and Tufts University for all that they have taught me over the years, and to Professor Robert Norton of the Worcester Polytechnic Institute for inspiring me to write this book. I would especially like to thank Professor Ashok Midha of the University of Missouri at Rolla and Professor Steven Kramer of the University of Toledo who kindly agreed to review the manuscript and bring any egregious errors to my attention.

I am deeply grateful to my children Gretchen, Julie, Jason, Kris, and Lili for all they have taught me about "life, death, and related subjects," and I am particularly grateful to my wife Beverly, who, in addition to teaching me so much, has shown monumental patience with my preoccupation during the preparation of this book.

Despite all the help from the above-mentioned people, I, through my own obstinate efforts, may have included some errors in this book. I apologize to readers for any such errors. I welcome readers' comments and suggestions, which may be sent to me at

27 Laurel Drive
Lincoln, Massachusetts
01773

Chapter

1

Basic Concepts and Definitions

1.1 Introduction

This chapter defines the concepts on which this discussion of the kinematic design of machines will be based. Consequently, the chapter starts by defining the terms *kinematics, kinematic design,* and *machine.* These definitions introduce the concept of rigid bodies (links) and connections (joints). The remainder of the chapter discusses how the positions and displacements of these bodies may be described and how the motions of these bodies are affected by joining them together.

1.2 Kinematics and Kinematic Design

Mechanics is the branch of science that deals with motions, time, and forces. As indicated in Fig. 1.1, mechanics may be considered to consist of two main branches: statics and dynamics. *Statics,* as would be expected from its name, deals with stationary systems, so it is concerned only with forces and the geometry of systems. Although it can involve static deflections, it is not concerned with motion or time. *Dynamics,* on the other hand, deals with systems that are in motion. Dynamics, in turn, can be subdivided into the disciplines of kinematics and kinetics.

Figure 1.1 indicates that *kinematics* consists of the study of the interaction between the geometry of a system and the motions of that system. That is, it is concerned with the magnitudes and directions of the displacements, velocities, and accelerations (and possibly higher motion derivatives also) of the parts of the system. It is also concerned with how these motion parameters change that geometry and with how these parameters are dependent on the geometry. The forces and torques that are required when the motions of a system are prescribed

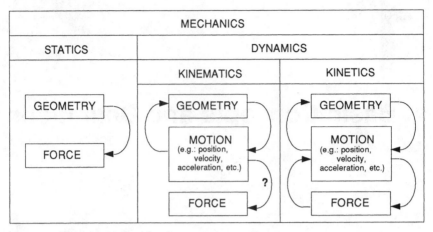

Figure 1.1 The science of mechanics.

can be computed from knowledge of the displacements, velocities, and accelerations in those prescribed motions and by treating inertial reaction forces (d'Alembert forces) and torques as static forces and torques. Such kinematic force and torque computations, then, involve both kinematics and statics and are often referred to as *kinetostatic analyses*. Such a kinetostatic analysis is discussed in Sec. 13.4.

Kinematics, however, does not consider how those forces might interact with the motions of the system. The distinction between kinematics and kinetics is that *kinetics* is concerned with how forces interact with the motions of the system.

One of the first phases in the design of a machine or mechanism is the choosing of the manner in which the parts of the machine must move in order to perform the functions for which the machine is being designed. Decisions must be made as to where the parts must be placed, how far and in what directions the parts must move, which parts must be connected to which other parts and how they must be connected, and what the critical dimensions of the parts must be. This phase will result in drawings or sketches of the general layout of the machine and will indicate how it will operate. This phase obviously involves the interactions between geometry and motions, so it and its resulting drawings will be referred to as the *kinematic design* of the machine. Although this effort may be performed simultaneously with other phases of the design, it is distinct from the design activities in which materials, strengths, responses to forces, wear, power, noise, vibration, costs, manufacturability, and so on are considered.

Although kinematic design is often taken for granted, great care should be exercised in its execution because it has a very powerful effect on all other aspects of the design process! The kinematics of a machine are its very soul.

1.3   Machines and Mechanisms

Each of us is familiar with what he or she considers to be machines, but there are as many published definitions of the term *machine* as there are authors writing about machines. The definition of the term *machine* that will be most useful in subsequent discussions in this book is a paraphrasing of one of the many such definitions given by Rouleaux in his pioneering book on kinematics (*Kinematics of Machinery: Outline of a Theory of Machines,* Dover Publications, New York, 1963, p. 35; originally published by Macmillan and Company in 1876):

> A machine is a combination of resistant bodies, so interconnected that by applying force or motion to one or more of those bodies, some of those bodies are caused to perform desired work accompanied by desired motions.

An example of a simple machine that illustrates the preceding definition is a pair of scissors (Fig. 1.2). The two movable parts of the scissors are connected to each other by a pivot in such a manner that by applying force or motion to the finger holes, the blades are caused to move relative to each other in such a manner as to do the desired cutting. Each of the two movable parts is a "resistant body" in the sense that it resists deformation sufficiently to allow it to move and work as desired when forces are applied to it.

An additional requirement is often imposed in the definition of a machine. This requirement is that one of the resistant bodies be rigidly fixed to a reference system known as *ground,* which for our purposes can be the surface on which the machine is mounted. A machine that is slightly more complicated than that in Fig. 1.2 and that satisfies this additional requirement and looks more like what we ordinarily think of as a machine is shown in Fig. 1.3. There it will be seen that the blade *B* is part of a resistant body that is attached to a stationary mounting surface or ground. Because that body does not move relative to ground, it will be considered to be part of the ground for purposes of our kinematic analyses. It also will be seen that by applying a force or motion *M* at the end of the body *P,* the fact that bodies, *P, L, B,* and *C* are suitably connected to each other (by pivots) causes the movable

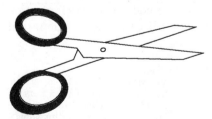

Figure 1.2   A very simple machine.

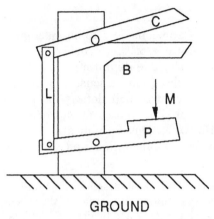

GROUND

Figure 1.3 Another simple machine (four-bar linkage).

body (blade) C to move relative to the fixed body (blade) B in such a manner as to perform the desired cutting.

In order to provide the desired motions in the machine of Fig. 1.3, it was necessary to connect or join the bodies like links in a chain. (In this case the chain is a closed loop.) Therefore, in the kinematic analysis and synthesis of machines, the bodies are referred to as *links,* and the connections are referred to as *joints.*

Note that the machine in Fig. 1.3 consists of four resistant bodies and of the connections between them. Even though one of the four bodies is fixed and does not move, in later chapters such a machine or mechanism will be referred to as a *four-bar linkage.*

Engineering literature contains many different definitions for the terms *machine* and *mechanism.* Although some of these definitions make distinctions between the two terms, this book will consider the preceding definition to apply to both terms, and they will be used interchangeably.

1.4   Properties of Rigid Bodies

Because the resistant bodies that are used in most machines are intended to be very stiff, they will be considered to be rigid bodies in these discussions. By *rigid* we mean that the body resists deformation so strongly that *no* deformation is possible.[1] It can be seen, then, that

---

[1]Although it is impossible to attain true rigidity in actual practice, for most machines the kinematic analysis can be performed using the assumption of rigidity. In those cases in which the deformations of the parts of the machine are important, the discipline known as *kinetoelastodynamics* can be used. However, discussion of this subject is beyond the scope of this book.

if no deformation of a rigid body is possible, a rigid body is a body in which the distance between any two points that are attached to that body remains constant, regardless of how the body is moved. In Fig. 1.4, for example, if points A and B are points that are attached to the rigid body L, the length of line segment AB will be constant, regardless of the motion of the body. (If that length were to change, stretching or compression of a portion of the body would have to take place, and the body would not be rigid.) Similarly, the lengths of line segments BC and CA will be constant. Because these three line segments completely define the triangle ABC, it may be seen that the angles between the lines AB and BC, between BC and CA, and between CA and AB will remain constant regardless of the motion of the body. We may then generalize these observations to establish two very useful properties of a rigid body:

> The distance between any two points that are fixed to a rigid body remains constant regardless of the motion of that body.
>
> The angle between any two intersecting lines that are attached to a rigid body remains constant regardless of the motion of that body.

1.5   Rigid Body Position and Degrees of Freedom

If discussion of the position of a body is to be quantitative, that position must be measurable relative to some reference frame. For example, we might say that a book (rigid body) is on the right front seat in a particular automobile. Such a statement implies that the reference frame with respect to which the book's position is measured is attached to the automobile frame. Use of the term *right front* implies that some two-axis coordinate system is being used. That is, right versus left is measured in one coordinate axis direction, and front versus back is measured in another coordinate axis direction

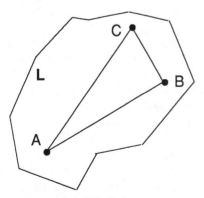

Figure 1.4 A rigid body.

that is perpendicular to the first. If the automobile happened to be in Midtown, U.S.A., we also might say that the book was at the corner of First Street and Avenue E in Midtown, U.S.A. In this description, the reference frame is attached to the U.S.A. or to Midtown, and the coordinate system is some system associated with the layout of the streets in that town. Note that the book is not moving relative to the automobile, but it will be moving relative to the reference frame that is attached to Midtown if the automobile is moving. Thus the measurement of position and motion will depend on the choice of reference frame.

Both the preceding reference frames have two attributes:

1. A coordinate system in terms of which the position of the object of interest is measured.

2. A convenient body to which that coordinate system is attached.

Although either of the preceding descriptions of the book's position might be useful in telling someone how to find the book, it is readily seen that similar, crude descriptions of the position of a machine part would be of little use in describing the operation of the machine. Not only would such descriptions indicate nothing of the *orientation* of the part (i.e., is it lying flat? standing on end? facing east? etc.), but they also would be too crude in terms of units of measurement. Obviously, a more rigorous definition of the reference frame will be necessary, and measurements relative to that frame in terms of lengths and/or angles must be used.

One very convenient coordinate system for our use is a three-axis orthogonal or Cartesian system. The body or surface to which the origin and axes of the coordinate system can be attached can be the floor or wall to which the machine is to be attached, the frame of the vehicle that contains the machine, the surface of the earth, or any other reference body from which it is convenient to make measurements. The position of the body whose position we wish to measure will then be measured or described in terms of distances and/or angles from the origin and/or axes of that coordinate system, and that coordinate system will be referred as the *reference frame*.

We will use three types of reference frames in subsequent discussions. The first type will be called an *absolute frame* or *ground*. It will be attached to some body that is considered to be not accelerating. It may be difficult or impossible to find such a body in the known universe. However, for our purposes, it is usually satisfactory to use any body that is not accelerating relative to the earth. Examples of such bodies are the earth itself, the floor of a building, and a vehicle that is traveling with constant velocity relative to the earth. Motions that are measured relative to this type of frame will be referred to as *absolute motions* or *positions,* as motions or positions relative to ground, or just simply as motions or positions.

The second type of reference frame is a *local reference frame,* which usually will be attached to some machine part that may be moving. Such a reference frame will consist of a coordinate system in which both the origin and the axes are rigidly attached to the reference body so that they all move in unison with the body. Motions or positions that are measured relative to this type of frame will be called *motions or positions relative to that reference body.*

The third type of reference frame is used when referring to *motion or position relative to a point.* In this type of frame, the origin of the co-ordinate system is rigidly attached to the point referred to (which usually will be a point that is rigidly attached to some moving body), but the axes of the system will remain parallel to (or at constant angles from) the ground system axes. That is, the moving axes will be non-rotating.

Consider, now, a method for describing the position of the rigid body L in Fig. 1.4 relative to the reference frame that is indicated by the axes X, Y, and Z and origin O in Fig. 1.5.[2] We might start by describing the position of a point such as A in body L. A very convenient way of doing so would be to give the X, Y, and Z coordinates x_A, y_A, and z_A of that point (i.e., the distances in the X, Y, and Z directions from the origin O to point A), as shown in Fig. 1.5. The position of body L is still not completely described because the body could be rotated about the point A without changing any of the co-ordinates of point A, so more than these three parameters must be specified.

Next, consider a coordinate system $X'Y'Z'$ whose origin is at A and whose axes are parallel to axes X, Y, and Z, respectively. Call the plane containing the axes X' and Y' plane $P1$ (indicated by the rectangular plate $P1$). Then, noting that points A, B, and C shown in body L in Figs. 1.4 and 1.5 define a plane $P2$ (indicated by rectangular plate $P2$ in Fig. 1.5), construct an axis Z'' perpendicular to plane $P2$ as shown. Construct an axis X'' along the segment AB. Complete the right-handed orthogonal axis system $X''Y''Z''$, and note that this system is fixed to the body L and must therefore move with it. The line of intersection of planes $P1$ and $P2$ is called the *line of nodes LN.* Call the angle between X' and LN α, that between the planes $P1$ and $P2$ β, and

---

[2]To be useful, this frame must be attached to some reference body, but for the present discussion, that body is not specified.

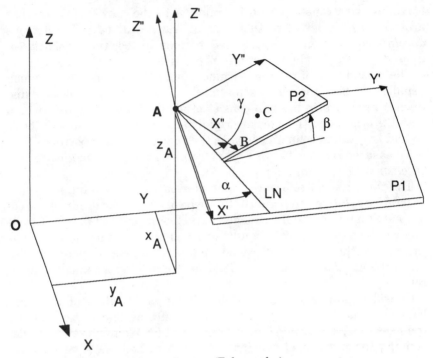

Figure 1.5 Rigid body position parameters (Euler angles).

that between LN and X'' γ. The angles α, β, and γ constitute what is known as a *Euler angle set*.[3]

Now carefully examine Fig. 1.5 and note that it would be impossible to move the body L that is represented by the axis system $X''Y''Z''$ in any way without changing the value of at least one of the parameters x_A, y_A, z_A, α, β, and γ. Thus we may conclude that a given set of values for these six parameters will uniquely describe the position of the body L relative to the reference frame XYZ, so the set is sufficient for such specification. Note further that the body may be moved in such a manner as to vary any one of the parameters without varying any of the

---

[3]Specifying relative orientation of two Cartesian coordinate systems by means of a Euler angle set consists of defining one plane attached to each of the two systems and then specifying the angles between those two planes and between the line of intersection of those planes and an axis in each coordinate system. Obviously, any one of three planes could be chosen in each axis system, so there are nine different Euler angle sets that could be used.

It should be noted that if the angle between the planes becomes zero, the line of nodes becomes undefined, and there is a singularity in the relationship between the relative orientation and the angle values. Because Euler angles are frequently encountered in articulated-arm robots, such singularities can seriously limit the range of motions that those robots can provide.

others so that if that varied parameter were ignored in the position description, the position of the body would not be completely described. The six parameters then form a necessary and sufficient set.

Obviously, there are many more sets of parameters that could be used to describe this relative position, but it will always be found that a necessary and sufficient set will consist of exactly six parameters. We therefore say that

> A system consisting of a three-dimensional reference frame and an isolated rigid body in space has six degrees of freedom.

1.6 Displacements in Three-Dimensional Space

Readers may, if they wish, skim quickly over Secs. 1.6 through 1.9 because the remainder of the book does not depend on understanding the material in them. However, it is important for practicing engineers to be aware of and to understand the concepts discussed in those sections. Sections 1.10 through 1.13 provide the foundations for the discussions in subsequent chapters.

The *displacement of a point* is simply the difference between its position after a motion and its position before that motion. It can be represented by a three-dimensional vector drawn from the initial position of the point to its final position. An example of such a displacement is indicated in Fig. 1.6, where A_1 represents the position of the point before the motion and A_2 represents the position of the point after the motion. The components of the displacement vector will be the changes in the coordinates of the point's position as measured in the reference coordinate system. We see then that not only are positions measured *relative* to some reference, but displacements are also relative phenomena.

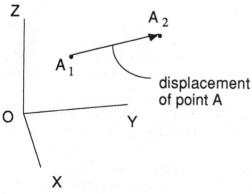

Figure 1.6 Displacement of a point.

Notice that no mention is made of the "path" followed by the point during the motion, nor is there mention of the time taken for the motion. Many different paths and times could be used to produce the same displacement. Displacement is strictly a "before versus after" phenomenon.

Rigid body displacements are more complicated than point displacements. A displacement of a rigid body is also a change in its position relative to some reference, but more than three parameters are needed to describe it. It is the difference between the position of that body after a motion has taken place and its position before that motion, all measured relative to the reference. Displacement of the body in Fig. 1.5, for instance, could be produced by moving it relative to the reference frame *XYZ*, thereby causing a change in one or more of the six parameters used to describe its position. The displacement could then be described by quoting the total change that occurred in each of the six parameters.

Two simple types of rigid-body displacement can be defined, and it will be found that all displacements can be considered to be combinations of these two types. These simple displacement types are *translation* and *rotation*.

An example of a translational displacement (or pure translation) would be a type of displacement of the body in Fig. 1.5 that would correspond to changes in the values of the coordinates x_A, y_A, and z_A and *no change* in the values of α, β, and γ. It will be seen that because these angle parameters did not change, the points A, B, and C would all have experienced displacements that were equal in length and parallel in direction. Then,

> It is an important property of a *pure translation* of a rigid body that the displacement vectors of all points in the body are identical and are nonzero.

This property can serve as a definition of a pure translation.

An example of a rotational displacement (or pure rotation) would be a displacement of the body in Fig. 1.5 in which the distance parameters x_A, y_A, and z_A did not vary but the angle parameters α, β, and γ changed. Such a displacement would involve *no displacement* of the point A. (Obviously, any other point in L could have been chosen as the undisplaced point.) Then,

> It is an important property of a *pure rotation* of a rigid body that although points in the body experience nonzero displacements, one point in that body experiences zero displacement.

This property can serve as a definition of a pure rotation.

In addition, Euler's theorem shows that in a pure rotation, all points along a particular line through that undisplaced point also experience zero displacement. This line will be called the *axis of rotational displacement* or simply the *axis of rotation,* and we speak of the rotation as being a rotation about that axis or line.

From the foregoing it may be seen that a translational displacement would change the three parameters x_A, y_A, and z_A by any desired amount and a rotational displacement would change the three parameters α, β, and γ by any desired amount. Obviously, then, a combination of a translation plus a rotation could change the six parameters by any desired amount and thus could achieve any desired displacement. Thus,

> Any displacement in three dimensions is equivalent to a translation plus a rotation.

This combination of translation plus rotation could be performed as a translation followed by rotation, a rotation followed by translation, or as simultaneous translation and rotation.

A theorem that is attributed to Chasles (pronounced somewhere between "shall" and "shahl") further states that any displacement of a rigid body can be accomplished by a translation along a line parallel to the axis of rotation that is defined by Euler's theorem plus a rotation about that same parallel line. Such a combination of translation plus rotation is called a *screw displacement* because if the body were attached to an appropriate nut, and if that nut were threaded on an appropriate stationary screw that was centered on the line described above, the nut carrying the body could be screwed along the screw and would displace the body in the desired manner. Then,

> Any displacement in three dimensions is equivalent to a screw displacement.

1.7   Joined Rigid Bodies in Three Dimensions: Degrees of Freedom

The machines that we will be interested in will all consist of more than one rigid body, so let us now consider a system consisting of two bodies, such as indicated in Fig. 1.7. A local reference frame is shown rigidly attached to each body. To describe the position of each body relative to the ground frame, it would be necessary to use six parameters. Therefore, to describe the positions of both bodies, twelve parameters would be required, so the system consisting of these two unconnected bodies and the ground is said to have twelve degrees of freedom.

Describing the positions of both bodies could be done by describing each absolute position independently or by describing the absolute position of one body such as body L_1 using six parameters and then

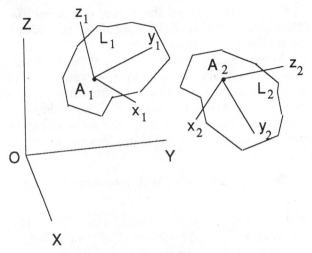

Figure 1.7 Positions of two rigid bodies.

describing the position of the other such as L_2 *relative to* the first. In the latter method, the absolute position of L_1 would be measured in terms of coordinates in the XYZ frame in a manner similar to that shown in Fig. 1.5. The position of L_2 relative to L_1 would then be measured in terms of coordinates in the $x_1y_1z_1$ frame in a manner similar to that shown in Fig. 1.5, where now the reference system is not grounded but is attached to the body L_1. Again, 12 parameters would be required to describe both positions.

As stated in the definition of a machine in Sec. 1.3, a machine consists of a combination of suitably *connected* bodies. The connections between the bodies serve to constrain the motions of the bodies so that they are not free to move with what would otherwise be six degrees of freedom for each body. For example, if the body L were to be attached to the reference frame XYZ in Fig. 1.5 by a spherical joint (see Fig. 1.8) centered at point A, that body L would then be free to move relative to the reference frame in such a manner as to vary the parameters α, β, and γ but not such as to vary x_A, y_A, or z_A. Use of the spherical joint would have reduced the freedom of body L relative to the reference frame from six degrees of freedom to three degrees of freedom. If we now define the number of degrees of freedom that a joint removes as the number of *degrees of constraint* that it provides, then we say that the spherical joint provides three degrees of constraint.

Now return to the two bodies in Fig. 1.7 and consider a situation in which body L_2 is connected to body L_1 by the spherical joint described in Fig. 1.8 in a manner similar to that described in the preceding paragraph. The body L_2 is no longer able to move relative to body L_1 with

NAME OF JOINT	FORM OF JOINT	SYMBOL	DEGREES OF FREEDOM		DEGREES OF CONSTRAINT
			rotary	trans-lational	
REVOLUTE (pivot, pin joint)	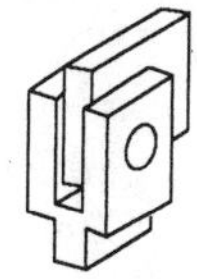	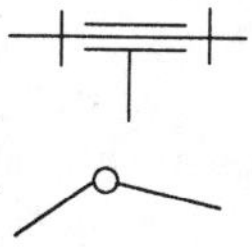	1	0	5
PRISMATIC (slide)	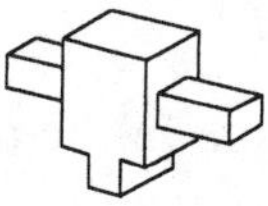	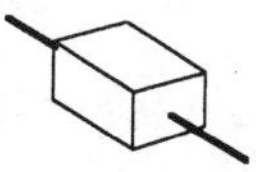	0	1	5
HELICAL (screw & nut)	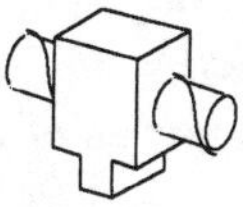		1	or 1	5
CYLINDRICAL	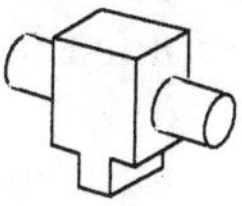	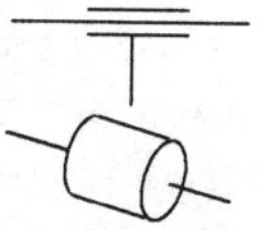	1	1	4
SPHERICAL (ball joint, ball & socket)			3	0	3
PLANAR	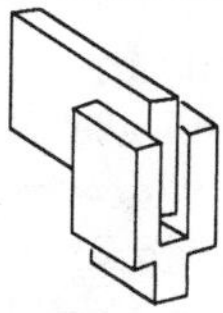	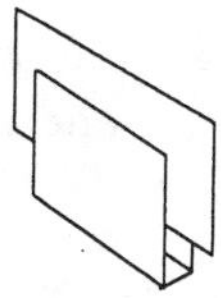	1	2	3

Figure 1.8 Kinematic joints in three dimensions.

six degrees of freedom; the spherical joint allows it only three degrees of freedom relative to body L_1. Nine parameters would now be adequate to describe the positions of the two bodies: six to describe the position of body L_1 relative to ground and three angle parameters to describe the position of body L_2 relative to body L_1. The three degrees of constraint provided by the spherical joint has removed three degrees of freedom from the total system.

There are many different types of joints that may be used to connect two bodies together. The number of degrees of freedom that each possesses and the number of degrees of constraint that each provides varies with the type of joint, but for a given joint in three-dimensional space, the sum of these two numbers is always 6. Several commonly used types of three-dimensional (spatial) joints are indicated in Fig. 1.8 together with the associated numbers of degrees of freedom and constraint.

It will be found that every time a joint is added to a system, the number of degrees of freedom in that system is reduced by the number of degrees of constraint provided by that joint. Then we may write

$$F = 6(L - 1) - 5J_1 - 4J_2 - 3J_3 - 2J_4 - J_5 \tag{1.1}$$

where F = the number of degrees of freedom in the machine (system of connected links)
   L = the number of links in the machine, including the ground link (which, being grounded, possesses no freedom)
   J_n = the number of joints having n degrees of freedom each.

Notice in each of the terms of the form NJ_n that $N + n = 6$ because the sum of the number N of degrees of constraint plus the number n of degrees of freedom of the joint must be equal to 6.

Equation (1.1) will be referred to as the spatial form of *Gruebler's formula*, which will be discussed in the Sec. 1.12 on planar motion. This formula can be very useful in the synthesis of machines because it tells how many input motions and output motions can be incorporated in a particular machine configuration. Some examples of its use and significance follow.

Figure 1.9 shows two schematic representations of the front suspension of an automobile, viewed from the front. In each of these views, two horizontal arms or links are shown joined at their left ends to the car frame, which is considered to be the reference frame or ground, by revolute joints and joined at their right ends by spherical joints to a link that carries the wheel. Thus it is seen that $L = 4$, $J_1 = 2$, and $J_3 = 2$, and Eq. (1.1) shows that the system has two degrees of free-

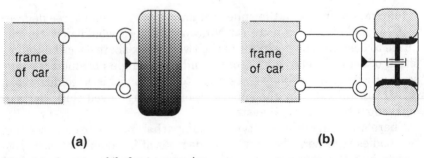

(a)   (b)

Figure 1.9   An automobile front suspension.

dom. One degree of freedom is obviously the freedom of the link that carries the wheel to swing up and down relative to the car frame to accommodate bumps in the road while remaining essentially vertical. The other degree of freedom is the freedom of the link carrying the wheel to rotate about a vertical axis to steer the wheel. Note also that the position of the system can be described completely by specifying values for two parameters such as an angle between the car frame and one of the horizontal links and some measure of the rotational displacement about the vertical axis of the wheel-carrying link.

As seen in Fig. 1.9*b*, an additional degree of freedom is actually provided on an automobile by a revolute joint that also allows the wheel (which constitutes a fifth link) to spin.

Figure 1.10 indicates a simple robot arm that consists of four links (including ground) joined by three revolute joints. Equation (1.1)

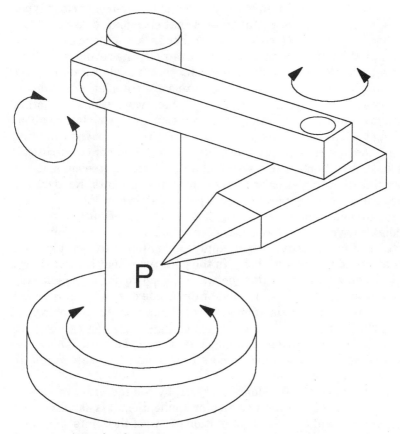

Figure 1.10   A simple robot arm.

shows that this system possesses three degrees of freedom because $L = 4$ and $J_1 = 3$, so

$$F = 6(4 - 1) - 5(3) = 18 - 15 = 3$$

To be a useful three-degree-of-freedom system, this system would require three motors or actuators to drive it: one at each of its revolute joints. It would then be able to position the tip P of the end link anywhere in the three-dimensional space that could be reached by the links, given their limited lengths. For example, the tip P could be positioned such that its three coordinates x, y, and z, as measured in some grounded coordinate reference frame, could each have any desired value within the reach of the links.

Note, however, that the orientation of the tip P could not be controlled independently of its location. For example, once the tip is positioned at a particular location xyz, it could not then be rolled onto its side. To allow such additional mobility, additional degrees of freedom would be required.

Returning to consideration of the degrees of freedom of various types of joints, refer to Fig. 1.8. It will be seen that each joint consists of two parts that may move relative to each other in a manner that is constrained by the nature of the joint. For example, the revolute joint allows the parts to rotate relative to each other but allows no relative translation. The cylindrical joint allows relative translation as well as relative rotation about the direction of that translation. The other joints allow various combinations of rotation and translation about and along various directions (axes). Realizing, then, that the combinations of freedoms provided by the joints in Fig. 1.8 do not constitute all possible combinations of translational and rotational freedoms, it can be seen that there must exist many more types of joints. Readers are encouraged to invent several of the types not shown in Fig. 1.8.

When the joints indicated in Fig. 1.8 are examined, it is seen that each joint shown involves contact of an appreciable area of joint surface attached to one body with an appreciable area of the joint surface that is attached to the other body. In the joint types that are *not* shown in Fig. 1.8, however, the contact between these two surfaces generally occurs only at one or two points and/or along one or two lines. If an appreciable force or torque is applied to the joints that are not shown in Fig. 1.8, the unit pressure at the point or line of contact can be very large, thereby contributing to wear or failure of the joint. The joints shown in Fig. 1.8, then, tend to be stronger and less susceptible to wear than those not shown.

Therefore, when the freedoms provided by the individual joints in Fig. 1.8 are not adequate for a particular application, it is usually found to be more practical to use a combination of two of the joints shown together with a small added link between them than to use any of the joint types not shown. Examples of two very common types of such joint

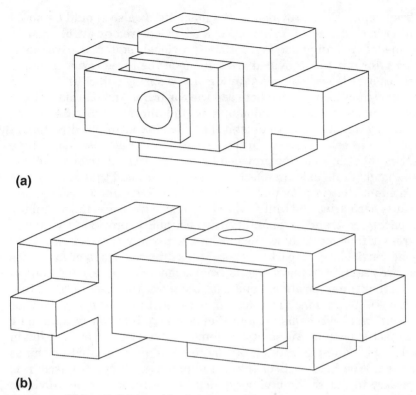

(a)

(b)

Figure 1.11 Kinematic joint combinations.

combinations are given in Fig. 1.11. The combination in Fig. 1.11*a* is usually called a *universal joint,* a *Hooke's joint,* or a *Cardan joint.* Although it is called a joint, it is actually a combination of two revolute joints and an intermediate body. The joint combination shown in Fig. 1.11*b*, the pivoted slider, is actually a combination of a revolute joint, a prismatic joint, and an intermediate body.[4] In both these joint combinations, the joint surfaces make contact over appreciable areas.

1.8   Number of Degrees of Freedom to Be Used in a Machine

The discussion in the preceding section has shown that for each machine position parameter that is to be varied independently, one degree

---

[4]Note that this pivoted slider joint has one rotational freedom and one translational freedom, just as does the cylindrical joint in Fig. 1.8. However, in the cylindrical joint, the translational freedom allows translation parallel to the axis about which rotation is possible. In the pivoted slider, the translational freedom is at an angle to the rotational freedom axis.

of freedom must be provided in the machine. It also should be noted that for each degree of freedom provided, one drive or control must be provided. For example, in Fig. 1.10, as previously stated, a drive motor or actuator must be provided at each one of the three revolute joints. Perhaps a little less obviously, in Fig. 1.9, a spring (which is a form of actuator) must be provided between the car frame and one of the horizontal links to keep it from flopping around uncontrollably. Some sort of steering linkage or drive must be provided to control the direction in which the wheel is steered. In the suspension in Fig. 1.9b, some drive and/or brake also must be provided to cause the wheel to rotate or stop rotating about its axle. In effect, then, it may be said that each degree of freedom that is provided also must be driven and/or controlled. Because each drive and control that is provided increases the machine's complexity, cost, and unreliability, it is desirable to provide only enough degrees of freedom to allow the machine to perform the desired tasks.

One particularly important class of machines consists of machines that are required to move objects from some first (initial) positions to some other (final) positions (and possibly to return them to the initial positions). These machines are often called *pick-and-place units* or *pick-and-place mechanisms,* and they are very common in automatic manufacturing and automatic assembly systems. Robots are often used as pick-and-place units, but more often, more limited machines are used. Whether robots or the more limited machines are used, it is necessary to choose the number of degrees of freedom to be provided, and for the reason stated in the preceding paragraph, it is desirable to minimize that number.

A pick-and-place machine is intended to change the position (location and orientation) of an object or objects. As shown in Sec. 1.6, a body (object) can be displaced from any position to any other position by a translation plus a rotation. This implies that a two-degree-of-freedom machine could be used. Actually, the translation and rotation could be performed simultaneously by a one-degree-of-freedom mechanism. An easily visualized example of such a one-degree-of-freedom mechanism may be inferred from Chasles' theorem (Sec. 1.6). That theorem implies that a one-degree-of-freedom machine consisting of a nut and screw (or any other combination of components that also would give a screw displacement) could be used to give *any* desired displacement.[5] For practical reasons, such as limitations on the available space, it may not be de-

---

[5]Note that a nut and screw mechanism would cause all points on the object to follow helical paths as the object is being displaced. However, as defined previously, a displacement is a "before versus after" phenomenon and does not imply anything about the path followed by the object. Thus there are many different one-degree-of-freedom mechanisms that will produce a given displacement of an object, although the paths followed may not be helical.

sirable to use a one-degree-of-freedom machine. However, the apparent complicated nature of the required displacement in a particular case does not, in itself, dictate the use of a multi-degree-of-freedom pick-and-place machine or robot. Let us consider, briefly, other factors that influence the choice of degrees of freedom to be used.

Often it is desired to move an object from one position to another, but the initial position in which it is to be found may vary and/or the final position in which it is to be placed may vary. The number of position parameters that must be used to describe any such variations must be combined with the theoretical minimum of one degree of freedom discussed in the preceding paragraph to give the total theoretical minimum number degrees of freedom required. For example, if an object is presented in its initial position on a table top, always right-side up and always facing east, but such that it could be anywhere on the table top, then its initial position would have two degrees of freedom because it would be necessary to specify values for two parameters (such as x and y) to describe its position variation. Then, in order to accommodate these position variations in the initial position only, two degrees of freedom would be required in the machine in addition to the theoretical minimum of one, giving a total of three. A similar situation pertains to the case in which only the final position varies.

In cases in which variations must be accommodated in both the initial and final positions, the situation is more complicated because some of the degrees of freedom provided to accommodate one position variation also may accommodate the variations in the other position. Therefore, the degrees of freedom cannot simply be added.[6]

As variations in the required initial and final positions increase, the required number of degrees of freedom increases. However, the theoretical required number will never exceed six, although practical considerations may dictate the use of more.

> Note that it is not the apparent complexity of the required displacement that dictates that more than one degree of freedom be used in a pick-and-place machine, but it is the *variations* in the initial and final positions that dictate the additional degrees of freedom. (Some practical considerations also may suggest the use of additional degrees of freedom. Such practical considerations may include constraints on the available space and may include requirements that the machine be adaptable to varied uses.)

---

[6]This problem usually can be solved by hypothesizing a "gripper" coordinate system that is rigidly attached to the theoretical "nut" that performs the nominal screw displacement. Then place the "nut" at the nominal initial position and note the position parameters that must be used to describe the variations in the object's initial position relative to the "gripper" coordinate system. Then do the same at the nominal final position, and any duplications will become evident.

Note that some objects (such as spheres, cylinders, etc.) possess symmetries that reduce the number of parameters required for describing their orientations.

1.9 Practical Degrees of Freedom versus Computed Degrees of Freedom

Although Eq. (1.1) accurately predicts the number of degrees of freedom that a system possesses, there are many practical cases that appear to contradict those predictions. These are cases in which special dimensional or angular relationships provide a freedom where the computations indicate that one does not exist. Equation (1.1) cannot predict such "practical" degrees of freedom because it contains no considerations of dimensions. One example of a mechanism with fewer computed degrees of freedom than "practical" degrees of freedom is indicated schematically in Fig. 1.12.

Figure 1.12 schematically depicts a side view of the kinematics of a crank-and-piston system such as would be found in an internal combustion engine or a compressor. This system contains four bodies or links: the frame F, the crankshaft S, the connecting rod R, and the piston P. These bodies are connected by the four joints: a revolute crankshaft bearing at A, a revolute connecting rod bearing at B, a cylindrical wrist-pin bearing at C, and a cylindrical bearing surface between the piston and cylinder at D. Then, $L = 4$, $J_1 = 2$, and $J_2 = 2$. Using Eq. (1.1) gives

$$F = 6(4 - 1) - 5(2) - 4(2) = 0$$

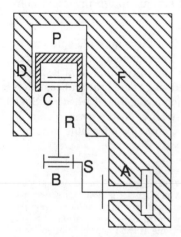

Figure 1.12 A crank-and-piston system.

The existence of zero degrees of freedom means that theoretically, *no position parameters could be varied;* the machine would be "locked up." However, there are millions of such machines operating successfully because the lengths and angles in their designs are such as to allow "practical" degrees of freedom.

When such a case is encountered, understanding of the paradox usually can be facilitated by hypothetically perturbing or distorting one or more of the dimensions or angles by some perceptible amount. Such a perturbation can show up the actual conflict that the computations indicate would prevent motion. In the case of the mechanism of Fig. 1.12, the axis of the revolute joint B is parallel to that of revolute joint A. To see the theoretical conflict, disturb that parallelism as shown in Fig. 1.13. When the piston is at the top of its stroke, the cylindrical wrist-pin bearing at the top of the connecting rod is not only displaced to the right on the piston but is also *tilted* to the right relative to the piston. When the piston is at the bottom of its stroke, the displacement and the tilt would be in the opposite directions, as shown by the dashed lines in the figure. Obviously, although the cylindrical wrist-pin bearing could accommodate the lateral displacements, it could not accommodate the back-and-forth tilting. Making the axes of revolute joints A and B parallel cures this problem in the practical machines.

The paradox also could be understood by just tilting the cylinder.

It must be remembered that manufacturing accuracies are not perfect, so there is always *potential* for physical conflicts in machines in which the computed number of degrees of freedom is inadequate. These conflicts can cause excessive stresses and wear during machine

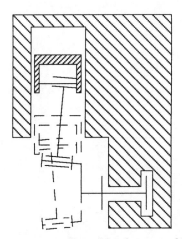

Figure 1.13 Loss of freedom caused by misalignment.

operation, and they can make the machine difficult or impossible to assemble when it is in some of its positions.

If it is desired to solve the conflict in the system of Figs. 1.12 and 1.13 by providing the needed number of *computed* degrees of freedom, the number of degrees of freedom of one of the joints could be increased by one. Suppose the two-degree-of-freedom cylindrical wrist-pin joint were changed to a three-degree-of-freedom spherical joint. Then $L = 4$, $J_1 = 2$, $J_2 = 1$, $J_3 = 1$, and $F = 6(4 - 1) - 5(2) - 4(1) - 3(1) = 1$.

This seems to indicate that the conflict is solved. Note, however, that in Fig. 1.13 a spherical joint at the wrist pin would not allow the lateral displacement shown. A conflict still persists. Where, then, is the degree of freedom indicated in the calculation? *Answer:* Joining the connecting rod to the piston by means of a spherical joint would allow the piston to rotate freely about a vertical axis. This is a useless freedom for purposes of this particular machine. The wrong freedom was provided.

The additional degree of freedom required at the wrist-pin joint is a rotational freedom about an axis more or less normal to the page while still preserving the rotational freedom and the lateral translational freedom of the cylindrical joint shown in Figs. 1.12 and 1.13. No such joint with this additional freedom is shown in Fig. 1.8. As discussed in Sec. 1.7, it is often desirable to use a combination of the joints shown when more degrees of freedom are required. Such a joint combination is shown in Fig. 1.14 as applied to the crank-and-piston system. There it is seen that the connecting rod has been divided into two links that are joined by means of a revolute joint D. This revolute joint D, combined with the cylindrical joint C and the intermediate body between them, provides the desired number and types of degrees of freedom. In this new configuration, $L = 5$, $J_1 = 3$, and $J_2 = 2$ so that Eq. (1.1) gives $F = 1$.

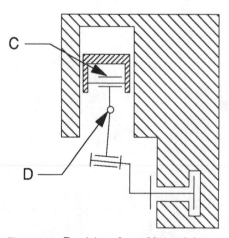

C

D

Figure 1.14 Provision of an additional degree of freedom.

The configuration in Fig. 1.14 provides a computed number of degrees of freedom of one, but in actual practice it is usually not necessary to use this additional joint unless the machine parts cannot be made accurately enough to provide the required angular alignments and the sufficiently accurate lengths.

Sometimes, when the computed number of degrees of freedom is inadequate but the practical number of degrees of freedom appears adequate, it is desirable to use in the joints bearings that "approximate" the appropriate numbers and types of degrees of freedom. This is to say bearings that approximate the types of joints that would give adequate computed degrees of freedom. For example, needle bearings running on hardened shafts should be used only as parts of revolute joints, but because they possess rather weak axial translational constraints, they sometimes can be used to approximate cylindrical joints sufficiently to allow easier assembly and to lower operating stresses in a machine. Similarly, some types of ball bearings (which also should be used only as parts of revolute joints) occasionally can be used to approximate spherical joints sufficiently well to accommodate very small misalignments. This is often a case in which a ball bearing is preferable to a roller bearing.

Use of these "approximating" bearings is not a substitute for careful tolerancing of dimensions and alignments.

From the foregoing, we see that

> Equation (1.1) can predict whether a given machine configuration will have enough freedom of motion so that it can perform the functions for which it is being designed.
>
> One or more of the freedoms predicted by Eq. (1.1) may be useless for performing the desired machine functions. The kinematics should be examined carefully to ensure that the desired types and locations of freedoms are provided.
>
> If Eq. (1.1) predicts too few degrees of freedom for the machine to perform the desired functions, the machine can still sometimes be made to have adequate "practical" degrees of freedom. This is done by using special length dimensions and/or angular dimensions in the design. If these dimensions are not held accurately enough, excessive operating stresses and wear and excessive assembly difficulty can result.

Further discussion of the ramifications of the relationships between computed and practical degrees of freedom is given by Goddard[7] and by Reshetov.[8]

---

[7]Goddard, D. L., Fixed and transient coincident mobilities in planar linkages and other mechanisms, in *Proceedings of the Second National Applied Mechanics and Robotics Conference, Cincinnati, Ohio, November 3–6, 1991* (for copies of this paper, contact Prof. A. H. Soni, University of Cincinnati, Cincinnati, OH 45221).

[8]Reshetov, L., Redundant constraints and mobilities in mechanisms, in *Self-Aligning Mechanisms,* Mir Publishers, Moscow, 1982, Chap. 1.

1.10   Planar Motion and Rigid Body
Degrees of Freedom

Although all real machines must exist in three-dimensional space, there is a very important class of motions in terms of which the action of most human-made machines can be analyzed. This class of motions is known as *planar* or *two-dimensional (2D) motion,* and the machines to which it applies are known as *planar machines* or *planar mechanisms. Planar motion* is motion in which all points in the system are assumed to move such that the distance from each point in the system to some reference plane remains constant. That is, all displacements in planar motion are parallel to that plane, and they can be represented by their projections onto that plane. Stated another way, a planar mechanism is one whose motions can be represented in a flat cardboard model in which all the mechanism parts are represented by flat cardboard pieces that are so interconnected that they are constrained to move by sliding on or just above the surface of the flat cardboard piece that represents the reference frame or system.

For example: the machine depicted in Fig. 1.3 may extend for a considerable distance in a direction perpendicular to the page, and obviously, some parts must pass in front of other parts as they move. However, all its important motions may be represented by motions of the projections of those parts onto the plane of the page, i.e., by motions as viewed from the viewpoint used in that figure.

To see how the position of a rigid body that is considered to have only planar motion can be described, refer to Fig. 1.15. In this figure, the

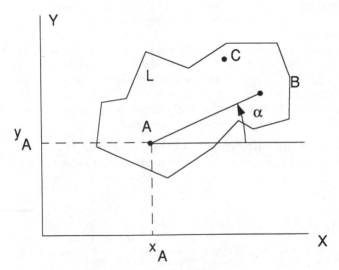

Figure 1.15   Position of a planar rigid body.

reference frame or coordinate system is represented by the axes X and Y. The rigid body L and the points A, B, and C attached to it are represented by their projections onto the XY plane. In discussions of planar machines and motions in the remainder of this book, these representative projections will be referred to simply as the rigid body itself and as points in that body.

Now note that the position of point A may be described by giving the X and Y coordinates x_A and y_A of that point. If, then, a value is specified for the angle α between a line parallel to the X axis and the line AB that is attached to the body, it will be seen that the position of the body has been described completely. Three measurements or parameters have been used to describe this position. It can be seen that the body cannot be moved (planar moves) without varying at least one of these parameters, and the body can be moved in such a manner that any one of the parameters will vary without varying either of the other two. The position of the body also could be described completely by other sets of parameters, but it will be found that a necessary and sufficient set will always consist of three parameters. Then, because three measurement parameters are necessary and sufficient to describe the position of the body, we say that

> A system consisting of a two-dimensional reference frame and an isolated rigid body in planar or two-dimensional motion has three degrees of freedom.

As will be shown in Sec. 1.12, if the body is connected to another body, its number of degrees of freedom will be changed.

Note that the point C need not be used in describing the position of body L. Only the two points A and B were used. Therefore, in most of the discussions of planar motion in the remainder of this book, the rigid bodies or links will be represented by just two points and the line segment connecting them.

1.11   Displacements in Planar Motion

In a manner similar to that discussed in Sec. 1.6, we note that the displacement of a point in planar motion can be represented by a vector from the point's position before the displacement to the point's position after that displacement, measured relative to some reference system. In planar motion, the vector is a two-dimensional one, as shown in Fig. 1.16. The displacement could be expressed in terms of the two components Δx_A and Δy_A of the displacement vector.

The displacement of a rigid body in planar motion is more complicated than that of a point, and it involves an additional parameter. Figure 1.17 shows a body L before and after a planar displacement.

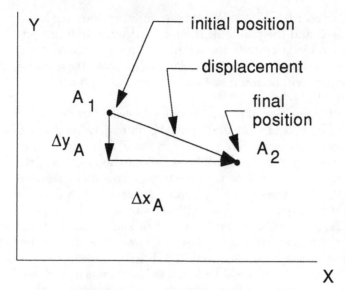

Figure 1.16 Displacement of a point in a plane.

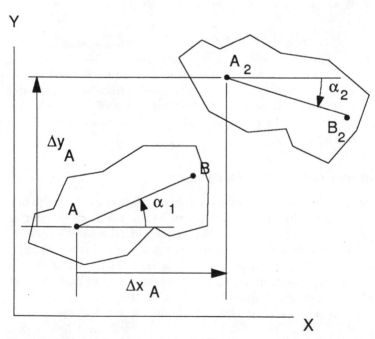

Figure 1.17 Planar displacement of a rigid body.

Because these two positions could each be described in terms of parameters such as those shown in Fig. 1.15, the displacement could be described in terms of the changes in those variables. That is, the displacement could be described in terms of Δx_A, Δy_A, and $\Delta \alpha$.

In planar motion as in three-dimensional motion, two simple types of rigid-body displacement can be defined, and it will be found that all displacements can be considered to be combinations of these two types. These simple displacement types are *translation* and *rotation*.

An example of a translational displacement (or pure translation) would be a type of displacement of the body in Fig. 1.15 that would correspond to changes in the values of the coordinates x_A and y_A and *no change* in the value of α. It will be seen that because that angle parameter did not change, the points A and B would have experienced displacements that were equal in length and parallel in direction. Then,

> It is an important property of a *pure translation* of a rigid body that the displacement vectors of all points in the body are identical and are nonzero.

This property can serve as a definition of a pure translation.

An example of a rotational displacement (or pure rotation) would be a displacement of the body in Fig. 1.15 in which the distance parameters x_A and y_A did not vary but the angle parameter α changed. Such a displacement would involve *no displacement* of the point A. (Obviously, any other point in L could have been chosen as the undisplaced point.) Then,

> It is an important property of a *pure rotation* of a rigid body that although points in the body experience nonzero displacements, *one* point in that body experiences zero displacement.

This property can serve as a definition of a pure rotation.

From the foregoing it may be seen that a translational displacement would change the two parameters x_A and y_A by any desired amounts and a rotational displacement would change the parameter α by any desired amount. Obviously, then, a combination of a translation plus a rotation could change the three parameters by any desired amount and thus could achieve any desired displacement. Thus,

> Any displacement in planar motion is equivalent to a translation plus a rotation.

This combination of translation plus rotation could be performed as a translation followed by rotation, a rotation followed by translation, or as simultaneous translation and rotation.

Consider now a general displacement of a body as shown in Fig. 1.18. In this figure, the body L has been displaced from position L_1 to position L_2, and consequently, the points A and B have been displaced from A_1 to A_2 and from B_1 to B_2, respectively. Construct the line A_1A_2 and its perpendicular bisector as shown. Also construct the line B_1B_2 and its perpendicular bisector as shown. Label the intersection of these two perpendicular bisectors C. It can then be seen (and proven, if you wish) that triangle A_1A_2C is isosceles so that length A_1C equals length A_2C. By similar reasoning, it is seen that B_1C equals B_2C. Body L is a rigid body, so A_1B_1 is equal to A_2B_2. Therefore, triangle A_1B_1C is congruent to triangle A_2B_2C. Thus we see that the triangle ABC has merely been rotated from its initial position A_1B_1C to its final position A_2B_2C about the point C, which has not moved. It can therefore be concluded that

> Any planar displacement of a rigid body is equivalent to a pure rotation of that body. We will call the point about which this equivalent rotation takes place the *center of rotational displacement* or simply the *center of rotation* or, even more briefly, the *pole* (for that displacement).

The center of rotation is not necessarily in the body itself and very frequently is at some appreciable distance from the body. The foregoing geometric construction involving the perpendicular bisectors is used to locate that center of rotation, and it will be found to be power-

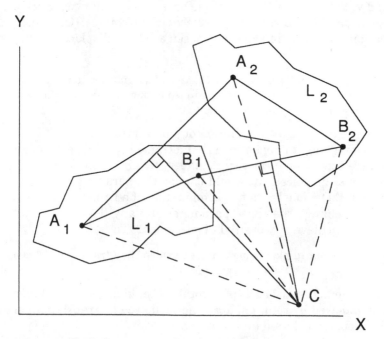

Figure 1.18 Equivalent pure rotation of a rigid body.

ful in analyzing motions and in synthesizing and analyzing mechanisms in succeeding chapters.

1.12 Joints and Their Effects on Planar Motion: Degrees of Freedom

The planar machines in which we will be interested will all consist of more than one rigid body, so let us now consider a system consisting of two bodies such as indicated in Fig. 1.19. A local reference frame is shown rigidly attached to each body. To describe the position of each body relative to the ground frame, it would be necessary to use three parameters. Therefore, to describe the positions of both bodies, six parameters would be required, so the system consisting of these two unconnected bodies and the ground would have six degrees of freedom.

Describing the positions of both bodies could be done by describing each absolute position independently or by describing the absolute position of one body such as body L_1 using three parameters and then describing the position of the other such as L_2 *relative* to the first. In the latter method, the absolute position of L_1 would be measured in terms of coordinates in the XY frame in a manner similar to that shown in Fig. 1.15. The coordinates might be the X and Y coordinates of point A_1 angle α_1 relative to the XY coordinate system. The position of L_2

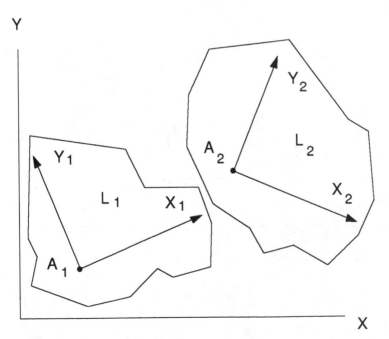

Figure 1.19 A system with two bodies.

relative to L_1 would then be measured in terms of coordinates in the $X_1 Y_1$ frame in a manner similar to that shown in Fig. 1.15, where now the reference system is not grounded but is attached to body L_1. These coordinates might be the X_1 and Y_1 coordinates of point A_2 and an angle α_2 measured from the X_1 axis to the X_2 axis. Again, six parameters would be required to describe both positions.

As stated in the definition of a machine in Sec. 1.3, a machine consists of a combination of suitably *connected* bodies, and of course this definition applies to planar machines as well. The connections between the bodies serve to constrain the planar motions of the bodies so that they are not free to move with what would otherwise be three degrees of freedom for each body. For example, if the body L in Fig. 1.15 were to be attached to the reference frame XY by a revolute joint or pivot (see Figs. 1.8 and 1.20 for definitions of joint types) centered at point A and with its axis perpendicular to the page, that body L would then be free

NAME OF JOINT	FORM OF JOINT	SYMBOL	DEGREES OF FREEDOM		DEGREES OF CONSTRAINT
			ROT.	TRANS.	
REVOLUTE (pivot, pin joint)	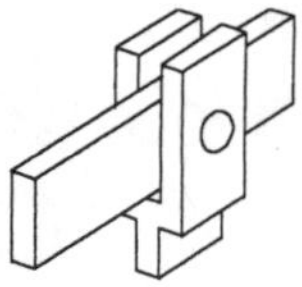	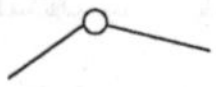 two bodies 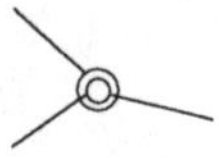 three bodies	1	0	2
SLIDER (prismatic)	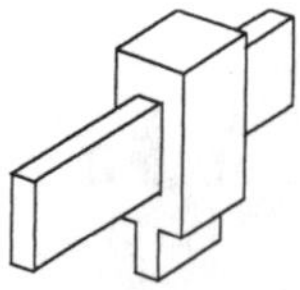	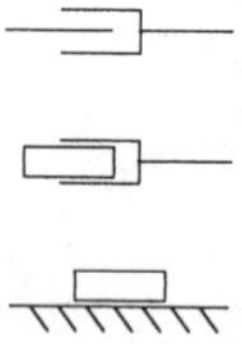	0	1	2
SLIDING PIN	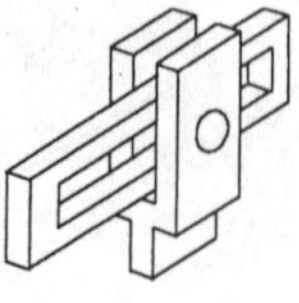	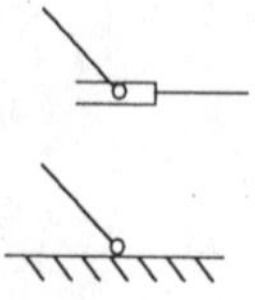	1	1	1

Figure 1.20 Kinematic joints in two dimensions.

to move relative to the reference frame in such a manner as to vary the parameter α but not such as to vary x_A or y_A. Use of the revolute joint would have reduced the freedom of body L relative to the reference frame from three degrees of freedom to one degree of freedom. If we now define the number of degrees of freedom that a joint removes as the number of degrees of constraint that it provides, then we say that in planar motion the revolute joint provides two degrees of constraint.

Now return to the two bodies in Fig. 1.19 and consider a situation in which body L_2 is connected to body L_1 at point A_2 by the revolute joint described in Figs. 1.8 and 1.20 in a manner similar to that described in the preceding paragraph. Body L_2 is no longer able to move relative to body L_1 with three degrees of freedom. That is true because the revolute joint prevents variation of the coordinates X_1 and Y_1 of point A_2 and allows L_2 only one degree of freedom relative to body L_1, i.e., the freedom to vary α_2. Four parameters would now be adequate to describe the positions of the two bodies: three to describe the position of body L_1 relative to ground and one angle parameter to describe the position of body L_2 relative to body L_1. The two degrees of constraint provided by the revolute joint have removed two degrees of freedom from the system. That is, the X_1 and Y_1 coordinates of point A_2 are no longer variables but are fixed dimensions of the two-body system or machine.

There are several different types of joints that may be used to connect two bodies together to make a planar machine. The number of degrees of freedom that each possesses and the number of degrees of constraint that each provides vary with the type of joint, but for a given joint in two-dimensional space, the sum of these two numbers is always 3. The commonly used types of planar joints are indicated in Fig. 1.20 together with the associated numbers of degrees of freedom and constraint. It should be noted that for planar motion the axes of rotation of these joints are perpendicular to the reference plane and the translational freedoms of the joints are parallel to the reference plane. Figure 1.20 also shows the symbols that are used to represent the joints in planar kinematic diagrams.

Note that, occasionally, more than two bodies will be joined to each other by revolute joints at a single location. In this case, the joints are drawn concentrically with each other, as shown in the figure for the example of three bodies. Remember that they remain separate and distinct joints even when they are coincident and are drawn concentrically with each other. The number of joints at a particular location will always be smaller by one than the number of bodies that are joined at that location.

It will be found that every time a joint is added to a system, the number of degrees of freedom in that system is reduced by the number of degrees of constraint provided by that joint. Then we may write

$$F = 3(L - 1) - 2J_1 - J_2 \qquad (1.2)$$

where F = the number of degrees of freedom in the planar machine
(system of connected links)

L = the number of links in the planar machine, including the
ground link (which, being grounded, possesses no freedom)

J_n = the number of joints having n degrees of freedom each

Notice in each of the terms of the form NJ_n that $N + n = 3$.

Equation (1.2) is referred to as *Gruebler's formula* or *Kutzbach's formula*. This formula can be very useful in the synthesis of planar machines because it tells how many input motions and output motions can be incorporated in a particular machine configuration. Some examples of its use and significance follow in the discussion of Figs. 1.22 and 1.23.

From the foregoing it is seen that the number of joints used in connecting links together is important. In subsequent chapters it will be seen that the locations of those joints on the links is important. However, the detailed physical shapes of the links themselves are not of kinematic significance. We will therefore use simplified symbols to represent the links in our planar kinematic diagrams. The symbol conventions used are indicated in Fig. 1.21 for links that possess from one to four joints. In the figure it is seen that the symbols indicate only the locations of the joints on the bodies (links), not the physical shapes of the bodies themselves.

Figure 1.22 indicates a very simplified front-end loader such as is used in earth excavation. In discussing this machine, the frame of the vehicle itself will be used as the ground reference. An arm 1 is pivoted to this frame at A by a revolute joint, and the bucket 2 is pivoted to the end of the arm at B by a revolute joint. Figure 1.22*b* shows a kinematic diagram of this system using the convention of a shaded line to indicate ground. This system consists of three links (including ground) and two single-degree-of-freedom joints, so $L = 3$ and $J_1 = 2$. Then, $F = 3(3 - 1) - 2(2) = 2$.

Therefore, the system has two degrees of freedom, so it will be necessary to use two parameters to describe its position at any time. Two such parameters could be the angle that the arm makes with the vehicle frame and the angle that the bucket makes with the arm. Being able to vary these two parameters is useful in allowing the bucket to be raised or lowered and allowing the bucket to be tilted so as to either retain the scooped-up earth or dump it. The provision of two degrees of freedom allows these two functions to be performed independently of each other.

LINK TYPE NAME	SOME EXAMPLES OF SCHEMATIC REPRESENTATIONS
	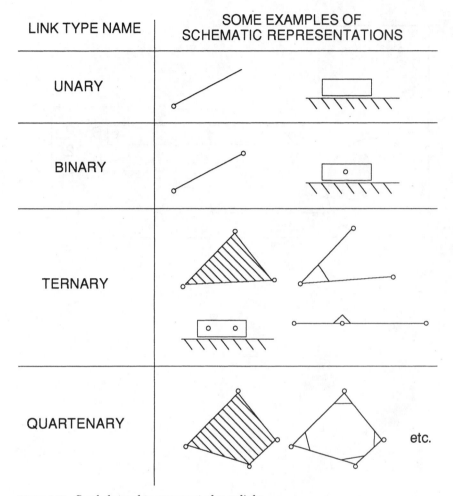

Figure 1.21   Symbols used to represent planar links.

In order to perform these two functions, it is necessary to provide a drive or actuator for each degree of freedom. These drives in actual practice usually consist of appropriately placed and connected hydraulic cylinders. Figure 1.22c indicates these hydraulic cylinders as prismatic joints C and D, and indicates schematically how they might be connected in the system. The system in this figure consists of seven links, six revolute joints, and two prismatic joints. Then, $F = 3(7 - 1) - 2(8) = 2$, as before.

The position of the system in Fig. 1.22c can be described by specifying the amount of extension of the hydraulic cylinders C and D, and therefore these two drives can be used to completely control the

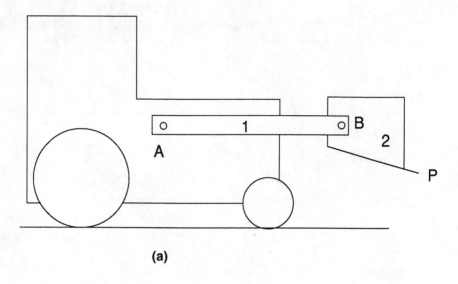

(a)

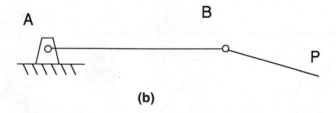

(b)

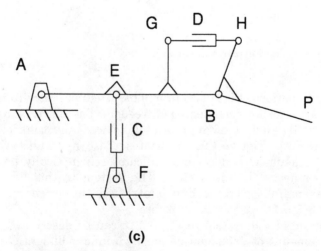

(c)

Figure 1.22 A simplified front-end loader.

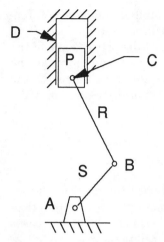

Figure 1.23 A planar crank-and-piston system.

positions of the system. These two drives are also *necessary* for complete control.

Figure 1.23 is a planar schematic diagram of a crank-and-piston system such as would be found in an internal combustion engine or compressor. It is an end view of the system that corresponds to the side view shown in Fig. 1.12. This planar system consists of four links (including ground), three revolute joints A, B, and C, and a slider or prismatic joint D.[9] Then, $L = 4$ and $J_1 = 4$, so $F = 3(4 - 1) - 2(4) = 1$.

The position of this single-degree-of-freedom system can be described completely by one parameter such as the angle of the crankshaft S or the position of the piston (slider) P. The system therefore can be driven by rotating the crankshaft S to produce a motion of the piston P as an output in the case of a compressor, or it can be driven by the motion of the piston to produce a rotation of the crankshaft as an output in the case of an engine.

In the foregoing examples, the number of degrees of freedom was a positive, nonzero number. In cases in which the number of degrees of freedom is zero, the number of position parameters that can be varied to represent variations in the position of the system is also zero. The position cannot be varied. The system is "locked up." If the system is to be made capable of moving, more degrees of freedom must be provided by changing the link arrangement and/or by changing some of the joints.

---

[9]In Fig. 1.12 the joints C and D were cylindrical joints. However, for planar system analysis and synthesis, the extra degrees of freedom are ignored.

In some systems it will be found that the number of degrees of freedom is a negative number. Such a system will not only be locked up,
but it may be impossible to assemble it without distorting some of the
links. This also means that if the system is to be made movable, the
number of degrees of freedom must be increased by changes in the
arrangement and/or joints until the appropriate positive number of degrees of freedom is obtained.

1.13 Computed Degrees of Freedom versus Practical Degrees of Freedom

Although Eq (1.2) can accurately predict the number of degrees
of freedom that a planar system possesses, there are practical
cases that appear to contradict these predictions. These are cases
in which special dimensional or angular relationships provide a
freedom where the computations indicate that one does not
exist. Equation (1.2) cannot predict such actual or "practical" degrees of freedom because it contains no considerations of dimensions. One example of a mechanism with fewer computed degrees
of freedom than "practical" degrees of freedom is given schematically in Fig. 1.24.

In Fig. 1.24, the computed number of degrees of freedom is zero,
indicating that the linkage is locked up. However, if the link lengths
and pivot locations are chosen carefully so as to form parallelograms
as shown in the figure, motion is possible. There is one practical degree of freedom. To see why Eq. (1.2) indicates that there are zero
degrees of freedom, perturb one of the lengths and/or one of the pivot
locations. For example, change the length of link 3 and the location

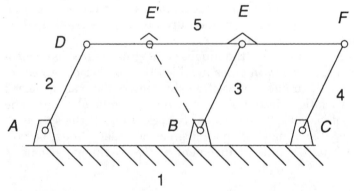

Figure 1.24 Practical degrees of freedom versus computed degrees of
freedom.

of its upper pivot so that they would be located as indicated by the dashed lines and not at the former location. It becomes obvious then that the new link 3 becomes a diagonal brace, preventing any motion of the linkage. To move the rest of the linkage, link 3 would have to stretch or contract (or other links would have to distort). Although the stretching, contracting, and/or distortion required for motion with link 3, in the position indicated by the dashed lines, is severe, similar but smaller effects will be produced even by very small deviations from the ideal dimensions that produce the parallelograms. These required distortions produce associated stresses in the machine parts, and the greater the deviation from the ideal dimensions, the greater are the stresses. To avoid such stresses, it is therefore necessary to build the machine quite accurately so that the practical freedom is provided, or it is necessary to change the configuration so as to provide an adequate number of computed degrees of freedom.

In summary, then,

> Equation (1.2) can be used to predict whether a planar machine will have the appropriate number of degrees of freedom so that it can perform the functions for which it is being designed.
>
> If Eq. (1.2) predicts too few degrees of freedom for the machine to perform the desired functions, the machine can still sometimes be made to have adequate "practical" degrees of freedom. This is done by using special length dimensions and/or angular dimensions in the design. If these dimensions are not held accurately enough, excessive operating stresses and wear and excessive assembly difficulty can result.

1.14   Kinematic Inversion

Figure 1.25 is a simplified drawing of a balance or scale for weighing objects or materials. It is a common application of the linkage shown in Fig. 1.24, and it is so designed that links 2 and 4 do not rotate as they rise and fall. Therefore, the weight that is placed on the pan on each of those links can be placed anywhere on that pan without affecting the balance. The links and pivots in Fig. 1.25 are labeled the same as those in Fig. 1.24, so it can be seen that the two linkages consist of the same arrangement of links and pivots. However, in Fig. 1.24 link 1, which is a ternary link, is grounded, whereas in Fig. 1.25 link 3, which is a binary link, is grounded. Thus, although both linkages consist of the same parallelogram arrangements of five links each, the motions of the two linkages will be different. This procedure of changing the choice of link (or links) to be

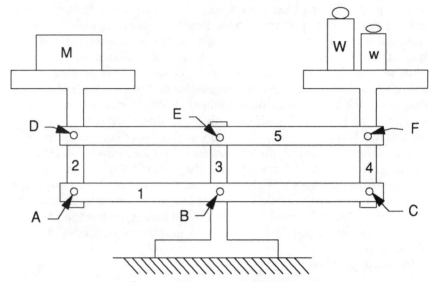

Figure 1.25 A kinematic inversion of the mechanism depicted in Fig. 1.24.

grounded while otherwise leaving the linkage unchanged is known as *kinematic inversion*. This procedure will be seen to be very useful in synthesizing, analyzing, and understanding mechanism motions later in this book.

The linkage in Fig. 1.25 obviously has zero computed degrees of freedom, just as was the case for Fig. 1.24. If a detailed analysis of the tolerances and performance of this balance mechanism were to suggest that the deficiency in degrees of freedom would unduly degrade performance, an added degree of freedom could be provided by breaking either link 1 or 5 (but not both) into two parts by adding a pivot somewhere along its length. Readers are encouraged to make such a modification and to investigate the resulting degrees of freedom.

Figure 1.26 is a schematic diagram of a linkage that is intended to cause an output shaft to be driven through a full 360° of rotation in unison with the rotation of an input shaft. This linkage is topologically the same as that in Fig. 1.25; i.e., it uses two ternary links and three binary links to form a set of parallelograms. This can be seen by imagining links 1 and 5 in Fig. 1.25 to be bent into 90° V shapes (without breaking them) and by then rotating the entire linkage as a unit clockwise through 90°. To emphasize this equivalence, the corresponding links in Figs. 1.25 and 1.26 are la-

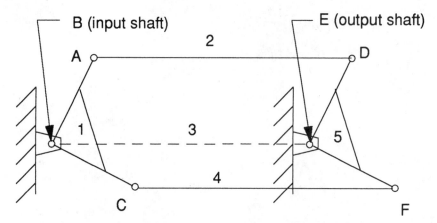

Figure 1.26 Another kinematic inversion of the mechanism in Fig. 1.24.

beled the same. Therefore, this machine encounters the same deficiencies in degrees of freedom as those in Figs. 1.24 and 1.25. The reader can verify the existence of these difficulties by distorting the machine in Fig. 1.26 in a manner analogous to that described in the discussion of Fig. 1.24.

2

Rigid Body Planar Motion

2.1 Introduction

This chapter presents a few simple principles of rigid body planar kinematics. These principles are useful in visualizing and understanding the motions of systems of interconnected rigid bodies (machines or mechanisms) that are studied in subsequent chapters. Readers who do not wish to engage in analytic treatment (i.e., algebra and calculus) of kinematics or who are already familiar with complex vector notation may wish to skip over Secs. 2.2, 2.3, and 2.4, the portions of Sec. 2.5 prior to the subsection entitled "Use of the velocity difference equation," Sec. 2.6, and the portions of Sec. 2.7 prior to the subsection entitled "Use of the acceleration difference equation." These skipped sections present a brief review of vector addition and differentiation, they introduce the representation of vectors by complex numbers, and they relate these operations to graphical depiction of vectors. The analysis and synthesis techniques described consist of analytical and graphical methods for describing the position, displacement, velocity, and acceleration of rigid bodies in planar motion. The useful concepts of instant centers, centrodes, and velocity-distribution triangles are introduced in the sections not skipped.

2.2 Position and Displacement of a Point

As discussed in Secs. 1.5 and 1.11, the quantitative description of the position of a point or body requires the use of a reference coordinate system in which to measure that position. Figure 2.1 shows such a coordinate frame and a point A that is positioned relative to that frame. The coordinate frame might be a ground reference frame, a local reference frame, or a reference frame attached to a point such as the

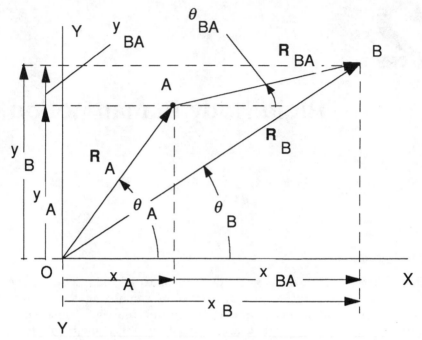

Figure 2.1 Positions of two points or displacement of a single point.

origin O (see Sec. 1.5). The position of point A obviously can be described relative to this frame by giving the Cartesian coordinates x_A and y_A. This position also can be described relative to this frame by stating that its position relative to the origin is given by the vector $\mathbf{R}_A$, noting that a vector possesses both magnitude and direction. In this case, the magnitude is the length R_A (a scalar) and the direction is described by the angle θ_A, which is measured at the "tail" of the vector and which is defined as being positive when measured counterclockwise from the positive X axis to the vector. The Cartesian coordinate description and the vector description of the position of point A are related by the scalar equations

$$x_A = R_A \cos \theta_A \tag{2.1}$$

$$y_A = R_A \sin \theta_A$$

A second point, B, is also shown in Fig. 2.1, and its position could be described by the vector $\mathbf{R}_B$, which has a magnitude of R_B and a direction represented by the angle θ_B, or its position might be described by the coordinates x_B and y_B.

The difference in the positions of points A and B can be described by the vector $\mathbf{R}_{BA}$, where the subscript BA should be read as meaning "to B from A" or "B relative to A."

It is seen that tracing along $\mathbf{R}_B$ from "tail to head" results in moving from the origin to point B, and that tracing along $\mathbf{R}_A$ from "tail to head" and then along $\mathbf{R}_{BA}$ from "tail to head" also results in moving from the origin to point B. The three position vectors in Fig. 2.1 are therefore related by the vector equation

$$\mathbf{R}_B = \mathbf{R}_A + \mathbf{R}_{BA} \tag{2.2}$$

Equations of the form of Eq. (2.2) will be referred to as *position-difference equations* because they relate the positions of the two separate points to the difference in their positions.

Examination of Fig. 2.1 shows that

$$x_B = x_A + x_{BA}$$
$$y_B = y_A + y_{BA} \tag{2.3}$$

Then, expressing these six Cartesian components in the form of the examples in Eq. (2.1) gives

$$R_B \cos \theta_B = R_A \cos \theta_A + R_{BA} \cos \theta_{BA}$$
$$R_B \sin \theta_B = R_A \sin \theta_A + R_{BA} \sin \theta_{BA} \tag{2.4}$$

Equations (2.4) expresses the same relationship as does Eq. (2.2). *Thus it is seen that a single vector equation is equivalent to two scalar algebraic equations.* If values are known for all but two of the six variables (three magnitudes and three angles) in the two scalar equations, the equations may be solved for values of the two remaining variables.

Equation (2.2) has the advantage that it is very compact, and thus it is easy to manipulate. However, it does not explicitly contain symbols for the six separate variables that are implied and which *do* appear explicitly in the equivalent Eq. (2.4). A technique that combines advantages of Eqs. (2.2) and (2.4) is based on the use of Euler's formula, which states that

$$e^{j\theta} = \cos \theta + j \sin \theta \tag{2.5}$$

where $j = \sqrt{-1}$.

Now, if the vector $\mathbf{R}_A$ is expressed in complex exponential form by stating that $\mathbf{R}_A = R_A e^{j\theta_A}$, Euler's formula shows that such an expression is equivalent to $\mathbf{R}_A = R_A \cos \theta + jR_A \sin \theta$. By comparing the terms on the right of this expression with Eq. (2.1), it is seen that now $\mathbf{R}_A = x_A + jy_A$. If, in a Cartesian reference coordinate system, the X axis is considered to be the real axis and the Y axis is considered to be the imaginary axis, this complex notation leads to a graphical

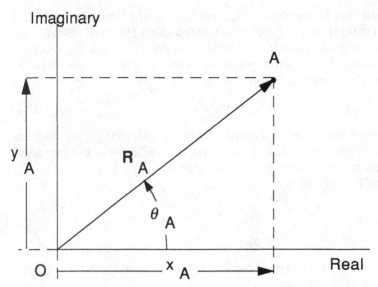

Figure 2.2 Graphic representation of complex vector notation.

representation of the vector $\mathbf{R}_A$ in the manner shown in Fig. 2.2 (called an *Argand diagram*), where again R_A is the magnitude and θ_A indicates the direction of the vector. Thus the complex notation "automates" the relationships between the vector notation and the Cartesian form.

If in Eq. (2.2) the vector $\mathbf{R}_A$ is expressed in complex exponential form as before by stating that $\mathbf{R}_A = R_A e^{j\theta_A}$ and the analogous thing is done for $\mathbf{R}_B$ and $\mathbf{R}_{BA}$, then Eq. (2.2) becomes

$$R_B e^{j\theta_B} = R_A e^{j\theta_A} + R_{BA} e^{j\theta_{BA}} \qquad (2.6)$$

The terms in Eq. (2.6) can each be expanded by using Euler's formula. The resulting equation will consist of three real terms and three imaginary terms. In a properly balanced equation, the real terms must balance only real terms and the imaginary terms must balance only imaginary terms. Therefore, the equation that results from expanding Eq. (2.6) implies two equations: one equation in the three real terms and one equation in the three imaginary terms. If each of the terms in the equation containing imaginary terms is divided by the factor j, the two equations become identical to Eq. (2.4). Readers are encouraged to perform the steps just mentioned in order to become familiar with the concepts.

The complex exponential form of vector equations such as Eq. (2.6) will be used frequently in subsequent sections because of the ease with which it can be manipulated and the ease with which it can be converted to the Cartesian form.

The foregoing discussion dealt with the positions of two individual points or particles A and B. Their positions were described by the vectors $\mathbf{R}_A$ and $\mathbf{R}_B$, and their position difference was described by the vector $\mathbf{R}_{BA}$. The positions of points A and B could just as easily have been considered to represent two different positions of the same point. That is, position A could have represented the initial position of a point or particle and position B could have represented the position of that point or particle after a displacement. Then the vector $\mathbf{R}_{BA}$ would have represented the displacement of that point or particle.

2.3   Position and Displacement of a Rigid Body

In Sec. 1.10 it was shown that the position of a rigid body in planar motion can be described by giving the Cartesian coordinates of a point that is attached to that body plus the angular orientation of a line segment that is attached to that body, as shown in Fig. 1.15. That figure is repeated as Fig. 2.3, but in Fig. 2.3 the body position is described by using two vectors, $\mathbf{R}_A$ and $\mathbf{R}_{BA}$. Because each of these vectors can be described by giving its Cartesian components or by giving its magnitude and direction, it is seen that the vector description in Fig. 2.3 is equivalent to that in Fig. 1.15.

In Fig. 2.3 it will be seen that the position of point B can be described by the vector sum $\mathbf{R}_B = \mathbf{R}_A + \mathbf{R}_{BA}$, where $\mathbf{R}_{BA}$ represents the position

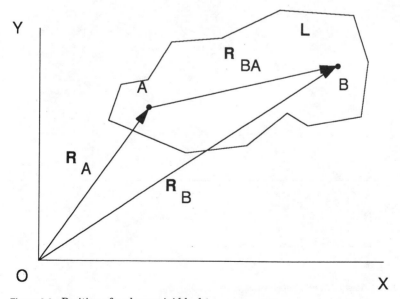

Figure 2.3   Position of a planar rigid body.

difference between points A and B or the position of point B relative to point A.

In Sec. 1.11 it was shown that a displacement of a rigid body in planar motion can be considered to be a combination of a translation and a rotation. Such a displacement is shown in Fig. 2.4. Point A has been displaced from position A_1 to position A_2, and point B has been displaced from position B_1 to position B_2 and then from position B_2 to position B_3. Now refer to Fig. 2.4 and note that the displacements of these points may be written in vector form as follows:

Displacement of point A is

$$\Delta \mathbf{R}_A = \mathbf{R}_{A_2 A_1} \tag{2.7}$$

Displacement of point B is

$$\Delta \mathbf{R}_B = \mathbf{R}_{B_3 B_1} = \mathbf{R}_{B_2 B_1} + \mathbf{R}_{B_3 B_2} \tag{2.8}$$

However, because the first displacement was a pure translation, $\mathbf{R}_{B_2 B_1} = \mathbf{R}_{A_2 A_1} = \Delta \mathbf{R}_A$, so

$$\Delta \mathbf{R}_B = \mathbf{R}_{B_3 B_1} = \Delta \mathbf{R}_A + \mathbf{R}_{B_3 B_2} \tag{2.9}$$

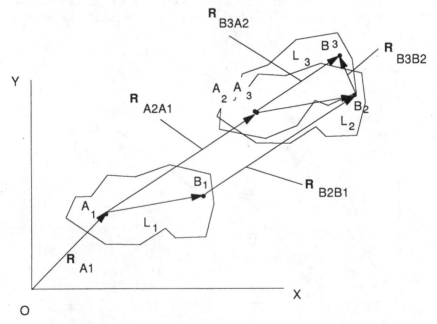

Figure 2.4 Translation and rotation of a rigid body.

Note now that the rotational displacement that produces $\mathbf{R}_{B_3B_2}$ just rotates the vector $\mathbf{R}_{BA}$ to produce a change $\Delta\mathbf{R}_{BA}$ in that vector $\mathbf{R}_{BA}$. Then Eq. (2.9) may be written

$$\Delta\mathbf{R}_B = \Delta\mathbf{R}_A + \Delta\mathbf{R}_{BA} \qquad (2.10)$$

Because this equation expresses the fact that the displacement of point B differs from that of point A only by the vector $\Delta\mathbf{R}_{BA}$, Eq. (2.18) is known as the *displacement-difference equation*.

2.4 Velocity of a Point and Some Vector Differentiation

Figure 2.5 depicts a point or particle A that is traveling along the path P which is fixed to the reference coordinate system. Point A_1 is the location of the moving point at time T_1, and point A_2 is the position of the moving point at a later time $T_2 = T_1 + \Delta t$. The displacement of point A that occurred during the time interval Δt is represented by the vector $\Delta\mathbf{R}_A$. Because velocity is defined as the time rate of change of position or as the displacement per unit of time, the average velocity $\mathbf{V}_{ave}$ of point A during the time interval Δt is given by

$$\mathbf{V}_{ave} = \frac{\Delta\mathbf{R}_A}{\Delta t} \qquad (2.11)$$

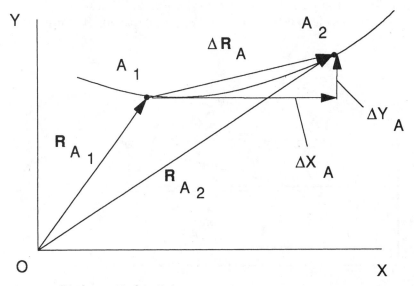

Figure 2.5 Displacement of a point.

It is seen that because the velocity is given by dividing the vector quantity $\Delta \mathbf{R}_A$ by a scalar Δt, the velocity is a vector quantity having the same direction as the displacement $\Delta \mathbf{R}_A$.

If the time interval Δt is made very small, the displacement vector $\Delta \mathbf{R}_A$ becomes very small also and approaches tangency with the path along which point A is traveling. Thus the average velocity approaches tangency with the path for very small displacements. Although $\Delta \mathbf{R}_A$ becomes very small, Δt also becomes very small, and because velocity is the ratio of these small quantities, the average velocity does not necessarily become very small for small displacements.

The instantaneous velocity of point A is defined as

$$\mathbf{V}_A = \lim_{\Delta t \to 0} \mathbf{V}_{\text{ave}} = \lim_{\Delta t \to 0} \frac{\Delta \mathbf{R}_A}{\Delta t} = \frac{d\mathbf{R}_A}{dt} \tag{2.12}$$

The vector differentiation implied by Eq. (2.12) can be performed quite conveniently by using the complex notation described in Sec. 2.2 and noting that in that notation, $\mathbf{R}_A = x_A + jy_A$. Then, differentiating both sides of this equation with respect to time gives

$$\mathbf{V}_A = \dot{\mathbf{R}}_A = \dot{x}_A + j\dot{y}_A \tag{2.13}$$

where the dot above a variable denotes differentiation of that variable with respect to time. It is thus seen that the velocity of the point consists of a rate of change of the X coordinate and a rate of change of the Y coordinate; i.e., $\dot{x}_A$ is the X component of the velocity and $\dot{y}_A$ is the Y component of the velocity, as indicated in Fig. 2.6. These components are frequently referred to as $\mathbf{V}_{A_x}$ and $\mathbf{V}_{A_y}$, respectively. It can be seen

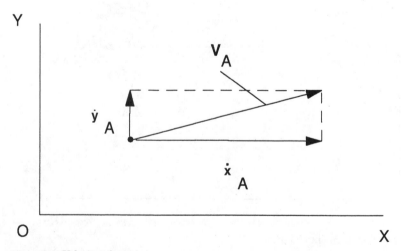

Figure 2.6 Velocity of a point.

that these components of velocity also could be derived by dividing Δx_A and Δy_A in Fig. 2.6 by Δt and then allowing Δt to approach zero.

It is often convenient to differentiate with the vector $\mathbf{R}_A$ expressed in the complex exponential form $\mathbf{R}_A = R_A e^{j\theta_A}$. Differentiation of this expression gives

$$\mathbf{V}_A = \dot{\mathbf{R}}_A = \dot{R}_A e^{j\theta_A} + j\dot{\theta}_A R_A e^{j\theta_A} \qquad (2.14)$$

By noting that Euler's formula gives $e^{j(\pi/2)} = j$, substitution of this equivalence into Eq. (2.14) followed by consolidation of terms gives

$$\mathbf{V}_A = \dot{R}_A e^{j\theta_A} + \dot{\theta}_A R_A e^{j(\theta_A + \pi/2)} \qquad (2.15)$$

The first term on the right side of Eq. (2.15) represents a vector with a direction indicated by the angle θ_A; i.e., it represents a vector in the same direction as the position vector $\mathbf{R}_A$. The magnitude of the vector represented by this first term is equal to the rate at which the *length* of the position vector is changing. This component is shown as $\mathbf{V}_{A_r}$ extending beyond the tip of $\mathbf{R}_A$ in Fig. 2.7. The velocity would consist of only this component if the point were moving strictly radially from or toward the origin, i.e., if θ_A were constant at the instant being considered.

The second term on the right side of Eq. (2.15) represents a vector with a direction indicated by the angle $(\theta_A + \pi/2)$. That is, it represents a vector that is rotated $\pi/2$ radians (90°) counterclockwise from the direction of the vector $\mathbf{R}_A$. The magnitude of the vector represented by this second term is equal to the length of the position vector times the

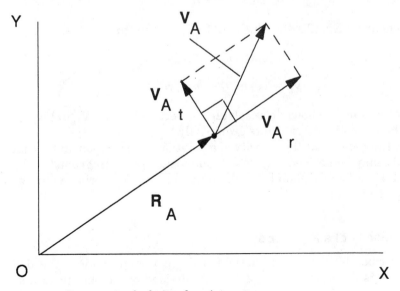

Figure 2.7 Components of velocity of a point.

rate of change of the angle of that position vector. This component is shown as V_{A_t} in Fig. 2.7. The velocity would consist of only this component if the length R_A of the position vector as measured in this reference system were constant for the instant being considered, i.e., if the point were moving along a path that was instantaneously perpendicular to the position vector.

This second component is shown in Fig. 2.7 to be directed upward and to the left, as would be the case for a positive value of rate of change of θ_A, i.e., for a counterclockwise changing direction of $\mathbf{R}_A$. If $\mathbf{R}_A$ were rotating in a clockwise direction, the rate of change of θ_A would be negative, so the length of this vector V_{A_t} would be negative and the vector would be directed downward and to the right.

In Eq. (2.15) the velocity has been resolved into two orthogonal components that are, respectively, parallel to and perpendicular to the position vector. In contrast, in Eq. (2.13) the velocity has been resolved into two orthogonal components that are, respectively, parallel to and perpendicular to the X axis. Obviously, Eq. (2.15) can be expanded into Cartesian form by applying Euler's formula and using the trigonometric formulas for sine and cosine of $(\theta_A + \pi/2)$ to obtain

$$\mathbf{V}_A = \dot{R}_A(\cos\theta_A + j\sin\theta_A) + \dot{\theta}_A R_A(-\sin\theta_A + j\cos\theta_A) \qquad (2.16)$$

Collecting real and imaginary terms gives

$$\mathbf{V}_{A_x} = (\dot{R}_A\cos\theta_A - \dot{\theta}_A R_A\sin\theta_A) + j(\dot{R}_A\sin\theta_A + \dot{\theta}_A R_A\cos\theta_A) \qquad (2.17)$$

By comparing Eq. (2.13) with Eq. (2.9), it is seen that

$$\dot{x}_A = \dot{R}_A\cos\theta_A - \dot{\theta}_A R_A\sin\theta_A \qquad (2.18)$$

$$\dot{y}_A = \dot{R}_A\sin\theta_A + \dot{\theta}_A R_A\cos\theta_A$$

This also could be shown by resolving the vectors $\mathbf{V}_{A_r}$ and $\mathbf{V}_{A_t}$ in Fig. 2.5 into their X and Y components graphically.

It is thus seen that the velocity of a point and the components of that velocity may be derived, expressed, and visualized both geometrically as in Figs. 2.5 and 2.6 and Eq. (2.13) and in complex variable form as in Eq. (2.15) and Fig. 2.7.

2.5   Velocity of a Rigid Body

To describe the velocity of a rigid body in planar motion, the displacements discussed in Sec. 2.3 can be divided by Δt, the time interval during which the displacements took place, and by then letting Δt approach zero. The discussion of displacements in Sec. 2.3 resulted in the development of equations for relationships between the displacements of points. The initial discussions in this section thus will start with a discussion of the velocities of points that are attached to rigid bodies.

Equation (2.10) expresses the displacement of point B in terms of the displacement of point A and the amount by which the displacement of point B differs from the displacement of point A. If each displacement in that equation is divided by the time increment Δt during which the displacements occur, Eq. (2.19) results:

$$\frac{\Delta \mathbf{R}_B}{\Delta t} = \frac{\Delta \mathbf{R}_A}{\Delta t} + \frac{\Delta \mathbf{R}_{BA}}{\Delta t} \qquad (2.19)$$

This will be recognized as an equation in average velocities. If, then, the time interval Δt is allowed to approach zero, the fractions in Eq. (2.19) become time derivatives of the displacements and, as shown in Sec. 2.3 and Eq. (2.8), these fractions become expressions for the instantaneous velocities of the points. Thus, in the limit, Eq. (2.19) becomes

$$\mathbf{V}_B = \mathbf{V}_A + \mathbf{V}_{BA} \qquad (2.20)$$

where the vector $\mathbf{V}_{BA}$ represents the amount by which the velocity of point B differs from that of point A. Equations of the form of Eq. (2.20) therefore are known as *velocity-difference equations*. If a nonrotating observer were attached to and traveling with point A, point B would appear to that observer to be traveling with a velocity of $\mathbf{V}_{BA}$.

The vectors $\mathbf{V}_B$ and $\mathbf{V}_A$ represent just the velocities of points, as described in Sec. 2.4. The vector $\mathbf{V}_{BA}$, however, has some special properties that are useful in analyzing the motion of rigid bodies. These special properties of $\mathbf{V}_{BA}$ can be visualized by noting that because points B and A are on the same rigid body, the distance between them must remain constant so that B cannot be moving toward or away from A. Therefore, $\mathbf{V}_{BA}$ cannot have a component parallel to the line connecting A and B; i.e., $\mathbf{V}_{BA}$ must be perpendicular to the line AB. Also, the rotation of line AB about A will give point B a velocity relative to A that is equal to the distance from A to B times the rate of rotation (in radians per second) of line AB.

To show this a little more rigorously, consider the difference $\Delta \mathbf{R}_{BA}$ in the displacements of points A and B. This difference is the result of the rotational displacement of the rigid body L. Such a displacement is

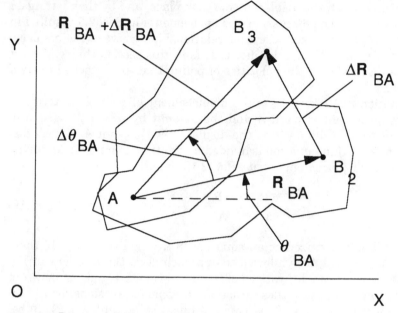

Figure 2.8 Rotational displacement of a rigid body.

shown in Fig. 2.8. It will be noted that because $\mathbf{R}_{BA}$ is a line segment attached to a rigid body, its length is constant, so the *magnitudes* of $\mathbf{R}_{BA}$ and $\mathbf{R}_{BA} + \Delta\mathbf{R}_{BA}$ are equal. The triangle formed by the vectors $\mathbf{R}_{BA}$, $\mathbf{R}_{BA} + \Delta\mathbf{R}_{BA}$, and $\Delta\mathbf{R}_{BA}$ is therefore isosceles. Then ΔR_{BA}, the magnitude of the vector $\Delta\mathbf{R}_{BA}$, is given by

$$\Delta R_{BA} = 2R_{BA} \sin\left(\frac{\Delta\theta_{BA}}{2}\right)$$

As Δt is caused to approach zero, $\Delta\theta_{BA}$ becomes very small, so

$$\sin\left(\frac{\Delta\theta_{BA}}{2}\right) \to \left(\frac{\Delta\theta_{BA}}{2}\right)$$

Therefore,

$$\Delta R_{BA} \to R_{BA}\Delta\theta_{BA} \tag{2.21}$$

Also, as Δt approaches zero, the vector $\Delta\mathbf{R}_{BA}$ approaches perpendicularity with the vector $\mathbf{R}_{BA}$. Therefore, the instantaneous velocity difference $\mathbf{V}_{BA}$, which is the limiting value of $\Delta\mathbf{R}_{BA}/\Delta t$ as Δt approaches zero, has the same direction as $\Delta\mathbf{R}_{BA}$ and is therefore perpendicular to $\mathbf{R}_{BA}$. The magnitude of this velocity difference vector is given by

$$V_{BA} = \lim_{\Delta t \to 0}\left(\frac{\Delta R_{BA}}{\Delta t}\right) = \lim_{\Delta t \to 0}\left(\frac{R_{BA}\Delta\theta_{BA}}{\Delta t}\right)$$

$$= R_{BA}\lim_{\Delta t \to 0}\left(\frac{\Delta\theta_{BA}}{\Delta t}\right) = R_{BA}\frac{d\theta_{BA}}{dt} \tag{2.22}$$

The derivative $d\theta_{BA}/dt$ is the rate at which the angle θ_{BA} is changing. It is denoted by the symbol ω_{BA} and is known as the *angular velocity* of the line BA or of the vector $\mathbf{R}_{BA}$.

As stated in Sec. 1.4, "the angle between any two intersecting lines that are attached to a rigid body remains constant regardless of the motion of that body." Therefore, if line segment AB that is attached to a rigid body in planar motion is rotated through an angle θ_{AB}, all other line segments that are attached to that body must rotate through the same angle θ_{AB}. It is therefore seen that the angular velocity of all line segments that are attached to a given rigid body must have the same angular velocity. This angular velocity is referred to as the *angular velocity of the entire rigid body*. Therefore,

> The velocity difference between two points A and B on a rigid body is a vector that is perpendicular to the line AB connecting these two points, and this velocity vector has a magnitude equal to the product of the length of the line segment connecting these two points times the angular velocity of the rigid body. (The sense of this vector is given by the sense of the component of $\mathbf{V}_B$ that is perpendicular to line AB minus the component of $\mathbf{V}_A$ that is perpendicular to that line.)

This relationship also may be derived by differentiating the complex exponential expression for the vector $\mathbf{R}_{BA}$:

$$\mathbf{V}_{BA} = \frac{d\mathbf{R}_{BA}}{dt} = \frac{d\left(R_{BA}e^{j\theta_{BA}}\right)}{dt}$$

which, when expanded, becomes

$$\mathbf{V}_{BA} = \dot{R}_{BA}e^{j\theta_{BA}} + j\dot{\theta}_{BA}R_{BA}e^{j\theta_{BA}} \tag{2.23}$$

Because the points A and B are attached to a rigid body, the length of R_{BA} is constant, and therefore, $\dot{R}_{BA}$ is zero and the first term on the right of Eq. (2.23) is zero. Then, remembering that $e^{j(\pi/2)} = j$ and that $\dot{\theta}_{BA} = \omega_{BA}$, Eq. (2.23) becomes

$$\mathbf{V}_{BA} = \omega_{BA}R_{BA}e^{j(\theta_{BA} + \pi/2)} \tag{2.24}$$

It is seen that the vector on the right-hand side of Eq. (2.24) is perpendicular to vector $\mathbf{R}_{BA}$, and that its magnitude is equal to the product of the length of the line segment connecting points A and B times the angular velocity of the rigid body. This is the same as the result that was obtained in the discussion of Fig. 2.8. However, the complex number method has the advantage that it automatically indicates the magnitude and direction of the velocity-difference vector.

Use of the velocity-difference equation

Examination of Eqs. (2.20) and (2.24) shows that if the velocities of two points on a rigid body are known, their velocity difference can be determined and the angular velocity of that rigid body can be calculated. Conversely, if the velocity of one point on a rigid body is known, and if the angular velocity of that rigid body is known, the velocity of any other point on that body can be calculated. The remainder of Sec. 2.5 describes manners in which these velocity-difference relationships can be used.

Velocity-distribution triangles

Because the velocity difference between two points on a rigid body is proportional to the distance between those points and is perpendicular to a line connecting them, the distribution of velocity differences for

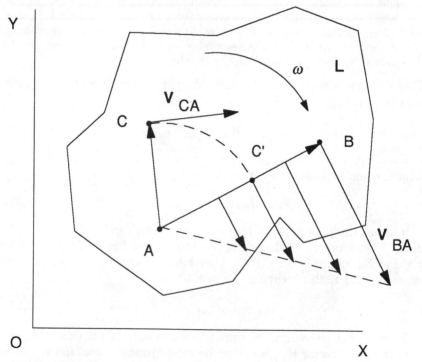

Figure 2.9 The velocity-distribution triangle.

points along that line may be drawn as shown in Fig. 2.9. Each velocity vector drawn attached to line AB in this figure is the difference between the velocity of the corresponding point along line segment AB and the velocity of point A. Or alternatively stated, it is the velocity of that point relative to point A. By considering similar triangles, it can be seen that the velocity-difference vector shown for each point along line segment AB is proportional to the distance to that point from point A. A triangle such as that formed by the vector $\mathbf{V}_{BA}$, the line $\mathbf{R}_{BA}$, and a line connecting the tip of $\mathbf{V}_{BA}$ with point A will be referred to as a *velocity-distribution triangle*. Such triangles are very useful in visualizing the relationship between the velocities of various points on a rigid body. Remember, however, that these triangles pertain to *relative* velocities or to velocity *differences* only for points on the *same rigid body*.

Consider a point C, which is *not* on line segment AB but which is on the same rigid body. Its velocity *relative to point A* will be perpendicular to line segment AC. The magnitude of this relative velocity will be the same as that of a point C' on line segment AB that is at the same distance from point A as is point C. This is true because the magnitude is proportional to the distance of the points from A (which is the same for point C as for point C') times the angular velocity of the body to which both points are attached. It is thus seen that if the velocity difference $\mathbf{V}_{BA}$ is known, the velocity difference $\mathbf{V}_{CA}$ can be found graphically by means of the following steps:

1. Construct the relative velocity-distribution right triangle having right-angled sides AB and $\mathbf{V}_{BA}$.
2. On that triangle, find the velocity of point C', which is at the same distance from point A as is point C.
3. Construct at point C the vector $\mathbf{V}_{CA}$ perpendicular to line segment AC and with a magnitude equal to the magnitude of the velocity of point C'.

Longitudinal and perpendicular velocities

$\mathbf{V}_{BA}$ and $\mathbf{V}_{CA}$, as discussed above, are velocity differences (or relative velocities). In general, point A will have an absolute velocity. Then, according to velocity-difference Eq. (2.20), the absolute velocities of points B and C can be obtained simply by adding the absolute velocity of point A to the respective velocity differences. Such an addition is shown for another general motion in Fig. 2.10. It can be seen that the velocity $\mathbf{V}_{BA}$ of point B relative to point A is perpendicular to the line connecting that point to point A, as is required by rigid body constraints. However, when $\mathbf{V}_A$, the absolute velocity vector of point A, is

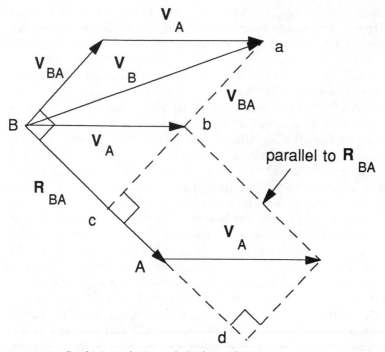

Figure 2.10   Combining relative and absolute velocities.

added to $\mathbf{V}_{BA}$, the resulting vector $\mathbf{V}_B$ is not necessarily perpendicular to line AB. In general, the velocities of points on a line will not be perpendicular to that line; the perpendicularity constraint applies only to the velocity *differences*.

Note that in the parallelogram that represents the addition of $\mathbf{V}_A$ to $\mathbf{V}_{BA}$ to give $\mathbf{V}_B$, line segment ab is parallel to $\mathbf{V}_{BA}$ and is therefore perpendicular to line segment AB. If line segment bc is constructed perpendicular to AB, then abc is a straight-line segment perpendicular to AB. Thus it is seen that line segment Bc represents the component of $\mathbf{V}_B$ along AB as well as the component of $\mathbf{V}_A$ along AB (that is, $Ad = Bc$). Then it can be concluded that

> The velocity vector of each of any two moving points can be resolved into a "longitudinal" component that is parallel to the line connecting the two points and a "perpendicular" component that is perpendicular to that line. *If these two points are attached to the same rigid body*, the longitudinal component of the velocity of each point is equal to the longitudinal component of the velocity of the other point.

Note that the difference between the perpendicular components of the two velocity vectors (i.e., length ba) is, of course, just the velocity difference. This velocity difference can be used with Eq. (2.24) and knowledge of the distance between the points to find the angular velocity of the rigid body.

Instant centers and velocity distributions

As discussed in the foregoing text, the velocity of a point can be considered to be an infinitesimal displacement vector of that point divided by a scalar that is the infinitesimal time during which that displacement takes place. Alternatively, in an infinitesimal time interval, each point on a moving rigid body can be considered to be displaced by an infinitesimal amount. This infinitesimal displacement vector of each point will be parallel to the velocity vector of that point, and the magnitude of the displacement will be proportional to the velocity magnitude of the point. These infinitesimal displacements of the points correspond to an infinitesimal displacement of the rigid body to which the points are attached.

In Sec. 1.11 it was shown that any planar displacement of a rigid body (even an infinitesimal displacement) is equivalent to a pure rotation of that body about some point called a *pole*. In that section it was shown further that the pole is located on the perpendicular bisectors of the displacements of the points that are attached to the rigid body. In Fig. 1.18, consider the limiting case in which the displacements become extremely small. The perpendicular bisectors remain perpendicular to the displacements (and thus they approach being perpendicular to the velocity vectors), but in the limit as the displacements become infinitesimal, the bisectors pass through the points themselves.

> For an infinitesimal displacement of a moving body, the location of the pole therefore can be determined by constructing at each of the two points on the body the perpendicular to the velocity vector at that point. The pole lies at the intersection of these two perpendiculars (see Fig. 2.11). This pole pertains only to the instant at which the particular velocity values exist. At any other instant, the velocities generally will be different, and the pole location also will be different. Because this pole pertains only to a particular instant, it is referred to as an *instant pole,* an *instant center of rotation,* a *velocity center,* or usually simply as an *instant center* (IC).

In Sec. 1.11 it was shown that the pole of a displacement experiences zero displacement. Then it can be seen that an instant center experiences zero infinitesimal displacement at the instant being considered. Therefore, the instant center is a point on the rigid body (or on an extension thereof) that has zero velocity at that instant. For any instant, a rigid body can be considered to be rotating about its instant center as a pivot. If the velocity of a point on the body is known, then once the instant center has been located, the velocity of any other point on

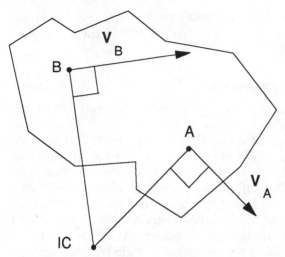

Figure 2.11   Locating an instant center.

the body can be determined by constructing velocity-distribution triangles, as was shown in Fig. 2.9 and as is shown in Fig. 2.12.

Note in Fig. 2.12 that the velocity of a point is proportional to the distance of that point from the instant center. That is, the velocity of a point is, as before, a vector perpendicular to a line connecting that point to the instant center, and its magnitude is equal to the distance from the instant center to the point times the angular velocity of the body. This relationship can be used "in reverse" to locate the instant center if the velocity of one point is known and the angular velocity of the body is known.

> That is, to find the *instant center,* construct through the point for which the velocity is known a line that is perpendicular to the velocity vector at that point. Then locate the instant center on this line such that the distance from the point to the instant center is equal to the magnitude of the velocity divided by the angular velocity of the body in radians per second. Note that the instant center also must be located such as to give the point a velocity *direction* that is consistent with ω.

It is thus seen that the location of the instant center can be determined if

1. The direction of the velocity of each of two points is known, or if

2. The direction and magnitude of the velocity of one point is known and the angular velocity of the body is known.

In the foregoing discussion of instant centers and velocity distributions, the velocities were spoken of as though they were absolute

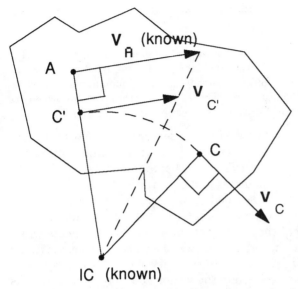

Figure 2.12   The instant center, velocity-distribution triangles, and velocities.

velocities. That is, they were treated as though they were measured relative to the ground reference frame or, in other words, as though they were seen by an observer who was attached to ground. The instant centers and velocity distributions that were derived from these velocities, then, were instant centers relative to ground and velocity distributions relative to ground. As observed earlier, the rigid body can be considered to be rotating about its instant center as though that center were a pivot (revolute joint). The instant center, then, represents two coincident points: one on the rigid body and one on the ground reference frame, both located at the pivot point. The point on the rigid body at the instant center will have the same velocity as the point on the ground at the instant center; i.e., they will both have zero velocity.

Consider now that, instead of absolute velocities, the velocities of points on the rigid body being considered are velocities relative to a moving reference frame (i.e., velocities as seen by an observer attached to the moving reference frame). This observer on the reference frame could perform the same graphic constructions as described above and obtain an instant center and sets of velocity distributions. This instant center would be a point at which the rigid body could be connected to the observer's moving reference frame by a revolute joint for just the instant being considered. In this case, again, the instant center represents two coincident points: one on the rigid body at the instant center and one attached to the moving reference frame at the instant center.

This moving reference frame could be attached to some second moving rigid body. Thus it may be seen that

> There is an instant center associated with any pair of moving bodies. This instant center represents two coincident points, one fixed to each body. Because the two bodies could be considered to be momentarily joined together by a revolute joint at these points, the velocities of these two points are equal at the instant for which the instant center pertains.

This equality of velocities at the instant center is an important property of an instant center that is useful in identifying instant centers and in calculating velocities. Two moving rigid bodies are shown in Fig. 2.13 along with the velocities of points C and D on those bodies. There is an instant center relative to the two bodies; it is located at point A on body L_1 and at point B on body L_2, and points A and B are coincident at the instant shown. Note that the velocity of point C on body L_1 is not equal to the velocity of point D on body L_2 but that the velocity of point A at the instant center on body L_1 is equal to the velocity of point B on body L_2.

Readers are encouraged to study the vector addition parallelograms at points C and D to appreciate these relationships.

The Aronhold-Kennedy theorem

At any instant there will be an instant center associated with any pair of rigid bodies that are moving relative to each other. Thus, if at some

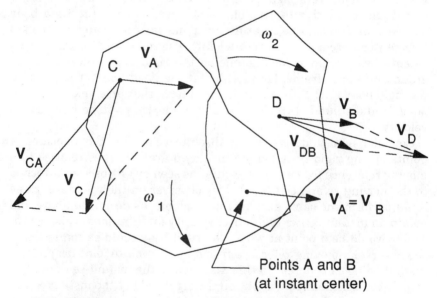

Figure 2.13 The instant center relative to the motions of two bodies.

instant three moving bodies L_1, L_2, and L_3 all have different velocities, there will be an instant center associated with pair L_1L_2, another instant center associated with pair L_1L_3, and a third instant center associated with pair L_2L_3. The Aronhold-Kennedy theorem states that

> The three instant centers associated with three rigid bodies that are moving relative to each other all lie on the same straight line.

The truth of this theorem may be demonstrated by referring to Fig. 2.14. In this figure, body L_2 is moving relative to body L_1 in such a manner that their mutual instant center is at IC_{12}. The point IC_{12} is the point at which body L_2 could be joined to body L_1 by a revolute joint for the instant being considered. Similarly, IC_{13} is the mutual instant center for bodies L_1 and L_3. Consider, then, a point P_2 located on body L_2 and a point P_3 located on body L_3 such that P_2 is coincident with P_3 at that instant. The velocity of P_2 relative to body L_1 must be perpendicular to the line $IC_{12}P_2$, and the velocity of point P_3 relative to body L_1 must be perpendicular to the line $IC_{13}P_3$ as shown. *If* the location of the coincident points P_2 and P_3 were to be the instant center IC_{23} associated with bodies L_2 and L_3, the velocities of these two points relative to *any* reference such as L_1 would have to be equal. These velocities could be equal only if they were parallel. They could be parallel only if line $IC_{12}P_2$ were parallel to line $IC_{13}P_3$. These conditions could exist only if points P_2 and P_3 were to lie on the line $IC_{12}IC_{13}$, i.e., if all three instant centers lie on the same straight line.

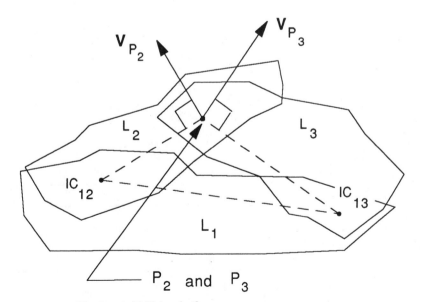

Figure 2.14   The Aronhold-Kennedy theorem.

Centrodes

As stated previously, an instant center pertains only to the instant being considered. At a different instant, the instant center will be located at a different point on each body. This phenomenon may be illustrated by considering the simple motion of a circle that is rolling *without slipping* along a nonmoving straight line, as indicated in Fig. 2.15. The rolling circle is shown in one position at the left of the figure and in another, later position toward the right of the figure. In the left-hand position, the point of nonslipping contact between the circle (body L_2) and the straight line on body L_1 is the instant center IC_{12}. This point of contact is the instant center because it is the point at which the velocities of the coincident points P_1 on body L_1 and P_2 on body L_2 are equal. (These point velocities are zero if one of the bodies is considered stationary, as is body L_1 in this illustration.)

As the body L_2 rolls to the position to the right, point P_2 moves to the new position P_2', as shown on the right-hand circle, and the point of nonslipping contact becomes the instant center IC_{12}'. This instant center corresponds to the point P_2' on body L_2 and the coincident point P_1' on body L_1. As would be seen by an observer moving with body L_2, the instant center has moved along the circular arc from P_2 to P_2'. As seen by an observer moving with body L_1 (i.e., a stationary observer), the instant center has moved along the straight-line segment from P_1 to P_1'. These two loci of the instant center are known as the *centrodes* of the motion. The locus of the instant center relative to the moving body is

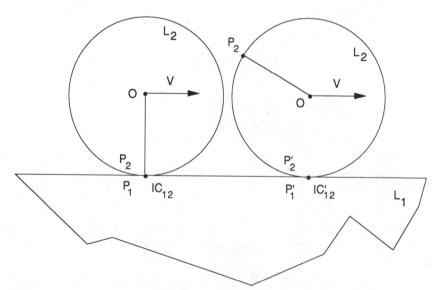

Figure 2.15 Centrodes for a circle rolling on a straight line.

known as the *moving centrode,* and the locus relative to the stationary body is known as the *stationary* or *fixed centrode.* The fixed centrode is attached to the stationary body and the moving centrode is attached to the moving body, and the motion can be considered to be generated by the moving centrode rolling without slipping on the fixed centrode.

In the foregoing simple example, the centrodes are simple curves: a circle and a straight line. In general, for two bodies moving relative to each other, the centrodes will not be simple curves; they will *not* depend on the shapes of the bodies but will depend only on the motion of the bodies. The centrodes will, however, be attached to the bodies and will roll without slipping on each other.

Now consider a more extended rigid body L_2 that is rigidly attached to the circle in Fig. 2.15 as shown at the left in Fig. 2.16. The points A, B, and C are attached to body L_2, and the circular centrode is rolling to the right without slipping along the straight-line centrode on L_1, carrying with it the attached body L_2. By following the path of point B as the circle rolls to the right, it can be seen that at the instant shown at the center of Fig. 2.16, point B and the center of the circle will have velocity vectors that are directed to the right. By using velocity-distribution triangles, it can be seen that point C will have a velocity that is directed to the left as shown. Further study of the motion will show that point A, which is located on the moving centrode, will follow a path that is a *cycloid,* and that that path will have a cusp

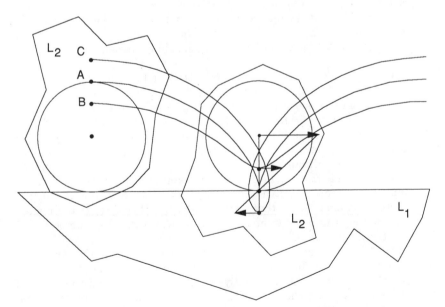

Figure 2.16 Centrodes and the paths followed by points in the bodies.

at each place where point A contacts the fixed centrode. At each such cusp the velocity of point A is zero for that instant. Point B will follow a smooth undulating curve known as a *curtate cycloid* with no cusps, and at no time will it experience zero velocity. Point C will follow a path that crosses itself and forms loops, and which is known as a *prolate cycloid*. Point C also will never experience zero velocity as long as there is relative velocity between the bodies.

In the general case of a body that is moving relative to some reference frame, if it is considered that a plane is attached to the moving body and is moving with it, then note that the moving centrode (the circle in the foregoing case) divides the moving plane into two regions. One such region contains (at least locally) the centrode that is fixed to the reference frame, and the other does not contain (at least locally) that fixed centrode. In general, if a point on the moving body is in the region of the moving plane that contains the *fixed* centrode, then the path followed by that point will tend to cross itself and contain one or more loops. If the point is in the region that does not contain the *fixed* centrode, the path followed by that point will not contain loops. If the point lies right on the *moving* centrode, the path followed by the point will contain one or more cusps.

Example 2.1   Finding the Centrodes of the Tchebycheff Straight-Line Mechanism
Figure 2.17a schematically depicts a mechanism consisting of four rigid bodies (links) that are connected to each other by revolute joints. This particular mechanism is known as the *Tchebycheff straight-line linkage*, and its link lengths are in the proportions $L_2 = 1$ unit, $L_1 = L_3 = 2.5$ units, and $L_4 = 2$ units. The problem in this example is to find and plot the centrodes for the motion of link L_2 relative to the ground link L_4.

In the preceding section on instant centers and velocity distributions, it was shown that if the directions of the velocities of two points on a rigid body (L_2 in this case) are known (relative to a second body, L_4 in this case), the instant center for motion of the first body relative to the second body may be found. Because the link L_3 is constrained to rotate about the joint O_2, the velocity of point A relative to L_4 must be perpendicular to a line (shown dashed) through points A and O_2, and the instant center of motion of L_2 relative to L_4 must lie on that line. Similarly, the velocity of point B must be perpendicular to a line through points B and O_1, and the instant center also must lie on that dashed line BO_1. Therefore, at the instant shown, the instant center must lie at point C, which is the intersection of the two dashed lines.

This instant center also could be located by using the *Aronhold-Kennedy theorem*. To do this, consider the three rigid bodies L_1, L_2, and L_4. The instant center IC_{12} relative to bodies L_1 and L_2 must lie at point B because these bodies are permanently pivoted to each other at that point. By the same reasoning, IC_{14} must lie at point O_1. Then, according

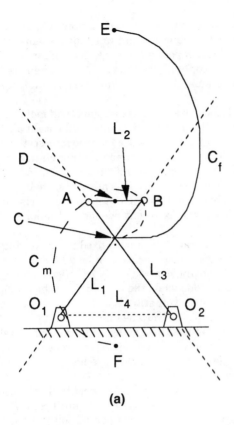

(a)

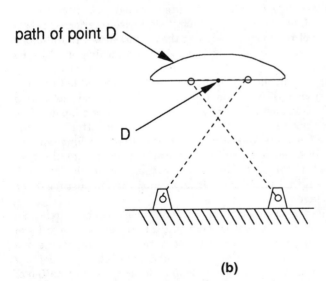

(b)

Figure 2.17 Centrodes of the Tchebycheff straight-line linkage.

to the Aronhold-Kennedy theorem, the instant center IC_{24} relative to bodies L_2 and L_4 must lie on the same straight line that contains IC_{12} and IC_{14}, i.e., the dashed line through points B and O_1. By similar reasoning, IC_{24} also must lie on the line containing points A and O_2. Then, as in the preceding paragraph, IC_{24} must lie at the intersection of these two dashed lines.

As the link L_2 is rotated clockwise, the intersection C will move initially from the point shown at C in the figure horizontally to the right along the curve C_f. Further clockwise rotation of L_2 will cause the intersection to follow a path upward along the curve C_f toward point E. For clarity, the portion of the curve C_f that is shown is only half the fixed centrode for the motion of L_2 relative to L_4; the remainder is a mirror image of the part shown. Together, the part shown and its mirror image form a closed, egg-shaped curve, and this curve is the fixed centrode for the motion of L_2 relative to ground (L_4).

The moving centrode for this motion is a path fixed to the moving body L_2, and it is the path of the instant center as seen by an observer who is attached to that body. This moving centrode may be constructed by considering the link L_2 to be grounded while the remainder of the linkage rotates about it. The instant center will still lie at the intersection of the lines AO_2 and BO_1, but these lines will now be rotating about points A and B, respectively. The resulting path is the moving centrode, and it is shown as curve C_m. Again, for clarity, only half the moving centrode is shown; the other half is a mirror image of the first part. Together the parts form a closed path that loops once inside itself.

As the link L_2 is rotated clockwise, the moving centrode C_m will roll clockwise without slipping along the fixed centrode C_f until point F on the moving centrode is brought into contact with point E on the fixed centrode, at which time link L_2 will have rotated through 180° from the position shown in the figure. The next 180° of clockwise rotation of L_2 will cause the portion of the moving centrode that is not shown to roll along the portion of the fixed centrode that is not shown, returning link L_2 to its original position.

It will be seen that in the region directly to the right and left of point C, the fixed centrode C_f is approximately a straight line, and the moving centrode C_m is almost a circular arc with its center at the midpoint D of link L_2. Then, because link L_2 moves as though it were attached to the moving centrode (which is almost a circular arc) that is rolling without slipping on the fixed centrode (which is almost a straight line), the path of point D (which is the center of this approximately circular arc) will move along a path that is almost a straight line, thereby justifying the name of this linkage.

Because this motion is very similar to the rolling motion of a circle on a straight line, the ratio of the velocity of point D to the angular velocity of link L_2 is very nearly constant over the useful range of straight-line motion. Thus, for a constant angular velocity of L_2, the velocity of point D will be nearly constant in this region. As may be seen, as the rotation of link L_2 continues beyond certain limits, the centrodes no longer possess the desired characteristics, and the path of point D no longer approximates a straight line as the moving centrode C_m continues to roll around inside the fixed centrode C_f. The actual path followed by the point D is shown in Fig. 2.17b.

In this example, the centrodes are closed, finite curves. Frequently, this is not the case, and the curves can extend to infinity and each curve can have more than one branch.

Probably no one other than Tchebycheff knows or knew by exactly what thought process Tchebycheff invented this mechanism, although Chap. 9 may provide some hints. However, consider a circle of radius DC with its center at D rolling along a horizontal line that is drawn through point C. If points A and B were attached to that circle's diameter and inside its circumference, as shown in Fig. 2.17a, those points would follow curtate cycloidal paths, as shown in Fig. 2.16. It also can be seen in this latter figure that those curtate cycloids could be approximated by circular arcs over appreciable portions of their lengths. Points A and B can be made to follow such circular arc paths by attaching them to bodies that are pivoted to ground at appropriate points such as O_2 and O_1, respectively. Evidently, Tchebycheff chose the locations of these points A and B on the moving body very well, and chose the proportions of the remainder of the linkage very well, because this linkage provides a motion that approximates a straight line for an appreciable distance in terms of the size of the linkage.

2.6 Acceleration of a Point

The discussion in Sec. 2.4 dealt with instantaneous velocities of points. Obviously, in general, instantaneous velocity will vary from instant to instant. Figure 2.18 indicates the velocity of a moving point A at time T when it is at location A_1 and at time $T + \Delta t$ when it is at location A_2. It is seen that the velocities at the two different instants differ in both magnitude and direction. As shown in Fig. 2.19, the difference between these velocities can be shown by moving vector $\mathbf{V}_2$ until its "tail" coincides with the tail of vector $\mathbf{V}_1$ without changing the magnitude or direction of vector $\mathbf{V}_2$. The difference vector is $\Delta\mathbf{V}$.

Acceleration is defined as the time rate of change of velocity or as the change in velocity per unit of time. The average of the point A in Figs. 2.18 and 2.19 is then $\mathbf{A}_{\text{ave}} = \Delta\mathbf{V}/\Delta t$. It is seen that because the acceleration is given by dividing the vector quantity $\Delta\mathbf{V}$ by a scalar Δt, the acceleration is a vector quantity having the same direction as the velocity increment $\Delta\mathbf{V}$.

If the time interval Δt is made very small, $\mathbf{V}_2$ approaches equality with $\mathbf{V}_1$, and the velocity difference $\Delta\mathbf{V}$ and its components also become very small. Although $\Delta\mathbf{V}$ and its components become very small, Δt also becomes very small, and because acceleration is

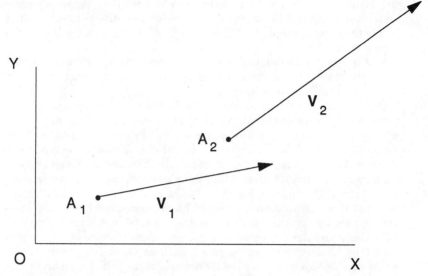

Figure 2.18 A change in the velocity of point A.

the ratio of these small quantities, the average acceleration and its components do not necessarily become very small for small time increments. As the time increment is decreased, the average acceleration approaches the instantaneous acceleration, which is defined as

$$\mathbf{A} = \lim_{\Delta t \to 0} \mathbf{A}_{\text{ave}} = \lim_{\Delta t \to 0} \frac{\Delta \mathbf{V}}{\Delta t} = \frac{d\mathbf{V}}{dt} \tag{2.25}$$

Note that $\Delta \mathbf{V}$ can be resolved (Fig. 2.19) into a component $\Delta \mathbf{V}^t$ parallel to $\mathbf{V}_1$ and another component $\Delta \mathbf{V}^n$ perpendicular to $\mathbf{V}_1$. Thus the average acceleration and therefore also the instantaneous acceleration (which is merely the limiting case) can be resolved into a component parallel to $\mathbf{V}_1$ and a component perpendicular to $\mathbf{V}_1$. The resulting components of this instantaneous acceleration are, respectively, $\mathbf{A}^t$ parallel to $\mathbf{V}_1$ and a component $\mathbf{A}^n$ perpendicular to (normal to) $\mathbf{V}_1$. Because the vector $\mathbf{V}_1$ is tangent to the path being followed by the moving point, $\mathbf{A}^t$ is a vector tangent to that path and the vector $\mathbf{A}^n$ is a vector perpendicular to (normal to) that path.

It can be seen that for very small time increments, $\mathbf{V}_1$ and $\mathbf{V}_2$ become almost equal to each other. Then the magnitude of $\Delta \mathbf{V}^t$ approaches the small-magnitude difference $V_2 - V_1$, and the magnitude of $\Delta \mathbf{V}^n$ approaches $V_1 \Delta \theta_V$. The magnitudes of the instantaneous acceleration components $\mathbf{A}^t$ and $\mathbf{A}^n$ are obtained by dividing these magnitudes by

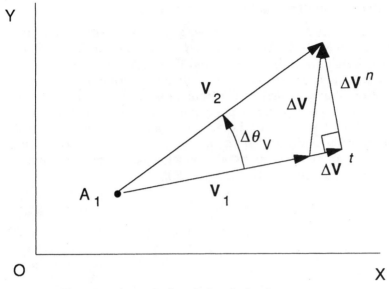

Figure 2.19 The vector change in the velocity of point A.

Δt and letting Δt approach zero. The magnitudes of these components then become the following.

Tangential acceleration magnitude:

$$A^t = \lim_{\Delta t \to 0}\left(\frac{V_2 - V_1}{\Delta t}\right) = \frac{dV}{dt} \tag{2.26}$$

Normal acceleration magnitude:

$$A^n = \lim_{\Delta t \to 0}\left(\frac{V_1 \Delta \theta_V}{\Delta t}\right) = V_1\left(\frac{d\theta_V}{dt}\right) = V_1 \omega_V \tag{2.27}$$

where ω_V is defined as the rate of rotation $d\theta_V/dt$ of the velocity vector.

The tangential acceleration is seen to be the result of changes in the speed with which the point is traveling along the path, and the normal acceleration is seen to be the result of changes in the direction of the path.

These components also may be obtained by differentiating $\mathbf{V} = Ve^{j\theta_V}$:

$$\dot{\mathbf{V}} = \dot{V}e^{j\theta_V} + j\dot{\theta}_V Ve^{j\theta_V} = \dot{V}e^{j\theta_V} + \dot{\theta}_V Ve^{j(\theta_V + \pi/2)} \tag{2.28}$$

where it is seen that the first term on the right-hand side of the equation is the tangential acceleration term and the second term is the normal acceleration term.

2.7   Acceleration of a Rigid Body

Figure 2.20 shows a rigid body L in two successive positions: L at time T and L' at time $T + \Delta t$. Points A and B are attached to that body and have velocities $\mathbf{V}_A$ and $\mathbf{V}_B$ at time T and $\mathbf{V}'_A$ and $\mathbf{V}'_B$ at time $T + \Delta t$, respectively. Because the velocities have changed during the interval Δt, accelerations have occurred. The individual average and instantaneous accelerations of points A and B can be treated in the manner discussed in Sec. 2.6. However, because the points are attached to the same rigid body, additional relationships relating their accelerations are available and useful. These relationships may be derived from the velocity-difference equations (see Eq. 2.20):

$$\mathbf{V}_B = \mathbf{V}_A + \mathbf{V}_{BA} \qquad (2.29)$$

$$\mathbf{V}'_B = \mathbf{V}'_A + \mathbf{V}'_{BA} \qquad (2.30)$$

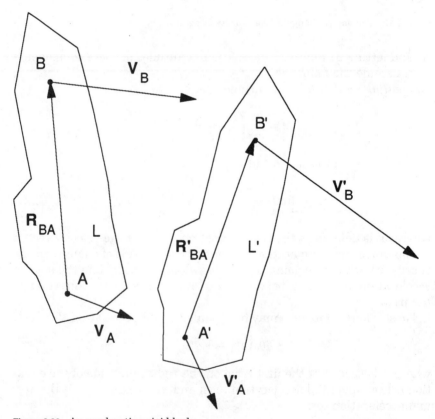

Figure 2.20   An accelerating rigid body.

These two equations are presented graphically in the upper two vector triangles in Fig. 2.21, where the corresponding vectors from Fig. 2.20 have been magnified to make Fig. 2.21 clearer. By comparing these two vector triangles, it may be seen that not only do $\mathbf{V}_A$ and $\mathbf{V}_B$ change during Δt, but $\mathbf{V}_{BA}$ also changes.

Subtracting Eq. (2.29) from Eq. (2.30) gives

$$\mathbf{V}'_B - \mathbf{V}_B = (\mathbf{V}'_A - \mathbf{V}_A) + (\mathbf{V}'_{BA} - \mathbf{V}_{BA})$$

which may be written

$$\Delta\mathbf{V}_B = \Delta\mathbf{V}_A + \Delta\mathbf{V}_{BA} \tag{2.31}$$

That is, the change in the velocity of B is equal to the change in the velocity of A plus the change in the velocity of B *relative to A*.

Dividing Eq. (2.31) by Δt and letting Δt approach zero gives

$$\mathbf{A}_B = \mathbf{A}_A + \mathbf{A}_{BA} \tag{2.32}$$

This is an *acceleration-difference equation*. It states that the acceleration of point B is equal to the acceleration of point A plus the acceleration of point B relative to point A. This relative acceleration $\mathbf{A}_{BA}$ is the acceleration of point B as it would appear to a nonrotating observer traveling with point A. As in the case of any acceleration, it is a limiting case of an average acceleration as the length of the time interval approaches zero. That is,

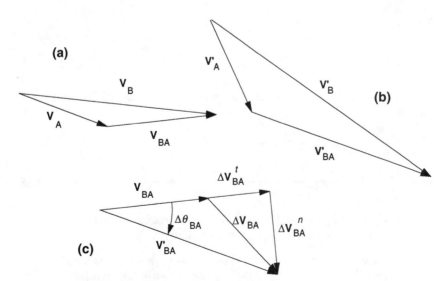

Figure 2.21 Vector changes in the velocities in Fig. 2.20.

$$\mathbf{A}_{BA} = \lim_{\Delta t \to 0}(\mathbf{A}_{BA})_{ave} = \lim_{\Delta t \to 0}\frac{\Delta \mathbf{V}_{BA}}{\Delta t} = \lim_{\Delta t \to 0}\left(\frac{\mathbf{V}'_{BA} - \mathbf{V}_{BA}}{\Delta t}\right) \qquad (2.33)$$

Now let us look at the nature of $\Delta \mathbf{V}_{BA}$ in this equation. The change, $\Delta \mathbf{V}_{BA}$, in $\mathbf{V}_{BA}$ can be seen by comparing $\mathbf{V}_{BA}$ and $\mathbf{V}'_{BA}$ in Fig. 2.21a and Fig. 2.21b and replotting them as done in Fig. 2.21c. As was shown previously (in Figs. 2.8 and 2.9 in Sec. 2.5), the *velocity* difference for two points on a rigid body is a vectors perpendicular to the line connecting those points, and it has a magnitude equal to the product of the distance R_{BA} between the two points times the angular velocity ω_{BA} of the body. Thus, for this case, the vectors $\mathbf{V}_{BA}$ and $\mathbf{V}'_{BA}$ as shown in Fig. 2.21c are perpendicular, respectively, to the lines AB and $A'B'$ in Fig. 2.20. In these figures it is noted that $\mathbf{V}_{BA}$ changes in both magnitude and direction. The change vector $\Delta \mathbf{V}_{BA}$ is shown resolved into two components: $\Delta \mathbf{V}_{BA}^{t}$ parallel to $\mathbf{V}_{BA}$ and $\Delta \mathbf{V}_{BA}^{n}$ normal to $\mathbf{V}_{BA}$ and directed from point B toward point A. We may then write Eq. (2.33) as

$$\mathbf{A}_{BA} = \lim_{\Delta t \to 0}\left(\frac{\mathbf{V}'_{BA} - \mathbf{V}_{BA}}{\Delta t}\right) = \lim_{\Delta t \to 0}\left(\frac{\Delta \mathbf{V}_{BA}^{n}}{\Delta t} + \frac{\Delta \mathbf{V}_{BA}^{t}}{\Delta t}\right) = \mathbf{A}_{BA}^{n} + \mathbf{A}_{BA}^{t} \qquad (2.34)$$

By considering Fig. 2.21 for the condition of very small changes Δt in time and very small changes in $\mathbf{V}_{BA}$, it can be seen that the magnitudes of the components on the far right-hand side of Eq. (2.34) are

$$A_{BA}^{n} = \lim_{\Delta t \to 0}\left(\frac{\Delta V_{BA}^{n}}{\Delta t}\right) = V_{BA}\lim_{\Delta t \to 0}\left(\frac{\Delta \theta_{BA}}{\Delta t}\right) = V_{BA}\left(\frac{d\theta_{BA}}{dt}\right) = V_{BA}\omega_{BA} \qquad (2.35)$$

and

$$A_{BA}^{t} = \lim_{\Delta t \to 0}\left(\frac{\Delta V_{BA}^{t}}{\Delta t}\right) = \lim_{\Delta t \to 0}\left(\frac{V'_{BA} - V_{BA}}{\Delta t}\right) = \lim_{\Delta t \to 0}\left(\frac{\Delta V_{BA}}{\Delta t}\right) = \left(\frac{dV_{BA}}{dt}\right) \qquad (2.36)$$

From Eq. (2.24) we see that $V_{BA} = \omega_{BA}R_{BA}$, so Eqs. (2.35) and (2.36) become

$$A_{BA}^{n} = V_{BA}\omega_{BA} = \omega_{BA}{}^{2}R_{BA} \qquad (2.37)$$

and $$A_{BA}^{t} = \frac{dV_{BA}}{dt} = R_{BA}\left(\frac{d\omega_{BA}}{dt}\right) = R_{BA}\alpha_{BA} \qquad (2.38)$$

where $\alpha_{BA} = d\omega_{BA}/dt$ is the *angular acceleration* of the rigid body. These results may be summarized as follows:

> The acceleration difference between two points attached to a rigid body is a vector that can be resolved into two components: One component, called the *centripetal acceleration* or *normal acceleration,* is parallel to a line

connecting the two points and is directed such as to indicate that the points are accelerating toward each other. Its magnitude is given by Eq. (2.37), and it represents the acceleration due to the changing *direction* of the velocity difference. The other component, called the *tangential acceleration,* is perpendicular to a line connecting the two points and is directed so as to indicate the direction in which the corresponding velocity difference is increasing. Its magnitude is given by Eq. (2.38), and it represents the acceleration due to the changing *magnitude* of the velocity difference.

These two types of acceleration components will be encountered frequently. Their magnitudes will each always contain two factors. The distance between the points will always be one factor. The magnitude of the centripetal component will always contain ω^2 as its other factor. The magnitude of the tangential component will always contain α as its other factor.

Figure 2.22 shows an example in which a rigid body has an angular velocity and an angular acceleration, both of which are positive (i.e., counterclockwise). Resulting centripetal and tangential acceleration components are indicated. Because ω_{BA} is counterclockwise, $\mathbf{V}_{BA}$ is directed leftward. Because α_{BA} is counterclockwise, $\mathbf{A}_{BA}^{t}$ is directed leftward. Reversal of α_{BA} and/or ω_{BA} would reverse $\mathbf{A}_{BA}^{t}$ and/or $\mathbf{V}_{BA}$, respectively. The vector $\mathbf{A}_{BA}^{n}$ will always be directed toward A, as indicated, regardless of the directions of the angular velocity and angular acceleration.

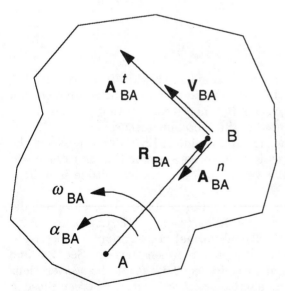

Figure 2.22 Velocity of point B relative to point A and components of acceleration of point B relative to point A.

Use of the acceleration-difference equation

Equation (2.32) relates the accelerations of two points on a rigid body to the acceleration difference for those two points. Equations (2.37), (2.38), and (2.39) relate the acceleration difference for the two points to the angular velocity and angular acceleration of the body. A machine must be analyzed to find its velocities before it can be analyzed to find its accelerations. Therefore, angular velocity in Eqs. (2.37) and (2.39) will be known when the acceleration analysis is attempted. Then it can be seen that if the acceleration of one of the two points is known, and if the angular acceleration of the body is known, the acceleration of any other point on the body can be calculated by using Eqs. (2.32), (2.37), (2.38), and (2.39). Conversely, if the accelerations of two points on the body are known, the angular acceleration can be calculated.

Acceleration difference using complex number notation

Differentiating Eq. (2.24) with respect to time gives

$$\mathbf{A}_{BA} = \frac{d\mathbf{V}_{BA}}{dt} = \frac{d\left(\omega_{BA}R_{BA}e^{j(\theta_{BA} + \pi/2)}\right)}{dt}$$

or

$$\mathbf{A}_{BA} = j\omega_{BA}R_{BA}e^{j(\theta_{BA} + \pi/2)}\left(\frac{d\theta_{BA}}{dt}\right) + \left(\frac{d\omega_{BA}}{dt}\right)R_{BA}e^{j(\theta_{BA} + \pi/2)}$$

so

$$\mathbf{A}_{BA} = -\omega_{BA}{}^2 R_{BA}e^{j(\theta_{BA})} + \alpha_{BA}R_{BA}e^{j(\theta_{BA} + \pi/2)} \tag{2.39}$$

The first term on the right-hand side of Eq. (2.39) represents the centripetal acceleration. It is seen to be a vector that is directed in a direction opposite to the vector $\mathbf{R}_{BA}$, and it is seen that its magnitude is $\omega_{BA}{}^2 R_{BA}$, just as was found in the preceding discussion.

The second term on the right-hand side of Eq. (2.39) represents the tangential acceleration. It is seen to be a vector that is rotated 90° counterclockwise from the vector $\mathbf{R}_{BA}$, and its magnitude is $\alpha_{BA}R_{BA}$, just as found previously.

The tangential acceleration and the centripetal acceleration described above in the derivations of Eqs. (2.34), (2.37), (2.38), and (2.39) are similar to the tangential and normal accelerations, respectively, described in Sec. 2.6. The difference between those in Sec. 2.6 and those just discussed is that those in Sec. 2.6 refer to the accelerations of a point that is traveling along an *arbitrary curve*, whereas those in the present section refer to the acceleration of a point (e.g., point B) relative to another point (e.g., point A) on the same rigid body. In this latter case, the path that is followed by the point of interest (B) relative to the reference point (A) is circular.

Acceleration-distribution triangles

In Fig. 2.22 the tangential and centripetal acceleration difference vectors for motion of point B relative to point A are shown normal and parallel, respectively, to vector $\mathbf{R}_{BA}$. Such vectors are shown again in Fig. 2.23, together with the total resulting acceleration difference $\mathbf{A}_{BA}$. The resultant is seen to make an angle ϕ with the vector $\mathbf{R}_{BA}$, where

$$\tan \phi = \frac{A_{BA}^t}{A_{BA}^n} \qquad (2.40)$$

Then, from Eqs. (2.37) and (2.38),

$$\tan \phi = \frac{R_{BA}\alpha_{BA}}{R_{BA}\omega^2_{BA}} = \frac{\alpha_{BA}}{\omega^2_{BA}} \qquad (2.41)$$

The variables α_{BA} and ω_{BA} are properties of the rigid body motion. Therefore, it is seen that for a given instantaneous motion condition, the angle ϕ is constant as it applies to all point pairs in the body. Equations (2.37) and (2.38) also show that both $\mathbf{A}_{BA}^n$ and $\mathbf{A}_{BA}^t$ are

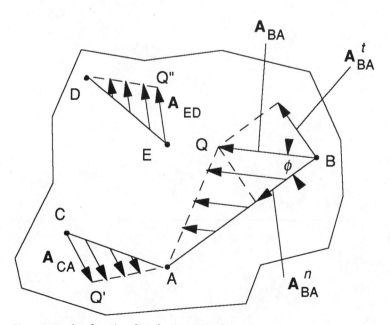

Figure 2.23 Acceleration-distribution triangles.

proportional to R_{BA}. Therefore, triangle ABQ in Fig. 2.23 may be drawn, and the vectors parallel to $\mathbf{A}_{BA}$ may be inscribed in it as shown. Each of these inner vectors is seen to be inclined to the vector $\mathbf{R}_{BA}$ by the angle ϕ, and its magnitude is proportional to the distance from point A. Therefore, each of these vectors represents the acceleration of a point on the line AB relative to point A. They are acceleration-difference vectors.

In a manner similar to that used with the velocity-distribution triangles in Sec. 2.5, these *acceleration-distribution triangles* may be used to find the acceleration difference between any two points in a rigid body, once the acceleration difference between one pair of points is known. Figure 2.22 shows example triangles ACQ' and DEQ'', which are similar to triangle ABQ and which can be used in that manner.

Whereas velocity-distribution triangles are always right triangles, acceleration-distribution triangles are not necessarily right triangles. This can somewhat complicate and limit their use, but they are useful in visualizing acceleration relationships within a rigid body.

Chapter

3

Crank-Slider Mechanisms

3.1 Introduction

The motions of the very useful linkages in this class known as *crank-slider mechanisms* may be analyzed and synthesized by considering each linkage to consist of only two moving rigid bodies (plus ground). Treatment of these mechanisms in this chapter, then, provides an easily followed introduction to the analysis and synthesis of mechanisms. Procedures for their design are given.

Crank-slider linkages are used

- For conversion of rotational motion to translational motion, and vice versa

- In reciprocating engines and compressors

- In presses for printing, stamping, cutting, punching, and shearing

- For providing timing of motions, including timing of quick-return motions (Although the timing characteristics that can be provided are somewhat limited, the crank-slider mechanism is usually the simplest way to provide them.)

In actual practice, much of the analysis and synthesis described in this chapter will be done with the assistance of a computer. However, when using a computer, engineers can easily lose sight of the significance of the interactions between the various kinematic phenomena involved. The investigation can then become one of wasteful trial and error. Therefore, this chapter emphasizes graphical analyses and syntheses that vividly depict those interactions and which are therefore an essential supplement to computerized techniques. These graphical procedures can be performed quite easily on a CAD (computer-aided drafting or design) system.

This chapter starts by analyzing and synthesizing the simple and easily visualized motion of the scotch yoke. It is then shown that the motions of the crank-slider and of the offset crank-slider mechanisms are merely distortions of or perturbations from the motion of the scotch yoke. Graphical and trigonometric methods are then applied to crank-slider motion analysis and synthesis. The concept of instant centers is used to assist in velocity analysis, and transmission angle effects are discussed. The loop-closure equation is introduced, and graphical methods for its solution and for velocity and acceleration analysis are presented. The concept of mechanical advantage is introduced, and simple methods for computing it are presented. Analytical methods using complex variables are then used to assist in manipulation of the equations and in an understanding of velocity and acceleration phenomena.

3.2   The Scotch Yoke

Figure 3.1a is a schematic drawing of a mechanism known as a *scotch yoke mechanism*. A link L_1 is pivoted to ground at point O. This link has a length L_1 and a pin P at its free end. This pin is constrained in such a manner that it can only slide up and down in a slot in the T-shaped body L_2. This body, in turn, is constrained such that it can slide only horizontally in a grounded slot or prismatic joint at the right. The pivoted link L_1 at the left is allowed to rotate through 360° and therefore is referred to as a *crank*. It can be seen that if the crank link L_1 is rotated continuously counterclockwise (for example), the slotted T-shaped body L_2 will be caused to slide horizontally back and forth. The ground constitutes a third body that we will call L_3.

Gruebler's formula [Eq. (1.2)] may be used to determine the mobility (number of degrees of freedom) of this mechanism. Note that there are three links, so $L = 3$. The pivot at O and the grounded prismatic joint at the right each possess one degree of freedom, so $J_1 = 2$. The sliding-pin joint connecting link L_1 to link L_2 at point P possesses two degrees of freedom because it allows both a rotation and a translation between links L_1 and L_2. Thus $J_2 = 1$. Then, $F = 3(3 - 1) - 2(2) - 1(1) = 1$, so the mechanism has a single degree of freedom. Because this mechanism has one degree of freedom, its position is completely determined if the position of either link L_1 or link L_2 is specified.

The sliding pin at P contacts the slot only at the two points where it is tangent to that slot. Thus any forces that are transmitted through

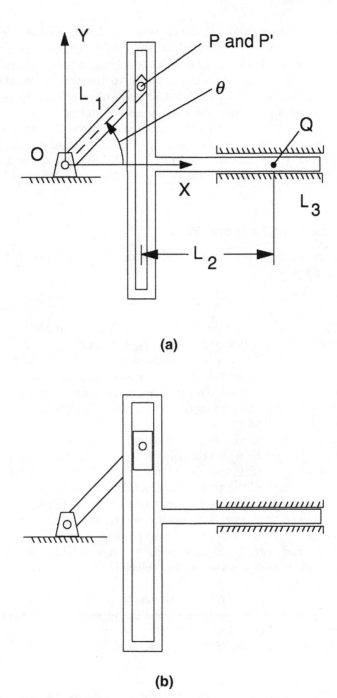

(a)

(b)

Figure 3.1   A scotch yoke.

the joint are concentrated at these points rather than being distributed over extended areas.[1] Therefore, the load-bearing capacity of the joint is quite limited. This type of joint is appropriate for use in small, lightly loaded mechanisms such as found in toasters, magnetic disk drives, etc. However, for more demanding applications, it is more appropriate to use a configuration such as shown in Fig. 3.1b, in which a block is pivoted to the crank at its end, and the block slides up and down in the slot. In this mechanism, $L = 4$, $J_1 = 4$, and $J_2 = 0$, so F is again equal to 1. By examination of Fig. 3.1a and Fig. 3.1b it can be seen that the two configurations of the mechanism perform identical motions.

3.3   Synthesis of the Scotch Yoke

In Fig. 3.1a it may be seen that the X coordinate of the point Q is given by the equation

$$x_Q = L_1 \cos \theta + L_2 \tag{3.1}$$

from which it may be seen that as θ varies from $0°$ to $180°$, the position of point Q varies between $x_{Q_{max}} = L_2 + L_1$ and $x_{Q_{min}} = L_2 - L_1$ for a total travel of $2L_1$. Therefore, to design a scotch yoke to give a desired total travel, it is necessary to choose a crank that has a length equal to one-half the total travel. The placement of point Q on the output-slide link can then be chosen to give an L_2 that will give the desired extreme positions of that point.

Although the foregoing discussion seems to assume that the mechanism is being driven by a rotation of link L_1 and that the motion of link L_2 is the output, Eq. (3.1) merely relates the positions of the two links. Regardless of which of the two links is considered to be the input link, the position of that input link can be used as the independent variable in Eq. (3.1), and that equation can be solved for the position of the other link as the dependent variable. In general, this interchangeability between which variable is independent and which is dependent will pertain for all subsequent position, velocity, and acceleration equations in this book.

The very simple reasoning used to arrive at Eq. (3.1) and to arrive at the expressions for the extreme positions of point Q constitutes position analysis of this linkage. Trivial as this particular analysis may seem, it led to relationships that can be used in synthesis of the link-

---

[1]Actually, these are line contacts in three-dimensional space, but the end views of the lines appear as points in the figure, which considers the mechanism to be a planar mechanism.

age. It will be found in subsequent sections that position analysis is useful in generating synthesis procedures.

3.4 Velocity and Acceleration Analysis of the Scotch Yoke

The velocity analysis begins by applying the velocity-difference equation (Eq. 2.20). Applying this equation to link L_1 gives

$$\mathbf{V}_P = \mathbf{V}_O + \mathbf{V}_{PO} \tag{3.2}$$

Because point O is fixed to ground, $\mathbf{V}_O = 0$, so the vector $\mathbf{V}_P = \mathbf{V}_{PO}$. Then, noting that points P and O are both attached to the same rigid body (link L_1), the velocity-difference vector $\mathbf{V}_{PO}$ and thus also $\mathbf{V}_P$ are perpendicular to L_1. The vector $\mathbf{V}_P$ is shown perpendicular to link L_1 in Fig. 3.2. There, $\mathbf{V}_P$ is also resolved into components $\mathbf{V}_{Px}$ in the X direction and $\mathbf{V}_{Py}$ in the Y direction. By analogy with Eq. (2.22), $\mathbf{V}_P$ has a magnitude of $\omega_1 L_1$, where ω_1 is the angular velocity of link L_1. Then, by inspection of Fig. 3.2, we may write

$$V_{Px} = -V_P \sin \theta_1 = -\omega_1 L_1 \sin \theta_1 \tag{3.3}$$

$$V_{Py} = V_P \cos \theta_1 = \omega_1 L_1 \cos \theta_1 \tag{3.4}$$

The same result can be obtained by expressing the vector $\mathbf{L}_1$ in complex notation as $\mathbf{L}_1 = L_1 e^{j\theta_1}$. Such notation represents link L_1 as a vector $\mathbf{L}_1$ with its tail at point O and its tip at point P, as shown in

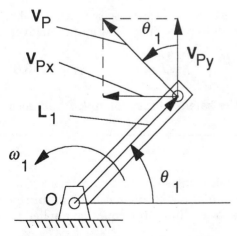

Figure 3.2 Velocity components for point P.

Fig. 3.2. Differentiating this vector $\mathbf{L}_1$ with respect to time in the manner used to derive Eq. (2.24) gives

$$\mathbf{V}_P = \mathbf{V}_{PO} = \frac{d\mathbf{L}_1}{dt} = L_1 j\left(\frac{d\theta_1}{dt}\right)e^{j\theta_1} = \omega_1 L_1 e^{j(\theta_1 + \pi/2)} \qquad (3.5)$$

When Euler's formula is applied to Eq. (3.5), it is found that the real and imaginary components of $\mathbf{V}_P$ are equal, respectively, to the X and Y components given by Eqs. (3.3) and (3.4).

Velocity of Link L_2

In Fig. 3.1 consider a point P' that is attached to link L_2 and which is coincident at the instant being considered with point P, which is attached to link L_1. Because the slot in link L_2 is vertical, any vertical component of motion of point P has no effect on the motion of point P'. However, if point P is to remain in the slot, the horizontal motion of point P' must be identical to the horizontal component of motion of point P. That is, the horizontal motion of point P is *transmitted* to point P' (or vice versa). The motion that is transmitted to point P' from point P (horizontal in this case) is known as the *motion of transmission*, and the motion of point P relative to point P' (vertical in this case) is known as the *motion of slip*.

Because link L_2 is constrained such that it cannot move vertically nor can it rotate, it can only translate horizontally, so its velocity and the velocities of all points attached to it (including point P') are equal to the *velocity of transmission*. (The velocities of all points in a rigid body in pure translation are identical to each other.) This velocity is given by Eq. (3.3).

The velocity of link L_2 also can be related to the velocity of crank L_1 by differentiating Eq. (3.1) with respect to time. Such differentiation gives

$$V_{Qx} = \frac{dx_Q}{dt} = -L_1 \sin\theta_1\left(\frac{d\theta_1}{dt}\right) = -\omega_1 L_1 \sin\theta_1 \qquad (3.6)$$

It is seen that Eq. (3.6) gives the same velocity for link L_2 as does Eq. (3.3).

Acceleration

The acceleration-difference equation such as Eq. (2.32) provides a powerful first step in the acceleration analysis of any mechanism. In the case of the scotch yoke, the acceleration-difference equation applied to link L_1 becomes

$$\mathbf{A}_P = \mathbf{A}_O + \mathbf{A}_{PO} \qquad (3.7)$$

Because point O is fixed to ground, $\mathbf{A}_O = 0$, so the vector $\mathbf{A}_P = \mathbf{A}_{PO}$. Then, noting that points P and O are both attached to the same rigid body (link L_1), the acceleration-difference vector $\mathbf{A}_{PO}$ (and thus vector $\mathbf{A}_P$) has a tangential component that is perpendicular to L_1 and a centripetal component that is directed from P toward O, as was shown in Sec. 2.7. These vector components are shown relative to link L_1 as $\mathbf{A}_P{}^t$ and $\mathbf{A}_P{}^n$, respectively, in Fig. 3.3. By analogy with Eqs. (2.38) and (2.37), $\mathbf{A}^t_P$ has a magnitude of $\alpha_1 L_1$, where α_1 is the angular acceleration of link L_1, and $\mathbf{A}^n_P$ has a magnitude of $\omega_1{}^2 L_1$. Then, by inspection of Fig. 3.3, we may write

$$A_{P_x} = -\omega_1{}^2 L_1 \cos \theta_1 - \alpha_1 L_1 \sin \theta_1 \qquad (3.8)$$

$$A_{P_y} = -\omega_1{}^2 L_1 \sin \theta_1 + \alpha_1 L_1 \cos \theta_1 \qquad (3.9)$$

Note now that the link L_2 is constrained such that it can move only horizontally. Then, as described in the preceding velocity discussion, the motion of transmission between links L_1 and L_2 is horizontal, and so the pin P *acceleration of transmission* is horizontal and given by Eq. (3.8). The acceleration of each and every point on link L_2 is therefore given by Eq. (3.8).

It can be seen that the first term on the right of Eq. (3.8) represents the contribution of the centripetal acceleration of P to the acceleration of link L_2, and the second term represents the contribution of the tangential acceleration of P to the acceleration of link L_2.

The same result may be obtained by using complex notation, as may be seen by differentiating Eq. (3.5) with respect to time to give

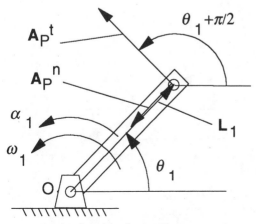

Figure 3.3 Acceleration of point P.

$$\mathbf{A}_P = \left(\frac{d\mathbf{V}_P}{dt}\right) = \alpha_1 L_1 e^{j(\theta_1 + \pi/2)} - \omega_1^2 L_1 e^{j(\theta_1)} \tag{3.10}$$

Applying Euler's formula to Eq. (3.10) produces real and imaginary components that are, respectively, equal to the X and Y components of acceleration of P given by Eqs. (3.8) and (3.9). The reader is encouraged to perform this differentiation and to then use Euler's formula.

A third method for obtaining the acceleration of link L_2 consists of differentiating Eq. (3.6) with respect to time to give

$$A_{Qx} = \left(\frac{dV_{Qx}}{dt}\right) = -\alpha_1 L_1 \sin\theta_1 - \omega_1^2 L_1 \cos\theta_1 \tag{3.11}$$

which is, of course, the same result as obtained from Eq. (3.8) because points P and Q must have the same horizontal component of acceleration.

3.5   The In-Line Crank-Slider Linkage

Figure 3.4a depicts a crank-slider mechanism schematically. A crank link L_1, represented by the vector $\mathbf{L}_1$, is pivoted to ground at point O.

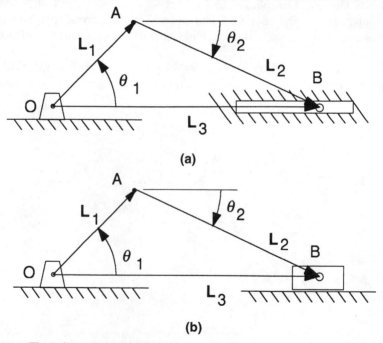

(a)

(b)

Figure 3.4 The in-line crank-slider linkage.

A link L_2, represented by the vector $\mathbf{L}_2$, is pivoted to crank link L_1 at point A. The other end of link L_2 is connected to ground by the sliding-pin joint at point B on link L_2. That is, the pin at B on L_2 slides in the grounded slot. Link L_2 is called a *coupler link* because it "couples" the input of the mechanism to the output of the mechanism. The translational freedom of the joint at B is along an axis that passes through the crank-to-ground pivot point O, so such a linkage is called an *in-line crank-slider linkage*. Rotation of crank L_1 will cause the sliding-pin joint B to slide back and forth in its slot. Alternatively, sliding the pin at B back and forth horizontally will cause link L_1 to rotate. The crank-slider mechanism in Fig. 3.4a is a single-degree-of-freedom mechanism.

Because of the limited area of contact between the pin at point B and the slot, the sliding-pin joint at point B is subject to the same load-carrying limitations that were discussed in Sec. 3.2. As a consequence, the crank-slider mechanism is often implemented in the form indicated in Figs. 3.4b, 1.12 and 1.23. Despite the presence of an additional body in these latter configurations, their motions are the same as those of the configuration in Fig. 3.4a, and their analyses and syntheses are the same as for the configuration in Fig. 3.4a.

A simple procedure for synthesizing these linkages is presented as Procedure 3.1.

3.6 Position Analysis and Synthesis of the In-Line Crank-Slider Mechanism

In Fig. 3.4a, place the origin of a stationary XY coordinate system at point O and place the X axis horizontal so that it passes through the slide at B. The X coordinate of point B is then the sum of the horizontal components of the vectors $\mathbf{L}_1$ and $\mathbf{L}_2$. The Y component of point B is the sum of the vertical components of these two vectors, noting that, as shown, θ_2 is negative (i.e., clockwise), so the vertical component of $\mathbf{L}_2$ is directed downward. Then the X and Y coordinates of point B may be related to the position of the link L_1 by

$$x_B = L_1 \cos \theta_1 + L_2 \cos \theta_2 \qquad (3.12a)$$

$$y_B = L_1 \sin \theta_1 + L_2 \sin \theta_2 = 0 \qquad (3.12b)$$

For a given set of linkage dimensions L_1 and L_2, Eqs. (3.12) contain three variables: x_B, θ_1, and θ_2. If a value is specified for one of these variables, the other two may be solved for. Equations (3.12a) and (3.12b) are easily solved if either θ_1, or θ_2 is known. However, if only x_B is given, it is more convenient to use Eq. (3.13), which is obtained simply by applying the law of cosines to the triangle OAB.

$$\cos \theta_1 = \frac{(L_1{}^2 + x_B{}^2 - L_2{}^2)}{2L_1 x_B} \tag{3.13}$$

Then, Eq. (3.12b) can be used to calculate the value of θ_2. Readers should be aware that Eq. (3.13) gives two values for θ_1. One of these solutions corresponds to the linkage position in which L_1 and L_2 are above L_3 in Fig. 3.4, and the other corresponds to the linkage position in which they are below L_3.

An alternative method for arriving at Eqs. (3.12a) and (3.12b) consists of using the vector equation

$$\mathbf{L}_3 = \mathbf{L}_1 + \mathbf{L}_2 \tag{3.14}$$

the truth of which can be seen by examining Fig. 3.4. Equation (3.14) is of a form known as a vector *loop-closure equation* because it expresses the relationship between all the vectors that represent the links that form a closed loop in the linkage.

Expressing the vectors $\mathbf{L}_1$, $\mathbf{L}_2$, and $\mathbf{L}_3$ in complex notation, Eq. (3.14) can be written as

$$L_3 e^{j\theta_3} = L_3 e^0 = L_3 = L_1 e^{j\theta_1} + L_2 e^{j\theta_2} \tag{3.15}$$

Readers should apply Euler's formula to Eq. (3.15) and note that the real equation that results is equivalent to Eq. (3.12a) and the imaginary equation that results is equivalent to Eq. (3.12b).

Synthesis of the in-line crank-slider linkage

Two special positions of the in-line crank-slider mechanism are of particular interest: the position with $\theta_1 = 0$ and the position with $\theta_1 = 180°$. For each of these positions, Eq. (3.12b) gives $\theta_2 = 0$. It can be seen that these positions are the positions in which the sliding point B is at the extremes of its travel. Then Eq. (3.12a) shows that $x_{B_{max}} = L_1 + L_2$ and $x_{B_{min}} = L_1 + L_2$, so the maximum travel (known as the *stroke length*) of point B is $x_{B_{max}} - x_{B_{min}} = 2L_1$. Therefore, in the synthesis of the crank-slider mechanism, just as in the synthesis of the scotch yoke, the crank length must be chosen to be one-half the slider stroke length.

Comparison of Eqs. (3.1) and (3.12a) suggests that the motion of the crank-slider will be similar to that of the scotch yoke, differing only because of the presence of the factor $\cos \theta_2$ in Eq. (3.12a). If θ_2 is kept small enough, $\cos \theta_2$ will be very close to unity, and the two equations will be almost identical and the two motions which they represent will be almost identical. That is, the motion of in-line crank-slider point B will be essentially a sinusoidal function of θ_1.

Equation (3.12b) may be solved for θ_2 to give

$$\sin \theta_2 = -\frac{L_1}{L_2} \sin \theta_1 = \frac{-\sin \theta_1}{\rho} \qquad (3.16)$$

where $\rho = L_2/L_1$ = the ratio of the coupler length to the crank length.

Equation (3.16) shows that if the coupler L_2 is made large compared with crank L_1, θ_2 will be small regardless of the value of θ_1, and the value of x_B will be almost a sinusoidal ("co-sinusoidal"?) function of θ_1. As L_2 is made *less large* compared with L_1, x_B, when plotted as a function of θ_1, will become less and less sinusoidal. This effect is shown in Fig. 3.5. The uppermost curve corresponds to a case in which the coupler is many times longer than the input crank. This curve is essentially sinusoidal. As the ratio of coupler length to crank length decreases, the distortion from a sine wave becomes progressively more pronounced. Note that the total range of travel remains the same if L_1 remains the same. However, as L_2/L_1 becomes smaller, the points where the curve crosses the midpoint of the travel move further from the $\theta_1 = 180°$ region. That is, the bottom of the curve becomes more horizontally spread out and blunt, and the top peaks becomes more sharp.

There are frequent occasions in which such distortion of the sine wave is useful for the timing and/or coordination of various functions in a machine. A common example of such use is found in the

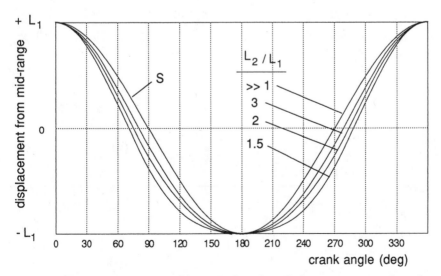

Figure 3.5 Slider displacement as a function of crank angle for various ratios of coupler length to crank length.

sewing machine needle motion shown in Fig. 3.6a. In this figure the drive shaft S rotates a disk D that constitutes the crank of length L_1. The coupler L_2 is pivoted to this crank at point A and to the slider N at point B. The slider N is caused to slide up and down, and thus causes the needle to alternately pierce the cloth and then move up and out of the way so that the cloth can be advanced horizontally for the next stitch. A typical value for the ratio of coupler length to crank length is about 2:1. Then, if the needle penetrates the cloth at the midpoint in the needle travel, Fig. 3.6b shows that the needle will be inserted in the cloth for only 151° of the machine cycle and will be out of the way for cloth movement for about 209° of the machine cycle. This linkage thus provides about 30° more cycle time out of the way of the cloth than would be available if a scotch yoke were used for this application. Also, the crank-slider mechanism avoids the added friction and wear problems associated with the slider in the yoke.

Procedure 3.1   Graphical Synthesis of an In-Line Crank-Slider Linkage

The lengths of the crank and of the coupler for a crank-slider linkage can be

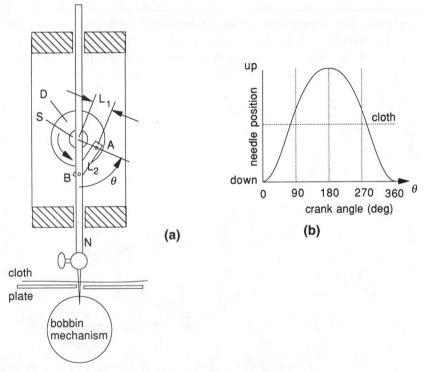

(a)   (b)

Figure 3.6 Crank-slider timing action in a sewing machine.

determined directly by referring to Fig. 3.5 or by using the simple graphical procedure illustrated in Fig. 3.7 and described below. In Fig. 3.5 the ordinate extends from $-L_1$ to $+L_1$, where L_1 is the length of the crank. The designer chooses the curve that provides the desired timing of slider position versus crank angle, and reads off the corresponding ratio L_2/L_1 of coupler length to crank length. The graphical procedure described below allows the designer to synthesize a linkage with an amount of distortion that will place

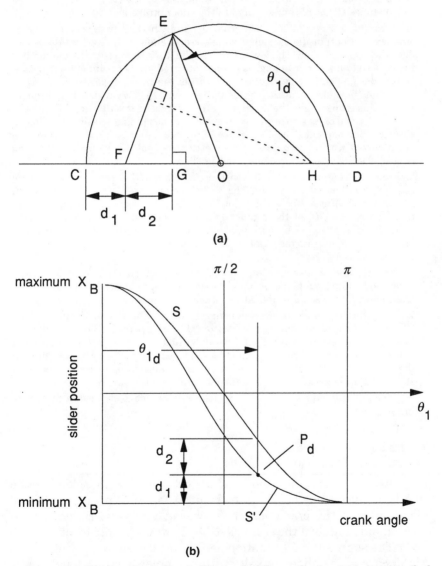

(a)

(b)

Figure 3.7 Graphical synthesis of an in-line crank-slider mechanism to give a prescribed timing.

the slider at a prescribed position in its stroke for a prescribed value of the crank input angle. The procedure is as follows:

1. Calculate the crank length $L_1 = 0.5$ times the desired output stroke length.

2. On a horizontal line, locate a point O as shown in Fig. 3.7a. This point will represent the fixed pivot point of the crank.

3. With point O as a center, draw a semicircle of radius L_1 as shown in the figure. This semicircle will represent the path of A, the moving pivot of the crank. Label as C and D the points where this semicircle intersects the horizontal line. The distance CD is, of course, equal to the slider stroke length.

4. Referring to Figs. 3.5 and 3.7b, consider that the top curve represents a cosine wave of output displacement versus crank angle θ_1 such as would be produced by a scotch yoke. The lower curve in Fig. 3.7b represents the motion that the desired crank-slider linkage is intended to produce. Let point P_d on this curve be the "design point" of particular interest. Let the crank angle at which it is desired that the slider be at the chosen design point P_d be θ_{1d} as labeled in the figure. At this crank angle, the desired location of the slider is at a desired distance d_1 from the end of its stroke that corresponds to the crank angle of 180°. This desired location of point P_d is also "distorted" by a distance d_2 from the cosine wave position that would be produced by a scotch yoke that had a crank of the same length. These distances are shown in Fig. 3.7b.

5. From the center O in step 3, construct a radius OE of the semicircle at the angle θ_{1d} as shown in Fig. 3.7a. This radius OE represents the design position of the crank. (*Alternatively,* the point P_d can be plotted to scale in Fig. 3.5, and then the ratio L_2/L_1 of the coupler length to the crank length can be determined from that figure by noting which curve the design point lies closest to. Then the complete linkage can be drawn, and the reader would proceed directly to step 8.)

6. Place a point F between points C and D at a distance d_1 from point C. Or, drop a perpendicular from point E to the horizontal line to locate point G, and then place point F between C and G at a distance d_2 from point G.

7. Draw line EF and its perpendicular bisector. The perpendicular bisector (shown dashed) will intersect the horizontal line at point H. The length of the line EH is the desired length L_2 of the coupler link. Therefore, OE represents the crank, EH represents the coupler, and OH represents the horizontal position of the slider, all at the design condition. As the crank rotates through 360°, the slider will move back and forth to the left and right of point H.

8. Check the extreme values of the transmission angle as defined in the following section.

The transmission angle
in a crank-slider linkage

As shown previously, the amount of distortion of the sinusoidal motion that is produced may be increased by decreasing the ratio ρ of the coupler length to the crank length. However, this ratio cannot be decreased indefinitely if the crank is to be allowed to rotate through a full 360°. The source of the limitation on this ratio is shown in Fig. 3.8. In Fig. 3.8a a crank-slider linkage with a coupler-to-crank length ratio $\rho = 2.0$ is shown.

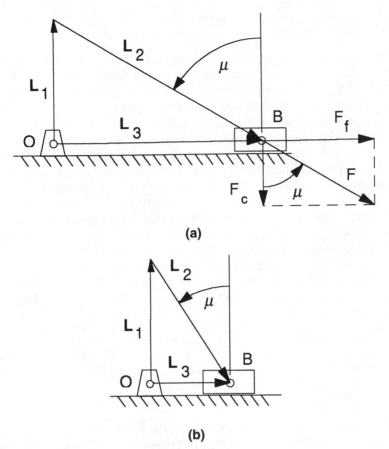

Figure 3.8 Extreme values of transmission angle.

Because link L_2 is pivoted at both ends, it is capable of transmitting forces only along and parallel to its length. For example, consider it to be in compression so that it is applying a force **F** at point **B** to whatever load may be associated with the output function of the mechanism. This force may be resolved into a component $\mathbf{F}_f$ in the direction in which point B is free to move and a component $\mathbf{F}_c$ in the direction in which point B is constrained from moving. Refer to the former direction as the *direction of freedom* and the latter as the *direction of constraint*.

The component $\mathbf{F}_f$ can act usefully on the load on which the mechanism is intended to act. The component $\mathbf{F}_c$ acts only on the constraint and is thus wasted effort that only serves to increase friction. It is therefore desirable to minimize the ratio of the wasted $\mathbf{F}_c$ to the total force **F**. This ratio is equal to the cosine of the angle μ, which is known as the *transmission angle*.

The *transmission angle* is the angle between the coupler link and the direction of constraint at the output end of the coupler.

Then large values (approaching unity) of this ratio of wasted force to total applied force will be associated with small values of the transmission angle (i.e., values for which its cosine is large). It also will be found that large velocities and large accelerations are often associated with small values of the transmission angle. Thus it is seen that small values of the transmission angle should be avoided and, therefore, large absolute values of the cosine of the transmission angle should be avoided. The linkages in Fig. 3.8 are in the positions at which the transmission angles are minimum for the linkage dimensions shown. Figure 3.8b shows a crank-slider mechanism in which the ratio ρ of coupler length to crank length is only 1.2. The associated minimum value of the transmission angle is about 33°. This compares unfavorably with the minimum value of 60° associated with the linkage in Fig. 3.8a. It is therefore seen that the smallness of the ratio ρ, and therefore the amount of distortion that may be provided by an in-line crank-slider linkage, is limited by the smallness of the transmission angle that may be tolerated.

An often-used rule of thumb is that the transmission angle should not be allowed to become smaller than about 40°. However, the range of transmission angle that can be tolerated in each individual proposed linkage design should be decided on only after investigating to determine whether the resulting forces, velocities, and accelerations are satisfactory.

Example 3.1   Synthesis of an In-Line Crank-Slider Linkage   In this example it is required that an in-line crank-slider mechanism be designed which, as its input crank rotates through 360°, will cause its slider to travel back and forth over a total distance of 4 in. It is further required that when the crank is rotated 110° from a position in which it is pointed directly along the axis along which the slider slides, the slider will be at a position that is 0.6 in from its closest approach to the crank's ground pivot.

In accordance with step 1 of Procedure 3.1, the crank length L_1 will be 2.0 in. A rough sketch of the required slider position time history will appear similar to that shown in Fig. 3.7b (or Fig. 3.5), and the total height of that plot will be 4 in (i.e., $2L_1$). The design point P_d on the sketch will be at dimension $d_1 = 0.6$ in up from the bottom extreme travel of the slider. This design point also will be at a horizontal dimension $\theta_{1d} = 110°$ from the start of the cosine wave.

Using these dimensions, the graphic construction corresponding to Fig. 3.7a will have a semicircle radius of 2.0 in, the line OE will be drawn at an angle $\theta_{1d} = 110°$, and the point F will be placed 0.6 in to the right of point C. Completing the construction according to Procedure 3.1 results in a required coupler length $L_2 = EH = 2.8$ in. This corresponds to a coupler-to-crank-length ratio of about 1.4.

Using these dimensions, the slider will travel back and forth between a point $L_2 + L_1 = 4.8$ in from the crank's ground pivot and a point $L_2 - L_1 = 0.8$ in from the crank's ground pivot as the crank rotates through $360°$.

Noting that the transmission angle is at its minimum value when the crank is at $90°$, it can be seen from Fig. 3.8 that $\cos \mu_{min} = L_1/L_2 = 2.0/2.8 = 0.71$. Consequently, the minimum value of the transmission angle is approximately $45°$, which is adequate for most applications.

3.7 Synthesis and Position Analysis of Offset Crank-Slider Linkages

Figure 3.9 schematically depicts a crank-slider linkage in which, unlike that in Fig. 3.4, the grounded pivot of the crank does *not* lie on an extension of the line along which the slider pivot B is free to slide. In Fig. 3.9 the crank pivot is offset from that line by the vector $\mathbf{L}_0$, and the linkage is referred to as an *offset crank-slider linkage*. The motion produced by this linkage is in some respects quite different from that of the previously discussed in-line crank-slider linkage.

Consider first the positions of the crank, which correspond to the extreme positions of the point at B. Figure 3.10a represents the linkage position at which the slider is at its leftmost position and $\mathbf{L}_2$ lies right on top of $\mathbf{L}_1$. In Fig. 3.10b the slider is at its right-most position. The positions of Fig. 3.10a and Fig. 3.10b are repeated in Fig. 3.11 superimposed on each other.

First, it will be noticed in Fig. 3.11 that the position of $\mathbf{L}_1$ is not π radians from the position of $\mathbf{L}_1{}'$. Therefore, if crank L_1 were rotated counterclockwise from position $\mathbf{L}_1$ to position $\mathbf{L}_1{}'$, the crank would rotate through an angle that is seen to be more than π radians by an

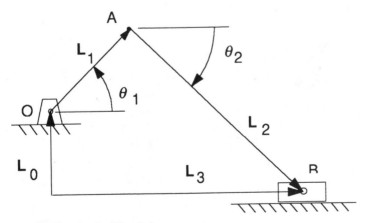

Figure 3.9 An offset crank-slider linkage.

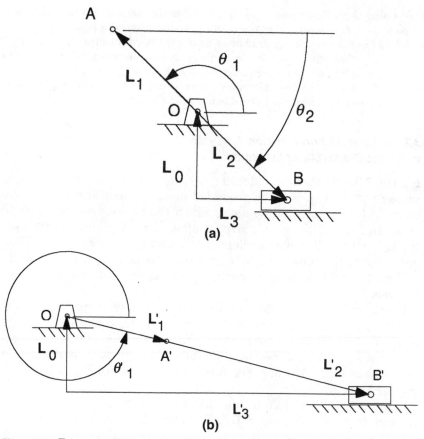

Figure 3.10   Extreme positions in an offset crank-slider linkage.

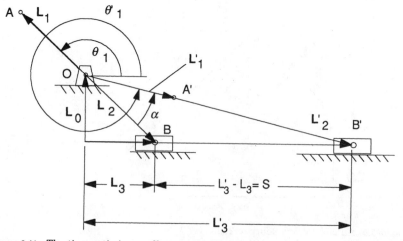

Figure 3.11   The time ratio in an offset crank-slider linkage.

amount α as shown. If the rotation of the crank were to continue counterclockwise from position L_1' back to position L_1, this second rotation
would be through an angle of less than π radians by the amount α
as shown.

If the input crank is caused to rotate counterclockwise at a constant
angular velocity ω_1, it is seen that as the crank moves from location L_1
to location L_1' and the slider moves from location B to position B', the
time elapsed will be $\Delta t_1 = (\pi + \alpha)/\omega_1$. Then, as the crank continues to
rotate and so returns from location L_1' back to location L_1 and the
slider returns from location B' back to position B, the time elapsed will
be $\Delta t_2 = (\pi - \alpha)/\omega_1$. The time increment Δt_2 is smaller than the time
Δt_1 required to make the stroke from B to B'.

An offset crank-slider mechanism used in this manner is known as
a *quick-return mechanism* because if the portion of the cycle in which
the slider travels from B to B' is considered to be the "work" stroke
portion, the time to *return* to the start of that portion is shorter than
the time consumed during the work portion. The ratio of "work" time
to return time is known as the *time ratio* and is given by

$$TR = \frac{\Delta t_1}{\Delta t_2} = \frac{(\pi + \alpha)}{(\pi - \alpha)} \tag{3.16}$$

This equation may be solved for α to give

$$\alpha = \pi \frac{(TR - 1)}{(TR + 1)} \tag{3.17}$$

so the angle α can be determined if the desired time ratio is
known.

Graphical synthesis of the offset crank-slider linkage

The offset crank-slider linkage can be synthesized to produce a desired output stroke length and desired time ratio by a graphic construction that is based on features of Fig. 3.11. In the figure it is seen
that the distance BB' is the output stroke length. It is also seen that
the crank-to-ground pivot O is at the apex of the triangle BOB', in
which the angle BOB' is of magnitude α. For each combination of desired stroke and α [which is determined by the desired time ratio, as
shown in Eq. (3.17)], there are an infinite number of triangles that
can be constructed with these characteristics. Each one of the resulting locations of the pivot O is a valid location for the crank pivot, and
the distance from each O to the extension of line BB' is the corresponding offset. In each case the length OB is the corresponding value
of $L_2 - L_1$, and the length OB' is the corresponding value of $L_2 + L_1$.

From these latter relationships, the lengths L_1 and L_2 may be computed as

$$L_1 = OB' - \frac{OB}{2} \quad \text{and} \quad L_2 = OB' + \frac{OB}{2} \tag{3.18}$$

Not all of the linkage designs that can be arrived at by constructing a triangle BOB' as described above will give a satisfactory performance, however. To visualize the variety of triangles that can be constructed having a given combination of values for α and stroke length B', and to visualize the performance limitations associated with each triangle, consider Fig. 3.12.

In this figure a circular arc has been constructed through points O, B, and B'. All points O that are the vertexes of angles of the same size α drawn through points B and B' will lie on this same circular arc. This is true because all angles that are inscribed in a given circle and which subtend the same chord of that circle are equal. (Remember high-school plane geometry?)

Then the graphical synthesis becomes as follows.

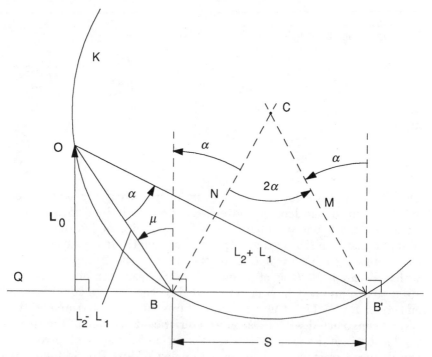

Figure 3.12 Locus of possible crank pivot points for an offset crank-slider linkage for a given time ratio.

Procedure 3.2   Synthesis of an Offset Crank-Slider Linkage to Give a Prescribed Stroke Length and a Prescribed Time Ratio

1. Choose the desired stroke length S and time ratio TR.
2. Calculate the angle α using Eq. (3.17).
3. On a horizontal line Q, place two points B and B' a distance S apart, with B on the left (see Fig. 3.12).
4. Construct a line from each of points B and B', and these lines should intersect above line Q at an angle α with each other. Label the point of intersection of these lines O. Construct a circular arc K through points B', B, and O. Alternatively, at point B' construct an upward-directed line M that is inclined leftward from a line that is perpendicular to line BB' by the angle α. At point B construct an upward-directed line N that is inclined rightward from a line that is perpendicular to line BB' by the angle α. Label as C the point of intersection of M and N. With point C as a center, construct circular arc K through points B and B'.
5. Anywhere along arc K, choose a point O.
6. From point O, draw lines to points B and B'.
7. The perpendicular distance from point O to line Q is the required offset distance L_0.
8. Measure lengths OB and OB'. Use Eqs. (3.18) to calculate the crank length L_1 and the coupler length L_2.

Steps 7 and 8 give all the dimensions that define a linkage that will produce the desired motion. However, note that as point O is moved further to the left of and upward from point B, the line OB becomes increasingly close to the vertical. Then, because this line represents the angular position of coupler link L_2 when the slider is at the leftmost end of its stroke, the transmission angle μ at that end of the stroke will tend to become smaller as O is moved leftward and upward along the arc. The actual *minimum* value that the transmission angle will reach during a cycle will be smaller than the angle between line OB and the vertical. Therefore, point O should be chosen tentatively such that line OB is inclined from the vertical by a bit more than the allowable minimum transmission angle. Then, when the resulting tentative link lengths have been computed, the linkage should be drawn to scale to check the actual value of the minimum transmission angle. This minimum value occurs when the link L_1 is directed vertically upward. If the resulting minimum value of the transmission angle is not satisfactory, a different position along the arc should be tried for point O. It will be found that when point O is in the region of the arc close to point B, the minimum value of the transmission angle also becomes small.

Example 3.2   Graphical Synthesis of an Offset Crank-Slider Linkage

Figure 3.13 illustrates an example synthesis of an offset crank-slider mechanism. The objective in this example is to synthesize an offset crank-slider mechanism with a stroke length of 100 units and a time ratio of 1.4. Equation (3.17) gives a value of $\alpha = \pi(0.4/2.4) = \pi/6$ rad $= 30°$.

Then, in Fig. 3.13, points B and B' are placed 100 units apart on the horizontal line Q. The upward-directed lines M and N are then constructed at points B' and B, respectively, such that each is inclined from the vertical by

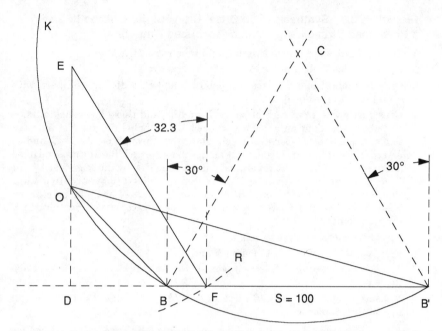

Figure 3.13 Example graphic synthesis of an offset crank-slider linkage for a prescribed time ratio and a prescribed stroke length.

the angle $\alpha = 30°$ toward the other, as shown. Using the point of intersection C of lines M and N as a center and length CB' as a radius, the circular arc K is drawn through points B and B'.

A point O is placed on arc K at a tentative position such that line OB is inclined upward and to the left at an angle of about 45° in the hope that the resulting minimum value of the transmission angle will not be less than about 40°.

The vertical distance OD is 37 units, so the offset for the mechanism is 37 units. The length OB is about 52.1 units and the length OB' is about 141.5 units, so Eqs. (3.18) give $L_1 = 44.8$ units and $L_2 = 96.8$ units.

In order to find out what the actual minimum value of the transmission angle will be, the position of the linkage at its worst transmission angle is constructed by drawing crank line L_1 vertically up from point O as shown as line OE. Then, using point E at the upper end of L_1 as a center and length of $L_2 = 96.8$ units as a radius, a circular arc R is drawn, and this arc intersects line Q at a point F.

The line OE represents the crank and line EF represents the coupler, both in the position that gives the worst transmission angle. Measuring this angle as shown, it is found to be only about 32.3°.

In an attempt to improve the minimum value of the transmission angle, other positions were tried for point O. It was found that positions on either side of that shown in Fig. 3.13 gave minimum transmission angles smaller than 32°, so the tentative position originally chosen (shown in the figure) represents about the best possible set of dimensions.

It is then concluded that an offset crank-slider quick-return mechanism with a time ratio of 1.4 will have a minimum transmission angle of no better than about 33°. Crank-slider mechanisms with larger time ratios will experience minimum transmission angles of worse than 33°, so the use of this type of mechanism for quick-return applications is quite limited. However, if the requirements for quickness of return are not very severe, the offset crank-slider mechanism offers the simplest implementation.

3.8 Position Analysis of the Crank-Slider Linkage

Once a crank-slider linkage has been synthesized, whether of the in-line type synthesized in Secs. 3.5 and 3.6 or of the offset type synthesized in Sec. 3.7, the linkage should be analyzed to determine whether the positions of its parts are satisfactory throughout the range of operation required of it. That is, a position analysis should be performed before building the mechanism. If more than a very small number of positions are to be analyzed, such analyses are most conveniently performed using a computer program such as *Working Model* or *Lincages* or any of many others on the market. However, understanding of the results of such computerized analyses is enhanced by familiarity with analytical and graphical techniques such as will now be described.

The position, velocity, and acceleration analyses are the same for in-line crank-slider linkages as they are for offset crank-slider linkages because a crank-slider linkage with no offset is merely a special case of an offset crank-slider linkage but in which L_0 is zero. The analyses in the following sections therefore will deal with the general case of the offset crank-slider linkage.

If the value θ_1 of the crank angle is given for the analysis, the positions of the coupler and the slider or sliding joint may be found graphically very simply by drawing to scale the offset vector $\mathbf{L}_0$ and the crank vector $\mathbf{L}_1$, connected tail to head and at the specified angles relative to the slide axis, as shown in Fig. 3.14a. Then an arc of radius equal to coupler length L_2 is constructed with its center at the tip of the crank vector. This arc intersects the slide axis at a point B, which corresponds to the position of the slider joint, and the positions of all links have been determined. Notice that there are two points at which the arc intersects the slide axis, corresponding to two possible positions for the slider.

If the position of the slider is given, the vectors $\mathbf{L}_3$ and $\mathbf{L}_0$ can be drawn to scale and perpendicular to each other, as in Fig. 3.14b. Then an arc of radius equal to crank length L_1 can be constructed with the tip of vector $\mathbf{L}_0$ as its center, and an arc of radius equal to coupler length L_2 can be constructed with tip of vector $\mathbf{L}_3$ as its center. The intersection of these two arcs is the location of the joint at A that joins the crank to the coupler, as shown in the figure, and the complete position of the linkage has been determined. Notice that the arcs intersect at two points, corresponding to two possible positions for links L_1 and L_2 as shown.

These graphical analyses can be performed using a CAD system, and very accurate results can be obtained. However, if the answers obtained by graphical analysis are not sufficiently precise, or if many positions must be analyzed, position analysis also may be performed analytically on a programmable pocket calculator, can be programmed into a spread sheet, or can be programmed into a computer. The equations used in such analytic means can be derived from the *vector loop closure equation:*

$$L_0 + L_1 + L_2 = L_3 \tag{3.19}$$

The truth of this vector equality may be seen by referring to Fig. 3.9 and by following the corresponding vectors in that figure sequentially from tail to head. Each of these vectors may be expressed in complex exponential form, so Eq. (3.19) becomes

$$L_0 e^{j\theta_0} + L_1 e^{j\theta_1} + L_2 e^{j\theta_2} = L_3 e^{j\theta_3} \tag{3.20}$$

In Figure 3.9 it is seen that $\theta_0 = 90°$ and $\theta_3 = 0$. Then Eq. (3.20) becomes

$$L_0 e^{j\pi/2} + L_1 e^{j\theta_1} + L_2 e^{j\theta_2} = L_3 e^{j0} \quad \text{or} \quad jL_0 + L_1 e^{j\theta_1} + L_2 e^{j\theta_2} = L_3 \tag{3.21}$$

Euler's formula may be applied to each term in Eq. (3.21), and the resulting equation can be separated into an equation in only real terms and an equation in only imaginary terms. If each of the terms in the equation in imaginary terms is divided by j, the two equations become

$$L_1 \cos \theta_1 + L_2 \cos \theta_2 = L_3 \tag{3.22a}$$

$$L_0 + L_1 \sin \theta_1 + L_2 \sin \theta_2 = 0 \tag{3.22b}$$

Equations (3.22) are seen to be very similar to Eqs. (3.12) when it is realized that L_3 represents the X distance from the grounded crank pivot to the slider position at B. If a value for either of the angles θ_1 or θ_2 is known, Eq. (3.22b) can be solved for the value of the other. Then the two angle values can be inserted into Eq. (3.22a), and the value of L_3 may be calculated. Because of the presence of the trigonometric functions in these equations, two solutions exist, corresponding to the two solutions shown in Fig. 3.14a.

If the value of neither θ_1 nor θ_2 is known but the position of the slider as represented by the value L_3 is known, Eqs. (3.22) are not easily solved for values for those angles. In this case it is more appropriate to use straightforward trigonometry. For the trigonometric method, refer

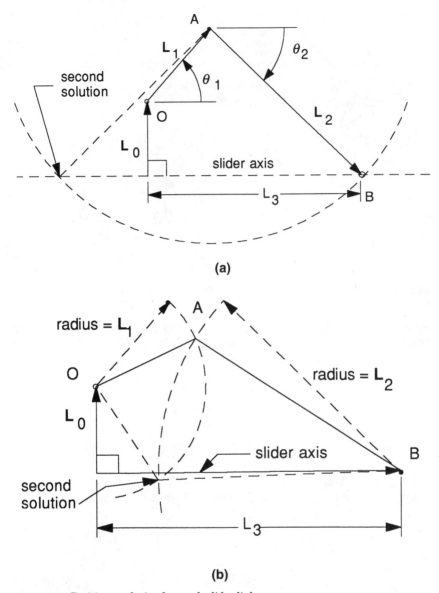

(a)

(b)

Figure 3.14 Position analysis of a crank-slider linkage.

to Fig. 3.15. In this figure a diagonal L_d is drawn in the quadrilateral formed by the four vectors $\mathbf{L}_0$, $\mathbf{L}_1$, $\mathbf{L}_2$, and $\mathbf{L}_3$, dividing that quadrilateral into two triangles.

In the lower triangle, it is seen that L_0 and L_3 are known, so $L_d = (L_0^2 + L_3^2)^{0.5}$ and $\tan \gamma = -L_0/L_3$, being careful to place γ in the

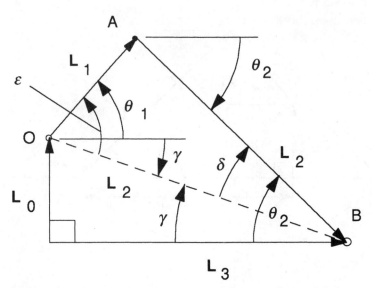

Figure 3.15   Trigonometric position analysis of a crank-slider linkage.

appropriate quadrant. For the vector values used in Fig. 3.15, γ is in the fourth quadrant. Then the upper triangle can be solved for values for angles δ and ϵ by using the law of cosines. This gives

$$\cos \delta = \frac{L_2^2 + L_d^2 - L_1^2}{2L_2L_d} \tag{3.23a}$$

$$\cos \epsilon = \frac{L_1^2 + L_d^2 - L_2^2}{2L_1L_d} \tag{3.23b}$$

Using the values thus found for γ, δ, and ϵ, and being careful with signs, values for θ_1 and θ_2 may be found from the relationships

$$\theta_1 = \epsilon + \gamma \tag{3.23c}$$

and

$$\theta_2 = \delta + \gamma \tag{3.23d}$$

as may be seen in Fig. 3.15. Again, it should be noted that each of Eqs. (3.23a) and (3.23b) gives two solutions, corresponding to the two positions shown in Fig. 3.14b. Care must be taken that the sign of δ be made the opposite of the sign of ϵ. (Note their directions in Fig. 3.15.)

Readers are encouraged to show that use of the complex variable method of analysis *automatically* controls the signs of the angle values and produces the same relationships as contained in Eqs. (3.23). This can be done by placing the terms in L_3 and L_0 in Eq. (3.21) on the right-hand side and expressing the right-hand side as $L_d e^{j\gamma}$. Then multiply all terms by $e^{-j\gamma}$. Use Euler's formula to obtain scalar equations, and then solve those equations for $\cos(\theta_1 - \gamma)$ and $\cos(\theta_2 - \gamma)$.

3.9 Velocity Analysis of Crank-Slider Linkages

Figure 3.16 depicts an offset crank-slider linkage, and this figure will be used to illustrate relationships in the velocity analysis.

Velocity analysis of the crank-slider linkage starts with application of the velocity-difference equation, just as did velocity analysis of the scotch yoke. The velocity relationships are illustrated in Fig. 3.16. In this figure crank link L_1 is shown pivoted to ground at point O, and it

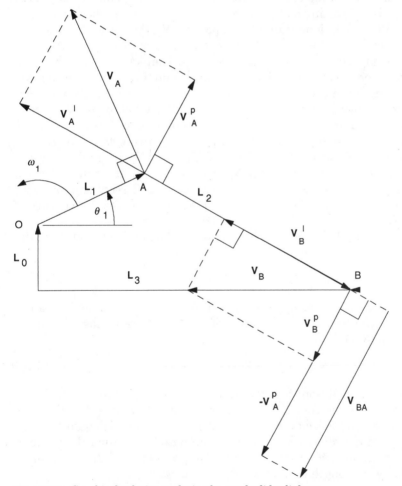

Figure 3.16 Graphical velocity analysis of a crank-slider linkage.

is rotating counterclockwise at an angular velocity of ω_1 rad/s. The applicable velocity-difference equation for the crank is

$$V_A = V_O + V_{AO} \qquad (3.24)$$

Then, because link L_1 is a rigid body, V_{AO} is perpendicular to link L_1 and has a magnitude of $\omega_1 L_1$, and because $V_O = 0$, the vector $V_A = V_{AO}$.

Graphical velocity analysis of crank-slider linkages

For a graphical analysis of the velocities of the linkage if the angular velocity of the crank is known, first calculate the vector V_A in the manner just described. Then plot V_A to some convenient scale on a scale drawing of the linkage, such as shown in Fig. 3.16. This velocity V_A of point A is shown in the figure resolved into a perpendicular component V_A^p and a longitudinal component V_A^l that are, respectively, perpendicular to line AB and along line AB. Because points A and B are both attached to the same rigid body (coupler L_2), the longitudinal component of the velocity of one must equal that of the other. We may then plot V_B^l equal to V_A^l but located at point B as shown. Now, V_B^l is only one of two components of V_B, the other being V_B^p, which is perpendicular to line AB. Because point B is constrained by the sliding joint to move only horizontally, these two orthogonal components must add to produce a velocity V_B that is horizontal. Then the vector resolution rectangle for V_B may be drawn at point B (shown as dashed lines in the figure). Construction of this rectangle produces the vectors V_B and V_B^p and thus completely determines the velocity of point B.

The velocity-difference equation may now be applied to link L_2 in the form $V_B = V_A + V_{BA}$, which may be rewritten as

$$V_{BA} = V_B - V_A = (V_B^p - V_A^p) + (V_B^l - V_A^l) \qquad (3.25)$$

Because points A and B are attached to the same rigid body (coupler L_2), the longitudinal components of their velocities must be equal. Then Eq. (3.25) becomes

$$V_{BA} = (V_B^p - V_A^p) + 0 = V_B^p + (-V_A^p) \qquad (3.26)$$

This vector addition is shown to the right of point B in Fig. 3.16, and it is seen that the resulting V_{BA} is perpendicular to link L_2. We know from rigid body velocity analysis that its magnitude is equal to $\omega_2 L_2$, so the angular velocity of L_2 may be computed by scaling the value of the magnitude of V_{BA} from the vector diagram and dividing that magnitude by the length of link L_2.

Example 3.3 Graphical Velocity Analysis of an Offset Crank-Slider Linkage Figure 3.16 is a reproduction of a schematic scale drawing of a linkage whose dimensions (in arbitrary units) are $L_0 = 5.5, L_1 = 10, L_2 = 20$, and $\theta_1 = 25°$. Position analysis gives $L_3 = 26.5$ and $\theta_2 = -30°$. For this analysis, the angular velocity of the crank is assumed to be $\omega_1 = 3$ rad/s, which, as shown, is counterclockwise.

Then the velocity of point A is plotted to some chosen scale at point A, perpendicular to line OA, and with a magnitude of $V_A = \omega_1 L_1 = (3)(10) = 30$ units/s. The vector $\mathbf{V}_A$ is then resolved into its longitudinal and perpendicular components, respectively, along line AB and perpendicular to line AB as shown.

The longitudinal component $\mathbf{V}_B^l$ is then laid out along line AB and equal to $\mathbf{V}_A^l$ as shown. From the tip of this vector $\mathbf{V}_B^l$ a perpendicular is drawn to intersect the horizontal. The total velocity $\mathbf{V}_B$ of point B is then a vector drawn from point B to that intersection. This is seen to be true because $\mathbf{V}_B$ is the sum of the orthogonal components $\mathbf{V}_B^l$ and $\mathbf{V}_B^p$ as drawn in the figure.

Measuring the lengths of vectors on the diagram from which Fig. 3.16 was reproduced gives $V_B = 28$ units/s, $V_B^p = 13.7$ units/s, and $V_A^p = 17.5$ units/s. Then, using Eq. (3.26), the construction at point B shows that $V_{BA} = 31.2$ units/s. Then, dividing V_{BA} by L_2 gives $\omega_2 = 31.2/20 = -1.56$ rad/s. The angular velocity ω_2 is negative because it is seen that the velocity $\mathbf{V}_{BA}$ is downward, indicating that point B is moving downward relative to point A, thereby causing a clockwise angular velocity of the coupler.

Graphical velocity analysis of crank-slider linkages using instant centers

Once the velocity of point A has been determined, the velocity of any point on coupler L_2 may be determined graphically by using the properties of instant centers. Figure 3.17 is a schematic scale drawing of a crank-slider mechanism with the velocity of the crank point A drawn at A to some chosen scale. The coupler L_2 is shown as an extended body, to which is attached its representative vector $\mathbf{L}_2$.

The instant center of L_2 relative to ground can be found by constructing at each of two points a perpendicular to the velocity (relative to ground) of that point. It is known that $\mathbf{V}_A$ is perpendicular to vector $\mathbf{L}_1$, so the instant center must lie somewhere along vector $\mathbf{L}_1$ or along an extension of vector $\mathbf{L}_1$. It is known that the velocity of point B on the slider must be horizontal, so the instant center must lie somewhere along a vertical line through point B. The instant center of the coupler relative to ground (L_2 relative to L_3) is therefore at the intersection of these two lines and is indicated as IC_{23} in Fig. 3.17.

A velocity-distribution triangle may then be drawn using $\mathbf{V}_A$ and the line segment from A to IC_{23} as its two perpendicular sides as shown. Point B' may be placed on the line from A to IC_{23} such that the distance from B' to the instant center is equal to the distance from B to

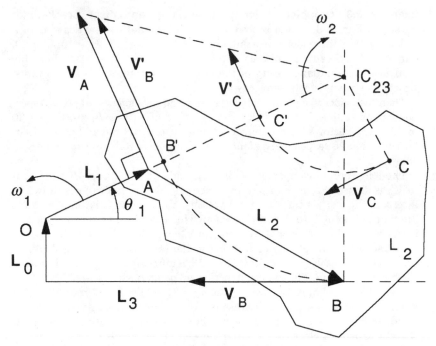

Figure 3.17 Graphical velocity analysis of a crank-slider linkage using instant centers.

that instant center. The velocity of point B' may then be determined by using the velocity-distribution triangle as described in Chap. 2. The magnitude of the velocity $\mathbf{V}_B$ of point B is then equal to that of point B', but $\mathbf{V}_B$ is perpendicular to the line from B to IC_{23}; i.e., it is horizontal, as shown in the figure.

The velocity of any other point that is attached to the coupler may be determined by a similar procedure, as shown in Fig. 3.17, e.g., for point C.

As described in Chap. 2, the angular velocity of any rigid body such as the coupler may be calculated by dividing the magnitude of the velocity of any point by the distance from that point to the instant center. That is, because $V_A = \omega_2 r$, where r is the distance from point A to IC_{23}, the angular velocity of the coupler is simply $\omega_2 = V_A/r$.

Mechanical advantage in a crank-slider linkage

Although, as discussed in Sec. 1.2 (also see Fig. 1.1), kinematics is not concerned with forces and torques, kinematic analysis (particularly velocity analysis) can be used very effectively to determine relationships between forces and torques that occur in machines. Such a procedure is particularly straightforward for the case of relating static forces and torques when the friction losses in the mechanism being analyzed are small and when energy storage in the mechanism is small. In such cases, very close approximations to the force and torque relationships may be obtained by assuming that the conservation of energy dictates that the work that is put into the mechanism is equal to the work that the mechanism produces at its output. Therefore, in such cases, because power is simply the rate at which work is done, the power input to the mechanism is assumed equal to the power output of the mechanism.

If power is input to a single-degree-of-freedom mechanism by means of applying a force F_{in} to a point on an input link while that point on the input link is moving with a velocity component V_{in} in the direction of that force, the input power can be expressed as $P_{in} = F_{in}V_{in}$. If the power is being input to the mechanism by means of applying a torque T_{in} to an input link while that link is rotating at an angular velocity ω_{in} about an axis parallel to the input torque axis, the input power can be expressed as $P_{in} = T_{in}\omega_{in}$.

Similar relationships pertain to the output power, so we may write $P_{in} = P_{out}$ as

$$
\left.\begin{array}{c} F_{in}V_{in} \\[6pt] or \\[6pt] T_{in}\omega_{in} \end{array}\right\} = \left\{\begin{array}{c} F_{out}V_{out} \\[6pt] or \\[6pt] T_{out}\omega_{out} \end{array}\right. \tag{3.27}
$$

where it must be realized that an input *force* can produce either an output force or an output torque and also that an input *torque* can produce either an output force or an output torque.

Mechanisms often are used to magnify input forces or torques. Therefore, the ratio of the output force or torque to the input force or torque is referred to as the *mechanical advantage* of the mechanism. This ratio usually is considered to be a nondimensional ratio of force to force or torque to torque. In such cases, Eq. (3.27) is seen to give the mechanical advantage as

$$MA = F_{out}/F_{in} = V_{in}/V_{out} \tag{3.28a}$$

$$\text{or}\quad MA = T_{out}/T_{in} = \omega_{in}/\omega_{out} \tag{3.28b}$$

A more general concept of mechanical advantage does *not* require it to be a nondimensional ratio. For this concept we may write

$$MA = F_{out}/T_{in} = \omega_{in}/V_{out} \tag{3.29a}$$

$$\text{or}\quad MA = T_{out}/F_{in} = V_{in}/\omega_{out} \tag{3.29b}$$

The dimensions of the velocities in these latter equations should be radians per second for angular velocities and either feet per second or meters per second for translational velocities. The resulting dimensions of the mechanical advantage will then be pounds per pound-foot or newtons per newton-meter, respectively, for Eq. (3.29a) and pound-feet per pound or newton-meters per newton, respectively, for Eq. (3.29b).

It can be seen, then, that a complete velocity analysis provides all the velocity values that allow Eqs. (3.28) or (3.29) to be used to compute the value of the mechanical advantage of a single-degree-of-freedom mechanism. Therefore, the simple relationships of Eqs. (3.28) and (3.29) allow easy evaluation of force and torque relationships. For all but the very simplest mechanisms, such evaluations would otherwise require complicated computations involving moment arms, angles of application of forces, and so on.

By examining Fig. 3.16, it may be seen that as the crank vector $\mathbf{L}_1$ approaches parallelism with the coupler vector $\mathbf{L}_2$, the vector $\mathbf{V}_A$ approaches perpendicularity with the coupler vector $\mathbf{L}_2$ and therefore the longitudinal component of $\mathbf{V}_A$ along $\mathbf{L}_2$ approaches zero magnitude. Therefore, in this position, known as a *toggle position,* the velocity of point B must be zero, regardless of the angular velocity of the crank. Figures 3.10 and 3.11 show that, in general, there are two such positions for a crank-slider linkage and that they constitute the slider's extreme positions.

If the crank motion is considered to be the input motion, Eq. (3.29a) shows that at the toggle positions, the mechanical advantage is infinite. In the *vicinity* of that position, the mechanical advantage is seen to be very large. Such large values of mechanical advantage are used in punch presses, printing presses, and some types of clamps to convert reasonable input torques to very large output forces.

If, conversely, the motion of the slider is considered to be the input motion and the crank rotation is considered to be the output motion, as in the case of piston and crank type engines, Eq. (3.29b) shows that the mechanical advantage is zero at these toggle positions. In such mechanisms, the toggle position is known as the "dead center" position, and no amount of force applied to the piston (slider) will cause the crank to rotate from that position. Auxiliary means such as a flywheel and/or an additional piston acting at another angle are used to cause the crank to continue to rotate from such a "dead center" position.

Analytical velocity analysis of crank-slider linkages

Velocity analysis of a crank-slider linkage by analytical means could proceed very much as discussed in the graphical analysis, but with the vectors and their components expressed in algebraic form (either trigonometrically or in complex variables). The manipulations would then be algebraic rather than graphical. However, rather than using an algebraic analogue of the graphical method, it is more compact and convenient to start with the vector loop closure equation and then to differentiate that equation.

For the crank-slider linkage as shown in Fig. 3.9, the vector loop closure equation, as given by Eq. (3.19), is

$$\mathbf{L}_0 + \mathbf{L}_1 + \mathbf{L}_2 = \mathbf{L}_3 \tag{3.30}$$

When expressed in terms of complex variables, this becomes Eq. (3.21), which is

$$jL_0 + L_1 e^{j\theta_1} + L_2 e^{j\theta_2} = L_3 \tag{3.31}$$

Noting that the only quantities in Eq. (3.31) that vary with time are θ_1, θ_2, and L_3, differentiation of Eq. (3.31) with respect to time gives

$$\omega_1 L_1 e^{j(\theta_1 + \pi/2)} + \omega_2 L_2 e^{j(\theta_2 + \pi/2)} = \left(\frac{dL_3}{dt}\right) = V_B \tag{3.32}$$

In Eq. (3.32), note that the first term on the left-hand side represents $\mathbf{V}_A$ and the second term represents $\mathbf{V}_{BA}$. By using Euler's formula, Eq. (3.32) can be separated into an equation in only real terms and an equation in only imaginary terms. The equation in only real terms is

$$\omega_1 L_1 \cos(\theta_1 + \pi/2) + \omega_2 L_2 \cos(\theta_2 + \pi/2) = V_B$$

which reduces to

$$-\omega_1 L_1 \sin(\theta_1) - \omega_2 L_2 \sin(\theta_2) = V_B \tag{3.33}$$

and the equation in only imaginary terms is

$$j\omega_1 L_1 \sin(\theta_1 + \pi/2) + j\omega_2 L_2 \sin(\theta_2 + \pi/2) = 0$$

which reduces to

$$\omega_1 L_1 \cos(\theta_1) + \omega_2 L_2 \cos(\theta_2) = 0 \tag{3.34}$$

Position analysis must be completed before velocity analysis can be performed. Position analysis provides values for θ_1 and θ_2, so Eqs. (3.33) and (3.34) become linear equations in the variables ω_1, ω_2, and V_B. If a value for any one of these three variables is known or given, these two equations may be solved for values of the other two variables. For simplicity in evaluation, these equations may be reduced to two equations, each of which contains only two variables. This can be done by solving Eq. (3.34) for ω_2 and substituting the result in Eq. (3.33), which then becomes

$$\omega_1 L_1(\tan \theta_2 \cos \theta_1 - \sin \theta_1) = V_B \qquad (3.35)$$

Equations (3.35) and (3.34) may then be used as alternates to Eqs. (3.33) and (3.34).

An alternative means for obtaining the relationship implied by Eq. (3.35) uses a technique which, in subsequent chapters, will be found very convenient for separating variables in vector equations that are expressed in complex notation. This technique consists of multiplying each term in the equation by some factor that causes the term involving one of the variables to become completely real or completely imaginary. This variable then appears in only one of the two scalar equations obtained by using Euler's formula. For example, multiplying each term in Eq. (3.32) by $e^{-j\theta_2}$ produces

$$\omega_1 L_1 e^{j(\theta_1 - \theta_2 + \pi/2)} + \omega_2 L_2 e^{j(\pi/2)} = V_B e^{-j\theta_2}$$

The corresponding scalar equation that contains only real terms is

$$\omega_1 L_1 \cos(\theta_1 - \theta_2 + \pi/2) + \omega_2 L_2 \cos(\pi/2) = V_B \cos(-\theta_2) \qquad (3.36)$$

It will be noted that the term containing ω_2 is identically zero, so the equation contains only the variables ω_1 and V_B. Then, solving for ω_1 and using trigonometric identities for angles in different quadrants gives

$$\omega_1 = \frac{V_B \cos \theta_2}{L_1 \sin(\theta_2 - \theta_1)} \qquad (3.37)$$

By applying trigonometric identities, it can be shown that Eq. (3.37) is identical to Eq. (3.35).

Analytical determination of velocity of a coupler point

Figure 3.18 depicts a crank-slider linkage in which the coupler consists of an extended rigid body that is pivoted to the crank at point A and whose point B slides along a horizontal path. Attached to the coupler body is a point C whose velocity is to be calculated. The position of point C can be related to a grounded origin at point O by the vector $\mathbf{R}_C$ from that origin to point C. The vector $\mathbf{R}_{CA}$ is fixed to the rigid coupler body at an angle β with respect to line AB and extends from point

A to point C. As can be seen in Fig. 3.18, $\mathbf{R}_C = \mathbf{L}_1 + \mathbf{R}_{CA}$. Expressing this vector equation in complex variable form gives

$$R_C e^{j\theta_c} = L_1 e^{j\theta_1} + R_{CA} e^{j(\theta_2 + \beta)} \tag{3.38}$$

where β is the fixed inclination of line AC relative to line AB. Then, because the velocity of point C is equal to the time rate of change of the position vector $\mathbf{R}_C$, the right-hand side of Eq. (3.38) may be differentiated to give the velocity of point C as

$$\mathbf{V}_C = j\omega_1 L_1 e^{j\theta_1} + j\omega_{CA} R_{CA} e^{j(\theta_2 + \beta)}$$

$$\text{or} \qquad \mathbf{V}_C = \omega_1 L_1 e^{j(\theta_1 + \pi/2)} + \omega_{CA} R_{CA} e^{j(\theta_2 + \beta + \pi/2)} \tag{3.39}$$

It is seen that Eq. (3.39) is a velocity-difference equation of the form $\mathbf{V}_C = \mathbf{V}_A + \mathbf{V}_{CA}$. Because line AC is attached to the coupler, $\omega_{CA} = \omega_2$. Then the corresponding scalar equations in real and imaginary variables are, respectively,

$$(V_C)_x = \omega_1 L_1 \cos(\theta_1 + \pi/2) + \omega_2 R_{CA} \cos(\theta_2 + \beta + \pi/2)$$

$$\text{and} \qquad j(V_C)_y = j\omega_1 L_1 \sin(\theta_1 + \pi/2) + j\omega_2 R_{CA} \sin(\theta_2 + \beta + \pi/2)$$

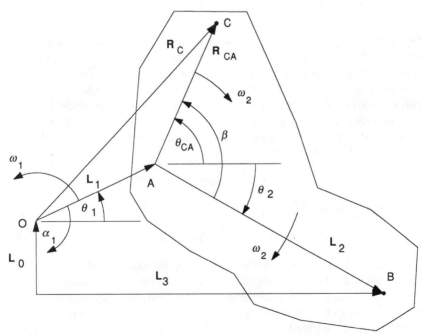

Figure 3.18 Graphical determination of velocity of a coupler point in a crank-slider linkage.

Eliminating the $\pi/2$ terms from the angle arguments and canceling out the factors j gives

$$(V_C)_x = -\omega_1 L_1 \sin \theta_1 - \omega_2 R_{CA} \sin(\theta_2 + \beta) \qquad (3.40a)$$

and $(V_C)_y = \omega_1 L_1 \cos \theta_1 + \omega_2 R_{CA} \cos(\theta_2 + \beta) \qquad (3.40b)$

3.10 Acceleration Analysis of Crank-Slider Linkages

It was seen in Sec. 3.9 that velocity analysis can be used not only for finding relationships between various velocities in a linkage but also for determining relationships between various forces and torques in that linkage. If the mechanism that is being designed or analyzed involves or interacts with any appreciable masses, it may be important to know how those masses are being accelerated and therefore what inertial reaction forces are being generated. Analysis of these accelerations can be performed once the position and velocity analyses have been completed.

Graphical acceleration analysis of crank-slider linkages

Often a graphical acceleration analysis of a mechanism at a particular stage of its motion is helpful in finding the source of some troublesome force or acceleration and in choosing means for improving mechanism performance. Graphical acceleration analysis vividly shows which acceleration vectors contribute prominently to overall acceleration.

Acceleration analysis starts with the use of acceleration-difference equations, which, when applied to the crank vector $\mathbf{L}_1$ and to the coupler vector $\mathbf{L}_2$ in Fig. 3.19, become

$$\mathbf{A}_A = \mathbf{A}_O + \mathbf{A}_{AO} \qquad \text{and} \qquad \mathbf{A}_B = \mathbf{A}_A + \mathbf{A}_{BA}$$

which combine to give

$$\mathbf{A}_B = \mathbf{A}_O + \mathbf{A}_{AO} + \mathbf{A}_{BA} \qquad (3.41)$$

in which it is seen that because point O is fixed to ground, $\mathbf{A}_O = 0$. Then, because the crank and the coupler are both rigid bodies, Sec. 2.7 shows that each of the remaining terms on the right-hand side of Eq. (3.41) may be resolved into a tangential component and a normal component. This gives

$$\mathbf{A}_B = \mathbf{A}_{AO}^t + \mathbf{A}_{AO}^n + \mathbf{A}_{BA}^t + \mathbf{A}_{BA}^n \qquad (3.42)$$

Section 2.7 shows that the vector $\mathbf{A}_{AO}^{t}$ (i.e., the tangential component of the acceleration of A) is directed perpendicular to line OA at point A and in the direction of the angular acceleration of the crank and has a magnitude of $L_1\alpha_1$. Note that, as drawn, Fig. 3.19a shows a *negative* (clockwise) value for α_1. From Sec. 2.7 it is seen that the vector $\mathbf{A}_{AO}^{n}$ (i.e., the normal or centripetal component of the acceleration of A) is directed toward point O from point A and has a magnitude of $L_1\omega_1^2$.

Similarly, the vector $\mathbf{A}_{BA}^{t}$ is directed perpendicular to line AB at point B and in the direction of the angular acceleration of the coupler and has a magnitude of $L_2\alpha_2$. The vector $\mathbf{A}_{BA}^{n}$ is directed toward point A from point B and has a magnitude of $L_2\omega_2^2$.

Now, because the position analysis and the velocity analysis must be completed before the acceleration analysis is attempted, the directions of lines OA and AB are known and the values of ω_1 and ω_2 are known. Therefore, the directions of all the vectors in Eq. (3.42) are known, and the magnitudes of the vectors $\mathbf{A}_{AO}^{n}$ and $\mathbf{A}_{BA}^{n}$ can be calculated from

$$A_{AO}^{n} = L_1\omega_1^2 \quad \text{and} \quad A_{BA}^{n} = L_2\omega_2^2 \tag{3.43}$$

If, for the acceleration analysis, the angular acceleration α_1 of the crank is given, then the magnitude of the vector $\mathbf{A}_{AO}^{t}$ can be calculated from

$$A_{AO}^{t} = L_1\alpha_1 \tag{3.44}$$

Then we may write Eq. (3.42) as

$$\overset{ud}{\mathbf{A}_{B}} = \overset{md}{\mathbf{A}_{AO}^{t}} + \overset{md}{\mathbf{A}_{AO}^{n}} + \overset{ud}{\mathbf{A}_{BA}^{t}} + \overset{md}{\mathbf{A}_{BA}^{n}} \tag{3.45}$$

where the notation directly above each vector denotes the status of knowledge of the magnitude and direction of that vector. The left-hand character refers to magnitude and the right-hand character refers to direction. For instance, the *magnitude* of the vector on the left-hand side of the equation is unknown, as indicated by the u (for unknown) as the left-hand character above that vector, whereas the *direction* of that vector *is* known, as indicated by the d (for direction) as the right-hand character above that same vector. Both the magnitude and the direction of each of the first two vectors on the right-hand side of the equation are known, as indicated by the presence of both an m and a d above each of the vectors. If the *direction* of one of the vectors had been unknown, a u (for unknown) would have been substituted for the d in the right-hand space above that vector.

Examination of Eq. (3.45) shows that this vector equation contains two unknowns: the magnitudes of the vectors $\mathbf{A}_B$ and $\mathbf{A}_{BA}^{t}$. The two scalar equations that Eq. (3.45) represents may therefore be solved for these unknowns. Because there is a solution to this vector equation, it also may be solved graphically. (The analytical solution will be described later.)

Graphical analysis of a linkage starts with writing an acceleration-difference equation such as Eq. (3.41) and expanding the acceleration vectors in terms of their normal and tangential components to produce an equation such as Eq. (3.45). Then a vector diagram is drawn to scale in such a manner that the vectors in that diagram add up in the same way that the vectors add up in the expanded acceleration-difference equation. In drawing that diagram, it will be found that not all the vectors are known. Therefore, those vectors which are completely known should be drawn first and connected tail to head as indicated by the additions shown in the equation. Then, in the manner shown in the following example, the unknown vectors are added to the diagram to the extent to which their characteristics are known. This will produce a closed vector loop in which the unknown vector quantities will be determined automatically.

Example 3.4 Graphical Acceleration Analysis of a Crank-Slider Linkage

Figure 3.19a is a reproduction of a diagram of the same linkage that was analyzed in the example graphical velocity analysis in Sec. 3.9. An acceleration-difference equation can be written for this linkage, and it will be identical to Eq. (3.45). From the results of the velocity analysis, the magnitudes A_{AO}^{n} and A_{BA}^{n} of the vectors $\mathbf{A}_{AO}^{n}$ and $\mathbf{A}_{BA}^{n}$ in Eq. (3.45) can be calculated as

$$A_{AO}^{n} = L_1\omega_1^2 = (10)(3)^2 = 90 \text{ units/s}^2$$

$$\text{and} \quad A_{BA}^{n} = L_2\omega_2^2 = (20)(-1.56)^2 = 48.7 \text{ units/s}^2$$

If, then, the angular acceleration of the input crank L_1 is given as -3 rad/s^2 (i.e., clockwise), the tangential component of the acceleration of point A will be downward and to the right, as shown in Fig. 3.19a, and it will have a magnitude of

$$A_{AO}^{t} = L_1\alpha_1 = (10)(-3) \text{ or } 30 \text{ units/s}^2$$

Examination of Eq. (3.45) shows that the *magnitudes* of all vectors except $\mathbf{A}_B$ and $\mathbf{A}_{BA}^{t}$ are known and that the *directions* of *all* the vectors are known. The graphical construction is then started by choosing an origin such as is shown as point O in Fig. 3.19b. Then, starting at that origin, the completely known vectors ($\mathbf{A}_{AO}^{t}$, $\mathbf{A}_{AO}^{n}$, and $\mathbf{A}_{BA}^{n}$) are plotted first, starting from O. These vectors are connected tail to head to represent their addition as indicated on the right-hand side of Eq. (3.45). The remaining term on the right-hand side of Eq. (3.45) is $\mathbf{A}_{BA}^{t}$, and it must be added to the other three vectors by placing its tail at the tip of vector $\mathbf{A}_{BA}^{n}$. However, all that is known about this vector $\mathbf{A}_{BA}^{t}$ is that it must be perpendicular to coupler L_2. Therefore, a line p perpendicular to L_2 is drawn through the tip of vector $\mathbf{A}_{BA}^{n}$, and the magnitude of $\mathbf{A}_{BA}^{t}$ will be determined later.

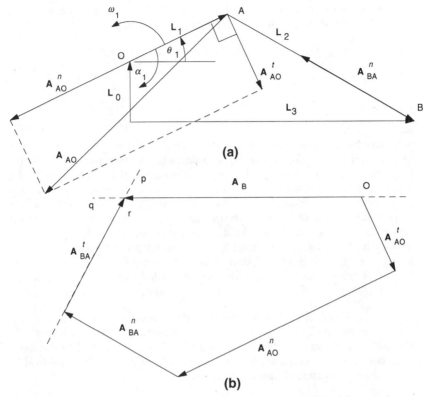

Figure 3.19 Graphical acceleration analysis of a crank-slider linkage.

Now, because no more is known about the right-hand side of the equation, attention is shifted to the left-hand side. When the vector(s) on the left-hand side are drawn, starting at point O and followed tail to head, they must add up to the same result as obtained by following the vectors on the right-hand side tail to head from point O. However, all that is known about $\mathbf{A}_B$, the only vector on the left-hand side, is that it must be horizontal because point B can only slide horizontally. Therefore, a line q is drawn horizontally through the origin O, and the magnitude of $\mathbf{A}_B$ will be determined later.

Note now that lines p and q intersect at a point r. If the magnitudes of $\mathbf{A}_{BA}^t$ and $\mathbf{A}_B$ were such as to place their heads at point r, the vector additions on the two sides of the equation would produce equal results, and the equation would be satisfied. These magnitudes can then be scaled off the diagram, and it is found that

$$A_B = 88.8 \text{ units/s}^2 \quad \text{and} \quad A_{BA}^t = 47.2 \text{ units/s}^2$$

Because $\mathbf{A}_{BA}^t$ is seen to be directed upward and to the right, point B is accelerating upward and to the right *with respect to* point A, so the coupler has a positive (counterclockwise) angular acceleration. The magnitude α_2 of this angular acceleration is computed from the relationship

$$A_{BA}^t = L_2\alpha_2$$

which gives

$$47.2 = 20\alpha_2$$

so $\alpha_2 = 2.4$ rad/s^2.

In the foregoing example the directions of the unknown vectors were known and their magnitudes were unknown, so they were represented temporarily just by lines in the known directions. If the magnitude of a vector had been known but its direction had been unknown, it could have been represented temporarily by a circular arc with a radius equal to the known magnitude and with its center at the origin O or at the tip of the vector to which it was to be added. If only one vector had been unknown, neither its direction nor its magnitude would have been known. Then, after all the other vectors had been drawn, it would be found that the unknown vector would be simply the vector needed to close the loop in the diagram.

Example 3.5 Coupler Point Accelerations Consider a point C on a coupler as shown in Fig. 3.20a, which is a reproduction of Fig. 3.19a but with the addition of point C. Because points C and A are attached to the same rigid body (i.e., the coupler), the acceleration $\mathbf{A}_C$ of point C can be found by using the acceleration-difference equation:

$$\mathbf{A}_C = \mathbf{A}_{AO} + \mathbf{A}_{CA} = \mathbf{A}_{AO}^n + \mathbf{A}_{AO}^t + \mathbf{A}_{CA}^n + \mathbf{A}_{CA}^t \qquad (3.46)$$

The first two vector components $\mathbf{A}_{AO}^t$ and $\mathbf{A}_{AO}^n$ on the right-hand side of Eq. (3.46) are already plotted in Fig. 3.19b, and they are replotted to a larger scale in Fig. 3.20b. The remaining two terms on the right-hand side must then be added graphically by placing them at the tip of the resultant of the first two terms, as shown in Fig. 3.20b. The magnitude A_{CA}^t of the vector $\mathbf{A}_{CA}^t$ is given by $A_{CA}^t = a_2 R_{CA} = (2.4)(12) = 28.8$ units/s^2, and it is directed perpendicular to $\mathbf{R}_{CA}$ and in a direction consistent with the counterclockwise angular acceleration of the coupler (as indicated by the direction of $\mathbf{A}_{BA}^t$). The magnitude A_{CA}^n of the vector $\mathbf{A}_{CA}^n$ is given by $A_{CA}^n = \omega_2^2 R_{CA} = (1.56)^2(12) = 29.2$ units/s^2, and it is directed parallel to $\mathbf{R}_{CA}$ and from point C toward point A.

By scaling values off Fig. 3.20b, it is found that $\mathbf{A}_{CA}$ is a vector of magnitude about 41 units/s^2 at an angle of about 57.5°. The total acceleration $\mathbf{A}_C$ that is found by graphically adding all the vectors in the diagram is a vector of approximate magnitude 55.4 units/s^2 at an angle of approximately 214°.

It can be seen that once the accelerations $\mathbf{A}_A$ and $\mathbf{A}_{BA}$ have been obtained, the acceleration-distribution triangle method that was described at the end of Chap. 2 also could be used to determine values for $\mathbf{A}_{CA}$ and $\mathbf{A}_C$.

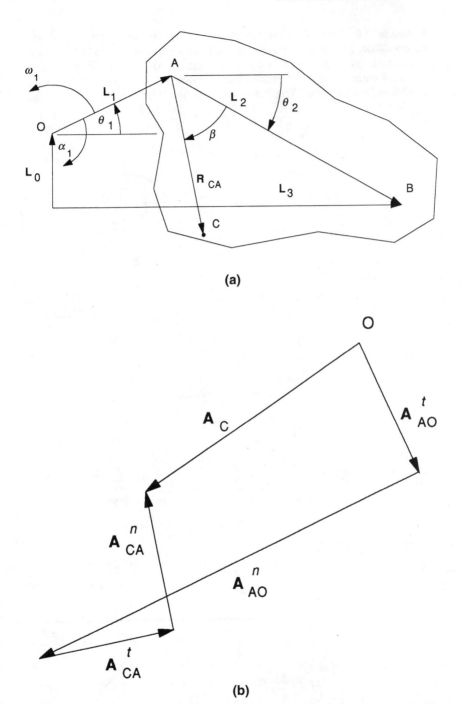

(a)

(b)

Figure 3.20 Graphical determination of acceleration of a coupler point in a crank-slider linkage.

Example 3.6 Peak Accelerations and Drive Torque Fluctuations in a Crank-Slider Linkage Figure 3.21a schematically depicts an in-line crank-slider linkage with a crank length L_1 of 10 cm and a coupler length L_2 of 30 cm. Assume a constant crank input angular velocity ω_1 of positive (counterclockwise) 10 rad/s. The velocities and accelerations of various points

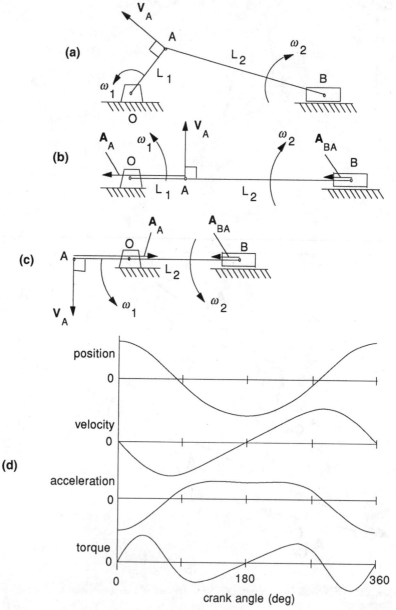

Figure 3.21 Peak accelerations and drive torque fluctuations in a crank-slider linkage.

on the bodies L_1 and L_2 can be computed for various positions of the linkage by the methods presented in previous sections. If it is desired to obtain values for those velocities and accelerations for more than a very few positions, a computer probably will be used. However, values for a couple of important positions can be calculated very simply.

Consider the extreme positions shown in Fig. 3.21b and c. These are positions at which the acceleration of the slider is at its peak values, and these values may be computed very simply. To do so, use the acceleration-difference equation:

$$\mathbf{A}_B = \mathbf{A}_A + \mathbf{A}_{BA}$$

Because the angular velocity of the crank is constant, the tangential component of $\mathbf{A}_A$ is zero, so $\mathbf{A}_A$ is purely centripetal (directed from A toward O), and its magnitude is $A_A = \omega_1^2 L_1 = (10)^2(10) = 1000$ cm/s^2. This vector $\mathbf{A}_A$ is drawn to scale with its tail at point A on Fig. 3.21b and c.

The velocity of point A is vertical, and its magnitude is given by $V_A = \omega_1 L_1 = (10)(10) = 100$ cm/s. Because $\mathbf{V}_A$ has no longitudinal component along L_2 and thus $\mathbf{V}_B$ must have no longitudinal component along L_2, and because $\mathbf{V}_B$ is constrained to be horizontal, $\mathbf{V}_B = 0$. Then the entire velocity difference between points A and B is $\mathbf{V}_A$. Thus $\omega_2 = -V_A/L_2 = -100/30 = -3.33$ rad/s^2.

The acceleration of point B must be horizontal, so we are only interested in the horizontal components of the vectors in the acceleration-difference equation. The horizontal component of $\mathbf{A}_{BA}$ is merely the centripetal component of $\mathbf{A}_{BA}$, and its magnitude is given by $A_{BA} = \omega_2^2 L_2 = (3.33)^2(30) = 332.7$ cm/s^2. The vector $\mathbf{A}_{BA}$ is plotted in Fig. 3.21b and c pointing toward point A and with its tail at point B.

Now note that in Fig. 3.21b, adding the acceleration vectors according to the acceleration-difference equation gives a net A_B of $1000 + 332.7 = 1332.7$ cm/s^2 directed leftward. In the case of Fig. 3.21c, however, the acceleration-difference equation dictates the addition of two vectors that are in *opposite* directions, giving a net A_B of $1000 - 332.7 = 667.3$ cm/s^2 directed rightward. It is thus seen that by means of a very simple calculation, it is found that even with a ratio of coupler length to crank length of 3, the peak accelerations of the slider at the extremes of the motion differ by a factor of 2. This is consistent with the fact that the curvature of the displacement curve in Fig. 3.5 is sharper at one extreme of its displacement than it is at the other extreme.

Figure 3.21d shows the results of computer calculations of the displacement, velocity, and acceleration of point B throughout a full rotation of the crank. The differences in the accelerations at the extremes of the travel (at 0° and at 180°) are manifest in the sharpness of the curvatures in the displacement curve and in the slopes of the velocity curve at the extremes.

Consider now the effects that accelerations of a slider mass will have on the crank shaft torque needed to drive this mechanism at constant shaft speed. To determine these effects, it is useful to employ the concept of mechanical advantage described in Sec. 3.9. Mechanical advantage can be used to relate the output force applied to the slider to the input torque applied at the crank shaft. That is,

$$F_{out}/T_{in} = \omega_{in}/V_{out}$$

or solving for T_{in},

$$T_{in} = F_{out} (V_{out}/\omega_{in}) = F_{out} (V_B/\omega_1)$$

Then, because F_{out} is the force required to accelerate the slider, $F_{out} = m_B A_B$. where m_B, is the mass of the slider block. Therefore,

$$T_{in} = \frac{m_B A_B V_B}{\omega_1}$$

Now note that for constant slider mass (and comparatively negligible coupler mass) and constant input shaft angular velocity, the required input torque is proportional to $A_B V_B$. A plot of this latter product (to an arbitrary scale) is given as the bottom curve in Fig. 3.21d. To obtain a curve of input torque, this curve would be multiplied by slider mass and divided by the input angular velocity (in radians per second). In cases in which a large slider mass is driven by such a mechanism and the mechanism is driven by an electric motor, these fluctuations in drive torque can cause substantial motor heating, even when the *average* torque is small. Such heating effects are added to whatever heating effects result from the average torque requirement, and the motor should be sized accordingly.

It is seen that this input torque waveform is distorted from a sine wave that has a frequency of twice the input rotational frequency. If the ratio of coupler length to crank length were large, this curve would approach a sinusoidal form. In cases in which the waveform is sufficiently sinusoidal, it is sometimes possible to cancel out this fluctuation by adding a second slider mechanism that is driven by the same shaft but which operates 90° out of phase with the subject mechanism. This may be seen by visualizing a waveform like that shown but shifted to the right by 90° from the original and then noting that the two waves tend to cancel. If they were true sinusoids, they would cancel exactly.

In order to gain further insight into the sources of these fluctuations, it is useful to return to consideration of the scotch yoke that was discussed at the beginning of this chapter. Such a scotch yoke is shown schematically for two positions of the input crank in Fig. 3.22. For a constant crank angular velocity, the acceleration of the pivot at point A at the end of the crank will always be pointed from that point toward the fixed (grounded) pivot, as shown in Fig. 3.22a and b. Because the pivoted block at point A must slide only up and down in the slot of the output member L_2 and that output member can slide only horizontally, the acceleration of member L_2 must be equal to the horizontal component of the acceleration of point A.

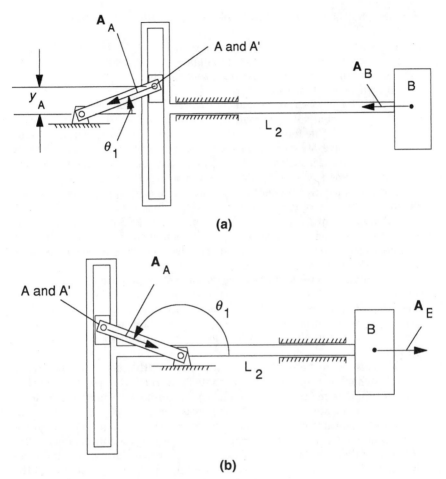

Figure 3.22 Accelerations in a scotch yoke mechanism.

Therefore, in the position shown in Fig. 3.22a, the acceleration of L_2 and a point B on it must be leftward, as shown. If a mass is attached to the end of L_2 at point B, the force needed to accelerate the mass leftward must come from a tension in member L_2. This tension will apply a clockwise torque to the crank because it acts horizontally at point A, and the effective moment arm relative to input crank shaft is the distance y_A. Although the acceleration of point B is large and therefore the tension in L_2 is large when θ_1 is small, the moment arm y_A is small, so the moment that this tension applies to the crank is small. As θ_1 increases, y_A increases, so the crank shaft torque increases. However, as θ_1 increases, $\mathbf{A}_A$ rotates toward a vertical orientation so that its horizontal component decreases and thus the acceleration of the mass at point B decreases and the tension in L_2 decreases. Then, even though the moment arm y_A is large when $\theta_1 = 90°$, the moment that is applied to the crank by the tension in L_2 drops to zero at that time.

In this first quadrant of crank motion (i.e., when $0° < \theta_1 < 90°$), it is seen that the driving torque that must be supplied to the crank shaft to counter this tension in L_2 must increase from zero to some counterclockwise value and then decrease to zero.

When θ_1 has increased beyond 90°, a condition such as that shown in Fig. 3.22b eventually will be reached, and it is seen that point B is accelerating rightward. The force required to produce this acceleration must come from compressive stresses in member L_2. This compression will apply a counterclockwise torque to the crank shaft. Because the compression in L_2 increases beyond $\theta_1 = 90°$, the counterclockwise torque will increase. However, because the moment arm y_A decreases as θ_1 approaches 180°, the counterclockwise torque will decrease to zero at $\theta_1 = 180°$.

In this second quadrant of crank motion (i.e., when $90° < \theta_1 < 180°$), it is seen that the driving torque that must be supplied to the crank shaft to counter this *compression* in L_2 must increase from zero to some *clockwise* value and then decrease to zero.

If the required crank shaft drive torque is computed and plotted, it is found that the plot consists of a complete cycle of a sine wave during the first 180° of crank rotation. The reader is encouraged to continue the foregoing reasoning through the remainder of the 360° of crank rotation to see that another cycle of sine wave occurs in the remaining 180° of crank rotation.

As discussed in the early part of this chapter, the motion of the crank-slider mechanism is quite similar to that of the scotch yoke mechanism. The varying angular velocity of the coupler in the crank-slider mechanism, however, causes an acceleration difference between point B and point A that varies in magnitude and direction in a manner such as to produce an acceleration of point B that differs from that in a scotch yoke mechanism. This difference plus the difference in the angle at which the tension and compression in the coupler in the crank-slider mechanism apply torque to the crank-shaft causes the drive torque waveform to deviate from a pure sine wave. For large ratios of coupler length to crank length, these deviations are small. As this ratio approaches unity, the distortions become large. The torque waveform shown in Fig. 3.21d is seen to be moderately distorted from a sine wave.

It may seem that the foregoing qualitative analysis is an unnecessary belaboring of a simple phenomenon in a simple mechanism. However, readers should keep in mind that this sort of analysis, cautiously applied, can be very helpful in attempts to understand what may otherwise seem to be mysterious phenomena in more complicated mechanisms.

Analytical acceleration analysis of crank-slider linkages

Acceleration analysis of a crank-slider linkage by analytical means could proceed very much as discussed in the graphical analysis, but with the vectors and their components expressed in algebraic form (either trigonometrically or in complex variables). The manipulations would then be algebraic rather than graphical. However, rather than using an algebraic analogue of the graphical method, it is more compact and convenient to start with the vector loop closure equation for Fig. 3.9 and then to differentiate that equation twice. The vector loop equation has already been differentiated once in Sec. 3.9 to give Eq. (3.32), which is repeated here as Eq. (3.47):

$$\omega_1 L_1 e^{j(\theta_1 + \pi/2)} + \omega_2 L_2 e^{j(\theta_2 + \pi/2)} = \left(\frac{dL_3}{dt}\right) = V_B \qquad (3.47)$$

Differentiating Eq. (3.47) with respect to time gives

$$\alpha_1 L_1 e^{j(\theta_1 + \pi/2)} + j\omega_1^2 L_1 e^{j(\theta_1 + \pi/2)} + \alpha_2 L_2 e^{j(\theta_2 + \pi/2)} + j\omega_2^2 L_2 e^{j(\theta_2 + \pi/2)} =$$
$$\left(\frac{d^2 L_3}{dt^2}\right) = A_B$$

Because $j = e^{j(\pi/2)}$ and $e^{j\pi} = -1$, this equation can be simplified to give

$$\alpha_1 L_1 e^{j(\theta_1 + \pi/2)} - \omega_1^2 L_1 e^{j\theta_1} + \alpha_2 L_2 e^{j(\theta_2 + \pi/2)} - \omega_2^2 L_2 e^{j\theta_2} = A_B \qquad (3.48)$$

Note that Eq. (3.48) is just the acceleration-difference equation expanded in terms of normal and tangential components of the accelerations. That is, it is merely

$$\mathbf{A}_A^t + \mathbf{A}_A^n + \mathbf{A}_{BA}^t + \mathbf{A}_{BA}^n = \mathbf{A}_B. \qquad (3.49)$$

Equation (3.48) can be expressed as two equations by using Euler's formula to give

$$\alpha_1 L_1 \cos(\theta_1 + \pi/2) - \omega_1^2 L_1 \cos\theta_1 + \alpha_2 L_2 \cos(\theta_2 + \pi/2)$$
$$- \omega_2^2 L_2 \cos\theta_2 = A_B \qquad (3.50)$$

and

$$j\alpha_1 L_1 \sin(\theta_1 + \pi/2) - j\omega_1^2 L_1 \sin\theta_1 + j\alpha_2 L_2 \sin(\theta_2 + \pi/2)$$
$$- j\omega_2^2 L_2 \sin\theta_2 = 0 \qquad (3.51)$$

By using trigonometric identities on these two equations and dividing Eq. (3.51) by j, we get

$$-\alpha_1 L_1 \sin\theta_1 - \omega_1^2 L_1 \cos\theta_1 - \alpha_2 L_2 \sin\theta_2 - \omega_2^2 L_2 \cos\theta_2 = A_B \qquad (3.52)$$

and $\quad \alpha_1 L_1 \cos\theta_1 - \omega_1^2 L_1 \sin\theta_1 + \alpha_2 L_2 \cos\theta_2 - \omega_2^2 L_2 \sin\theta_2 = 0 \qquad (3.53)$

The angular velocities ω_1 and ω_2 are determined by a preceding velocity analysis. Then it can be seen that Eqs. (3.52) and (3.53) are linear equations in the acceleration variables α_1, α_2, and A_B. If a value for any one of these acceleration variables is given, the values for the other two may be solved for.

Analytical determination of the acceleration of a coupler point

Once the acceleration of point A and the angular acceleration of the coupler in Fig. 3.20a have been determined, the acceleration of any point on the coupler such as point C can be calculated by using the acceleration-difference equation, which relates the acceleration of that coupler point to the known acceleration of point A. For this example point C, the acceleration-difference equation is simply $\mathbf{A}_C = \mathbf{A}_A + \mathbf{A}_{CA}$, which can be expanded into terms of the tangential and normal components to give

$$\mathbf{A}_C = \mathbf{A}_A^t + \mathbf{A}_A^n + \mathbf{A}_{CA}^t + \mathbf{A}_{CA}^n$$

The vector $\mathbf{A}_C$ will in general have both a horizontal (X or real) component and a vertical (Y or imaginary) component. The angular orientation of the vector $\mathbf{R}_{CA}$ from point A to point C is seen to be represented by the angle $\theta_2 + \beta$ (and in this particular figure both θ_2 and β are shown as negative angles). Then, noting that the tangential and normal components of the vectors on the right-hand side of this expanded equation may be written in their now familiar complex variable form, this acceleration-difference equation becomes

$$(A_C)_x + j(A_C)_y = \alpha_1 L_1 e^{j(\theta_1 + \pi/2)} - \omega_1^2 L_1 e^{j\theta_1} + \alpha_2 R_{CA} e^{j(\theta_2 + \beta + \pi/2)}$$

$$- \omega_2^2 R_{CA} e^{j(\theta_2 + \beta)} \quad (3.54)$$

Then, using Euler's formula, the two scalar equations that correspond to Eq. (3.54) become

$$(A_C)_x = -\alpha_1 L_1 \sin \theta_1 - \omega_1^2 L_1 \cos \theta_1$$

$$- \alpha_2 R_{CA} \sin(\theta_2 + \beta) - \omega_2^2 R_{CA} \cos(\theta_2 + \beta) \quad (3.55)$$

and

$$(A_C)_y = \alpha_1 L_1 \cos \theta_1 - \omega_1^2 L_1 \sin \theta_1$$

$$+ \alpha_2 R_{CA} \cos(\theta_2 + \beta) - \omega_2^2 R_{CA} \sin(\theta_2 + \beta) \quad (3.56)$$

The values of all variables on the right-hand side of Eqs. (3.55) and (3.56) are known from foregoing analyses, so the values of the horizontal and vertical components, respectively, of the acceleration of point C can be computed directly from these two equations.

3.11 Summary

This chapter has shown that crank-slider linkages produce slider motions that are distortions of sinusoidal motions. Graphical methods for synthesizing such linkages to give various degrees of such distortion were presented. These methods provide designs for mechanisms that will produce desired timing and quick-return motions. The amount of timing and quick return that may be provided was shown to be limited by the transmission angle that may be tolerated in a particular application.

Simple, direct graphical procedures for analyzing the position, velocities, and accelerations were presented. These procedures vividly show the interactions between the geometry and the velocities and accelerations. The use of instant centers in velocity analysis was demonstrated. The usefulness of velocity analysis in computing force relationships and mechanical advantage was shown.

Analytical methods for performing these same analyses also were presented. The analytical methods included the use of complex exponential vector notation, which facilitates the generation of expressions from which the values of the unknown velocities and accelerations may be computed.

4

Pin-Jointed Four-Bar Linkages

4.1 Introduction

This chapter presents graphical procedures for the synthesis and analysis of pin-jointed four-bar linkages. These procedures are valuable means for determining linkage geometry, which will provide motions required for many tasks that machines must perform. They also can be used to provide first estimates of linkage designs prior to conducting the more complete computer-aided syntheses of Chap. 9. Without such estimates, computer syntheses can involve extensive trial and error.

A four-bar linkage consists of four interconnected rigid bodies (including ground), and it may contain any combination of types of joints. However, those four-bar linkages containing four revolute joints (pin joints) are among the most commonly encountered.[1] Synthesis and analysis of these linkages require consideration of the motion of *three* moving rigid bodies rather than just the two considered in the case of crank-slider linkages in Chap. 3. Because of the presence of the added moving body, these linkages can be synthesized to provide more varied motions than is the case for crank-slider mechanisms.

Four-bar pin-jointed linkages can be synthesized to

* Cause the orientation of an output link (one end of which is pivoted to ground) to assume a prescribed orientation for each of a set of

---

[1]Strictly speaking, the crank-slider linkage is also a four-bar linkage. This is most easily seen in the case where a pivoted sliding block is used, as shown in Figs. 3.4b and 3.6a. In these configurations, the block or sliding rod constitutes the fourth link, and there are three revolute joints and one sliding (prismatic) joint.

orientations of an input link. This is known as *function generation* because the output is made to be a desired function of the input. Synthesis methods used for mechanisms to provide such operation are called *function synthesis methods.*

- Cause the *coupler link* to assume a sequence of prescribed positions in translation and rotation as the input link is rotated. This is known as *motion generation.* Synthesis methods used for mechanisms to provide such operation are called *motion synthesis methods.*

- Cause a point attached to the coupler body to pass through a sequence of prescribed points as the input link is rotated. This is known as *path generation.* Synthesis methods used for mechanisms to provide such operation are called *path synthesis methods.*

Some of the many uses to which pin-jointed four-bar linkages may be put are

- As pick-and-place units for automatic assembly machines
- As drivers for "walking beam" mechanisms for progressively transporting parts along an assembly line
- As mechanisms that provide prescribed timing between the motion of two shafts
- As mechanisms for causing machine components to move through a prescribed sequence of positions

In Chap. 3, *function synthesis procedures* were presented for use with crank-slider linkages. Although crank-slider linkages also can be used for motion generation and path generation, greater synthesis flexibility is available for four-bar linkages. In this chapter, then, procedures will be presented for *motion synthesis* and *path synthesis* of pin-jointed four bar linkages, as well as procedures for *function synthesis.*

This chapter starts by showing the similarities between the motions of the crank-slider mechanism and the crank-rocker mechanism. Graphical techniques for function synthesis and motion synthesis are presented, and the synthesis of pick-and-place mechanisms is illustrated. The Grashof mobility criterion is introduced, and double-crank and double-rocker linkages are introduced. Coupler-joint paths and cognate linkages and the uses of both are described. The velocity and acceleration analysis techniques used in Chap. 3 are then extended for use on pin-jointed four-bar linkages.

4.2   Crank-Rocker Linkages

Figure 4.1 schematically depicts a pin-jointed four-bar linkage in which the links $(L_1, L_2, L_3,$ and $L_4)$ are represented by the vectors $\mathbf{L}_1$, $\mathbf{L}_2, \mathbf{L}_3,$ and $\mathbf{L}_4,$ respectively. The links are joined by four revolute joints at points $O_1, A, B,$ and $O_2.$ For the proportions shown in the figure, the link L_1 can rotate continuously around point $O_1,$ as indicated by the dashed circle that represents the path that would be followed by point $A.$ Link L_1 is, as in the case of the crank-slider linkage, known as a *crank* because it can rotate continuously through 360°. For the link lengths shown, as L_1 rotates, it acts through coupler L_2 to cause link L_3 to rock back and forth between the positions indicated by the dashed lines L_3' and $L_3'',$ carrying point B back and forth along the dashed circular arc between points B' and B'' shown. Because link L_3 rocks back and forth, it is referred to as a *rocker,* and a linkage of this type is called a *crank-rocker linkage.*

General design considerations

The aim of *function* synthesis of a crank-rocker linkage is to provide an output position such as the position of point B or the position of link $L_3,$ which has a prescribed value for each of a prescribed set of positions of link $L_1.$ That is, the output of the mechanism is made to be some desired *function* of the input. The output will be a continuous function of the input, but it will be specified for synthesis purposes at

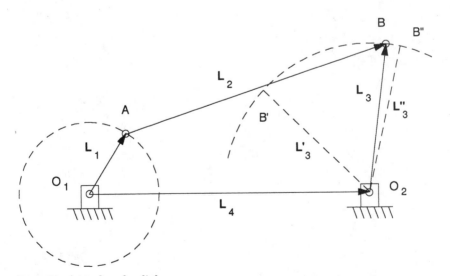

Figure 4.1   A crank-rocker linkage.

only a limited set of values of input. These specified pairs of input and output values are known as *precision points* on a plot of the function.

By referring to Fig. 4.2, it will be seen that if the rocker is long compared with the length of the crank and coupler, there are similarities between the motion of a crank-rocker linkage and a crank-slider linkage. In Fig. 4.2, link L_3 is very long (and point O_2 has been moved down on the page from its position in Fig. 4.1), so the circular arc path followed by point B as it travels back and forth approximates a straight line; i.e., the dashed arc has a large radius. Then the motion of point B for this linkage will be very much like the motion of a slider. Such a linkage can be synthesized to provide timing motions very similar to those provided by the crank-slider linkages that were synthesized in Chap. 3, and the same synthesis procedures can be used. That is, the crank-rocker linkage with a long rocker can be synthesized by first synthesizing a simple crank-slider or offset crank-slider linkage to provide a desired motion for point B. Then, a long rocker can be substi-

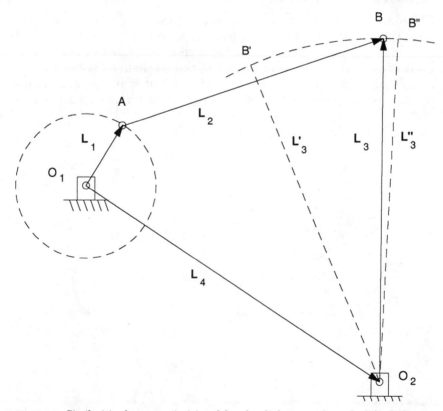

Figure 4.2 Similarities between pin-jointed four-bar linkages and crank-slider linkages.

tuted for the slider by placing the grounded rocker pivot at a point O_2 that is on the perpendicular bisector of the line $B'B''$. This grounded pivot is placed at some appreciable distance from that line $B'B''$ by then using a long rigid link extending from the end B of the coupler to the pivot point O_2. Such a synthesis will be referred to as a *crank-slider analogy synthesis*. Such a synthesis is useful when timing such as discussed in Chap. 3 is to be provided. In Sec. 4.3, two-point and three-point function synthesis procedures, in which the output rocker is *not* required to be long, are also presented.

The choice between a slider and a rocker

The similarities between the crank-slider linkage and the crank-rocker linkage raise the question of which type of linkage to choose for a particular application. Some of the advantages of each of these types of linkage are as follows.

Advantages of crank-slider linkages:

- Generally, the crank-slider linkage requires less space than does the crank-rocker linkage.

- There are occasions for which a very straight-line output motion must be produced and which can be produced only by a slider.

- Transmission-angle problems often can be less severe in crank-slider linkages than in crank-rocker linkages.

Advantages of crank-rocker linkages:

- Crank-rocker linkages use only revolute joints. Such joints can consist of bearings that are easily kept clean and well lubricated, whereas sliding joints (which are required in slider mechanisms) usually have large sliding surfaces that are exposed to possible contamination and damage.

- The ability to choose the amount of curvature of the path followed by the output end of the coupler (point B in the foregoing figures) allows the crank-rocker mechanism to be designed to produce a greater variety of output motions, not only for function generation but also for motion generation and path generation.

4.3 Crank-Rocker Function Synthesis Procedures

Three types of function synthesis procedures for crank-rocker linkages are presented in this section: crank-slider analogy function synthesis, three-point function synthesis, and overlay function synthesis. As explained in the procedures, each is applicable to a slightly different design requirement.

Procedure 4.1: Crank-Slider Analogy Function Synthesis (see also Examples 4.1 and 4.2) This type of synthesis can be used when an output body must be made to rock back and forth in response to the motion of an input crank that is free to rotate through a full 360°. This synthesis procedure provides an output motion of a quick-return nature or a motion in which the output member remains near one end of the output range during a specified portion of the input rotation.

As pointed out in Sec. 4.2, *if the output rocker of a crank-rocker linkage is long compared with the length of the crank,* the motion of the rocker's connection to the coupler will be very much like that of a slider, and a crank-slider linkage analogy synthesis can be used. Readers will understand this synthesis best by examining the quantitative Examples 4.1 and 4.2. First, however, the procedure is described in detail in the following 14 steps.

1. Choose where the endpoints of the motion of the output are to be located. These points correspond to the points B' and B'' in Fig. 4.2. Plot these points to a suitable scale on paper or on a computed-aided design (CAD) system, and draw an extended line through them.[2] Find the midpoint of the line segment between points B' and B'', and construct at that midpoint a line perpendicular to the line segment $B'B''$. Locate the grounded pivot O_2 of the output rocker on this perpendicular line. Draw lines O_2B' and O_2B''. The distance from this grounded pivot to either point B' or B'' will be the output rocker length. In order to use this type of synthesis, the rocker length usually should be at least 1.5 times the distance between the points B' and B''. The resulting figure will be a drawing of the limiting positions of the output rocker (see Fig. 4.4 in Example 4.1, where C corresponds to B' and D to B'').

2. Decide whether this is to be a quick-return mechanism or a mechanism in which it is intended that the output link will spend a greater portion of the rotation of the input crank toward one of its travel than toward the other. This latter case corresponds to an output motion similar to those indicated in Fig. 3.5.

3. If it is desired that the output link will spend a greater portion of the rotation of the input crank toward one end of the output travel than toward the other, go to step 4. If a quick-return mechanism is desired, go to Step 10.

4. Label the two points plotted in step 1 such that the one farthest from the input crank shaft is point D and the nearer one is point C, as in Fig. 4.4.

5. On the rocker position drawing from step 1, draw radius O_2F to represent a desired design position that will correspond to the rocker's spending the desired portion of its cycle nearer to the crank shaft. That is, this point F is analogous to the point F in Fig. 3.7a and corresponds to the point P_d in Fig. 3.7b. (For an illustration of the choice and use of this point, see Example 4.1.)

---

[2]The use of a CAD system has the advantages that not only is it easy to make accurate graphical synthesis drawings with such systems, but also, once the synthesis drawing is complete, it may be flipped horizontally or vertically, rotated, and/or scaled to suit the desired location of the final mechanism. By dimensioning the CAD drawing, very accurate answers can be obtained.

6. Drop a perpendicular from point F to the chord CD, and label the base of this perpendicular F'', as shown in Fig. 4.4. This point F'' is the desired design point for the crank-slider analogue that will be synthesized in the next few steps.

7. The design distances d_1 and d_2 (analogous to those shown in Fig. 3.7a) can then be measured along line CD as in Fig. 3.7a.

8. Apply Procedure 3.1 to the plotted points C, D, and F''. The result of such a synthesis is shown in Fig. 4.4, where the construction shown is similar to that in Fig. 3.7a. Note that the synthesis diagram in Fig. 3.7a can be reversed from left to right. Note also that the coupler-to-crank length ratio also can be obtained by locating point P_d on Fig. 3.5 as described in step 5 of Procedure 3.1, and the designer could then proceed directly to step 12 of the present procedure.

9. The length of line segment OE is the length of the required input crank, and the length of line segment EH is the length of the required coupler. The required location of the crank's grounded pivot will be on line CD, will be outside the semicircle, and will be a distance $(EH - OE)$ from point C. If a CAD system is used, very accurate values can be obtained for these lengths. Go to step 12.

10. If a quick-return mechanism is to be synthesized, choose the time ratio, and calculate the angle α as indicated in Procedure 3.2.

11. Label the two endpoints plotted in step 1 such that the one closest to the intended crank pivot is B and the other is B' (see Fig. 4.6 and Example 4.2). Then, using these points and the value of α from step 10, apply Procedure 3.2. In that procedure, note that point O can be located anywhere on the circular arc K, including at locations *to the right* of points B and B'. The center C of arc K also can be located *below* the line BB', as shown in Fig. 4.6, in which case the arc curves downward. It is usually desirable to locate the grounded pivot of the input crank (point O) on the same side of line BB' (extended) as the grounded pivot (point O_2) of the output rocker.

If a CAD system is used for this synthesis, very accurate values for link lengths and pivot locations may be computed from the dimensioned lengths on the synthesis graphic.

12. Using the link lengths and pivot locations arrived at in the synthesis, construct a scale drawing of the resulting linkage in some convenient part of its motion cycle. If a CAD system has been used for the foregoing, this scale drawing can be very easily constructed on another layer by duplicating, moving, rotating, and scaling all or parts of the synthesis construction.

13. Determine the extremes of the transmission angle in accordance with the discussion in the paragraph following this procedure. The transmission angle extremes may be found to be unsatisfactory as a result of (a) the fact that a nonoptimal link-length choice or pivot-location choice was made during the synthesis or (b) the fact that the time ratio chosen in step 10 or the design point chosen in step 6 cannot be accommodated by this type of linkage without encountering an unacceptable transmission angle.

14. If the transmission angles are not acceptable, return to the graphical construction and see whether some change can be made to alleviate the problem. If not, it may be necessary to use a quick-return mechanism such as those discussed in Chap. 5, use a more complicated function generator, or use a function generator synthesized in Procedure 4.2. The transmission angles produced in a four-bar linkage often can be worse than those produced by an analogous crank-slider mechanism. Perhaps, if the design deficiency is small, the desired function could be generated by a crank-slider mechanism or by just lengthening the rocker.

Transmission angles in pin-jointed four-bar linkages

In Sec. 3.6 the *transmission angle* in a crank-slider linkage was defined as "the angle between the coupler link and the direction of constraint at the output end of the coupler." To see how this definition applies to a pin-jointed four-bar linkage, refer to Fig. 4.3a, which shows a four-bar linkage in which the input motion or torque is applied to the

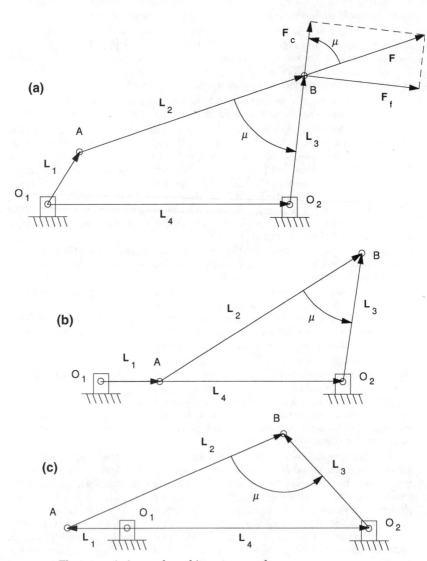

Figure 4.3 The transmission angle and its extreme values.

crank link L_1 and the output link (rocker) is link L_3. The output end of coupler L_2 is at point B. In the position shown, the input crank motion can cause point B to move *only* along a circular arc of radius L_3. At any instant, then, point B is therefore *constrained from* moving in a direction parallel to the instantaneous direction of link L_3.

> *The direction of constraint of point B is therefore parallel to the length of the output link L_3 at each instant.* The direction of *freedom* is perpendicular to that length. In accordance with the preceding definition of transmission angle, *the transmission angle in a pin-jointed four-bar linkage is the angle μ between the coupler L_2 and the output link L_3 as shown.*
>
> The transmission angle often is defined as the *acute* angle between these two links. As will be seen in the ensuing discussion, the important aspect of the transmission angle is the absolute value of its cosine. Therefore, it is not important whether the acute angle or its supplement is used. In either case, it is desired to avoid large values of the absolute value of the cosine of the transmission angle. That is, large values of the deviation of the transmission angle from 90° should be avoided.

The coupler link L_2 is pivoted at each end and is therefore able to experience only tension and compression. The coupler therefore can transmit to output link L_3 only forces parallel to the coupler's length. Such a force is indicated by the vector $\mathbf{F}$ in Fig. 4.3a. This force vector is resolved into a component $\mathbf{F}_c$ parallel to the direction of constraint and a component $\mathbf{F}_f$ parallel to the direction in which point B is free to move in response to input motion. The component $\mathbf{F}_c$ contributes nothing to the output and serves only to increase friction and stresses in the mechanism. The magnitude of $\mathbf{F}_c$ is given by

$$F_c = F \cos \mu \quad \text{which gives} \quad F_c/F = \cos \mu$$

The absolute value of the ratio F_c/F is indicative of the amount of wasted force F_c that is being produced by the coupler output force F. In order to avoid excessive values of this ratio, large values of the absolute value of cos μ should be avoided. For this purpose:

> A rule of thumb that is often used is that the transmission angle should not be allowed to differ from 90° by more than ±50°. This means that if the acute-angle definition of transmission is used, the angle should be kept larger than 40°.
>
> It can be shown that in a four-bar linkage containing an input crank, the transmission angle reaches its extreme values (maximum and minimum) when the input link (crank) is parallel to the ground link, as shown in Fig. 4.3b and c.

Example 4.1: Crank-Slider Analogy Synthesis of a Four-Bar Linkage
Objective: Synthesize a crank-rocker mechanism in which, in response to continuous rotation of the input crank, the 4.5-in-long output rocker is

caused to rock back and forth through an angle of 40°. In addition, the rocker must remain within 10° of one end of its travel while the input crank rotates through a total of 160°.

The first step is to lay out the desired positions in the motion of the output rocker as represented by the points O_2, C, D, and F shown in Fig. 4.4 (steps 1, 4, and 5 of Procedure 4.1). Note that the points C, D, and F lie on the arc of radius 4.5 in that is centered at point O_2 and the angle DO_2C is 40°. The problem statement requires the rocker to remain between the line O_2F and the line O_2C while the input crank rotates through 160°. The angle FO_2C equals 10°.

Then draw the chord CD and project point F down onto that chord to give point F' (step 6 in Procedure 4.1). Either by computation or measurement on Fig. 4.4, chord length CD is found to be 3.08 in. As shown in Procedure 3.1 and Example 3.1, a crank-slider linkage could by synthesized to cause a slider to move back and forth between points C and D. For

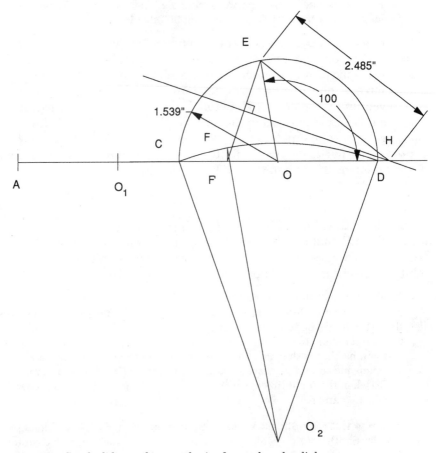

Figure 4.4 Crank-slider analogy synthesis of a crank-rocker linkage.

such a motion, the crank length would be one-half the distance CD, or $(0.5)(3.08) = 1.54$, and the ground pivot of the crank would be on line CD extended to the left of point C.

It remains, then, to determine the coupler length that would cause the slider to remain between points C and F' for 160° of crank rotation. Draw a semicircle centered at point O (i.e., the midpoint of line CD) and passing through points C and D as shown. The coupler length can then be determined by using Procedure 3.1, as illustrated in Fig. 3.7. For this example, the timing requires that θ_1 in that figure be 100°. Therefore, line OE in Fig. 4.4 is drawn at an angle of 100° from line CD. Then line $F'E$ and its perpendicular bisector are drawn. The intersection of this bisector and line CD is point H. The length of line EH is the required length of the coupler. By scaling the drawing, it is found that the coupler length must be 2.48 in.

As an alternative to the construction shown in Fig. 4.4, a horizontal line could be drawn at a distance of 0.25 of the total travel up from the bottom of the plot in Fig. 3.5 because the distance CF' is found by measurement in Fig. 4.4 to be 0.76 in or 0.25 of the total travel CD. Then, note that such a horizontal line would cross a curve for L_2/L_1 of about 1.6 in Fig. 3.5 at values of crank angle equal to about 100° and 260°. It is then seen that if a coupler length of $1.6L_1 = (1.6)(1.54) = 2.48$ in were chosen, the slider would remain between points C and F' in Fig. 4.4 (i.e., below the previously mentioned horizontal line in Fig. 3.5) for $260 - 100 = 160°$ of crank rotation.

The required location for the crank pivot O_1 may be determined in Fig. 4.4 by connecting a coupler of length 2.48 in to point C and extending to the left to point A as shown. Because this represents the coupler in its most leftward position, the crank of length 1.54 in extends rightward from point A to point O_1 as shown. This crank-slider mechanism, then, is the *crank-slider analogy* of the crank-rocker linkage that we wish to synthesize. It remains, then, to draw the corresponding crank-rocker linkage and to see how closely the design requirements have been met.

Using point O_1 as the grounded pivot of the crank, point O_2 as the grounded pivot of the output rocker, a crank length of 1.54 in, a coupler length of 2.48 in, and an output rocker length of 4.5 in, the linkage of Fig. 4.5a can be drawn. The solid lines represent the linkage when the crank angle is 100°, and the dashed lines show the position of the links when the crank angle is 260°. As actually drawn, the fixed pivot O_1 was moved upward from the position shown in Fig. 4.4 by distance FF' so that a line O_1F would be parallel to chord CD. This results in links L_1 and L_2 in Fig. 4.5a and their dashed counterparts in that figure being symmetrically located about a *horizontal* line. This ensures that rocker link BO_2 will pass through the prescribed timing positions at crank angles that are 160° apart and which are symmetrically located relative to 180°. As indicated in Fig. 4.5a, the resulting total travel of rocker BO_2 will not be exactly 40° (it is actually 40.11°), and the extremes of its travel will not occur at crank angles of exactly 0° and 180°, but the errors will be small.

It is seen that the motion of point B is very close to that of the slider in the crank-slider analogy, and thus, because point F is very close to point F', the synthesis objectives have been closely satisfied. If these objectives must

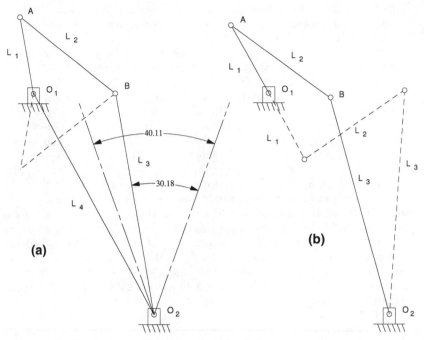

Figure 4.5 Synthesized example of a crank-rocker linkage.

be satisfied more precisely, some minor adjustment of link lengths can be tried, or Procedure 4.2 can be used.

The extremes of the transmission angle should then be checked by aligning the crank with ground link L_4 in the two positions shown in Fig. 4.5b. In this figure it is seen that the transmission angle extremes differ from 90° by about 40°.

Example 4.2: Crank-Slider Analogy Synthesis of a Quick-Return Crank-Rocker Linkage *Objective*: Synthesize a crank-rocker mechanism in which, in response to continuous rotation of the input crank, the 4.5-in-long output rocker is caused to rock back and forth through an angle of 40°. In addition, for a constant crank rotation rate of 1 revolution per second, the rocker must take 0.6 s to rotate clockwise from one extreme position to the other. It will therefore take 0.4 s to return by counterclockwise rotation to its original extreme position. That is, the time ratio is to be 0.6/0.4, or 1.5 : 1.

The first step is to lay out the desired positions in the motion of the output rocker as represented by the points O_2, B, and B' shown in Fig. 4.6. (Note that points B and B' lie on the arc of radius 4.5 in that is centered at point O_2.) As shown in Example 3.2, a crank-slider linkage could be synthesized to cause a slider to move back and forth between points B and B' along the dashed straight line BB' with a quick-return motion. Such a synthesis will now be performed.

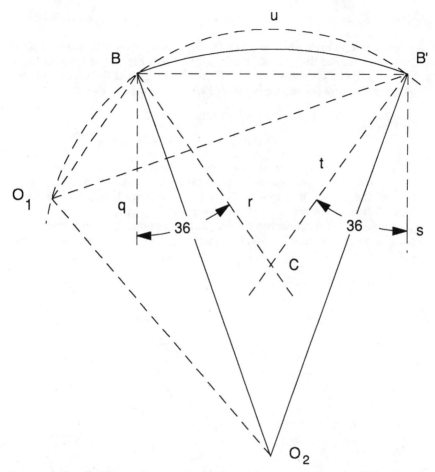

Figure 4.6 Crank-slider analogy synthesis of a quick-return crank-rocker linkage.

An angle α is calculated from Eq. (3.17) to be

$$\alpha = \pi(1.5 - 1)/(1.5 + 1) = 0.2\pi \text{ rad} = 36°$$

The subsequent synthesis follows the steps in Procedure 3.2. However, as cautioned in step 11 of Procedure 4.1 for crank-rocker synthesis, the grounded crank pivot and the rocker pivot usually should be placed on the same side of chord BB' (extended). Therefore, the construction procedure shown in Figs. 3.12 and 3.13 will be flipped vertically to give the geometry shown dashed in Fig. 4.6.

At points B and B' construct lines q and s perpendicularly downward from chord BB'. From points B and B' construct lines r and t downward and inclined toward each other by the angle $\alpha = 36°$ from the lines q and s. Lines r and t intersect at a point C. Using C as a center and length CB' as a radius, construct circular arc u through points B and B' and beyond.

Any point on arc u such as point O_1 can be chosen as the grounded pivot for the crank. For the point O_1 that was chosen in Fig. 4.6, the length of the ground link O_1O_2 is measured as 3.78 in. The lengths of the crank and coupler are given by the relationships of Eqs. (3.18), which are

$$L_1 = (O_1B' - O_1B)/2 = (4.28 - 1.69)/2 = 1.30 \text{ in}$$

$$L_2 = O_1B' + O_1B)/2 = (4.28 + 1.69)/2 = 2.99 \text{ in}$$

In Fig. 4.7 these link lengths were used to construct the resulting linkage in two positions, superimposed on portions of the construction of Fig. 4.6. In Fig. 4.7 the circle p of radius 1.30 in is the path that the tip of the crank L_1 would follow as the crank rotated about point O_1.

As pointed out in the subsection on transmission angle, when crank L_1 is parallel to ground link L_4, as at positions O_1D and O_1F, the transmission

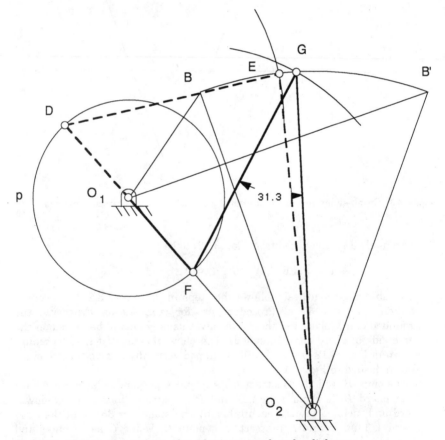

Figure 4.7 Motion of a synthesized quick-return crank-rocker linkage.

angle will be at its extreme values. The heavy dashed lines O_1DEO_2 show
the linkage in one such position, and the heavy solid lines O_1FGO_2 show it
in the other. The transmission angle (the angle between coupler DE and
rocker EO_2) is seen to be close to 90° in the first case. In the case of the heavy
solid lines, however, the transmission angle (between FG and GO_2) is seen
to be only a little more than 31°. This value is smaller than the minimum
limit suggested by the usual rule of thumb. Whether such a small value
could be tolerated in a particular application would have to be determined
by investigation of the loads induced in that particular mechanism.

The reader might wish to consider performing a synthesis in which the
dashed construction (except for line O_1O_2) in Fig. 4.6 is flipped vertically so
that arc u is concave upward. Such a synthesis would place the grounded
pivots O_1 and O_2 on opposite sides of chord BB'. It will be found that the
minimum value of the transmission angle in the resulting mechanism will
be only about 15°—much worse than the one just found.

More general function synthesis procedures

The outputs of the mechanisms synthesized by Procedure 3.1
(Examples 3.1 and 3.2) are rather special functions of the input crank
motion. It is sometimes desired to generate an output function that
has given values at only two or three chosen values of the input link
angle. As can be seen in Examples 4.1 and 4.2, synthesis Procedure 4.1
results in either only one set of possible linkage dimensions or (in the
case of the quick-return linkage) a quite limited choice of sets of linkage dimensions. In function synthesis Procedures 4.2 and 4.3, which
will be described next, however, even though requirements for output
function to be provided are rigidly specified, additional assumptions
must be made *before* the synthesis can proceed. The linkage dimensions that result from these syntheses depend on these assumptions.
These results must be checked to determine whether the overall motion parameters (such as transmission angles) of the linkage are satisfactory. One particular aspect of the linkage motion that may be of
concern is whether the input link can rotate through a full 360°, i.e.,
whether it will be a crank. To check this feature, Grashof's criterion as
described in Sec. 4.6 is useful.

**Procedure 4.2: Three-Point Function Synthesis Procedure for a Four-Bar
Linkage** This type of synthesis can be used when the angular position of
some output body that is pivoted to ground is to have a specified value for
each of two or three specified angular positions of an input body that is also
pivoted to ground. This type of synthesis is also frequently useful as a preliminary step to choose initial estimates for linkage proportions when using
the function synthesis procedures of Chap. 9 for more than three positions.
The synthesis will be performed in terms of the angular positions of the
input link (to which the input body is attached) and the output link (to
which the output body is attached). However, it should be noted that in most

mechanism designs the input *body* and the output *body* can be oriented at any convenient angle to the input *link* and output *link*, respectively. Therefore, the orientations of these links in these syntheses are arbitrary assumptions, *except* that the angular *displacement* of each of these links *from one position to the next* must be the same as that which is specified for the attached body. This will be demonstrated in the following example. This procedure will be described by using it on an example problem.

Example 4.3: Three-Point Function Synthesis *Objective*: Synthesize a four-bar pin-jointed linkage to coordinate the tilting of two mirrors such that the tilt of a pivoted mirror P is related to the tilt of a mirror Q in the manner indicated by the following table, where the tilt angles are counterclockwise (ccw) relative to horizontal:

Position	Tilt of mirror P	Tilt of mirror Q
1	0°	90° ccw
2	40° ccw	115° ccw
3	80° ccw	160° ccw

Each of these positions is referred to as a *precision point* because on a plot of output versus input for the synthesized mechanism, points corresponding to these positions will have coordinates that precisely match the coordinates specified in this table.

1. As mentioned above, the orientation of the links that are attached to the bodies (the mirrors) need not be the same as the orientations of their attached bodies, the important parameters being the displacements from one position to another. We therefore construct a second table from the preceding table:

Displacement	Link P	Link Q
From position 1 to position 2	40° ccw	25° ccw
From position 1 to position 3	80° ccw	70° ccw

2. To start the procedure, draw three positions of link P or Q. Such a drawing has been made for link Q of arbitrary length 2 in, in Fig. 4.8. In this figure position 2 of link Q has been chosen arbitrarily to be vertically upward at O_2B_2, and positions 1 and 3 are arranged to be consistent with the preceding table of displacements. As a result, link Q is inclined 25° to the right of vertical for position 1 (O_2B_1) and is inclined 45° to the left of vertical for position 3 (O_2B_3). It is generally best to avoid having any of the positions of this link be close to parallel to the ground link (at least in the first synthesis attempt).

3. Then, for this example, the ground link from O_2 to O_1 is chosen arbitrarily to be 4 in long and to extend horizontally to the left. The point O_1 has thereby been chosen as the ground pivot of link P. It remains then to determine the lengths and orientations of link P and of the coupler link from link P to link Q.

The synthesis procedure uses the principle of kinematic inversion. That is, one link of the two links whose orientations are to be coordinated is assumed to be fixed (in this example it is link P, the link whose length and orientation are to be determined), while the remainder of the linkage (including ground) is rotated about the grounded pivot of that fixed link.

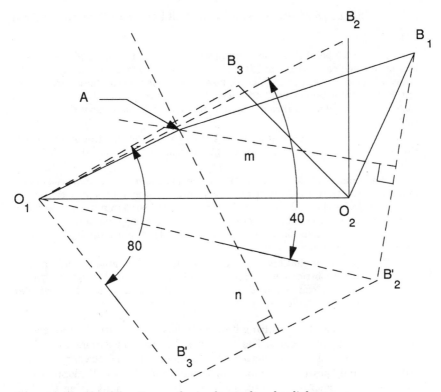

Figure 4.8 Graphical function synthesis of a crank-rocker linkage.

4. Note now that the tables indicate that in this example, link P rotates *counterclockwise* through 40° from position 1 to position 2. When the linkage is inverted by grounding link P, the remainder of the linkage must be rotated in the opposite direction. Therefore, for this synthesis, when link P is fixed at position 1, the remainder of the linkage is rotated *clockwise* about point O_1 through 40°. This is the kinematically inverted rotation that corresponds to the transition from position 1 to position 2. This rotation of the remainder of the inverted linkage "swings" point B_2 clockwise through 40° to position B'_2 as shown.

This could be done on a CAD system by rotating a duplicate (perhaps on a separate layer) of the solid lines in Fig. 4.8 through the required angle about point O_1.

5. Similarly, inversion of the link P rotation from position 1 to position 3 would "swing" point B_3 clockwise through 80° to position B'_3 as shown.

Now note that because link P has been assumed to be fixed, a point A (which is yet to be found) at which link P is pivoted to the coupler will be fixed, and the coupler will simply rotate about that point A. Therefore, because the length of the coupler cannot change, the point B at which it is pivoted to link Q must move along a circular arc that is centered at point A. That is, points B_1, B'_2, and B'_3 must all lie on a circular arc centered at point A.

6. The center point A is then found by constructing the perpendicular bisectors of line segments $B_1B'_2$ and $B'_2B'_3$, as shown by dashed lines m and n. The point A is at the intersection of these bisectors. Some CAD systems can pass an arc through three points and find the center of that arc automatically. In such a case, pass a circular arc through points B_1, B'_2, and B'_3, and label the center of that arc A.

7. The required linkage in position 1 then consists of the ground link O_1O_2, the link P that is the line O_1A, the coupler that is line AB_1, and the link Q that is line O_2B_1.

8. The completed linkage should then be checked for satisfactory performance in all positions that it is expected to experience. That is, its transmission angle range should be checked (see the subsection on transmission angle, earlier in this section). Also, the linkage should be checked to see whether any of the links that may have been specified to be cranks can indeed rotate through 360° (see Grashof's criterion in Sec. 4.4).

9. The dimensions that are determined by steps 1 through 8 actually merely determine the proportions of the required linkage. That is, the linkage synthesized in these steps can be scaled up or down and rotated to suit the particular machine into which the linkage is to be incorporated.

Figure 4.9 shows the resulting linkage in position 2, with the mirrors attached at appropriate angles. The position of the links in positions 1 and 3 are shown dashed (without the mirrors). The reader might wish to verify that the linkage actually positions the mirrors as required. This can be done by geometrically constructing the linkage at its design positions using the link lengths determined above or by using a CAD system to "animate" the linkage.

Comments and suggestions

Figure 4.10 is a plot of the angle of mirror Q versus the angle of mirror P for the range between positions 1 and 3. Such a plot can

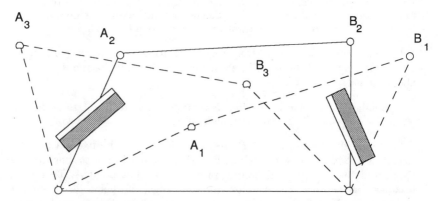

Figure 4.9 Coordinated positioning of two mirrors using a function-synthesized four-bar linkage.

be used to show whether the angular relationship between the mirrors is satisfactory in the regions between the design (precision) positions.

Examination of Fig. 4.9 shows that in position 1, point B and link Q are very near their rightmost extreme positions because link P is almost parallel to the coupler. Correspondingly, Fig. 4.10 shows that near this position, the angle of mirror Q is very insensitive to the position of mirror P because the curve is nearly horizontal in this region. Depending on the particular application, such a condition might constitute a marginal feature of the design. Indeed, the linkage barely meets the requirement for providing position 1. Also, if link Q were considered to be the input link, the transmission angle would be very bad in position 1.

Thus it is seen that although the linkage has been made to satisfy position requirements at three points, the curve of output versus input may have less than desirable characteristics *between* these points. The shape of the curve can be varied by changing the arbitrary assumptions that were made in steps 2 and 3. That is, a different orientation of the group of link Q positions could be chosen, and/or a different length of the ground link O_1O_2 could be chosen. For example, if the group of link Q positions were assumed to be rotated slightly clockwise from the position assumed in Fig. 4.8, the curve of output versus input would have a minimum point between positions 1 and 2. A counterclockwise change in the assumed position of the group of link Q positions would result in a steeper slope of the curve at position 1.

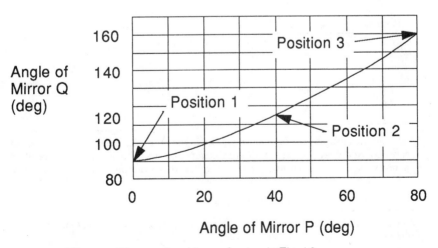

Figure 4.10 Mirror positions produced by mechanism in Fig. 4.9.

General rules for varying the assumptions to achieve given results cannot be given, but readers often will find that after a couple of variations have been tried, patterns of effects begin to appear and an iterative process of varying those assumptions can be developed for the particular problem being worked on.

The desired shape of the curve of output versus input could be specified more completely by specifying more precision points. However, it will be noted that if more than three points are specified, more than two perpendicular bisectors would appear in Fig. 4.8, and conflicting determinations of the location of point A would result. Therefore, if it is desired to specify more than three points, Procedure 4.3 (overlay procedure) should be used, or a computer-aided procedure such as described in Chap. 9 should be used.[3]

Procedure 4.3: Overlay Procedure for Function Synthesis   The overlay procedure is an iterative graphical procedure that can be used to synthesize a linkage whose plot of output versus input matches a specified function at several precision points. The synthesis may result in only *approximate* matching at some of the specified points, and the more points that are specified, the more difficult it is to obtain good matches. However, such a synthesis often can serve as a preliminary step to provide insight into the design of a function-generating linkage using the function synthesis methods of Chap. 9.

This procedure uses either a transparent overlay drawing or two layers of a CAD system. The steps are illustrated by the following example.

Example 4.4: Overlay Function Synthesis   *Objective*: To synthesize a four-bar linkage in which an output link Q that is pivoted to ground is to assume positions that are related to the positions of an input link P (which is also pivoted to ground) in the manner indicated by the following table:

Position	Angle of link P	Angle of link Q
1	90°	0°
2	75°	5°
3	60°	12°
4	45°	21°
5	30°	32°
6	15°	44°
7	0°	57°

1. On either a sheet of paper or on one layer of a CAD system file, choose an origin O_1, and draw line segments radiating from it at the orientations specified for one of the links such as P or Q above. (In this example, link P was chosen,

---

[3]As shown in Chap. 9, as many as five points may be specified for *exact* matching in a function synthesis. However, the resulting linkage is not always practical.

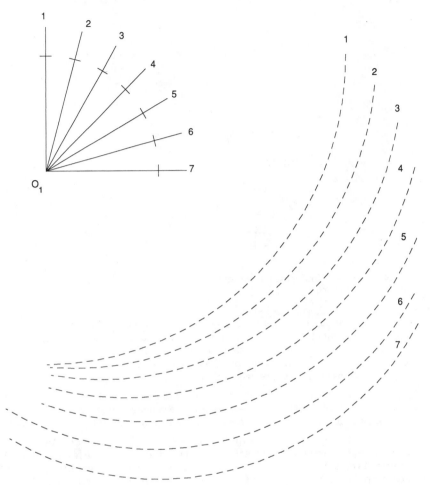

Figure 4.11 Input and coupler graphic constructions for overlay type of function synthesis.

and the resulting line segments are shown in Fig. 4.11.) Label each segment with the number of the corresponding position.

   2. On each of the line segments drawn in step 1, mark off a convenient length from the origin, as shown in Fig. 4.11. (In this example, the line segments represent the positions of link P, and the length marks represent the successive locations of the pivot connecting that link to the coupler.)

   3. Using each of the length marks from step 2 as a center, construct an arc having a convenient radius (which will represent the assumed coupler length), as shown dashed in Fig. 4.11. Label each such arc to correspond to the associated position of the line segment on which its center lies (e.g., in the example, arc 1 corresponds to line 1, arc 2 corresponds to line 2, etc.)

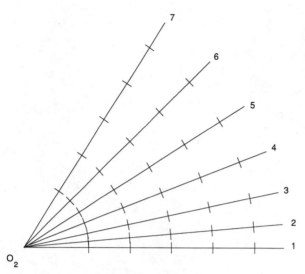

Figure 4.12 Output link position construction for overlay type of function synthesis.

4. On a transparent sheet or on a second layer in the CAD system file, choose an origin and label it O_2. From this origin, draw radiating line segments at the angular orientations of the positions of the other link (Q in this example), as shown in Fig. 4.12. In this example the spacings of the angles of link P are all equal, but those of link Q are relatively uneven.

5. On each of the line segments in step 4, place a series of length marks at a convenient spacing from the origin, making the spacing the same on all line segments, as shown in Fig. 4.12. These marks indicate potential lengths for link Q.

6. Place the transparent drawing from steps 4 and 5 over the drawing generated in steps 1 through 3. If a CAD system is used, *show* both the layer generated in steps 1 through 3 and the layer generated in steps 4 and 5. (This latter layer will be referred to as the *transparent drawing* or *transparency*.) Rotate and slide this transparent drawing around until a set of length marks that are all at the same distance from O_2 on the transparency fall on the corresponding numbered arcs on the underlying drawing, as shown in Fig. 4.13. In this figure a rather good (but not perfect) fit has been achieved by sliding and rotating the transparency until the third mark from O_2 on each line on the transparency has been made to lie almost on an arc on the underlying drawing that has the same number as the line on which the mark lies. It was found that these "third" marks can be made to lie more accurately on their corresponding arcs than can any of the other sets of marks. It often requires quite a bit of sliding the transparency around to find a good match. In this particular example it was necessary to rotate the transparency almost 180° to allow link Q to rotate counterclockwise while link P rotates clockwise.

7. Draw the resulting linkage in position 1. This can be done in Fig. 4.13 by noting that in position 1, link P is the line segment O_1A and point B is the position of the pivot connecting the coupler to link Q. Therefore, link P is O_1A, the coupler is AB, link Q is BO_2, and the ground link is O_1O_2.

8. Check the resulting linkage for transmission angle values and for the presence of a crank if required (see Sec. 4.4).

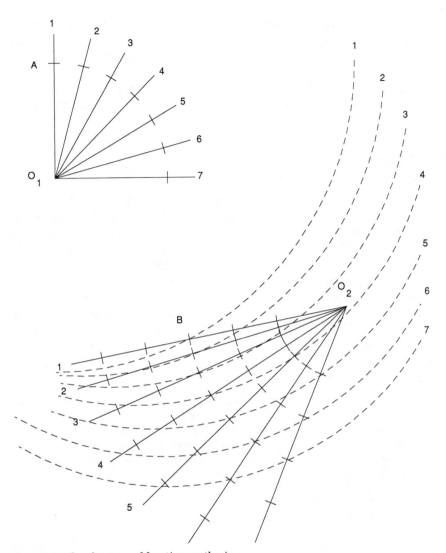

Figure 4.13 Overlay type of function synthesis.

Comments and suggestions

The linkage that results from the preceding steps will be based on the assumption of a length for link P and of a ratio of length of coupler to length of link P. Such linkage may be scaled up or down (as a unit) and rotated (as a unit) to suit any space requirement without altering the functional relationship between the angles of link P and link Q. Therefore, the choice of length for link P is not important to the quality of the fitting of the generated function to the precision points.

However, the assumed ratio of the coupler length to the length of link P has a large effect on the ability to match those points. It may be necessary to try various values of this ratio in order to obtain a satisfactory fit.

Note that the dashed arcs shown in Fig. 4.11 can cross each other. Therefore, functions that have slopes that change sign can be produced.

4.4   Grashof's Criterion and Kinematic Inversion

The linkages that were discussed in Chap. 3 and those which were discussed in connection with synthesis Procedures 4.1 and 4.2 contained cranks. That is, they each contained a link that was capable of rotating through 360° relative to ground, and they could therefore each be driven by a continuously rotating shaft. There are also other useful four-bar linkages in which no link is capable of full 360° rotation relative to any of the other three links in same linkage. It is possible to determine whether a given four-bar linkage contains a link that is capable of such 360° rotation by applying a criterion that is ascribed to Grashof (pronounced "grahz-hofe").

Grashof's criterion

If, in a pin-jointed four-bar linkage, the sum of the length of the longest link plus the length of the shortest link is *smaller than* the sum of the lengths of the other two links, then the shortest link (and only the shortest link) in the linkage can be rotated through 360° relative to all the other three links in that linkage. Such a linkage is referred to as a *Grashof linkage*.

If, in a pin-jointed four-bar linkage, the sum of length of the longest link plus the length of the shortest link is *greater than* the sum of the lengths of the other two links, then no link in the linkage can be rotated through 360° relative to any of the other three links in that linkage. Such a linkage is referred to as a *non-Grashof linkage*.

If, in a pin-jointed four-bar linkage, the sum of length of the longest link plus the length of the shortest link is *equal to* the sum of the lengths of the other two links, then at some stage in the motion of the linkage all four of the links can become parallel to each other and the motion subsequent to reaching that condition is indeterminate. Such a linkage has some of the characteristics of each a Grashof linkage and a non-Grashof linkage and thus might be called a *borderline Grashof linkage*.

We may write Grashof's criterion algebraically as

$$L_g + L_s > L_p + L_q \qquad \text{gives a Grashof linkage}$$

$$L_g + L_s < L_p + L_q \qquad \text{gives a non-Grashof linkage} \qquad (4.1)$$

$$L_g + L_s = L_p + L_q \qquad \text{gives a borderline Grashof linkage}$$

In these expressions, L_g is the length of the longest link, L_s is the length of the shortest link, and L_p and L_q are the lengths of the other two links.

Figure 4.14a shows a Grashof crank-rocker linkage. The link-length proportions are $L_1 = 1$ unit, $L_2 = 2$ units, $L_3 = 2$ units, and $L_4 = 2.5$ units. Then, according to Eq. (4.1), the shortest link (L_1) can rotate through 360° relative to the other links, including ground, so it is indeed a crank-rocker linkage.

In Fig. 4.14b the same linkage has been inverted kinematically by grounding link L_2 rather than L_4, and the shortest link L_1 can still rotate through 360° relative to ground. This too is a crank-rocker linkage.

In Fig. 4.14c the same linkage has been inverted kinematically by grounding link L_3 rather than L_4, and the shortest link L_1 can still rotate through 360° relative to the other three links. None of the other links can rotate through 360°, so links L_2 and L_4 are rockers and the linkage is a *double-rocker linkage.*

In Fig. 4.14d the same linkage has been inverted kinematically by grounding link L_1 rather than L_4. In this inversion, the rotation of the shortest link L_1 through 360° *relative* to the other three links actually

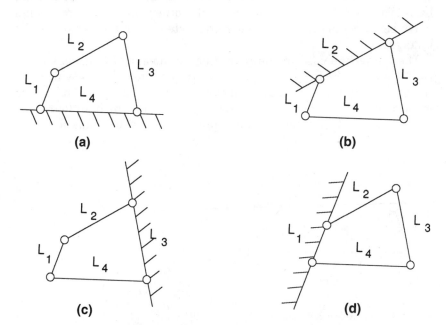

(a)

(b)

(c)

(d)

Figure 4.14   Kinematic inversions of a crank-rocker linkage (Grashof linkage).

consists of the rotation of the other three links through 360° *relative to* L_1 because L_1 is grounded. Because both L_2 and L_4 can rotate through 360° relative to ground, this is a *double-crank linkage.* It is frequently called a *drag-link linkage* and can be used to transmit rotation from one continuously rotating shaft to another continuously rotating shaft. This type of linkage is discussed in Sec. 5.5.

Figure 4.15 depicts a linkage with lengths $L_1 = 1$ unit, $L_2 = 2$ units, $L_3 = 2$ units, and $L_4 = 3.5$ units. According to Eq. (4.1), this is a non-Grashof linkage, and none of the links can be rotated through 360° relative to any other link. Therefore, links L_1 and L_3 are rockers, and this linkage is a double-rocker linkage. *Regardless of which link in a non-Grashof linkage is grounded, the linkage will be a double-rocker linkage.*

In a non-Grashof linkage, the two intermediate length links (e.g., L_2 and L_3 in Fig. 4.15) will become colinear when the shortest link (L_1) is at one of its extreme positions (to the left in Fig. 4.15). Rotating the shortest link from that extreme position would allow the intermediate links to move toward *either* of the two positions shown (one of the positions is shown dashed). Because the intermediate links can change from one of these positions to the other *only* by passing through this colinear position, this colinear position is known as a *change point.*

Such alternate positions for the intermediate links are also possible in Grashof linkages, but in order to move the intermediate links from one of these positions to the other, one of the pin joints must be disconnected and then reconnected after *reassembling* the linkage.

The borderline case in the preceding statement of Grashof's criterion, that in which $L_g + L_s = L_p + L_q$, results in linkages which, although useful, must be used with care. That is, either such a linkage must be prevented from moving to the position in which all four links become parallel, or some auxiliary provision must be made for guid-

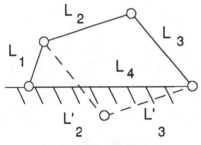

Figure 4.15 A non-Grashof linkage.

ing the motion of the links through that position. The need for these precautions is best illustrated by building and operating a planar cardboard and thumb tack model, and readers are encouraged to do so. Try a linkage with link lengths $L_1 = 1$ in, $L_2 = 2$ in, $L_3 = 3$ in, and $L_4 = 4$ in.

Borderline Grashof linkages include parallelogram linkages such as shown in Fig. 4.16a and deltoid or kite-form linkages such as shown in Fig. 4.16b (as well as many other configurations). It is readily seen that when L_1 becomes parallel to L_4 in either of the linkages in Fig. 4.16, all four of the links become colinear. This colinear position is a change point *from which* links L_2 and L_3 could either return to positions such as shown as solid lines *or* proceed to positions such as shown dashed. In the case of the kite-form linkage of Fig. 4.16a, there is also a change point when links L_2 and L_3 are parallel to each other but not coincident, while links L_1 and L_4 are not parallel to each other. Thus it is seen that borderline Grashof linkages contain change points just as do non-Grashof linkages. It is also seen that in these borderline linkages one of the links can be a crank, as was the case in Grashof linkages. (Actually, more than one of these links can be a crank.)

The reader will find it instructive to make cardboard and thumb-tack models of the linkages in Fig. 4.16 and to move the links to all their possible positions while considering various links to be grounded.

4.5 Motion Synthesis

As defined in Sec. 4.1, motion synthesis can be used to devise a linkage that will cause its coupler to assume a sequence of prescribed positions. Although computerized synthesis programs are available for performing such syntheses, graphical techniques such as described in this section are helpful in visualizing the geometric interactions involved. These graphical methods also can give first approximations to

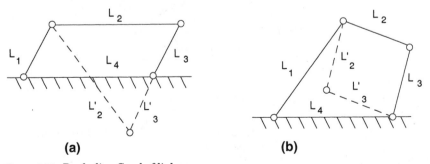

(a)   **(b)**

Figure 4.16 Borderline Grashof linkages.

the design, thereby minimizing the trial-and-error iterations required in the computerized process. CAD programs can be used to perform these graphical syntheses and provide very precise answers. In many cases these precise answers may be adequate for the final design, eliminating the need for further computerized synthesis.

In the preceding synthesis procedures the output motion was that of a link that was pivoted to ground. This motion, therefore, consisted of rotation only. When both rotation and translation must be provided, the motion must be that of a coupler because the coupler is the only link in a four-bar linkage that both rotates and translates. The task, then, consists of determining the lengths of the links that connect the coupler to ground, the locations of the two pivots on the coupler (the *moving pivots*), and the locations of the pivots on the ground (the *fixed pivots*). In Procedure 4.4 the locations of the moving pivots are assumed, and the remaining parameters are then determined. In Procedure 4.5 the locations of the fixed pivots are assumed, and the remaining parameters are then determined.

Procedure 4.4: Two- and Three-Position Motion Synthesis with Chosen Moving Pivots This procedure will be described using the example sequence of output body positions that are depicted in Fig. 4.17a. In this fig-

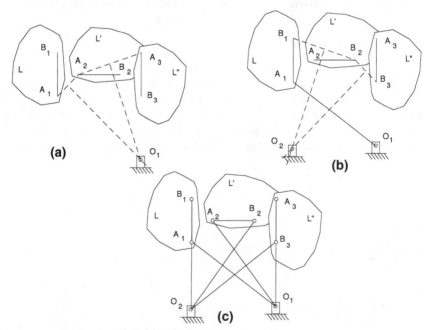

Figure 4.17 Motion synthesis with prescribed moving pivots.

ure a body L is required to pass through the sequence of three positions L, L', and L'' as shown. These required positions are known as the *precision positions*. Points A_1 and B_1 at the ends of the line segment shown attached to body L are chosen as the locations of the pivots on that moving body. As the body moves to positions L' and L'', the chosen pivots move to positions A_2 and B_2 and then to A_3 and B_3, respectively.

Because points A and B are each to be pivoted to a link that is in turn pivoted to ground, each of those points A and B must move along a circular arc that is centered at the fixed pivot for the corresponding connecting link. The synthesis procedure then consists of locating the centers of these arcs using the following steps:

1. Either on paper or on a CAD system, draw the body to scale in the precision positions that it is to be required to pass through, as shown at L, L', and L'' in the example in Fig. 4.17a.

2. At each of the required body positions, indicate the locations of the pivots that are to be attached to the body, as shown at A_1, B_1, A_2, B_2, A_3 and B_3 in Fig. 4.17a.

3a. If the synthesis is being performed manually on paper, draw the line segments A_1A_2 and A_2A_3 shown dashed in Fig. 4.17a. Construct the perpendicular bisectors of these two line segments as also shown dashed in the figure. The intersection of these two bisectors is the location O_1 of the fixed pivot of the link from moving pivot A to ground.

3b. If the synthesis is being performed using a CAD program, some shortcuts to finding the center of the arc through A_1, A_2, and A_3 may be available. Some such systems allow an arc to be passed through these points, and then the center O_1 of that arc is automatically determined. In others, the line segments A_1A_2 and A_2A_3 can be drawn, and then the segments can be rotated through 90° about their individual centers to form the perpendicular bisectors of the segments, as shown in Fig. 4.17a. The intersection of the perpendicular bisectors is the desired location of the fixed pivot O_1.

Note that if only two precision positions were specified, only one perpendicular bisector can be constructed, and the fixed pivot O_1 could lie *anywhere* along that bisector.

4. Repeat step 3a or 3b but using points B_1, B_2, and B_3 instead of points A_1, A_2, and A_3. This step is illustrated in Fig. 4.17b. The intersection of these perpendicular bisectors is O_2, the location of the fixed pivot of the link connecting point B to ground.

Note that if only two precision positions were specified, only one perpendicular bisector could be constructed, and the fixed pivot O_2 could lie *anywhere* along that bisector. In this case, the pivots O_1 and O_2 could be *chosen* to coincide (if desired) in a single pivot at the intersection of the bisectors from steps 3 and 4.

5. The resulting linkage consists of a ground link O_1O_2, a link O_1A, a link O_2B, and a coupler AB. This linkage is shown in the three precision positions in Fig. 4.17c.

6. Check the linkage throughout its expected range of motion for values of transmission angle and, if necessary, for the presence of cranks (Grashof's criterion).

Comments and suggestions

In Fig. 4.17c it will be noted that when the body (coupler) is in the first precision position L, the coupler AB is aligned exactly with the link O_2B. If the linkage were to be driven by trying to rotate the link O_1A, link O_2B would be the output link, and the angle between AB and O_2B would be the transmission angle. Because this transmission angle is zero in the first precision position, the linkage could not be driven through that position using O_1A as the input link.

An analogous condition occurs in the third precision position L''. In this position the linkage cannot be driven by using the link O_2B as the input link. Thus it is seen that this linkage as synthesized cannot be driven to and from the precision positions *without some auxiliary means*.

When unacceptable linkage characteristics result from a given synthesis such as the preceding, relief often can be obtained by changing the position of one or both of the moving pivots relative to the coupler body. In changing these pivot locations, note that fixed pivot O_1 was determined entirely by the precision locations of pivot A; i.e., the motion of pivot B had no effect on the location of O_1. Similarly, the motion of pivot A has no effect on the location of pivot O_2. Then O_1 and A constitute an interdependent pair (known as a *Burmester point pair*), and O_2 and B constitute another such pair. Each pair can be treated independently of the other.

If in Fig. 4.17c, with the body fixed in the position L, the pivot at A_1 were moved to the left relative to body L without moving pivot B_1, the line A_1B_1 would no longer be aligned with link O_2B because changing the location of pivot A does not change the location of pivot O_2 (O_2B would remain as shown). With this change, the transmission angle problem in position L could be improved. Thus it is seen that by changing the choice of moving pivot locations, some undesirable conditions can be improved.

In Fig. 4.18a the body L is shown in the same three positions as were shown in Fig. 4.17. However, in Fig. 4.18a a new set of moving pivots C and D has been chosen. (The original moving pivots A and B are also shown in the first precision position.) Using the same synthesis procedure described above, new positions for pivots O_1 and O_2 are determined in Fig. 4.18a and b. Figure 4.18c shows the resulting linkage in the three precision positions. There it is seen that the pivot and link positions are quite different from those in Fig. 4.17c, and the trans-

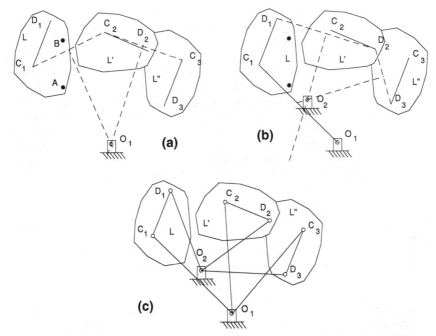

Figure 4.18 Motion synthesis with an alternative choice of prescribed moving pivots.

mission angle problems have been alleviated. This new linkage can be driven by rotating link O_1C or link O_2D.

It must be realized that although the linkages of Figs. 4.17c and 4.18c will cause the body L to reach the same three precision positions, the *motions of body* L *between these precision positions will be different* for these two linkages. Such motion should be checked.

In the examples just described, the three precision positions given allow moving pivot A to determine a circular arc that passes through the three successive positions of point A. Then the center of that arc is determined by constructing two perpendicular bisectors of the line segments joining these positions. If four precision positions of the body had been specified, an additional perpendicular bisector would have to be constructed in an attempt to find an arc that passes through the four positions of point A. In general, this third bisector will not intersect the other two in the same point, and, therefore, the three bisectors will in general give inconsistent indications of the location of an arc center. This is another way of saying that a circular arc cannot in general be passed through more than three arbitrarily chosen points.

However, it can be shown (see Chap. 9) that it is possible to find an infinite number of locations for point A in the body that will give successive positions through which a circular arc can be passed. To find such points and therefore perform four precision-position motion synthesis, it is usually most practical to use the computerized synthesis procedures described in Chap. 9.

Procedure 4.5: Two- and Three-Position Motion Synthesis with Chosen Fixed Pivots By comparing Figs. 4.17c and 4.18c it can be seen that the same set of precision positions can be provided by linkages having not only different sets of moving pivot locations but also different sets of fixed pivot locations. There are occasions on which the location of the fixed pivots is a more important requirement than the location of the moving pivots. In such cases, the location of the fixed pivots can be chosen, and a synthesis procedure using kinematic inversion can be used to find the locations of the moving pivots. This procedure is as follows:

1. On paper or on a CAD system, draw to scale the moving body in the three (or two) specific precision positions. Figure 4.19a shows such a drawing in which the precision positions L, L', and L'' are the same as those shown in Figs. 4.17 and 4.18. (To make the comparison with those figures more apparent, the line AB between the original moving pivots is also shown in Fig. 4.19a, although the shape of the body has been altered.)

2. On the drawing in step 1, indicate the desired locations of the fixed pivots. In the example used in this description, the fixed pivots O_1 and O_2 are chosen quite close to the prescribed precision positions of the body (presumably to make the resulting linkage more compact).

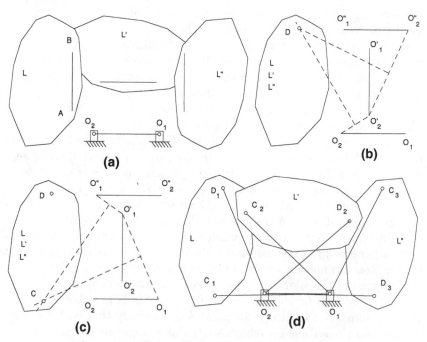

Figure 4.19 Motion synthesis with prescribed fixed (grounded) pivots.

3. The drawing produced in steps 1 and 2 will be called the *drawing*. (Label the CAD layer accordingly.)

4. Place a transparent sheet (or if a CAD system is used, a *new layer*) over the original drawing produced in steps 1 and 2. We will call this transparent sheet the *transparency*. (Label the new CAD layer accordingly.) On the transparency, trace or copy the locations of fixed pivots O_1 and O_2 and the moving body in the *first* precision position. These features are shown in Fig. 4.19b (along with some subsequent features).

The subsequent steps make use of kinematic inversion. That is, they pretend that the moving body is actually *fixed* in the first precision position and that the ground and other links move relative to that body.

5a. If the synthesis is being performed manually on paper, slide the drawing beneath the transparency until the second precision position L' on the drawing coincides with the precision position L on the transparency. Trace the locations of the fixed pivots O_1 and O_2 on the drawing onto the transparency, and label the tracings O_1' and O_2', respectively, as shown in Fig. 4.19b.

5b. If the synthesis is being performed on a CAD system, step 5a can be performed very accurately as follows: The location (in translation and rotation) of some feature, such as an edge of the body in the second precision position L' on the drawing, can be dimensioned relative to the location of that same feature in the first precision position L on the transparency. Then, knowing these dimensions, the entire drawing layer can be precisely rotated and translated to bring the body position L' on the drawing layer into exact coincidence with body position L on the transparency layer.

Be careful of the order in which you do the dimensioning and in which you make the translation and rotation, and be careful of the point around which the CAD program makes rotations. In the CAD system used to generate Fig. 4.19b, the following sequence was followed: (a) The angle between an edge of the body in position L' on the drawing layer and a horizontal line was dimensioned; (b) the angle between the corresponding edge of the body in position L on the transparency layer and a horizontal line was dimensioned; (c) the difference between these two angles was noted, and then the *entire drawing layer* was rotated through an angle equal to this angular difference to make the corresponding edges of the body at L' in the drawing layer and L in the transparency layer parallel; (d) the horizontal and vertical distances from a point on the body at this new position of body at L' in the drawing layer were dimensioned to the corresponding point on the body at L in the transparency layer; and (e) the *entire drawing layer* was moved by translation through a distance equal to the distances indicated by the distance dimensions, thereby bringing L' on the drawing layer into exact coincidence with L on the transparency layer.

6. Repeat step 5a or 5b for the third position L'' to give the location of points O_1'' and O_2'' on the transparency, as shown in Fig. 4.19b.

The drawing that appears on the transparency (layer) will now contain three successive positions each of the fixed pivots O_1 and O_2, as shown in Fig. 4.19b. This is a kinematic inversion of the situation in Fig. 4.19a, in which the relative motion between the moving body and the fixed ground in Fig. 4.19a has been portrayed as the same relative motion between the fixed body and moving ground. The remainder of this procedure is very much the same as Procedure 4.4.

7a. If the synthesis is being performed manually on paper, draw the line segments $O_2O'_2$ and $O'_2O''_2$ shown dashed in Fig. 4.19b. Construct the perpendicular bisectors of these two line segments as also shown dashed in the figure. The intersection of these two bisectors is the location D of the moving pivot that corresponds to the fixed pivot O_2.

7b. If the synthesis is being performed using a CAD program, some shortcuts to finding the center of the arc through O_2, O'_2, and O''_2 may be available. Some such systems allow an arc to be passed through these points, and then the center D of that arc is automatically determined. In others, the line segments $O_2O'_2$ and $O'_2O''_2$ can be drawn, and then these segments can be rotated through 90° about their individual centers to form the perpendicular bisectors of the segments, as shown in Fig. 4.19b. The intersection of the perpendicular bisectors is the desired location of the moving pivot D.

Note that if only two precision positions were specified, only one perpendicular bisector can be constructed, and the moving pivot D can lie anywhere along the bisector.

8. Repeat step 7a or 7b but using points O_1, O'_1 and O''_1 instead of points O_2, O'_2 and O''_2. This step is illustrated in Fig. 4.19c. The intersection of these perpendicular bisectors is C, the location of the moving pivot that corresponds to the fixed pivot O_1.

Note that if only two precision positions were specified, only one perpendicular bisector can be constructed, and the moving pivot C can lie anywhere along the bisector. In this case, the moving pivots C and D can be chosen to coincide (if desired) in a single pivot on the body at the intersection of the bisectors from steps 7 and 8.

9. The resulting linkage consists of a ground link O_1O_2, a link O_1C, a link O_2D, and a coupler CD, where each of these links can be drawn between the corresponding points on the transparency (layer) that is shown in Fig. 4.19c. This linkage is shown in the three precision positions in Fig. 4.19d.

10. Throughout its expected range of motion, check the linkage for values of transmission angle and, if necessary, for the presence of cranks (Grashof's criterion).

Comments and suggestions

Figure 4.19d shows that throughout the range of motion shown, and regardless of whether link O_1C or link O_2D is used to drive the linkage, the transmission angle is reasonable. Thus it is seen that Procedure 4.5 provides yet another approach that may be used to try to improve unacceptable linkage characteristics that result from an initial synthesis.

It should be further noted that because each Burmester point pair can be treated independently, the position of one moving pivot and the position of one fixed pivot could be chosen and then Procedures 4.4 and 4.5 could be used together to determine the location of the other moving pivot and of the other fixed pivot. For example, locations for O_1 and D could have been chosen arbitrarily, and Procedure 4.4 could then have been used to locate O_2 and Procedure 4.5 to locate C.

The same comments that apply to Procedure 4.4 relative to four-position synthesis also apply to this procedure.

4.6   A Synthesis Technique
for Pick-and-Place Mechanisms

A frequently encountered synthesis problem that does not readily fall into the category of two- or three-position motion synthesis is that of synthesis of a so-called pick-and-place mechanism. Such a mechanism is required to grasp (pick) an object that is presented to it at some initial location and in some initial orientation. The mechanism must then move the object to some specified final location and orientation, where it will release (place) that object. An example of the sort of operation that might be required of such a mechanism is given in Fig. 4.20a.

In Fig. 4.20a, an object such as a label indicated by line AB is shown in an initial position A_1B_1 in a feed tray at the left of the figure. The label is to be picked up out of the feed tray and transported to the right,

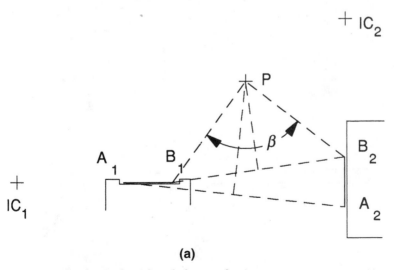

(a)

Figure 4.20   Synthesis of a pick-and-place mechanism.

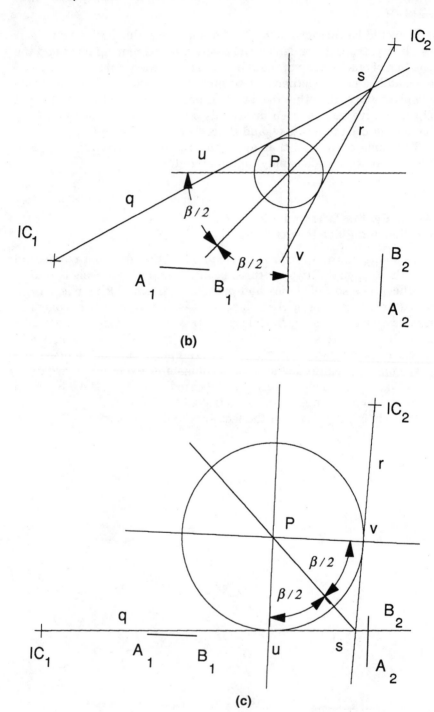

(b)

(c)

Figure 4.20 *(Continued)*

where it is to be placed at position A_2B_2 on the side of a box. Although the only positions specified are those at A_1B_1 and A_2B_2, it also is necessary to lift the label vertically out of the feed tray and bring it into contact with the box without sliding it along the vertical surface of the box. That is, as the label leaves the tray, its velocity must be vertical, and as it approaches the box, the label's velocity must be horizontal.

The direction of the velocity at the feed-tray end of the transfer could be specified by specifying, in addition to position A_1B_1, a position just barely above position A_1B_1. To get from position A_1B_1 to this second position, the label would have to travel vertically, and thus the initial velocity would have been specified indirectly as being vertical. In the limit, the displacement between these two positions would be made infinitesimal.

The direction of the velocity as the label approached the box similarly could be specified by specifying a third position just barely to the left of the position A_2B_2. Using the original two positions A_1B_1 and A_2B_2 plus the two additional positions just described, the problem becomes one of four-position motion synthesis. Such a synthesis can be performed using readily available computer programs such as Linkages. However, graphical Procedure 4.6 can be used not only to provide insight into the phenomena involved but also to provide a first approximation to the mechanism configuration, preparatory to using computer synthesis for refining the dimensions. In many cases this graphical procedure will provide a very good design by itself, without recourse to computerized synthesis.

Procedure 4.6: Graphical Synthesis of a Pick-and-Place Mechanism
This procedure will be described using the example indicated in Fig. 4.20a and described in the preceding text. The procedure itself is based on a graphical construction that was described in P. W. Jensen, Streamlined linkage design, *Machine Design* (March 21, 1991):169–172. Instead of explicitly using four positions to specify the desired motion, it uses just two positions and makes use of the concepts of *poles* (see Sec. 1.11) and *instant centers* (see Sec. 2.5). The following example will illustrate the procedure.

Objective: Determine the proportions of a four-bar pin-jointed linkage that will displace a label of specified size from a feed tray to a box that is a given horizontal distance from that tray. The label must lie horizontally in the tray and must be lifted vertically out of the tray. As the label approaches the box, it must be oriented vertically and be traveling horizontally (as shown in Fig. 4.20a).

1. Draw to scale the object that the mechanism is to pick and place, in the pick position and in the place position. In this example, these are positions A_1B_1 and A_2B_2, respectively, in Fig. 4.20a.

2. Locate the pole for the displacement from the pick position to the place position. This pole is the point around which the transported body *could* be rotated

to take it from the pick position to the place position. Any two points on the body can be used in the determination of this pole. In this example, points A and B will be used, and as shown in Fig. 4.20a, these points move from A_1 and B_1 to A_2 and B_2.

Draw line A_1A_2 and its perpendicular bisector, as shown dashed in Fig. 4.20a. Draw line B_1B_2 and its perpendicular bisector, as shown dashed in Fig. 4.20a. The intersection of these perpendicular bisectors is the pole P of the displacement from pick to place.

Draw lines from P to the two positions of any point on the transported body. In Fig. 4.20a these lines are the dashed lines PB_1 and PB_2. Measure the angle between these lines (shown as β). This is the angle through which the body (the label) could be rotated about point P to displace the body from A_1B_1 to A_2B_2. In the present example, β happens to be 90°.

3. Locate the instant center for the motion of the body (the label) at the pick location. As shown in Sec. 2.5, this instant center can be found by choosing a point on the body and noting the direction of the velocity of that point. Then draw a line through the point and perpendicular to the velocity of the point. The instant center will lie on that line, and the distance from the point to the instant center will depend on the angular velocity of the body. In most cases readers will not wish to specify a numerical value for the angular velocity. Usually it will be desired only that the angular velocity be small so that the body does not rotate rapidly as it departs the pick position or that it have a certain direction of rotation as it departs. In such cases the choice of the location of the instant center is somewhat arbitrary, as long as it lies on a line perpendicular to the velocity of a chosen point in the body.

In the present example, the body velocity at the pick position is desired to be vertical, so the velocity of all points in the label AB must be vertical. Drawing a line through either A_1 or B_1 and perpendicular to the velocity results in drawing a horizontal line. This dictates that the instant center lie horizontally to the right or left of A_1B_1. In Fig. 4.20a the instant center is chosen somewhat arbitrarily at point IC_1.

4. Repeat step 3 for the place position. In the place position, the body in this example is required to have a horizontal velocity, so the instant center has been chosen arbitrarily to be vertically above position A_2B_2 at IC_2.

5. Because the subsequent constructions will involve drawing several lines, which might result in confusion, it may be advisable to trace the points P, IC_1 and IC_2 and the lines A_1B_1 and A_2B_2 onto two transparencies or onto two new CAD layers.

6. On one of the transparencies or layers from step 5, draw a circle of convenient radius centered at point P, as shown in Fig. 4.20b. The convenience of the radius will become apparent in step 7.

7. From each instant center, draw a line tangent to the circle drawn in step 6. Such lines are shown as lines q from IC_1 and r from IC_2 in Fig. 4.20b. The location of a fixed pivot for the mechanism will lie at the intersection (labeled s) of these two tangent lines. Notice that two tangent lines can be drawn from *each* instant center, so there are four possible intersections of these lines and thus four possible locations for the fixed pivot. Also, the locations of these intersections will depend on the radius chosen for the circle. Therefore, choose a radius and a set of tangent lines that will give a practical location for a fixed pivot.

8. Draw a line through points P and s as shown in Fig. 4.20b. Through P, draw a line that is rotated from line Ps in a direction *opposite* to that in which the *body* would be rotated to move it from pick to place, and rotate this line from line Ps through an angle $\beta/2$. That is, in this example, the body would be rotated counterclockwise through β about P to move it from pick to place, as seen in Fig. 4.20a. Therefore, the line is drawn rotated clockwise from line Ps, and this new line intersects tangent line q at point u, as shown in Fig. 4.20b. This point u is the location of the pick position of the moving pivot, which is on the same link as the fixed pivot found in step 7. That is, in the pick position, the line segment from s to u represents the position of one link from ground to the coupler.

If it is desired to see where that moving pivot will be when the mechanism is in the place position, another line can be drawn that is rotated from line Ps in the *same* direction as that in which the *body* would be rotated to move it from pick to place and rotate this line from line Ps through an angle $\beta/2$. That is, in this example, the body would be rotated counterclockwise through β about P to move it from pick to place, as seen in Fig. 4.20a. Therefore, this new line is drawn rotated counterclockwise from line Ps, and this new line intersects the other tangent line r at point v, as shown in Fig. 4.20b. This point v is the location of the place position of the moving pivot.

9. On the second transparency or the second added CAD layer, repeat steps 6, 7, and 8 using a different radius and/or set of tangent lines. This will produce locations for another fixed pivot and another moving pivot. Such a construction is shown in Fig. 4.20c. That is, in the pick position, the line segment from s to u represents the position of a second link from ground to the coupler. In Fig. 4.20c the points u and v appear to be at the points of tangency of the lines and the circle. This is just coincidence, resulting from the choices of circle radius and the tangents in this example.

The locations of the four pivots for the pick-and-place four-bar linkage have now been determined.

10. Trace the locations of the pivots as determined above onto a single drawing or CAD layer. The resulting example mechanism is shown in its pick and place positions in Fig. 4.21, where points O_2 and D_1 are tracings of points s and u from Fig. 4.20b and points O_1 and C_1 are tracings of points s and u from Fig. 4.20c. Investigate the linkage in both pick and place positions and at positions between those positions to see whether the mechanism will operate without encountering bad transmission angles and without requiring disassembly and reassembly to reach all required positions.

Comments and suggestions

The dashed lines in Fig. 4.21 show that the links O_1C and O_2D, when extended, intersect at IC_1 in the pick position (Fig. 4.21a) and at IC_2 in the place position (Fig. 4.21b). As will be shown in Sec. 4.8, this indicates that the linkage does indeed produce the chosen instant center locations.

The mechanism configuration depicted in Fig. 4.21 resulted from several assumptions that were made before and during the 10 steps just described. In any such synthesis, it is important to consider the effects that such assumptions have on the result and to attempt to

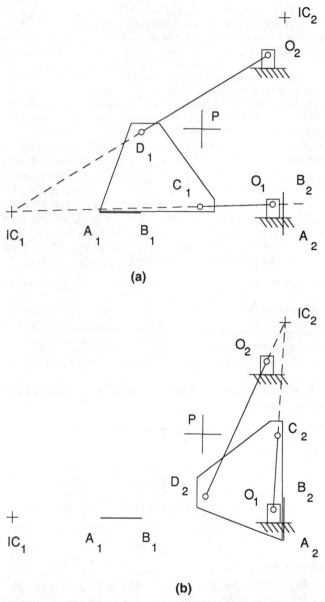

Figure 4.21 Pick-and-place mechanism synthesized from Fig. 4.20.

visualize how variations in these assumptions could be used to improve the result. Although these same assumptions also can be varied in the computerized syntheses described in Chap. 9, graphic Procedures 4.6 can provide aid in visualization of the interactions between variations in the assumptions and the results.

The assumptions that were made during steps 1 through 10 were the locations of the instant centers, the radii of the construction circles, and the choices of to which side of the circles the tangent lines should pass. Although it might not seem like an assumption, it was assumed before starting steps 1 through 10 that the resulting mechanism should meet the objectives stated just before step 1. The following are illustrations of how variations in these assumptions might be used in the particular example considered here.

By following the linkage through the range of positions from pick to place, it will be seen that if an attempt were made to drive the mechanism by rotating the link O_2D, the transmission angle would pass through zero. If the linkage is driven by link O_1C, the transmission angle is relatively well behaved. However, it is seen in Fig. 4.21b that near the place position it becomes a bit less than 30°. If this is deemed not tolerable, the choices of assumptions should be reexamined to see how the pivot locations might be changed. In Fig. 4.21b, if the pivots O_1 and C_2 were moved to the right without moving pivots O_2 and D_2, the transmission angle near the place position could be improved. Then, because the locations of pivots O_1 and C were determined by the construction in Fig. 4.20c, note that if the radius of the circle in that figure were increased, points s and v would be moved to the right. Because O_1 and C_2 are tracings of that s and that v, they would be moved to the right, and the transmission angle would be improved.

Figure 4.22a shows the positions of the label AB for successive 10° displacements of link O_1C for the linkage shown in Fig. 4.21. It will be

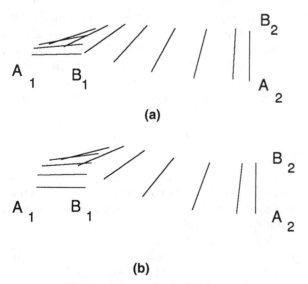

(a)

(b)

Figure 4.22 Motion of the synthesized pick-and-place mechanism.

noted that the label does indeed rise vertically from the feed tray at the left, rotates and translates, and then approaches the box at the right with a horizontal velocity. However, as the label rotates and translates rightward, it narrowly avoids colliding with the initial position A_1B_1. Depending on the nature of the label and of the device that transports it, this near-collision might be unacceptable.

In an effort to better avoid this near collision, consider the possibility of increasing the vertical velocity of the label as it leaves the pick position. Such increased velocity might raise the label further before it started to translate to the right, and thus it might be higher as it passes over the initial place position. To see how this vertical velocity might be increased, draw a velocity-distribution triangle (see Sec. 2.5) on Fig. 4.21a. This triangle should have its right-angle vertex at point C_1, and one side of the triangle should be line C_1IC_1. Draw an upward velocity vector of some convenient length at C_1 perpendicular to line C_1IC_1. Complete the velocity-distribution triangle by drawing a line from the tip of this vector to point IC_1. This triangle indicates that the velocity of the label at this position is about half that of point C_1. If the instant center IC_1 had been chosen to be much further to the left, the upper and lower sides of the triangle would have been more parallel, and the velocity of the label would have been much closer to being equal to the velocity of point C_1. That is, the vertical velocity of the label at the pick position could be increased by assuming a position of IC_1 that is further to the left. Such a choice does indeed improve the collision-avoidance problem in this example, as shown in Fig. 4.22b, which shows the label motion when IC_1 is moved further to the left and other parameters are left unchanged.

The collision avoidance also would be improved if the length AB were a smaller fraction of the horizontal distance from pick to place. This is easily visualized in Fig. 4.22 by considering AB to be smaller at each position. For a given size of label, this would mean that the translation distance would have to be greater than that shown. If the design constraints allow this greater displacement, such an approach could be used.

4.7   Coupler Point Paths and Cognate Linkages

The motion synthesis procedures of Procedures 4.4, 4.5, and 4.6 result in linkages in which successive locations *and orientations* of a body (the coupler) are caused to have specified values. There are many occasions in which it is required that a point pass through a sequence of prescribed locations or that the point follow a prescribed path but in which the *orientation* of the body to which it is attached is not important.

The path of any point on a link that is pivoted directly to ground is always a circle or circular arc. Points that are attached to the coupler of a pin-jointed four-bar linkage, however, can have a very large variety of shapes and sizes. The synthesis of four-bar linkages to provide a desired coupler point-path characteristic can take four forms:

1. A coupler curve and the associated linkage proportions can be chosen from an atlas in which many linkages and their associated coupler point curves are illustrated. Such an atlas by Hrones and Nelson[4] shows several thousand coupler point curves and the Grashof linkage proportions that produce them. The same information is also available in some CAD and kinematic analysis software. Engineers can search through these collections of curves until a curve with the desired characteristics is found, and then the curve and the associated linkage can be scaled to the desired size.

2. Because the variety of linkage proportions in any atlas is necessarily limited, and because such an atlas may not be readily available to the reader, qualitative synthesis techniques such as described in Chap. 8 may be used. Indeed, such techniques are useful adjuncts to the other techniques mentioned here. Readers should use qualitative techniques to obtain first approximations of linkage proportions. The dimensions of the linkages can then be refined by iterative use of simulation or by using any of the other three techniques mentioned here.

3. Graphical synthesis methods can be used when the coupler point path need only pass through a limited number of prescribed positions (preferably no more than three). These graphical methods are often useful in narrowing the range of values that must be assumed before the other types of syntheses are attempted.

4. Computer-aided synthesis such as described in Chap. 9 can be used to provide accurate linkage proportions that will produce a coupler path through a set of points.

The third of these four types of coupler point-path synthesis procedures will be described in this section.

Procedure 4.7: Graphical Synthesis of a Linkage to Produce a Coupler Point Path that Passes Through Three Prescribed Points This synthesis uses kinematic inversion, and it will be described using the requirements shown in Fig. 4.23a.

---

[4]Hrones, J. A., and Nelson, G. L., *Analysis of the Four-Bar Linkage,* Technology Press (MIT), Cambridge, MA, 1951.

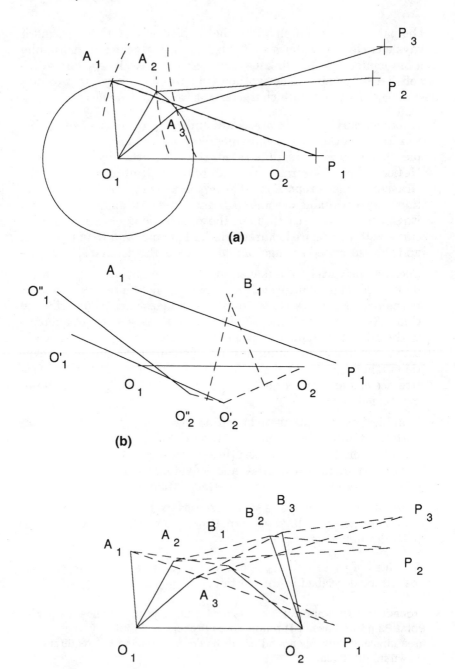

Figure 4.23 Graphical path synthesis.

Objective: Synthesize a four-bar linkage that will cause a point on a body to pass through points P_1, P_2, and P_3 shown in Fig. 4.23a. In addition, the mechanism that produces the required motion must be somewhat to the left of that set of points.

1. Make a drawing of the locations of points P_1, P_2, and P_3 on paper or on a CAD system layer called *drawing*. On that drawing, locate fixed pivots O_1 and O_2 in positions consistent with the requirements in the objective. In this example, these pivots were placed arbitrarily as shown in Fig. 4.23a.

2. Assume a convenient length for a link that will be pivoted to either O_1 or O_2. In this example it was chosen as the link pivoted at O_1. On the drawing, draw a circle with a radius equal to that convenient link length and centered at the chosen fixed pivot, as shown in Fig. 4.23a.

3. For one of the prescribed point positions, choose a convenient angular position for the link chosen in step 2. From the chosen fixed pivot to the circumference of the circle, draw a line representing the chosen link in that position. For this example, this line is shown as line O_1A_1 in Fig. 4.23a. This link was chosen to correspond to point position P_1, so the line A_1P_1 is drawn.

4. The points A and P will both be on the same rigid body (the coupler), so they must remain a constant distance apart as the linkage moves. Therefore, using a radius equal to length A_1P_1, draw an arc centered at P_2 intersecting the circle. This intersection will be the point A_2 so that O_1A_2 is the position of the chosen link that corresponds to the position of point P_2. Draw line A_2P_2. Repeat for points P_3 and A_3 as shown.

5. On a transparent sheet or a new CAD layer called *transparency,* trace lines O_1O_2 and A_1P_1, maintaining their relative positions from the drawing such as shown in Fig. 4.23b. This transparency will be used to perform a kinematic inversion in which the line AP will be assumed to be stationary and the ground represented by line O_1O_2 will be assumed to move.

6. Slide and rotate the original drawing or CAD drawing layer around beneath the transparency until line A_2P_2 coincides exactly with line A_1P_1 on the transparency. Then trace line O_1O_2 from the drawing to the transparency and label it $O_1'O_2'$ as shown.

7. Repeat step 6 but using line A_3P_3 instead of A_2P_2. Label this new traced position of line O_1O_2 as $O_1''O_2''$, as shown in Fig. 4.23b. The resulting drawing shows how the relative motion between the lines AP and O_1O_2 would look to an observer attached to line AP.

The coupler will have another moving pivot B that will be at one end of a link that is pivoted to ground at fixed pivot O_2. In the inversion that is shown in Fig. 4.23b, the coupler was held stationary, so the pivot B will be stationary and the pivot O_2 must travel along a circular arc centered at B. Therefore:

8. On the transparency, draw line O_2O_2' and its perpendicular bisector, both shown dashed in Fig. 4.23b.

9. On the transparency, draw line $O_2'O_2''$ and its perpendicular bisector, both shown dashed in Fig. 4.23b.

10. The intersection of the two perpendicular bisectors is the position of the moving pivot B relative to line AP, and line AP is, of course, shown in the first position in Fig. 4.23b. Therefore, label this point of intersection as B_1. The synthesized four-bar linkage in the position corresponding to point P_1 therefore consists of links O_1A_1, A_1B_1, B_1O_2 and the ground link O_2O_1. (For clarity, these actual links are not drawn in Fig. 4.23b. They are included in Fig. 4.23c, however, where the coupler is shown as the dashed triangle ABP in the three successive positions.)

11. Check the linkage over the range of positions expected to be used to see whether the transmission angle is satisfactory, whether the coupler point P moves from one position to the next in the desired order, and whether or not the linkage is a Grashof type of linkage (if the linkage input is to be a crank motion).

Comments and suggestions

The linkage synthesized in the preceding example is shown in the three positions that correspond to the precision positions P_1, P_2, and P_3. The coupler body is depicted as an obtuse angle ABP in the three positions.

The link lengths are such that this is a non-Grashof type of linkage, so none of the links can be rotated through 360°. It is a double-rocker linkage.

Regardless of which link, O_1A or O_2B, is used as an input driver, a transmission angle of less than 30° is encountered in the range of motion for which this mechanism is intended.

If either the Grashof condition or the transmission angle condition is undesirable for the intended application, remember that several assumptions were made. They were the locations of the fixed pivots O_1 and O_2, the length of link O_1A, and the initial angular position of link O_1A. Trial-and-error changing of these assumptions *one at a time*, combined with careful visualization of the effects of each change often can lead to much improved results.

Procedure 4.8: Graphical Synthesis of a Linkage to Produce a Coupler Point Path that Passes through Three Prescribed Points with Prescribed Coordination of Coupler Point Position with Input Body Angular Position

In some applications the coupler point must be located at each prescribed position while the input body is at a prescribed angle. Synthesis of a linkage to perform in this manner is known as *path synthesis with prescribed timing*.

Although links are often represented as straight lines connecting their pivots, the lines are attached to bodies that have both length and width and which can have any orientation relative to the line connecting the pivots. Therefore, when speaking of input body angular positions for timing purposes, the only important angular measurements are the *angular displacements from one position to another*.

The procedure for path synthesis with prescribed timing will be described using an example requirement that is depicted in Fig. 4.24a.

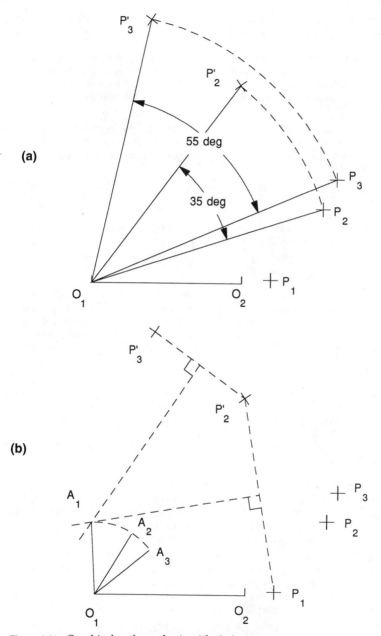

(a)

(b)

Figure 4.24 Graphical path synthesis with timing.

Objective: Synthesize a linkage in which the input link (or body) must rotate through 35° clockwise as a point *P* moves from position P_1 to position P_2 and must rotate through 20° clockwise as point *P* moves from position P_2 to position P_3, where these positions of point *P* are shown in Fig. 4.24*a* (together with some subsequent geometric constructions). The fixed pivots are to be located to the left of the prescribed point positions.

1. Make a drawing of the locations of points P_1, P_2, and P_3 on paper or on a CAD system layer. On that drawing, locate fixed pivots O_1 and O_2 in positions consistent with the requirements in the objective. In this example these pivots were placed at arbitrary positions to the left of P_1, P_2, and P_3, as shown in Fig. 4.24a.

2. Choose which of the fixed pivots is to be attached to the input link (body). In this example O_1 was chosen as the fixed pivot for the input link. From this fixed pivot, draw a line segment to the second prescribed coupler point P_2, as shown in Fig. 4.24a.

3. From that same fixed pivot, draw a line segment to the third prescribed coupler point P_3, as shown in Fig. 4.24a.

The next steps use kinematic inversion in which the input link is assumed to be fixed, and the ground (including the points P_2 and P_3) is rotated around the fixed pivot in a direction opposite to the prescribed rotation of the input link.

4. Rotate the line drawn in step 2 about the fixed pivot and through the angle through which the input link is specified to rotate from position 1 to position 2 but in the opposite direction. That is, in the present example the input link is specified to rotate clockwise through 35° from position 1 to position 2. Therefore, rotate line O_1P_2 counterclockwise through 35° about the fixed pivot O_1, as shown in Fig. 4.24a. Label the displaced end of this rotated line segment P_2'.

5. Rotate the line drawn in step 3 about the fixed pivot and through the angle through which the input link is specified to rotate from position 1 to position 3 but in the opposite direction. That is, in the present example the input link is specified to rotate clockwise through 55° from position 1 to position 3. Therefore, rotate line O_1P_3 counterclockwise through 55° about the fixed pivot O_1, as shown in Fig. 4.24. Label the displaced end of this rotated line segment P_3'.

Steps 4 and 5 have rotated points P_2 and P_3 about point O_1 through the prescribed timing displacement angles but in the opposite directions to give the "primed" point positions. The input link has not rotated in this inversion, and, therefore, the moving pivot A on that input link has not moved. The coupler has therefore rotated about the pivot A. Because the pivot A and point P are both on the same rigid body (the coupler), successive positions of P as represented by the points P_1, P_2', and P_3' must lie on a circular arc centered at A. Steps 6 and 7 locate this point.

6. Delete the lines O_1P_3 and O_1P_3'. Delete the lines O_1P_2 and O_1P_2'.

7. Draw line P_1P_2' and its perpendicular bisector as shown dashed in Fig. 4.24b.

8. Draw line $P_2'P_3'$ and its perpendicular bisector as shown dashed in Fig. 4.24b.

9. The intersection of the two perpendicular bisectors is the location of the moving pivot A of the input link in the first prescribed position.

10. The line O_1A can then be labeled O_1A_1, and lines O_1A_2 and O_1A_3 can be drawn with lengths equal to O_1A_1 and displaced from O_1A_1 by the prescribed timing angles, as shown in Fig. 4.24b.

Note now that the lines from step 10, plus line O_1O_2 and points P_1, P_2, and P_3, as they all appear in Fig. 4.24b, constitute the same sort of information as that in Fig. 4.23a. Therefore, Procedure 4.7 can be used to complete the synthesis. The same comments and suggestions for that procedure pertain to this synthesis.

Cognate linkages

In each of the two preceding synthesis procedures, a point on the coupler of the synthesized linkage was caused to pass through a sequence of prescribed points. In passing through these points, the coupler point would describe a continuous path which, if the linkage were a Grashof type of linkage and were to pass through all its possible positions without reassembly, would be a closed path. In the comments and suggestions accompanying the discussions of those procedures, it was shown that pin-jointed four-bar linkages with other dimensions also could be synthesized that would have coupler points that would pass through the same prescribed points. In general, though, the coupler points on those other linkages would *not* follow the same paths *between and beyond* the prescribed points.

However, for any four-bar pin-jointed linkage in which a coupler point follows a given path, it is possible to find two other different four-bar pin-jointed linkages, each of which has a coupler point that *exactly* follows that same path at *all* points. This fact is often stated as the *Roberts-Chebychev theorem,* and each of these two related linkages is known as a *cognate* of the original linkage. These cognates therefore occur in sets of three four-bar linkages that are cognates of each other. Characteristics of these sets are discussed in the Comments subsection and in the geared five-bar cognate linkages subsection, both of which follow Procedure 4.9. Means for finding the cognates of a given linkage are described in Procedure 4.9.

It is often useful to find one or both of the cognate linkages of a given linkage to substitute for the original linkage if the original linkage has some undesirable size or shape. A cognate of a linkage also can be very effective when a point on a body must be caused to follow a given path while that body is prevented from rotating. That is, when *parallel motion* must be provided. Such an application was described in A. H. Soni, *Mechanism Synthesis and Analysis* (McGraw-Hill, New York, 1974, pp. 381–383). Use of a cognate linkage in providing parallel motion is illustrated in the following procedure.

Procedure 4.9: Synthesis of a Cognate Linkage to Provide Parallel Motion
This procedure will be illustrated by means of the following example.

Objective: To design a mechanism that will progressively move small parts along the surface of a table. In the process of designing this mechanism, a linkage has already been synthesized to cause a coupler point to travel along a roughly elliptical path. The linkage and its coupler point path are shown in Fig. 4.25. The links in this linkage are an input crank O_1A, a coupler link AB, a rocker O_2B, the ground link O_1O_2, and the coupler point C. The link lengths are $O_1A = 1$ in, $AB = 5$ in, and $O_2B = 4$ in, and the length from point A to the coupler point C is $AC = 5$ in.

The purpose of this mechanism is to progressively move a series of small parts horizontally from right to left along a table surface indicated by dot-dashed line T. To do this, a link M is pivoted to link L_2 of the original linkage at coupler point C. The link M possesses a series of "fingers" on its upper surface as shown. As the crank rotates counterclockwise, these fingers must rise up through a slot in the table, move to the left while pushing the parts,

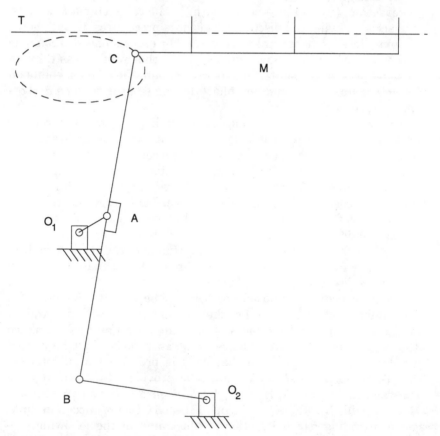

Figure 4.25 An approximate elliptical motion generator for a walking-beam mechanism.

retract downward through the slot, return rightward, and then repeat the cycle. A mechanism that operates in this manner is often called a *walking-beam mechanism*.

The link M must remain horizontal at all times if it is to perform its function. That is, M must not rotate; it must move in pure translation. Therefore, as stated in Sec. 1.11, all points in M must move in unison, and thus all such points must follow identical paths. To hold M horizontal, then, some means must be found for forcing a second point on M to follow a path identical to that followed by C. A cognate of the original linkage will be used for this purpose.

When the dimensions of the original linkage and its coupler point location are known, the dimensions and orientations of the parts of each of its two cognate linkages can be determined by the use of a diagram such as shown in Fig. 4.26a. It is important to remember that this diagram is intended to *qualitatively* represent the original linkage and its cognates. The shapes, lengths, and angles may not resemble those of the linkages concerned; the diagram is used as a *guide* in constructing the cognates.

1. Draw an arbitrary triangle such as ABC in Fig. 4.26a. The points A and B will represent the moving pivots A and B of the linkage in Fig. 4.25, and point C will represent coupler point C in that figure. This triangle will represent the coupler of the original linkage, and as mentioned above, it need not resemble the shape of the actual coupler.

2. In Fig. 4.26a and using side BC as a side, draw a parallelogram $BCDO_2$ and label the vertices accordingly. The line O_2B will represent the rocker O_2B in Fig. 4.25, and O_2 will represent the fixed pivot O_2 of Fig. 4.25.

3. In Fig. 4.26a and using side AC as a side, draw a parallelogram $ACGO_1$ and label the vertices accordingly. The line O_1A will represent the crank O_1A in Fig. 4.25, and O_1 will represent the fixed pivot O_1 of Fig. 4.25.

4. In Fig. 4.26a and on line CD construct a triangle CDE that is similar to triangle ABC as shown. Note that the angles α, β, and γ appear in corresponding places in ABC and CDE.

5. In Fig. 4.26a and on line GC, construct a triangle GCF that is similar to triangle ABC as shown. Note that the angles α, β, and γ appear in corresponding places in ABC and GCF.

6. In Fig. 4.26a complete the diagram by constructing the parallelogram $CEHF$ on sides CF of triangle GCF and CE of triangle CDE. The point H will represent a fixed pivot of the two cognate linkages.

Lines O_2D, DE, and EH will represent one cognate linkage with a coupler CDE such that point C traces the same curve as did C in the original linkage represented by O_1ABO_2. Lines O_1G, GF, and FH will represent another cognate linkage with a coupler GFC such that point C traces the same curve as did C in the original linkage. The points O_1, O_2, and H represent the fixed pivots for the three linkages. It remains to translate these representations into actual linkage dimensions and arrangements of the cognate linkages themselves using *relationships* from Fig. 4.26a but actual lengths and directions from Fig. 4.25.

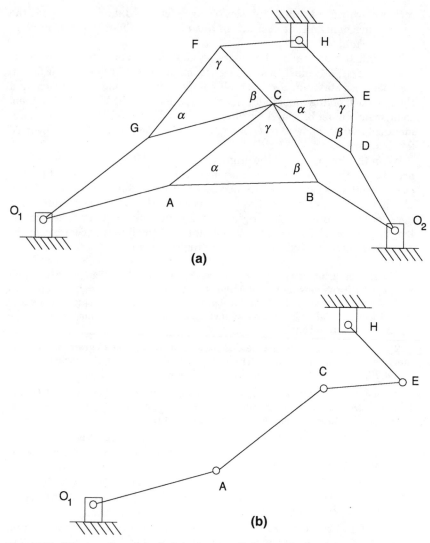

Figure 4.26 Diagram for use in deriving cognate linkages (Cayley diagram).

7. Let us choose to determine the cognate linkage that corresponds to the cognate represented on the right of Fig. 4.26a, i.e., the one represented by O_2DEH. We note that the diagram in Fig. 4.26a indicates that link O_2D is parallel and equal to the distance BC because they are opposite sides of a parallelogram. Figure 4.27 is a reproduction of the original linkage portion of Fig. 4.25. Therefore, on Fig. 4.27 draw dashed line O_2D parallel to and equal to the distance BC on that figure as shown.

8. Figure 4.26a indicates that DC is equal to and parallel to O_2B, so draw it thus in Fig. 4.27 as a dashed line as shown. It is seen that the parallelogram in Fig. 4.26a translates into a parallelogram in Fig. 4.27.

9. Figure 4.26a indicates that triangle DCE is similar to triangle BAC, so the corresponding construction is to be made in Fig. 4.27. However, note that in Fig. 4.27 the triangle BAC in this example is flattened into a straight line. That is, the angle $BAC = \alpha$ is 180°, and the angles $ABC = \beta$ and $BCA = \gamma$ are each zero. Therefore, construct the triangle DCE in Fig. 4.27 accordingly as dashed lines as shown. In this example, the triangle DCE is flattened into a straight line because the triangle BAC is a straight line. If triangle BAC had not been a straight line, triangle DCE would have been similar to but smaller than BAC and rotated so that DC was parallel to O_2B.

10. To complete the cognate linkage in Fig. 4.27, the link HE must be determined. In Fig. 4.26a it is shown that HE is parallel to FC. Also, in that figure FC differs in direction from GC by the constant angle β and GC is parallel to O_1A, the input crank of the original linkage. In Fig. 4.27 the angle β for this example, is, as indicated in step 9, equal to zero. Therefore, HE in Fig. 4.27 must be drawn parallel to O_1A and in the same direction (sense).

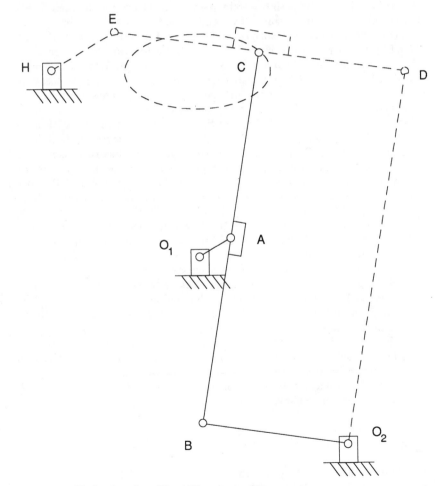

Figure 4.27 Mechanism from Fig. 4.25 and one of its cognates.

The length of HE is indicated in Fig. 4.26a to be equal to length FC, and the length of GC is indicated to be equal to length O_1A. However, because triangle GCF is similar to triangle ABC, then $FC/GC = CB/AB$. Therefore, $HE = O_1A(CB/AB) = 1(10/5) = 2$ in. The link HE is then drawn parallel to O_1A and with a length of 2 in as a dashed line in Fig. 4.27. The pivot H on link HE is a fixed pivot, as indicated in Fig. 4.26a.

In Fig. 4.26a, note that because GC must remain parallel to O_1A, the angular velocity of triangle GCF must be equal to the angular velocity of input crank O_1A of the original linkage. But HE must remain parallel to FC, which is a side of triangle GCF, so the angular velocity of HE must be equal at all times to the angular velocity of the original input crank O_1A. It can be seen that HE is the *only* link that is pivoted to ground and which will rotate at the same angular velocity as will the input crank O_1A.

The cognate linkage $HEDO_2$ therefore can be driven by rotating crank HE in unison with crank O_1A, and point C on the cognate linkage will trace a path that is identical to the path traced by point C on the original linkage. These paths will be traced in unison, so the points C on the two linkages in Fig. 4.27 will coincide continuously. Such coincidence is not useful for causing the link M in Fig. 4.25 to remain horizontal. To provide this function, simply translate the cognate linkage without distortion or rotation to a new location and pivot the link M to the new point C (relabeled C') on the link DE in its new location.

The new location for cognate linkage $HEDO_2$ can be chosen arbitrarily. However, because cranks O_1A and HE are to be driven in unison, it is convenient to move the cognate linkage until fixed pivot H coincides with fixed pivot O_1. This has been done in Fig. 4.28, resulting in the cognate linkage $H'E'D'O'_2$. Links $H'E'$ and O_1A rotate in unison, so they have been combined into a single ternary link. Also in this figure, link M has been pivoted to the original linkage at point C and to the cognate linkage at the point C'. Because these two pivot points move in unison, link M will move without rotating.

If we apply Gruebler's formula to this linkage in Fig. 4.28, $L = 7$, $J_1 = 9$, and $J_2 = 0$. Then $F = 3(7 - 1) - 9(2) = 0$; the linkage has zero degrees of freedom and thus is locked up. If the dimensions of the links and pivot locations were controlled extremely accurately, this linkage could be caused to have a *practical* degree of freedom, as discussed in Sec. 1.13. Note, however, that if the portion $C'D'O'_2$ of the cognate linkage were removed but the link $E'C'$ were still pivoted to link M at C', the motion of point C' would be the same as if these portions had not been removed. These portions are therefore seen to be unnecessary.

In Fig. 4.28 the unnecessary portions of the cognate linkage are indicated by dashed lines. If these dashed portions are removed, the application of Gruebler's formula gives $L = 6$, $J_1 = 7$, and $J_2 = 0$, so $F = 3(6 - 1) - 7(2) = 1$. The linkage can be driven by the cranks O_1A and $H'E'$, both mounted on the same shaft.

Comments

The other linkage that is a four-bar cognate of the original linkage could be determined using the same diagram shown in Fig. 4.26a but

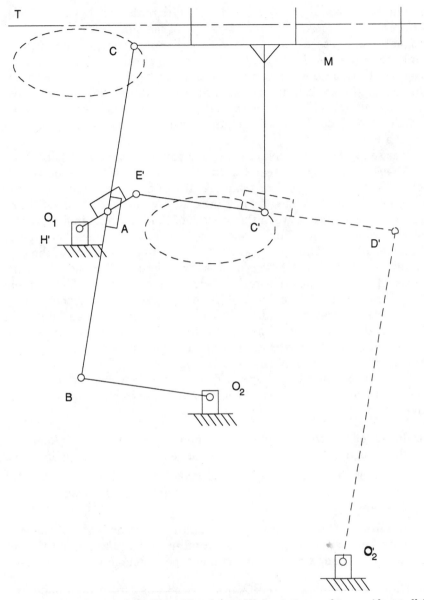

Figure 4.28 Mechanism from Fig. 4.25 and one of its cognates used to provide parallel motion.

using the relationships indicated by the left-hand linkage O_1GFH in that figure. This second cognate linkage would be quite different from the first cognate just synthesized, but a point C on it would nonetheless trace a path that would be identical to that traced by point C on the original linkage. In many applications this second cognate linkage might be of equal interest to the first cognate synthesized above. However, as pointed out following step 10, the only link in this second cognate that rotates in unison with the input crank of the original linkage is the *coupler GCF*. In order to cause point C on this second cognate to trace the same curve as traced by point C on the original linkage and trace it in unison with the original, the *coupler* of the second cognate would have to be caused to rotate in unison with the input crank O_1A. To do this would require considerable complication in the mechanism.

It is seen, then, that cognate four-bar linkages come in sets of three linkages that are cognates of each other. By using the properties of parallelograms and similar triangles, it can be shown that the ratios between the lengths of the four links in any one of the cognates are equal to the ratios between the lengths of the four links in each of the other two cognates. Thus, if one of the cognates is a Grashof type of linkage, all three linkages are Grashof-type linkages. If one of the cognates is a non-Grashof type of linkage, all three linkages are non-Grashof-type linkages. The *order* of arrangement of the links and their length ratios are different in the three cognates. By using the properties of parallelograms and similar triangles again, it can be shown that, as discussed following step 10 in Procedure 4.9, certain of the links in Fig. 4.26a must rotate at angular velocities that are equal to the angular velocities of certain other links.

Then, using the reasoning just discussed, adventurous readers will be able to show that in any possible set of three cognate four-bar linkages there are three possible combinations:

1. If one of the three cognates is a non-Grashof type of linkage, all three cognate four-bar linkages must be non-Grashof-type linkages.

2. If one of the three cognates is a drag-link (double-crank) type of linkage, all three cognate four-bar linkages must be drag-link-type linkages.

3. If neither of the two preceding conditions pertains, two of the cognate four-bar linkages must be crank-rocker linkages, and the remaining linkage must be a Grashof type of double-rocker linkage.

The original linkage O_1ABO_2 in Fig. 4.28 was synthesized using the qualitative principles described in Chap. 8. These same principles show that the path traced by coupler point C on cognate linkage $HEDO_2$ in that figure will be very similar to that traced by point C on linkage O_1ABO_2. The Roberts-Chebychev theorem tells us *more*, however. It tells us that by using the linkage lengths and angles according to the rules of the diagram of Fig. 4.26a, these two paths will be *identical*.

Geared five-bar cognate linkages

Return now to Fig. 4.26a, and note that a five-bar linkage consisting of ground link O_1H and moving links O_1A, AC, CE, and EH moves along with the four-bar cognates in such a manner as to cause point C to trace the same path as traced by the four-bar cognates. Note also that by tracing along the parallelograms and the triangle along the left side of Fig. 4.26a, link EH is seen to be oriented at the *constant angle* of β from link O_1A, so these two links must rotate at the same angular velocity. Figure 4.26b shows this five-bar linkage. If links O_1A and EH are "geared" together so that they *maintain the fixed angle β between each other,* the pivot C will trace the same path as traced by the coupler points C on the four-bar cognates.

Examination of Fig. 4.26a shows that there are also two other geared five-bar cognate linkages in this set. They are O_1GCDO_2 and O_2BCFH. Link O_1G must be geared to link DO_2, and link O_2B must be geared to link FH. This so-called one-to-one gearing can be accomplished by a gear train, by a chain and sprockets, by a timing belt and pulleys, or by adding another link to the five-bar linkage. This added link would connect suitable extensions on these links to each other in such a manner that, together with the ground link and the extensions, a parallelogram would be formed (see Figs. 1.24 and 1.25).

> Thus it is seen that a set of three one-to-one geared five-bar cognate linkages is associated with each set of three four-bar cognate linkages.

Care must be taken when using five-bar cognate linkages to be sure that the angle between the two links that are not pivoted to ground does not reach undesirable values. If one of the three five-bar cognates is found to be unsatisfactory, perhaps one of the others will be found to be more suitable for a given application.

4.8   Velocity Analysis of the Pin-Jointed Four-Bar Linkage

Knowledge of the velocities of various parts of a linkage is useful when the velocity of some part of that linkage must bear some prescribed relationship to the velocity of some other part of the machine being designed or when the mechanical advantage of the mechanism must be determined. Graphical methods of velocity analysis are very helpful in visualizing the relationships between the geometry of a linkage and the velocities in the linkage.

The velocity analysis of a four-bar linkage makes use of the velocity-difference equation, just as did the velocity analysis of the crank-slider linkage. That equation states that the velocity of a point on a rigid body is equal to the velocity of a second point on that same body plus the difference in the velocities of the two points. Velocity analysis will be illustrated using the linkage shown in Fig. 4.29. In this figure a link L_1 is shown pivoted to ground at point O_1, and it is rotating counterclockwise at an angular velocity of ω_1 rad/s. A coupler link L_2 is pivoted to L_1 at point A and to another link L_3 at point B. Link L_3 is pivoted to ground at fixed pivot O_2.

Graphical velocity analysis of pin-jointed four-bar linkages

For a graphical analysis of the velocities of the linkage if the angular velocity of one of the links that is pivoted to ground is known, first calculate the velocity vector $\mathbf{V}_A$ of the moving pivot on that link as follows: For example, assume that the angular velocity of link L_1 in Fig. 4.29 is known. The applicable velocity-difference equation for link L_1 is

$$\mathbf{V}_A = \mathbf{V}_{O_1} + \mathbf{V}_{AO_1}$$

where $\mathbf{V}_A$ is the velocity vector of point A, $\mathbf{V}_{O_1}$ is the velocity of fixed pivot O_1 (which is, of course, zero), and $\mathbf{V}_{AO_1}$ is the velocity of point A relative to point O_1 (i.e., the velocity difference between the two points).

Then, because link L_1 is a rigid body, $\mathbf{V}_{AO_1}$ is perpendicular to link L_1 and has a magnitude of $\omega_1 L_1$ (see Sec. 2.5). Also, because $\mathbf{V}_{O_1} = 0$, the vector $\mathbf{V}_A = \mathbf{V}_{AO_1}$.

Then plot $\mathbf{V}_A$ to some convenient scale on a scale drawing of the linkage such as shown in Fig. 4.29. This velocity $\mathbf{V}_A$ of point A is shown in

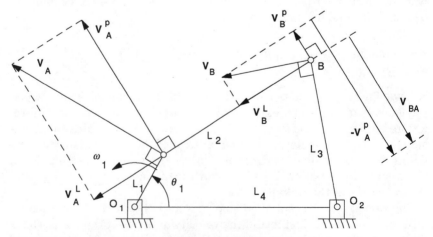

Figure 4.29 Graphical velocity analysis using longitudinal and perpendicular velocity components.

the figure resolved into a perpendicular component V_A^p and a longitudinal component V_A^l which are, respectively, perpendicular to line AB and along line AB. Because point A and B are both attached to the same rigid body (coupler L_2), the longitudinal component of the velocity of one must equal that of the other (see Sec. 2.5). We may then plot V_B^l equal to V_A^l but located at point B as shown. Now V_B^l is only one of two components of V_B, the other being V_B^p, which is perpendicular to line AB. Because point B is constrained by link L_3 to move only perpendicular to link L_3, these two orthogonal components of V_B must add to produce a velocity V_B that is perpendicular to L_3. Then the vector resolution rectangle for V_B may be drawn at point B as shown in the figure. Construction of this rectangle produces the vectors V_B and V_B^p and thus completely determines the velocity of point B.

The velocity-difference equation may now be applied to link L_2 in the form $V_B = V_A + V_{BA}$, which may be rewritten as

$$\mathbf{V}_{BA} = \mathbf{V}_B - \mathbf{V}_A = (\mathbf{V}_B^p - \mathbf{V}_A^p) + (\mathbf{V}_B^l - \mathbf{V}_A^l)$$

where $\mathbf{V}_B$ is the velocity of point B, $\mathbf{V}_A$ is the velocity of point A, and $\mathbf{V}_{BA}$ is the velocity of point B relative to point A.

Because points A and B are attached to the same rigid body (the coupler L_2), the longitudinal components of their velocities must be equal. Then this equation becomes

$$\mathbf{V}_{BA} = (\mathbf{V}_B^p - \mathbf{V}_A^p) + 0 = \mathbf{V}_B^p + (-\mathbf{V}_A^p)$$

This vector addition is shown to the right of point B in Fig. 4.29, and it is seen that the resulting $\mathbf{V}_{BA}$ is perpendicular to link L_2. We know from rigid body velocity analysis (see Sec. 2.5) that its magnitude is equal to $\omega_2 L_2$, so the angular velocity of L_2 may be computed by scaling the value of the magnitude of $\mathbf{V}_{BA}$ from the vector diagram and dividing that magnitude by the length of link L_2.

Example 4.5: Velocity Analysis of a Pin-Jointed Four-Bar Linkage Figure 4.29 is a reproduction of a schematic scale drawing of a linkage whose dimensions (in arbitrary units) are $L_1 = 1$, $L_2 = 3$, $L_3 = 2.5$, $L_4 = 3.5$, and $\theta_1 = 60°$. For this analysis, the angular velocity of the crank is assumed to be $\omega_1 = 3$ rad/s, which, as shown, is counterclockwise.

Then the velocity of point A is plotted to some chosen scale at point A, perpendicular to line OA and with a magnitude of $V_A = \omega_1 L_1 = (3)(1) = 3$ units/s. The vector $\mathbf{V}_A$ is then resolved into its longitudinal and perpendicular components, respectively, along line AB and perpendicular to line AB as shown.

The longitudinal component $\mathbf{V}_B^l$ is then laid out along line AB and equal to $\mathbf{V}_A^l$ as shown. From the tip of this vector $\mathbf{V}_B^l$, a perpendicular is drawn to intersect a line that is perpendicular to link L_3. The total velocity $\mathbf{V}_B$ of point

B is then a vector drawn from point B to that intersection as shown. This is seen to be true because $\mathbf{V}_B$ is the sum of the orthogonal components $\mathbf{V}_B^l$ and $\mathbf{V}_B^p$ as drawn in the figure.

Measuring the lengths of vectors on the diagram from which Fig. 4.29 was reproduced gives $V_B = 1.53$ units/s, $\mathbf{V}_B^p = 0.57$ units/s, and $\mathbf{V}_A^p = 2.65$ units/s. Then, using $\mathbf{V}_{BA} = \mathbf{V}_B^p + (-\mathbf{V}_A^p)$, the construction at point B shows that $V_{BA} = 2.07$ units/s. Then, dividing V_{BA} by L_2 gives $\omega_2 = -2.07/3.0 = -0.69$ rad/s. The angular velocity ω_2 is negative because it is seen that the velocity $\mathbf{V}_{BA}$ is downward, indicating that point B is moving downward relative to point A, thereby causing a clockwise angular velocity of the coupler.

Graphical velocity analysis of four-bar linkages using instant centers

Once the velocity of point A has been determined, the velocity of any point on the coupler L_2 may be determined graphically by using the properties of instant centers. Figure 4.30 is a schematic scale drawing of a crank-rocker mechanism with the velocity of the link moving pivot

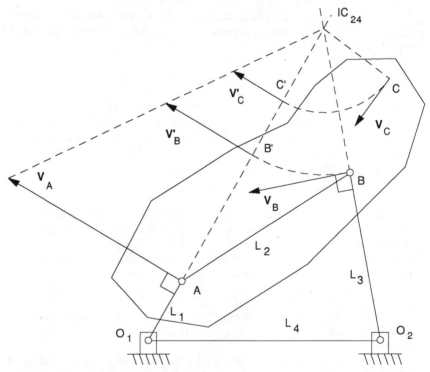

Figure 4.30 Graphical velocity analysis for coupler points using the coupler's instant center and velocity-distribution triangles.

A drawn at A to some chosen scale. The coupler L_2 is shown as an extended body.

The instant center of L_2 relative to ground can be found by constructing at each of two points a perpendicular to the velocity (relative to ground) of that point. It is known that $\mathbf{V}_A$ is perpendicular to vector link L_1, so the instant center must lie somewhere along line L_1 or along an extension of L_1. It is known that the velocity of point B on the link L_3 must be perpendicular to L_3, so the instant center must lie somewhere along L_3 or along an extension of L_3. The instant center of the coupler relative to ground (L_2 relative to L_4) is therefore at the intersection of these two lines and is indicated as IC_{24} in Fig. 4.30.

A velocity-distribution triangle may then be drawn using $\mathbf{V}_A$ and the line segment from A to IC_{24} as its two perpendicular sides as shown. Point B' may be placed on the line from A to IC_{24} such that the distance from B' to the instant center is equal to the distance from B to the instant center. The velocity of point B' may then be determined by using the velocity-distribution triangle as described in Chap. 2. The magnitude of the velocity $\mathbf{V}_B$ of point B is then equal to that of point B', but $\mathbf{V}_B$ is perpendicular to the line from B to IC_{24}, as shown in the figure.

The velocity of any other point that is attached to the coupler may be determined by a similar procedure, as shown, for example, for point C.

As described in Chap. 2, the angular velocity of any rigid body such as the coupler may be calculated by dividing the magnitude of the velocity of any point by the distance from that point to the instant center. That is, because $V_A = \omega_2 r$, where r is the distance from point A to IC_{24}, the angular velocity of the coupler is simply $\omega_2 = V_A/r$.

Readers will notice that some additional, potentially useful relationships are implied by Fig. 4.30. For example, the angular velocities of the coupler L_2 and the link L_1 are related by $V_A = \omega_1 L_1 = \omega_2 r$, where r is the distance from A to IC_{24}. Also, the angular velocities of the coupler L_2 and the link L_3 are related by $V_B = \omega_3 L_3 = -\omega_2 s$, where s is the distance from B to IC_{24}.

Angular velocity ratio

A further use of instant centers is illustrated in Fig. 4.31. The instant center IC_{13} is the point along which the link L_1 instantaneously rotates *relative to* link L_3. Its location can be determined by using kinematic inversion, in which L_3 is considered to be fixed and the remainder of the linkage moves relative to L_3. In this inversion, the velocity of point A relative to link L_3 would be constrained to be perpendicular to link L_2, and the velocity of point O_1 relative to link L_3 would be constrained

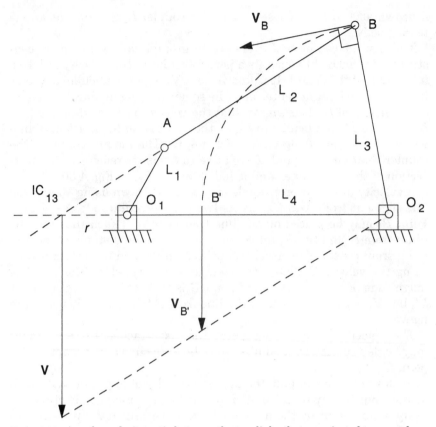

Figure 4.31 Angular velocity ratio between the two links that are pivoted to ground.

to be perpendicular to link L_4. The instant center of rotation of link L_1 relative to link L_3 must therefore lie on a line through point A that is perpendicular to the velocity of point A; i.e., it must lie on link L_2 or an extension of that link. It also must lie on a line through point O_1 that is perpendicular to the velocity of point O_1 and that is on link L_4 or an extension of that link. Thus IC_{13} is seen to lie at the intersection of dashed extensions of links L_2 and L_4, as shown in Fig. 4.31.

As pointed out in Sec. 2.5, the instant center is a point on *each* of the two bodies concerned, and these two coincident points have the same velocity at the instant considered. Therefore, with link L_4 fixed to ground and link L_1 rotating counterclockwise in Fig. 4.31, point IC_{13}, considered as a point on link L_1, would have a velocity **V** vertically downward as shown in that figure. The point IC_{13}, considered as a point on link L_3, would have an equal velocity vertically downward. The angular velocities of links L_3 and L_1 are then related by

$$V = \omega_1 r = \omega_3(L_4 + r) \quad \text{or} \quad \frac{\omega_3}{\omega_1} = \frac{r}{L_4 + r}$$

Analytical velocity analysis

Velocity analysis of a pin-jointed four-bar linkage by analytical means could proceed very much as discussed in the graphical analysis but with the vectors and their components expressed in algebraic form (either trigonometrically or in complex variables). The manipulations would then be algebraic rather than graphical. However, rather than using an algebraic analogue of the graphical method, it is more compact and convenient to start with the vector loop closure equation and then to differentiate that equation.

The loop closure equation expresses the fact that the vectors that represent the links form a closed loop. For the linkage shown in Fig. 4.32, this equation is

$$\mathbf{L}_1 + \mathbf{L}_2 = \mathbf{L}_4 + \mathbf{L}_3$$

When expressed in terms of complex variables (see Sec. 2.2), this becomes

$$L_1 e^{j\theta_1} + L_2 e^{j\theta_2} = L_4 e^{j\theta_4} + L_3 e^{j\theta_3}$$

Note that θ_4 is zero and the only quantities in this equation that vary with time are θ_1, θ_2, and θ_3. Then, differentiating this equation with respect to time gives

$$\omega_1 L_1 e^{j(\theta_1 + \pi/2)} + \omega_2 L_2 e^{j(\theta_2 + \pi/2)} = \omega_3 L_1 e^{j(\theta_3 + \pi/2)} \tag{4.2}$$

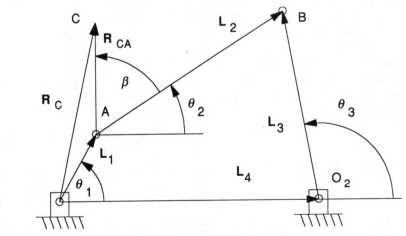

Figure 4.32 Analytical velocity analysis for a point on the coupler.

In this equation note that the first term on the left-hand side represents $\mathbf{V}_A$, the second term represents $\mathbf{V}_{BA}$, and the term on the right-hand side represents $\mathbf{V}_B$. It is therefore seen that this equation is just an alternative form of the velocity-difference equation $\mathbf{V}_B = \mathbf{V}_A + \mathbf{V}_{BA}$.

By using Euler's formula, Eq. (4.2) can be separated into an equation in only real terms and an equation in only imaginary terms. The equation in only real terms is

$$\omega_1 L_1 \cos(\theta_1 + \pi/2) + \omega_2 L_2 \cos(\theta_2 + \pi/2) = \omega_3 L_3 \cos(\theta_3 + \pi/2)$$

which reduces to

$$-\omega_1 L_1 \sin(\theta_1) - \omega_2 L_2 \sin(\theta_2) = -\omega_3 L_3 \sin(\theta_3) \qquad (4.3)$$

and the equation in only imaginary terms is

$$j\omega_1 L_1 \sin(\theta_1 + \pi/2) + j\omega_2 L_2 \sin(\theta_2 + \pi/2) = j\omega_3 L_3 \sin(\theta_3 + \pi/2)$$

which reduces to

$$\omega_1 L_1 \cos(\theta_1) + \omega_2 L_2 \cos(\theta_2) = \omega_3 L_3 \cos(\theta_3) \qquad (4.4)$$

Position analysis must be completed before velocity analysis can be performed. Position analysis provides values for θ_1, θ_2, and θ_3, so Eqs. (4.3) and (4.4) become linear equations in the variables ω_1, ω_2, and ω_3. If a value for any one of these three variables is known or given, these two equations may be solved for values of the other two variables. Then note that the three terms in Eq. (4.3) are the horizontal components of $\mathbf{V}_A$, $\mathbf{V}_{BA}$, and $\mathbf{V}_B$, respectively, and the terms in Eq. (4.4) are the vertical components of those same velocities. Therefore, the known values of the L's, θ's, and ω's may be substituted in these terms to calculate the values of the components of those velocities.

An alternative to solving Eqs. (4.3) and (4.4) simultaneously consists of multiplying each term in the equation by some factor that causes the term in one of the variables to become completely real or completely imaginary. That variable then appears in only one of the two scalar equations obtained by using Euler's formula. For example, multiplying each term in Eq. (4.2) by $e^{-j\theta_2}$ produces

$$\omega_1 L_1 e^{j(\theta_1 - \theta_2 + \pi/2)} + \omega_2 L_2 e^{j(\pi/2)} = \omega_3 L_1 e^{j(\theta_3 - \theta_2 + \pi/2)}$$

The corresponding scalar equation that contains only real terms is

$$\omega_1 L_1 \cos(\theta_1 - \theta_2 + \pi/2) + \omega_2 L_2 \cos(\pi/2) = \omega_3 L_3 \cos(\theta_3 - \theta_2 + \pi/2)$$

It will be noted that the term containing ω_2 is identically zero, so the equation contains only the variables ω_1 and ω_3. Then, solving for ω_3

and using trigonometric identities for angles in different quadrants gives

$$\omega_3 = \frac{\omega_1 L_1 \sin(\theta_2 - \theta_1)}{L_3 \sin(\theta_2 - \theta_3)} \tag{4.5}$$

If link L_1 is considered the input link and therefore ω_1 is known, Eq. (4.5) can be used to find ω_3. Then the terms on the right-hand side of Eqs. (4.3) and (4.4), which are the components of the velocity of point B, can be evaluated.

By multiplying all terms in Eq. (4.2) by $e^{-j\theta_3}$ and performing steps similar to the preceding, an expression for ω_3 that is analogous to Eq. (4.5) is obtained. That expression is

$$\omega_2 = -\frac{\omega_1 L_1 \sin(\theta_3 - \theta_1)}{L_2 \sin(\theta_3 - \theta_2)} \tag{4.6}$$

Analytical determination of velocity of a coupler point

Figure 4.32 depicts a linkage in which the coupler consists of an extended rigid body that is pivoted to the crank at point A. Attached to the coupler body is a point C whose velocity is to be calculated. The position of point C can be related to a grounded origin at point O by the vector $\mathbf{R}_C$ from that origin to point C. The vector $\mathbf{R}_{CA}$ is fixed to the rigid body at an angle β with respect to line AB and extends from point A to point C. As can be seen in Fig. 4.32, $\mathbf{R}_C = \mathbf{L}_1 + \mathbf{R}_{CA}$. Expressing this vector equation in complex variable form gives

$$R_C e^{j\theta_C} = L_1 e^{j\theta_1} + R_{CA} e^{j(\theta_2 + \beta)}$$

where β is the fixed inclination of line AC relative to line AB. Then, because the velocity of point C is equal to the time rate of change of the position vector $\mathbf{R}_C$, the right-hand side of Eq. (3.38) may be differentiated to give the velocity of point C as

$$\text{Velocity of } C = j\omega_1 L_1 e^{j\theta_1} + j\omega_{CA} R_{CA} e^{j(\theta_2 + \beta)}$$

or $\quad\quad \text{Velocity of } C = \omega_1 L_1 e^{j(\theta_1 + \pi/2)} + \omega_{CA} R_{CA} e^{j(\theta_2 + \beta + \pi/2)} \tag{4.7}$

It is seen that Eq. (4.7) is a velocity-difference equation $\mathbf{V}_C = \mathbf{V}_A + \mathbf{V}_{CA}$. Because line AC is attached to the coupler, $\omega_{CA} = \omega_2$, which can be calculated from Eq. (4.6). Then the corresponding scalar equations in real and imaginary variables are, respectively,

$$(V_C)_x = \omega_1 L_1 \cos(\theta_1 + \pi/2) + \omega_2 R_{CA} \cos(\theta_2 + \beta + \pi/2)$$

and $\qquad j(V_C)_y = j\omega_1 L_1 \sin(\theta_1 + \pi/2) + j\omega_2 R_{CA} \sin(\theta_2 + \beta + \pi/2)$

Eliminating the $\pi/2$ terms from the angle arguments and canceling out the factors j gives:

$$(V_C)_x = -\omega_1 L_1 \sin \theta_1 - \omega_2 R_{CA} \sin(\theta_2 + \beta) \qquad (4.8)$$

and $\qquad (V_C)_y = \omega_1 L_1 \cos \theta_1 + \omega_2 R_{CA} \cos(\theta_2 + \beta) \qquad (4.9)$

4.9 Acceleration Analysis of the Pin-Jointed Four-Bar Linkage

It was seen in Sec. 3.9 that velocity analysis can be used not only for finding relationships between various velocities in a linkage but also for determining relationships between various forces and torques in that linkage. Although that section dealt with crank-slider linkages, the same principles apply to pin-jointed four-bar linkages. If the mechanism that is being designed or analyzed involves or interacts with any appreciable masses, the forces and torques involved may consist largely of inertial effects. If so, it may be important to know how the masses are being accelerated and therefore what inertial reaction forces are being generated. Analysis of these accelerations can be performed once the position and velocity analyses have been completed.

Graphical acceleration analysis of pin-jointed four-bar linkages

Acceleration analysis makes use of acceleration-difference equations, which, like the velocity-difference equations, relate the motions of two points with the difference in their motions. Figure 4.33 shows the same linkage as shown in Fig. 4.29 but in which link L_1 is not only rotating counterclockwise with angular velocity ω_1 but in which L_1 has a clockwise angular acceleration α_1.

When the acceleration-difference equations are applied to link L_1 and to the coupler L_2 in Fig. 4.33, they become

$$\mathbf{A}_A = \mathbf{A}_{O_1} + \mathbf{A}_{AO_1} \qquad \text{and} \qquad \mathbf{A}_B = \mathbf{A}_A + \mathbf{A}_{BA}$$

which combine to give

$$\mathbf{A}_B = \mathbf{A}_{O_1} + \mathbf{A}_{AO_1} + \mathbf{A}_{BA}$$

where $\mathbf{A}_B$ is the acceleration of point B, $\mathbf{A}_{O_1}$ is the acceleration of point O_1 (which, because point O_1 is fixed to ground, is zero), $\mathbf{A}_{AO_1}$ is the acceleration of point A relative to point O_1, and $\mathbf{A}_{BA}$ is the acceleration of point B relative to point A.

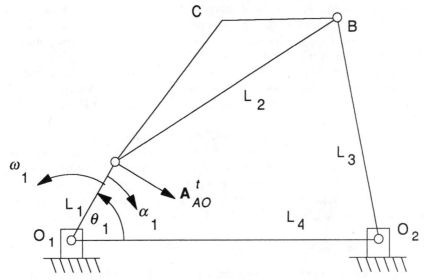

Figure 4.33 Acceleration analysis for a four-bar linkage.

Then, because link L_1, the coupler L_2, and link L_3 are all rigid bodies, Sec. 2.7 shows that each of the remaining terms in this equation may be resolved into tangential components and normal components. This gives

$$\mathbf{A}_B^t + \mathbf{A}_B^n = \mathbf{A}_{AO}^t + \mathbf{A}_{AO}^n + \mathbf{A}_{BA}^t + \mathbf{A}_{BA}^n$$

But, because point O_1 is fixed to ground, $\mathbf{A}_{AO}^t = \mathbf{A}_A^t$ and $\mathbf{A}_{AO}^n = \mathbf{A}_A^n$. Then,

$$\mathbf{A}_B^t + \mathbf{A}_B^n = \mathbf{A}_A^t + \mathbf{A}_A^n + \mathbf{A}_{BA}^t + \mathbf{A}_{BA}^n$$

Section 2.7 shows that the vector $\mathbf{A}_A^t$ (i.e., the tangential component of the acceleration of A) is directed perpendicular to line O_1A at point A and in the direction of the angular acceleration of link L_1, as shown in Fig. 4.33, and it has a magnitude of $L_1\alpha_1$. Note that, as drawn, Fig. 4.33 shows a *negative* (clockwise) value for α_1. From Sec. 2.7 it is seen that the vector $\mathbf{A}_A^n$ (i.e., the normal or centripetal component of the acceleration of A) will be directed toward point O_1 from point A and will have a magnitude of $L_1\omega_1^2$.

Similarly, the vector $\mathbf{A}_{BA}^t$ will be directed perpendicular to line AB at point B and in the direction of the angular acceleration of the coupler, and it will have a magnitude of $L_2\alpha_2$. The vector $\mathbf{A}_{BA}^n$ will be directed toward point A from point B and will have a magnitude of $L_2\omega_2^2$.

Also, the vector $\mathbf{A}_B^t$ will be directed perpendicular to link L_3 at point B and in the direction of the angular acceleration of L_3, and it will have a magnitude of $L_3\alpha_3$. The vector $\mathbf{A}_B^n$ will be directed toward point O_2 from point B and will have a magnitude of $L_3\omega_3^2$.

Thus it is seen that the direction of all the vectors are determined by the directions of the links, and the magnitudes of the vectors are determined by the angular velocities, angular accelerations, and the link lengths.

Now, because the position analysis and the velocity analysis must be completed before the acceleration analysis is attempted, the directions of the lines O_1A, O_2B, and AB will be known, and the values of ω_1, ω_2, and ω_3 will be known. Therefore, the *directions* of all the vectors in the preceding equation will be known, and the magnitudes of the vectors $\mathbf{A}_B^n$, $\mathbf{A}_{AO_1}^n$, and $\mathbf{A}_{BA}^n$ can be calculated from

$$A_B^n = L_3\omega_3^2 \qquad A_{AO_1}^n = L_1\omega_1^2 \qquad \text{and} \qquad A_{BA}^n = L_2\omega_2^2$$

If, for the acceleration analysis, the angular acceleration α_1 of the crank is given, then the magnitude of the vector $\mathbf{A}_{AO_1}^t$ can be calculated from

$$A_{AO_1}^t = L_1\alpha_1$$

Then, we may write the equation for the acceleration of point B as

$$\mathbf{A}_B^{u,d} + \mathbf{A}_B^{m_n d} = \mathbf{A}_{AO_1}^{m,d} + \mathbf{A}_{AO_1}^{m_n d} + \mathbf{A}_{BA}^{u,d} + \mathbf{A}_{BA}^{m_n d}$$

where the notation directly above each vector denotes the status of knowledge of the magnitude and direction of that vector. The left-hand character refers to magnitude, and the right-hand character refers to direction. For instance, the *magnitude* of the first vector on the left-hand side of the equation is unknown, as indicated by the u (for unknown magnitude) as the left-hand character above that vector, while the *direction* of that vector is known, as indicated by the d (for direction) as the right-hand character above that same vector. The magnitude and direction of each of the first two vectors on the right-hand side of the equation are known, as indicated by the presence of both an m and a d above each of these vectors. If the *direction* of one of the vectors had been unknown, a u (for unknown direction) would have been substituted for the d in the right-hand space above that vector.

Examination of this equation shows that this vector equation contains two unknowns: the magnitudes of the vectors $\mathbf{A}_B^t$ and $\mathbf{A}_{BA}^t$. The two scalar equations that this equation represents therefore may be solved for these unknowns. Because there is a solution to this vector equation, it also may be solved graphically. (The analytical solution will be described later.)

Graphical analysis of a linkage starts with writing an acceleration-difference equation such as discussed above and expanding the acceleration vectors in terms of their normal and tangential components to produce an equation such as the last equation above. Then a vector diagram is drawn to scale in such a manner that the vectors in that diagram add up in the same way that the vectors add up in the expanded acceleration-difference equation. In drawing this diagram, it will be found that not all the vectors are known. Therefore, those vectors which *are* completely known should be drawn first, connected tail to head as indicated by the additions shown in the equation. Then, in the manner shown in the following example, the unknown vectors are added to the diagram to the extent to which their characteristics are known. This will produce a closed vector loop in which the unknown vector quantities will be determined automatically. This graphical procedure will be illustrated by the following example.

Example 4.6: Graphical Acceleration Analysis of a Pin-Jointed Four-Bar Linkage

Figure 4.33 is a reproduction of a diagram of the same linkage that was analyzed in the graphic velocity analysis in Example 4.5. The linkage dimensions are as given in that example. In addition, the angular acceleration of link L_1 is given as clockwise at 4 rad/s². That is, $\alpha_1 = -4$ rad/s².

The first step in graphical acceleration analysis is to write the acceleration-difference equation that relates accelerations throughout the linkage. The expanded acceleration-difference equation for this linkage will be analogous to the last equation in the preceding discussion. That is,

$$\mathbf{A}_B^{u_t d} + \mathbf{A}_B^{m_n d} = \mathbf{A}_A^{m_t d} + \mathbf{A}_A^{m_n d} + \mathbf{A}_{BA}^{u_t d} + \mathbf{A}_{BA}^{m_n d}$$

From the results of the velocity analysis in Example 4.5, the magnitudes A_B^n, A_A^n, and A_{BA}^n of the vectors $\mathbf{A}_B^n$, $\mathbf{A}_{AO_1}^n$, and $\mathbf{A}_{BA}^n$ can be calculated as

$$A_B^n = L_3\omega_3^2 = (2.5)(0.612)^2 = 0.94 \text{ units/s}^2$$

$$A_A^n = L_1\omega_1^2 = (1)(3)^2 = 9 \text{ units/s}^2$$

and $\qquad A_{BA}^n = L_2\omega_2^2 = (3)(-0.69)^2 = 1.43 \text{ units/s}^2$

Because these accelerations are centripetal accelerations, the direction of $\mathbf{A}_B^n$ is from B toward O_2, the direction of $\mathbf{A}_{AO_1}^n$ is from A toward O_1, and the direction of $\mathbf{A}_{BA}^n$ is from B toward A.

The angular acceleration of input link L_1 is given as -4 rad/s² (i.e., clockwise), so the tangential component of the acceleration of point A will be downward and to the right, as shown in Fig. 4.33, and it will have a magnitude of

$$A_A^t = L_1\alpha_1 = (1)(-4) \text{ or } 4 \text{ units/s}^2$$

Examination of the acceleration-difference equation shows that the magnitudes of all the vectors except $\mathbf{A}_B^t$ and $\mathbf{A}_{BA}^t$ are known and that the directions of all the vectors are known. The graphical construction is then started by choosing an origin such as is shown as point O in Fig. 4.34. Then, starting at the origin, the completely known vectors on the right-hand side of the equation ($\mathbf{A}_A^t$, $\mathbf{A}_A^n$, and $\mathbf{A}_{BA}^n$) are plotted first, starting from O. These vectors are connected tail to head to represent their addition, as indicated on the right-hand side of the acceleration-difference equation. The remaining term on the right-hand side of that equation is $\mathbf{A}_{BA}^t$, and it must be added to the other three vectors by placing its tail at the tip of vector $\mathbf{A}_{BA}^n$. However, all that is known about this vector $\mathbf{A}_{BA}^t$ is that it must be perpendicular to coupler L_2. Therefore, a line p perpendicular to L_2 is drawn through the tip of vector $\mathbf{A}_{BA}^n$, and the magnitude of $\mathbf{A}_{BA}^t$ will be determined later.

Now, because no more is known about the right-hand side of the equation, attention is shifted to the left-hand side. When the vector(s) on the left-hand side are drawn, starting at point O and followed tail to head, they must add

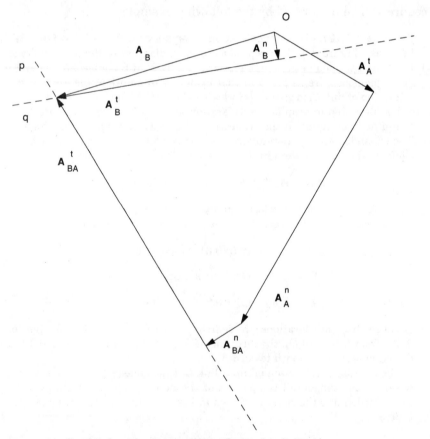

Figure 4.34 Graphical acceleration analysis of a four-bar linkage.

up to the same result as obtained by following the vectors on the right-hand side tail to head from point O. The magnitude and direction of $\mathbf{A}_B^n$ are both known, so it is plotted with its tail at the origin as shown. However, all that is known about $\mathbf{A}_B^t$ is that it must be perpendicular to L_3. Therefore, a line q is drawn through the tip of vector $\mathbf{A}_B^n$ and perpendicular to L_3, and the magnitude of $\mathbf{A}_B^t$ will be determined later.

Note now that lines p and q intersect at a point r. If the magnitudes of $\mathbf{A}_{BA}^t$ and $\mathbf{A}_B^t$ were such as to place their heads at point r, the vector additions on the two sides of the equation would produce equal results, and the equation would be satisfied. These magnitudes can then be scaled off the diagram, and it is found that

$$A_B^t = 7.81 \text{ units/s}^2$$

and
$$A_{BA}^t = 9.84 \text{ units/s}^2$$

Because $\mathbf{A}_{BA}^t$ is seen to be directed upward and to the right, point B is accelerating upward and to the right with respect to point A, so the coupler has a positive (counterclockwise) angular acceleration. The magnitude α_2 of this angular acceleration is computed from the relationship

$$A_{BA}^t = L_2\alpha_2$$

which gives

$$9.84 = 3.0\alpha_2$$

so $\alpha_2 = 3.28 \text{ rad/s}^2$.

Because $\mathbf{A}_B^t$ is seen to be directed leftward, point B is accelerating leftward with respect to point O_2, so link L_3 has a positive (counterclockwise) angular acceleration. The magnitude α_3 of this angular acceleration is computed from the relationship

$$A_B^t = L_3\alpha_3$$

which gives

$$7.81 = 2.5\alpha_3$$

so $\alpha_3 = 3.12 \text{ rad/s}^2$. The total acceleration of point B is the vector sum of $\mathbf{A}_B^t$ and $\mathbf{A}_B^n$, and this addition can be done graphically as shown.

Comments

In this foregoing example the directions of the unknown vectors were known and their magnitudes were unknown, so they were represented temporarily just by lines in the known directions. If the magnitude of a vector had been known but its direction were unknown, it could have been represented temporarily by a circular arc with a radius equal to the known magnitude and with its center at the origin O or at the tip of the vector to which is was to be added. If only one vector had been unknown, neither its direction nor its magnitude would have been known. Then, after all of the other vectors had been drawn, it would be found that the unknown vector would simply be the vector needed to close the loop in the diagram.

Example 4.7: Coupler Point Accelerations   Consider a point C on the coupler shown in Fig. 4.33. Point C is a distance $R_{CA} = 2$ units from point A, and the angle BAC is $20°$. Because points C and A are attached to the same rigid body (i.e., the coupler), the acceleration $\mathbf{A}_C$ of point C can be found by using the acceleration-difference equation:

$$\mathbf{A}_C = \mathbf{A}_A + \mathbf{A}_{CA} = \mathbf{A}_A^n + \mathbf{A}_A^t + \mathbf{A}_{CA}^t + \mathbf{A}_{CA}^n$$

The first two vector components $\mathbf{A}_A^t$ and $\mathbf{A}_A^n$ on the right-hand side of this equation are already plotted in Fig. 4.34, and they are replotted in Fig. 4.35. The remaining two terms on the right-hand side must then be added graphically by placing them at the tip of the resultant of the first two terms, as shown in Fig. 4.35. The magnitude A_{CA}^n of the vector $\mathbf{A}_{CA}^n$ is given by $A_{CA}^n = \omega_2^2 R_{CA} = (0.69)^2(2) = 0.95$ units/s², and it is directed parallel to $\mathbf{R}_{CA}$ and in a direction from point C toward point A. The magnitude A_{CA}^t of the vector $\mathbf{A}_{CA}^t$ is given by $A_{CA}^t = \alpha_2 R_{CA} = (3.28)(2) = 6.56$ units/s², and it is directed perpendicular to $\mathbf{R}_{CA}$ and in a direction consistent with the counterclockwise angular acceleration of the coupler (as indicated by the direction of $\mathbf{A}_{BA}^t$).

The preceding equation for $\mathbf{A}_C$ indicates that $\mathbf{A}_C$ equals the sum of all the terms on the right-hand side of the equation. Those terms have all been plotted tail to head from the origin O in Fig. 4.35. The vector $\mathbf{A}_C$ is then simply a vector drawn from the origin to the tip of the vector $\mathbf{A}_{CA}^t$ as shown. Scaling the length of that vector from the drawing gives a magnitude of $A_C = 9.49$ units/s² at the angle shown.

Analytical acceleration analysis

Acceleration analysis of a pin-jointed four-bar linkage by analytical means could proceed very much as discussed in the graphic analysis but with the vectors and their components expressed in algebraic form (either trigonometrically or in complex variables). The manipulations would then be algebraic rather than graphical. However, rather than using an algebraic analogue of the graphical method, it is more compact and convenient to start with the vector loop closure equation for Fig. 4.33 and then to differentiate that equation twice. The vector loop equation has already been differentiated once in Sec. 4.8 to give Eq. (4.2), which is repeated here as Eq. (4.10):

$$\omega_1 L_1 e^{j(\theta_1 + \pi/2)} + \omega_2 L_2 e^{j(\theta_2 + \pi/2)} = \omega_3 L_1 e^{j(\theta_3 + \pi/2)} \tag{4.10}$$

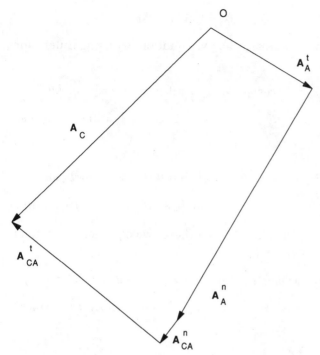

Figure 4.35 Graphical acceleration analysis for a coupler point on a four-bar linkage.

Differentiating Eq. (4.10) with respect to time gives

$$\alpha_1 L_1 e^{j(\theta_1 + \pi/2)} + j\omega_1^2 L_1 e^{j(\theta_1 + \pi/2)} + \alpha_2 L_2 e^{j(\theta_2 + \pi/2)} + j\omega_2^2 L_2 e^{j(\theta_2 + \pi/2)}$$
$$= \alpha_3 L_3 e^{j(\theta_3 + \pi/2)} + j\omega_3^2 L_3 e^{j(\theta_3 + \pi/2)}$$

Because $j = e^{j(\pi/2)}$ and $e^{j\pi} = -1$, this equation can be simplified to give

$$\alpha_1 L_1 e^{j(\theta_1 + \pi/2)} - \omega_1^2 L_1 e^{j\theta_1} + \alpha_2 L_2 e^{j(\theta_2 + \pi/2)} - \omega_2^2 L_2 e^{j\theta_2}$$
$$= \alpha_3 L_3 e^{j(\theta_3 + \pi/2)} - \omega_3^2 L_3 e^{j\theta_3} \quad (4.11)$$

Note that Eq. (4.11) is just the acceleration-difference equation expanded in terms of normal and tangential components of the accelerations. That is, it is merely

$$\mathbf{A}_A^t + \mathbf{A}_A^n + \mathbf{A}_{BA}^t + \mathbf{A}_{BA}^n = \mathbf{A}_B$$

Equation (4.11) can be expressed as two equations by using Euler's formula to give

$$\alpha_1 L_1 \cos(\theta_1 + \pi/2) - \omega_1{}^2 L_1 \cos\theta_1 + \alpha_2 L_2 \cos(\theta_2 + \pi/2) - \omega_2{}^2 L_2 \cos\theta_2$$
$$= \alpha_3 L_3 \cos(\theta_3 + \pi/2) - \omega_3{}^2 L_3 \cos\theta_3$$

and

$$j\alpha_1 L_1 \sin(\theta_1 + \pi/2) - j\omega_1{}^2 L_1 \sin\theta_1 + j\alpha_2 L_2 \sin(\theta_2 + \pi/2) - j\omega_2{}^2 L_2 \sin\theta_2$$
$$= j\alpha_3 L_3 \sin(\theta_3 + \pi/2) - j\omega_3{}^2 L_3 \sin\theta_3$$

By using trigonometric identities on these two equations and dividing the second one by j, we get

$$-\alpha_1 L_1 \sin\theta_1 - \omega_1{}^2 L_1 \cos\theta_1 - \alpha_2 L_2 \sin\theta_2 - \omega_2{}^2 L_2 \cos\theta_2$$
$$= -\alpha_3 L_3 \sin\theta_3 - \omega_3{}^2 L_3 \cos\theta_3 \quad (4.12)$$

and

$$\alpha_1 L_1 \cos\theta_1 - \omega_1{}^2 L_1 \sin\theta_1 + \alpha_2 L_2 \cos\theta_2 - \omega_2{}^2 L_2 \sin\theta_2$$
$$= \alpha_3 L_3 \cos\theta_3 - \omega_3{}^2 L_3 \sin\theta_3 \quad (4.13)$$

The angular velocities ω_1, ω_2, and ω_3 are determined by a preceding velocity analysis. Then it can be seen that Eqs. (4.12) and (4.13) are linear equations in the acceleration variables α_1, α_2, and α_3. If a value for any one of these acceleration variables is given, the values for the other two may be solved for.

Analytical determination of the acceleration of a coupler point

Once the acceleration of point A and the angular acceleration of the coupler have been determined, the acceleration of any point on the coupler (such as point C shown in Fig. 4.33) can be calculated by using the acceleration-difference equation that relates the acceleration of that coupler point to the known acceleration of point A. For this example point C, the acceleration-difference equation is simply $\mathbf{A}_C = \mathbf{A}_A + \mathbf{A}_{CA}$, which can be expanded into terms of the tangential and normal components to give

$$\mathbf{A}_C = \mathbf{A}_A^t + \mathbf{A}_A^n + \mathbf{A}_{CA}^t + \mathbf{A}_{CA}^n$$

The vector $\mathbf{A}_C$ will in general have both a horizontal (X or real) component and a vertical (Y or imaginary) component. The angular orientation of the vector $\mathbf{R}_{CA}$ from point A to point C is seen to be represented by the angle $\theta_2 + \beta$. Then, noting that the tangential and normal components of the vectors on the right-hand side of this expanded equation may be written in their now familiar complex variable form, this acceleration-difference equation becomes

$$(A_C)_x + j(A_C)_y$$

$$= \alpha_1 L_1 e^{j(\theta_1 + \pi/2)} - \omega_1^2 L_1 e^{j\theta_1} + \alpha_2 R_{CA} e^{j(\theta_2 + \beta + \pi/2)} - \omega_2^2 R_{CA} e^{j(\theta_2 + \beta)}$$

Then, using Euler's formula, the two scalar equations that correspond to this equation become

$$(A_C)_x = -\alpha_1 L_1 \sin \theta_1 - \omega_1^2 L_1 \cos \theta_1 - \alpha_2 R_{CA} \sin(\theta_2 + \beta) - \omega_2^2 R_{CA} \cos(\theta_2 + \beta)$$

and

$$(A_C)_y = \alpha_1 L_1 \cos \theta_1 - \omega_1^2 L_1 \sin \theta_1 + \alpha_2 R_{CA} \cos(\theta_2 + \beta) - \omega_2^2 R_{CA} \sin(\theta_2 + \beta)$$

The values of all variables on the right-hand side of these two equations are known from foregoing analyses, so the values of the horizontal and vertical components, respectively, of the acceleration of point C can be computed directly from these two equations.

5

Inverted Crank-Slider Mechanisms, Velocity-Matching Mechanisms, and Quick-Return Mechanisms

5.1 Introduction

Graphical, chart, and associated analytical methods for synthesizing inverted crank-slider linkages to give useful timing motions, velocity-matching motions, and large-angle output motions are presented in this chapter. These methods are then extended and used to develop synthesis techniques for the shaper quick-return linkage, the Whitworth linkage, the drag-link linkage, and the "washing machine" linkage. These latter two linkage types are four-bar, revolute-jointed linkages, but many of their characteristics are similar to those of inverted crank-slider linkages.

The motions of inverted crank-slider mechanisms are more complicated than those of the crank-slider mechanisms discussed in Chap. 3 because the motions of inverted crank-slider mechanisms involve slip and transmission components of velocity and acceleration, and they also involve Coriolis accelerations. These inverted mechanisms, however, can provide some very useful motions, and the techniques used in previous chapters are easily extended to include the extra terms involved in their analysis. Although computer-based methods probably will be used by most readers to analyze and/or simulate the motions of these linkages, better insight into the sources of some of the motion phenomena involved may be obtained from knowledge of the analysis methods described in Secs. 5.7 and 5.8.

5.2   The Inversions of the Crank-Slider Linkage and Their Applications

In the crank-slider linkages discussed in Chap. 3, the body or link containing the slide or slot was considered to be grounded, and the other links moved relative to it. In many applications it is useful to ground one of the other links, thereby producing an inversion of the crank-slider linkage. (Kinematic inversion was discussed in Sec. 4.4 for four-bar pin-jointed linkages.) Figure 5.1 shows schematically the four possible inversions of crank-slider linkages.

The linkage in Fig. 5.1a is an in-line crank-slider linkage in which the link CD that contains the slide or slot in which the slider B slides is grounded. This inversion, of course, was discussed in detail in Chap. 3. The link CA is shorter than the coupler AB, so the link CA can rotate through 360° while the coupler AB only translates and rocks back and forth.

Figure 5.1b depicts an inversion in which the link AC (which was the crank in Fig. 5.1a) is grounded, and links AB and CD are pivoted to it. In this inversion, because link CA is again shorter than AB, links AB and CD can rotate through 360°.

When the link AB is grounded, the inversion shown in Fig. 5.1c results. In this configuration, the link AC and the slider at B are pivoted to ground, and again, AC is shorter than AB. The link CD is pivoted to link AC, and it slides relative to the pivoted slider B. Thus, as AC rotates through 360°, link CD slides back and forth on the pivoted block at B while both link CD and that block rock back and forth.

In the inversion in Fig. 5.1d the slider at B is grounded, and the link CD must slide relative to it without rotating. The pivoted links AB and AC move relative to ground. Link AC, since it is shorter than AB, can rotate through 360° while link AB merely rocks back and forth about the pivot B.

Examples of common uses of these inversions of the crank-slider mechanism are depicted in Fig. 5.2. The inversion shown in Fig. 5.1b is shown in Fig. 5.2a as what is known as a *Whitworth mechanism,* and it is discussed in more detail in Sec. 5.4. In this configuration, the length of the input crank AB is greater than the length of the grounded link AC, so as the crank rotates through 360°, the path of point B encircles the pivot C, thereby causing link CD to rotate through 360°. If crank AB rotates at constant velocity, then when point B is at its lowest position, the angular velocity of link CD will be maximum, and when B is at its highest position, the angular velocity of link CD will be minimum. This mechanism can be used to provide quick-return action, as described in Sec. 5.4.

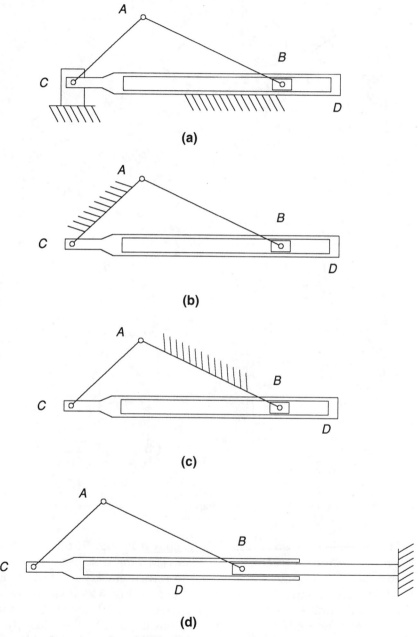

Figure 5.1   Inversions of crank-slider linkages.

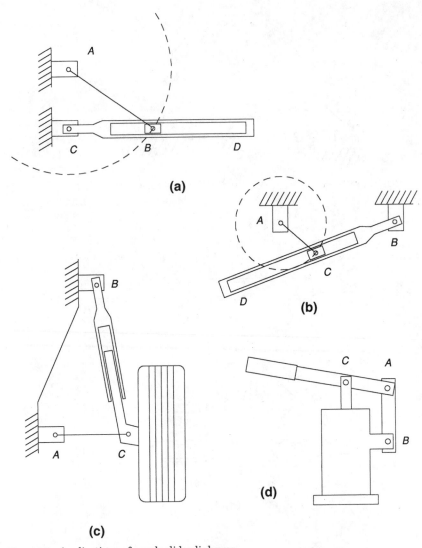

Figure 5.2 Applications of crank-slider linkages.

The inversion shown in Fig. 5.1c is shown in Fig. 5.2b as what is known as a *shaper quick-return mechanism*. This is so because it has been used to cause a tool in a shaper to return quickly to its starting position after making a slow cutting stroke. In Fig. 5.2b it can be seen that because link AC is shorter than ground link AB, the slide that is pivoted to ground at B will rock up and down, just like the slider that is pivoted to ground at B in Fig. 5.1c. Also, the sliding and rocking of the slider at C in Fig. 5.2b can be seen to be the same as the motions of the slide CD in Fig. 5.1c. It is thus seen that the slide and slider are interchangeable.

In Fig. 5.2b, as the input crank AC rotates at constant speed, the slider at C slides left and right in the slot in link BD and causes BD to rock up and down. When the slider C is at its right-most position, the angular velocity of the rocking link BD will be maximum, and when C is at its leftmost position, that angular velocity will be minimum. This varying angular velocity can be used for quick-return action, as discussed in the Sec. 5.3.

Figure 5.2c depicts an automotive application of the inversion shown in Fig. 5.1c. This configuration is known as a *MacPherson strut,* and it is used in the front suspension of many present-day automobiles. An arm AC is pivoted to the frame of the car at point A and is pivoted at C to a member that carries the axle of the wheel. To see the equivalence of Figs. 5.2c and 5.1c, note that the combination of the slider at B and the slotted link CD in Fig. 5.1c consists simply of an unbending but variable-length link that is pivoted to links AC and AB at its ends. Then note that the cylinder and piston assembly in Fig. 5.2c also constitutes such an unbendable but variable-length link.

The MacPherson strut is shown schematically as a *planar mechanism* in Fig. 5.2c, whereas it is actually a spatial (three-dimensional) mechanism. In actual practice, the joint at point C is a ball joint (spherical joint), the piston sliding in the cylinder is a cylindrical joint, and the joint at point B is in effect an elastomeric Hooke's joint. This combination of joints provides the additional degree of freedom required to allow the wheel to be steered.

The hand-actuated hydraulic pump depicted in Fig. 5.2d is an example application of the crank-slider inversion shown in Fig. 5.1d. The equivalence of these two figures can be seen by comparing the functioning of the like-labeled pivots and slides.

5.3 Synthesis of the Shaper Quick-Return Mechanism

A shaper quick-return mechanism such as shown in Fig. 5.2b is redrawn schematically in two extreme positions in Fig. 5.3a. The input crank is shown by solid lines at position AB, which places the output rocker CD at its rightmost position. A link DE is pivoted to link CD at point D, and a horizontally sliding block is pivoted to link DE at point E. When CD is in its rightmost position, the block at E is also in its rightmost position. In these positions of the links, link AB is perpendicular to link CD. If link AB is rotated counterclockwise through some angle β, it will reach a position shown dashed as AB', at which time link CD will be at its leftmost position, shown dashed as CD', and the block at E will be at its leftmost position (not shown). Further counterclockwise rotation of the crank through an angle α will return the output rocker and the slider to their rightmost positions. The motion of the mechanism while AB rotates through the angle β will be referred to as the *forward motion*, and the motion of the mechanism while AB rotates through the angle α will be referred to as the *return motion*. If the input crank AB is rotating counterclockwise at a constant angular velocity, the ratio TR of the time it would take to rotate through the angle β (forward time) to the time it would take to rotate through the return angle α (return time) would be given by

$$\text{Time ratio} = TR = \frac{\beta}{\alpha} = \frac{360 - \alpha}{\alpha} \tag{5.1}$$

where the angles α and β are given in degrees. This expression may be solved for α to give

$$\alpha = \frac{360}{1 + TR} \tag{5.2}$$

Now note that because triangles ABC and $AB'C$ are congruent right triangles,

$$\cos \frac{\alpha}{2} = \frac{L_1}{L_4} \tag{5.3}$$

where L_1 is the length of the input crank AB and L_4 is the length of the ground link AC. Consideration of triangles ABC and $AB'C$ also shows that the total angle δ through which the output rocker CD oscillates is given by

$$\delta = 180 - \alpha \tag{5.4}$$

Example 5.1: Synthesis of a Shaper Quick-Return Mechanism If a shaper quick-return linkage is to be used in a mechanism that must move some tool or part from one position to another and then return the mechanism to its starting position, and if the ratio of the times to make these two motions is specified, Eqs. (5.1) through (5.4) can be used to determine the proportions of that linkage. For example, if it is required that the mechanism move the slider shown in point E in Fig. 5.3a from its rightmost position (shown) to its leftmost position 8 in away in 2 s and that it return to the rightmost position in 1 s, the required time ratio would be $2/1 = 2$. Then Eq. (5.2) gives $\alpha = 360/(1 + 2) = 120°$. From Eq. (5.4) we find that $\delta = 180 - 120 = 60°$. Now note that the total travel of the slider

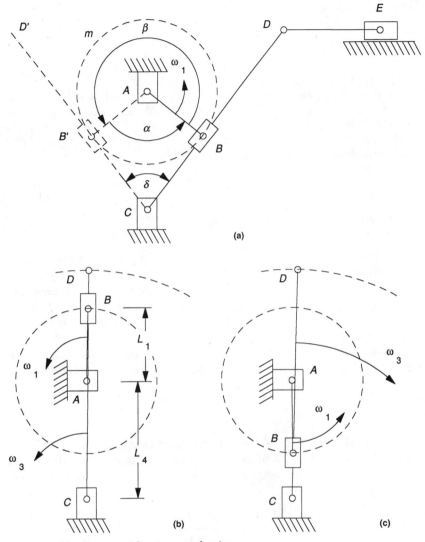

Figure 5.3 The shaper quick-return mechanism.

at point E is equal to the total horizontal travel s of point D, which is given by

$$s = 2(CD)\sin(\delta/2) = 8 \text{ in}$$

Then

$$CD = \frac{8}{2 \sin 30} = 8 \text{ in}$$

From Eq. (5.3) we obtain

$$\frac{L_1}{L_4} = \cos 60° = 0.5$$

where L_1 is length AB and L_4 is length AC.

Thus we have found that length CD must be 8 in and that it must oscillate through a total angular excursion of 60°. We also have found the required ratio of L_1 to L_4, but the choice of a value for either one of these lengths is arbitrary. For example, a convenient length L_4 for the ground link AC would be 4 in. Then the length L_1 would be

$$L_1 = 0.5L_4 = 2 \text{ in}$$

Use of a shaper quick-return mechanism for velocity matching

It is sometimes desired to cause the velocity of a point on a mechanism to periodically and momentarily match the velocity of some object. For example, it might be desired that a marking stylus or print head momentarily match the velocity of objects that regularly pass by on an assembly line. The shaper quick-return mechanism often can be used for such an application.

A velocity analysis of the mechanism in the position illustrated in Fig. 5.3b shows that the horizontal velocity of point D is given by

$$V_D = \frac{CD\omega_1 L_1}{L_1 + L_4}$$

where ω_1 is the angular velocity of the input crank AB in radians per second. A similar analysis of the mechanism in the position illustrated in Fig. 5.3c shows that the horizontal velocity of point D for that case is given by

$$V_D = \frac{CD\omega_1 L_1}{L_1 - L_4}$$

It is seen that this mechanism can give both a large output velocity and a small output velocity depending on the lengths of the links and the position of the crank. The direction of the output velocity obviously depends on the direction of rotation of the crank as well as its position.

5.4   Synthesis of a Whitworth Mechanism

In the design of machinery it is often necessary to provide means for causing one continuously rotating shaft to rotate more rapidly than another continuously rotating shaft during one portion of the machine cycle while maintaining cycle-to-cycle synchronism between the rotations of the two shafts. Such a requirement may arise from a need to provide a quick-return action (such as provided by the shaper quick-return mechanism just described), from a need to momentarily match the velocities of components that have different "pitches," or from a need to cyclically advance and retard the relative timing between the two shaft rotations. A Whitworth mechanism is an effective mechanism for providing this required action.

A schematic drawing of a Whitworth mechanism in two positions is given in Fig. 5.4. The fixed link AC has a length L_4. The input link of length L_1 is shown at position AB in the first position and dashed at AB' in the second position. Similarly, the output link is shown at CD in the first position and dashed at CD' in the second position. In the particular positions shown, CD' is oriented 180° from CD, and these two output positions can be considered to be the extreme positions in a quick-return mechanism, for which the Whitworth mechanism is often used.

Note that as the input link rotates counterclockwise from AB to AB', it traverses an angle β, and as it returns to AB by continued counterclockwise rotation, it traverses an angle α. During each of these input angle rotations, the output angle (the angle of CD) traverses 180°. If the input link AB rotates at constant counterclockwise angular velocity ω_1, the ratio of the time that it takes CD to go from CD to CD' to the time it takes it to *return* to CD is known as the time ratio TR, and this ratio will be given by

$$TR = \frac{\beta}{\alpha} = \frac{360 - \alpha}{\alpha} \tag{5.5}$$

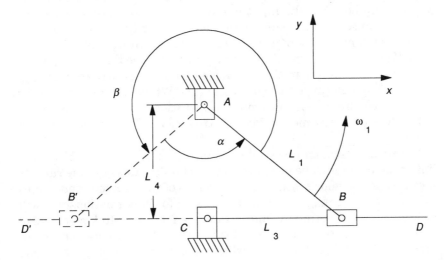

Figure 5.4 The Whitworth mechanism.

It also will be noted that

$$\frac{L_4}{L_1} = \cos\frac{\alpha}{2} \tag{5.6}$$

where L_1 is length AB and L_4 is length AC. It thus can be seen that these simple relationships may be used to design a Whitworth mechanism to give a chosen time ratio.

Readers can quite easily demonstrate that if the input link AB rotates at constant angular velocity ω_1, the output link angular velocity ω_3 of link L_3 will be in the same direction and will have a minimum value when the input link angle, measured counterclockwise from the x axis shown in Fig. 5.4, is 90° and will have a maximum value when that input link angle is 270°. These extreme values are given, respectively, by

$$(\omega_3/\omega_1)_{\min} = \frac{L_1}{L_1 + L_4} \tag{5.7}$$

and

$$(\omega_3/\omega_1)_{\max} = \frac{L_1}{L_1 - L_4} \tag{5.8}$$

It is thus seen that the synthesis of a Whitworth mechanism to give a desired minimum or maximum velocity ratio is also quite simple.

5.5   Synthesis of a Drag-Link Mechanism

Although a drag-link mechanism is not a crank-slider mechanism, the relationship between the input motion and output motion in a drag-link mechanism is quite similar to the corresponding relationship for a Whitworth mechanism such as was described in the preceding section. Therefore, a drag-link mechanism can be used to provide functions that are similar to those provided by a Whitworth (or turning-block) linkage. Because the synthesis of the Whitworth linkage is very simple, it can provide insight into the operation of the drag-link mechanism. The drag link mechanism is introduced in Sec. 4.4 and is shown in Fig. 4.14d.

The present section gives relationships by means of which a drag-link mechanism may be designed to give chosen values of time ratio or given values of velocity ratio. The extremes of the transmission angles experienced by linkages designed according to these relationships will be at or near the optimal values attainable for a drag-link mechanism with the chosen time ratio or velocity ratio. These relationships are presented as design charts into which readers may enter desired time

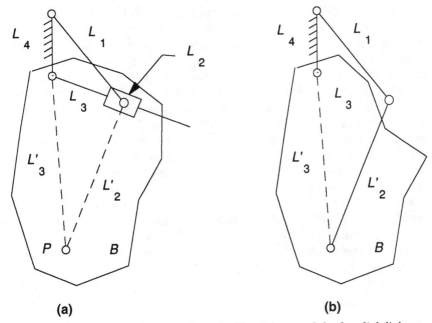

(a)   **(b)**

Figure 5.5 Equivalence of the inverted crank-slider linkage and the drag-link linkage.

ratios or desired angular velocity ratios and may then read the necessary linkage proportions, the initial angle conditions, and the worst expected values of the transmission angle. Use of these charts is described in Procedures 5.1 and 5.2. This material is based on work previously described in a paper by the author.[1]

In slider mechanisms (such as the Whitworth), the slider can be visualized as a very long pivoted link whose far pivot is far removed from the slider motion. Using this technique, a Whitworth mechanism may be visualized as shown in Fig. 5.5a. In this figure the fixed or ground link is labeled L_4, and the input link is labeled L_1. The slider is indicated as sliding block L_2, and the output link is indicated as the variable-length link L_3. If the output link L_3 were considered to be rigidly attached to the rigid body B shown in the figure, then the slider L_2 would move along a straight line *relative to body B,* as determined by link L_3. If the input link L_1 were attached by a revolute joint to a link L'_2 (shown dashed) instead of to the slider, and if link L'_2 were

[1]Eckhardt, H. D., Charts for the design of drag-link mechanisms with optimum or near-optimum transmission angle excursions, in *Procedings of the Second National Applied Mechanisms and Robotics Conference,* Cincinnati, OH, November 3–6, 1991. (These pamphlets were distributed privately. Contact the author at 27 Laurel Drive, Lincoln, MA 01773 for availability of these papers.)

attached to body B by a revolute joint at point P, the joint between L_1 and L'_2 would travel relative to body B along an arc centered at P. By making the length of L'_2 very long, that arc could be made to approximate the straight line determined by L_3. With a very long L'_2, the motion of body B would be the same in response to the motion of L_1, whether body B were attached to link L_3, as shown by solid lines in Fig. 5.5a, or were attached to a very long link such as L'_3 which in turn was connected by revolute joints to L_4 and L'_2, as shown dashed in Fig. 5.5a and shown solid in Fig. 5.5b. It can be seen that links L_1, L'_2, L'_3, and L_4 in Fig. 5.5b form a drag-link mechanism. Thus a drag-link mechanism having very long coupler and output links could be synthesized by using the simple relationships in Eqs. (5.5) to (5.8).

For practical reasons, a drag-link mechanism will not be designed with very long links. Therefore, the motion of a drag-link mechanism, although similar in many respects to that of a Whitworth mechanism, will be perturbed from that of a Whitworth mechanism. However, the ratio L_1/L_4 that is seen to be important in Eqs. (5.5) to (5.8) has similar effects

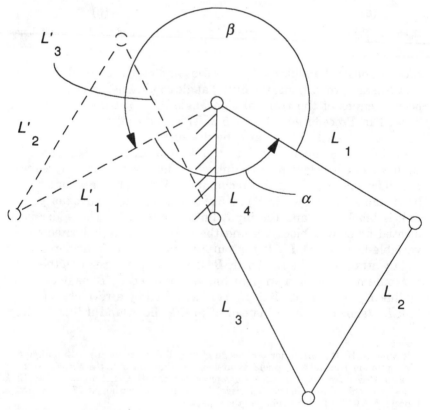

Figure 5.6 The drag-link linkage.

in the synthesis of a drag-link mechanism. The similarity between a Whitworth mechanism and a drag-link mechanism can be seen by comparing Figs. 5.4 and 5.6 and noting that in each figure the direction of the output link L_3 when it is in the dashed position is 180° from its direction when it is in the position indicated by the solid line. If the ratio L_1/L_4 were the same in the two figures, the value of α would be the same in the two figures. The synthesis therefore will start with an investigation of means for minimizing the deviation from optimum of the transmission angle in a linkage when a value is chosen for the length ratio L_1/L_4.

The transmission angle will experience its extreme values when the angle of the input link L_1 measured counterclockwise from the positive x axis is 90° and 270°, as shown in Fig. 5.7. The transmission angle

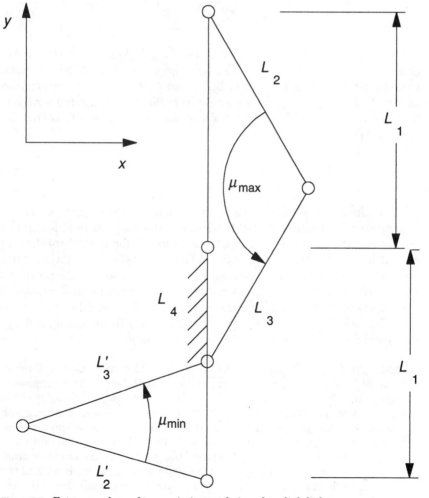

Figure 5.7 Extreme values of transmission angle in a drag-link linkage.

μ_{max} is shown for $\theta_1 = 90°$, and μ_{min} is shown for $\theta_1 = 270°$ (where θ_1 is the angle of the input link L_1 relative to the x axis). For the particular linkage proportions in these figures, the transmission angle extreme value μ_{max} is greater than 90°, and the other extreme value μ_{min} is smaller than 90°. For other linkage proportions, both extremes can be greater than 90° or both can be smaller than 90°. In any event, as discussed in the subsection on transmission angles in Sec. 4.3, it is desirable to minimize the extremes of deviation of the transmission angle from 90°. The paper by Eckhardt mentioned earlier shows that if a value for the ratio L_1/L_4 is chosen for a drag-link linkage design, then *for a linkage having that value for that ratio*, the maximum deviation of the transmission angle from 90° will be minimized if

$$L_2^2 = L_3^2 = \frac{L_1^2 + L_4^2}{2} \tag{5.9}$$

Then, to generate design charts for drag-link mechanisms, a sequence of values of the ratio L_1/L_4 was chosen, and Eq. (5.9) was used to compute values of the ratios L_2/L_4 and L_3/L_4. Next, the maximum deviation δ of the transmission angle from 90° was computed for those linkage proportions. The paper mentioned earlier shows that this deviation δ is given by

$$\delta = \sin^{-1} \left| 1 - \frac{(L_1 + L_4)^2}{2L_2^2} \right| \tag{5.10}$$

Then, using each link-length ratio set determined as just described, a computer program searched the linkage positions until it found the positions that gave the greatest time ratio TR for a 180° rotation of output link L_3 for that set of ratios. (The greatest value for time ratio for 180° of output motion does *not* necessarily occur for the positions shown in Fig. 5.6.) Then those ratios and the associated deviation δ were each plotted versus that maximum TR. The results are the design charts in Figs. 5.8 through 5.11. Their use in designing a drag-link mechanism is described in Procedure 5.1.

Procedure 5.1: Synthesis of a Drag-Link Mechanism to Give a Desired Value of Time Ratio   The values of maximum deviation of the transmission angle from 90°, values of L_1/L_4, values of L_2/L_4, and values of angles of the input and output links at the start of the forward motion in a drag-link quick-return application are each plotted versus the value of quick-return time ratio in Figs. 5.8, 5.9, 5.10, and 5.11. By choosing a value of desired time ratio on the charts in those figures, the design parameters for a drag-link mechanism that will give that time ratio may be determined, and the corresponding maximum deviation of the transmission angle from 90° may be determined. In mechanisms designed using these charts, the trans-

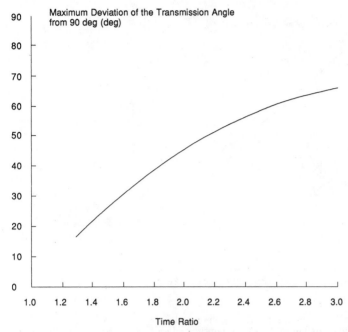

Figure 5.8 Optimized maximum deviations of transmission angle from 90° in a drag-link linkage versus time ratio provided by that linkage.

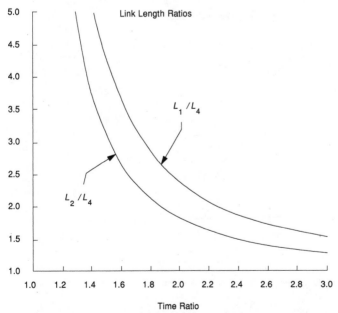

Figure 5.9 Link-length ratios versus time ratio provided in a drag-link linkage that has been optimized with respect to maximum deviations of transmission angle from 90°.

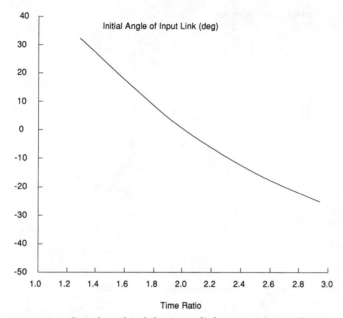

Figure 5.10 Initial angle of the input link versus time ratio provided in a drag-link linkage that has been optimized with respect to maximum deviations of transmission angle from 90°.

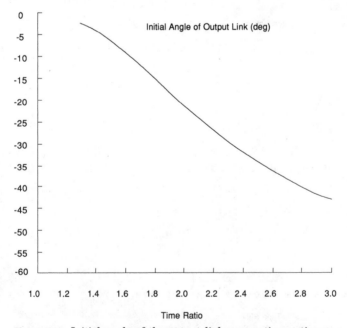

Figure 5.11 Initial angle of the output link versus time ratio provided in a drag-link linkage that has been optimized with respect to maximum deviations of transmission angle from 90°.

mission angle's maximum deviation in the positive direction from 90° will be equal in magnitude to its maximum deviation in the negative direction from 90°.

For example, consider synthesizing a drag-link mechanism to give a time ratio of 2.0. Figure 5.8 shows that the maximum deviation of the transmission angle from 90° will be about 45°. That is, the transmission angle will vary between 45° and 135° during a complete revolution of the input link.

Figures 5.9, 5.10, and 5.11 show that for the time ratio of 2.0, the link-length ratios would be $L_1/L_4 = 2.3$ and $L_2/L_4 = 1.8$, the interval of rotation through β for link L_1 would begin at an angle of $-1°$, and the corresponding angle interval for link L_3 would begin at an angle of $-22°$.

It will be noted that the maximum deviation of the transmission angle from 90° becomes quite large for large time ratios. Therefore, whether these large time ratios can be provided will depend on the loads and speeds that must be accommodated, on the looseness in the bearings used, and on the compliances in the linkage members.

It also will be noted that the extreme transmission angles differ very little from 90° when the time ratio is small. Therefore, for time ratios of less than 1.4, it is probably satisfactory to use a Whitworth analogue synthesis in which the ratio L_1/L_4 is determined using Eqs. (5.5), (5.6), and (5.9).

Procedure 5.2: Synthesis of a Drag-Link Mechanism to Give a Desired Value of Velocity Ratio

By using the instant centers method, it may be shown that the velocity ratio ω_3/ω_1 for a drag-link mechanism will have values given by Eqs. (5.7) and (5.8) when the input link angle is 90° and 270°, respectively. A drag-link mechanism therefore could be synthesized for a desired velocity ratio by using a Whitworth analogue synthesis based on Eqs. (5.7), (5.8), and (5.9).

However, although the velocity ratio values given by Eqs. (5.7) and (5.8) are the minimum and maximum values for a Whitworth mechanism, they are not the *extreme values for a drag-link mechanism*. It will be found that the extreme values for a drag-link mechanism occur at larger values of the input link angle. The previously referenced paper by Eckhardt presents charts that take this difference into account.

Those charts are reprinted here as Figs. 5.12 through 5.17, and they are used in a manner similar to that for Figs. 5.8 through 5.11. For example, if a particular maximum value is desired for the ratio of output angular velocity to input angular velocity, that value can be entered into the chart in Fig. 5.12, and that figure will indicate the maximum deviation of the transmission angle from 90° during operation of the required linkage. Entering the desired maximum value for velocity ratio into Fig. 5.13 gives the required values for the link-length ratios L_1/L_4 and L_2/L_4. Entering the desired maximum value for velocity ratio into Figs. 5.14 and 5.15 gives the input link angle and output angle, respectively, at which that maximum value of angular velocity ratio occurs.

In a similar manner, Figs. 5.16 and 5.17 can be used to find the maximum deviation of the transmission angle from 90° and the required link-length ratios for providing a desired minimum value of the velocity ratio.

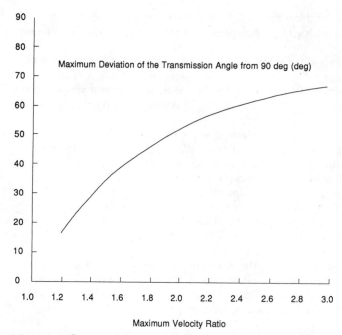

Figure 5.12 Optimized maximum deviations of transmission angle from 90° in a drag-link linkage versus maximum velocity ratio provided by that linkage.

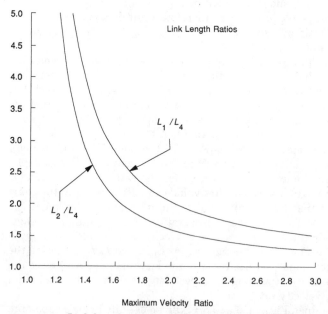

Figure 5.13 Link-length ratios versus maximum velocity ratio provided in a drag-link linkage that has been optimized with respect to maximum deviations of transmission angle from 90°.

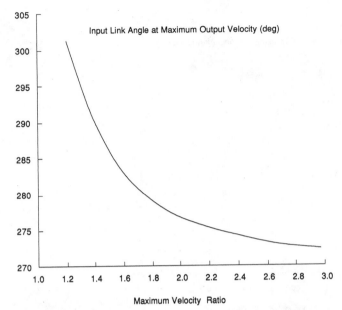

Figure 5.14 Input link angle at maximum output velocity versus maximum velocity ratio provided in a drag-link linkage that has been optimized with respect to maximum deviations of transmission angle from 90°.

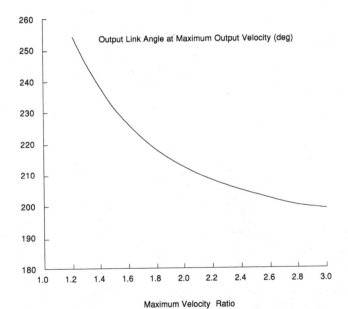

Figure 5.15 Output link angle at maximum output velocity versus maximum velocity ratio provided in a drag-link linkage that has been optimized with respect to maximum deviations of transmission angle from 90°.

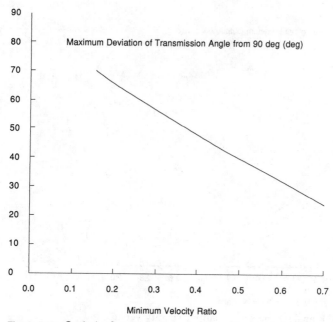

Figure 5.16 Optimized maximum deviations of transmission angle from 90° in a drag-link linkage versus minimum velocity ratio provided by that linkage.

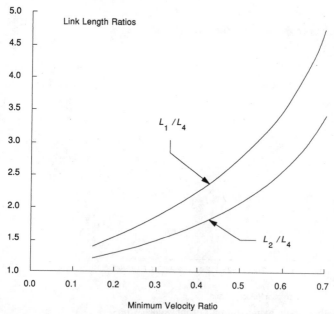

Figure 5.17 Link-length ratios versus minimum velocity ratio provided in a drag-link linkage that has been optimized with respect to maximum deviations of transmission angle from 90°.

By entering a desired value of maximum or minimum velocity ratio in the charts in Figs. 5.12 through 5.17, the design parameters for a drag-link mechanism that will provide those ratios may be determined, and the resulting linkage will experience extremes of transmission angle that are at or near optimal values.

It is seen that for small values of minimum velocity ratio and for large values of maximum velocity ratio, the maximum deviation of the transmission angle from 90° is quite large. Whether these velocity ratios can be provided will depend on the same considerations as mentioned previously for the time-ratio synthesis.

For minimum velocity ratios between 0.7 and 1.0 and for maximum velocity ratios between 1.0 and 1.4, the transmission angle is seen to be close to 90°, so for these ratios, a Whitworth analogue synthesis using Eqs. (5.6), (5.7), and (5.9) gives quite satisfactory results.

5.6   Synthesis of an Oscillating Linkage with Large Output Angular Excursions

It is often necessary to use a quite limited input angular motion to cause some part in a machine to rotate through a large angle (perhaps as much as 180° or more), although this output need not be caused to rotate through more than 360°. Frequently, this motion can be provided by a four-bar linkage of a type that is often called a *"washing machine" linkage* because of its occasional use in driving the large angular oscillations of a washing machine agitator. Perhaps a better name for it would be *backhoe mechanism* because it is frequently seen in the bucket actuation mechanism in a backhoe. A four-bar linkage of this type constitutes a simpler, more easily maintained mechanism than one using extra actuators, gears, belts, etc. The range of the output angular motion that can be provided by such a four-bar linkage, however, tends to be limited by the range of transmission angle that the designer is willing to tolerate. This section presents design charts by means of which four-bar linkages may be designed to provide desired output angle ranges for desired input angle ranges while avoiding undesirable values of transmission angle. Use of these charts is described in Procedure 5.3.

The relationships and the charts presented in this section are based on the approach presented in a paper by Eckhardt in 1993.[2] Over the chosen range of operation of the mechanism, the maximum value of the transmission angle in any linkage designed using the charts in the present section will differ from 90° by the same number of degrees as

---

[2]Eckhardt, H. D., Charts for the design of four-bar linkages with large output angular excursions and optimum transmission angle excursions, in *Proceedings of the Third National Applied Mechanics and Robotics Conference,* Cincinnati, OH, Nov. 8–10, 1993. (These pamphlets were distributed privately. Contact the author at 27 Laurel Drive, Lincoln, MA 01773 for availability of these papers.)

will the minimum value of the transmission angle. This transmission angle variation magnitude will be the optimal magnitude for the chosen mechanism input and output ranges.

Maximum and minimum values of the transmission angle

Figure 5.18 shows a four-bar linkage in a sequence of three positions that represent conditions in which a moderate angular displacement

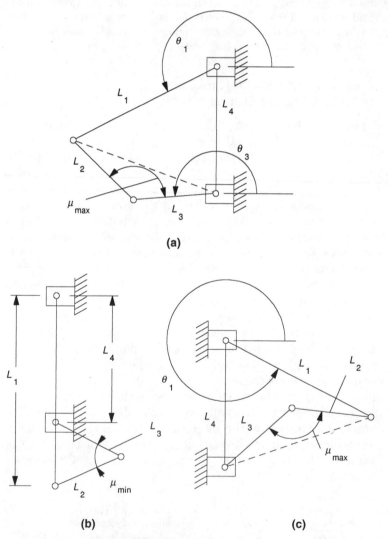

(a)

(b)   **(c)**

Figure 5.18 An oscillating linkage that produces large output angular motions.

of input link L_1 causes a large angular displacement of output link L_3. The transmission angle μ varies from a maximum value in Fig. 5.18a to a minimum value in Fig. 5.18b to another maximum value in Fig. 5.18c as input link L_1 and output link L_3 rotate counterclockwise. (Note that in Fig. 5.18c the linkage has moved past the condition of maximum counterclockwise rotation of L_3.) It can be seen that the range of input and output angles will be limited by the range of values allowed for the transmission angle. The referenced technical paper used a computer program to search for the maximum range of output angle obtainable for a given limit on the range of transmission angle and for given linkage proportions. Then, in a manner similar to that used for the drag-link mechanism design charts, the charts of

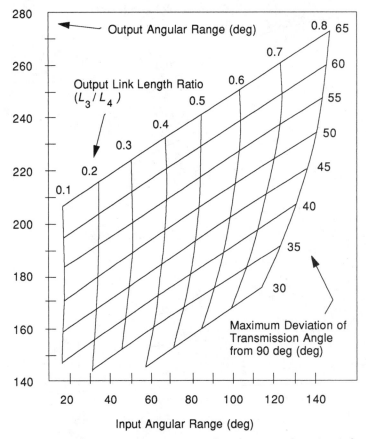

Figure 5.19 Output angular range, output length ratio, and maximum deviation of transmission angle versus input angular range for an oscillating linkage that produces large output angular motions and which has been optimized with respect to maximum deviations of transmission angle from 90°.

Figs. 5.19 through 5.23 were generated. Their use is described in Procedure 5.3.

Procedure 5.3: Synthesis of a Four-Bar Linkage to Produce a Large Output Angular Excursion   The synthesis of a linkage to produce a large output angular excursion usually will start with a statement of the required magnitude of that output excursion. If, then, a value is chosen for the magnitude of the input excursion, these two magnitude values can be entered into Fig. 5.19, and a value for the maximum deviation of the transmission angle from 90° and a value for the ratio of length L_3 of the output link to

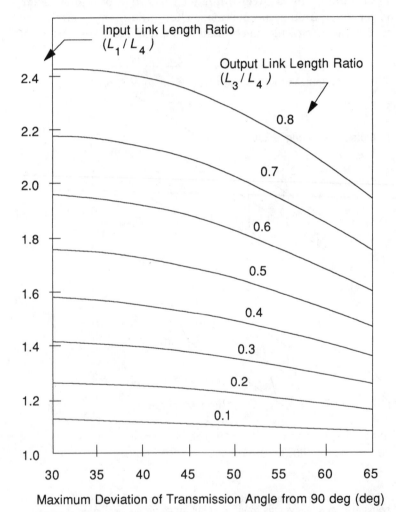

Figure 5.20   Input link-length ratio and output link-length ratio versus optimized maximum deviation of transmission angle from 90° for an oscillating linkage that produces large output angular motions.

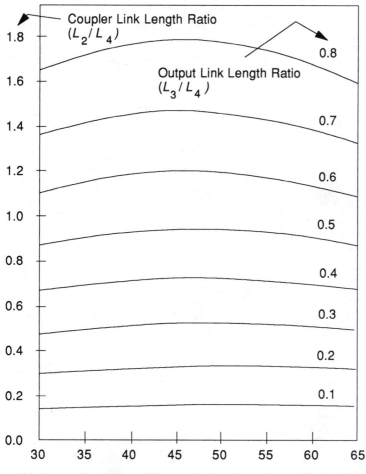

Maximum Deviation of Transmission Angle from 90 deg (deg)

Figure 5.21 Coupler link-length ratio and output link-length ratio versus optimized maximum deviation of transmission angle from 90° for an oscillating linkage that produces large output angular motions.

length L_4 of the ground link are read from the chart. (Actually, values for any two of the variables in Fig. 5.19 could be chosen, and the remaining two could then be determined from the chart.)

Once the four values in Fig. 5.19 are known, Figs. 5.20 through 5.23 can be used to determine the values for the remaining linkage proportions and the angular orientations of the input and output links at the end of their travels.

To illustrate the use of these charts, consider a design problem in which one component of a machine oscillates through an angular range of 100°, and it is desired to cause some other component to oscillate through a range of

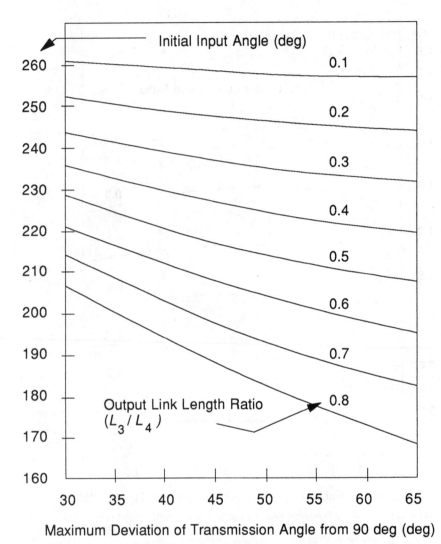

Figure 5.22   Initial input link angle and output link-length ratio versus optimized maximum deviation of transmission angle from 90° for an oscillating linkage that produces large output angular motions.

250° in synchronization with the oscillation of the first component. Using a value of 100° on the horizontal axis of Fig. 5.19 and a value of 250° on the vertical axis, it is seen that the optimal linkage would have a ratio of L_3 to L_4 of about 0.58, and the maximum transmission angle deviation from 90° would be about 65°. Then, using Fig. 5.20 with $L_3/L_4 = 0.58$ and with the maximum transmission angle deviation from 90° of 65°, it is seen that L_1/L_4 is approximately 1.59. In a similar manner, Figs. 5.21, 5.22, and 5.23 indicate that L_2/L_4 is about 1.05, and the angular orientations of the input and

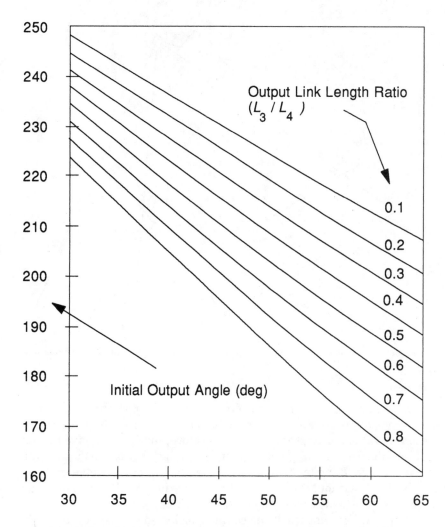

Maximum Deviation of Transmission Angle from 90 deg (deg)

Figure 5.23 Initial output link angle and output link-length ratio versus optimized maximum deviation of transmission angle from 90° for an oscillating linkage that produces large output angular motions.

output links at one end of their excursions will be approximately 198° and 177°, respectively.

Figure 5.24 depicts the motion of the resulting linkage, where links L_1 and L_3 are pivoted to ground at points A and B, respectively. It will be noted that while the input link (the longest link) rotates from about 198° to about 298° (a range of 100°), the output link rotates from about 177° to about 427° for a range of 250°. During this motion, the transmission angle varies between 155° and 25°.

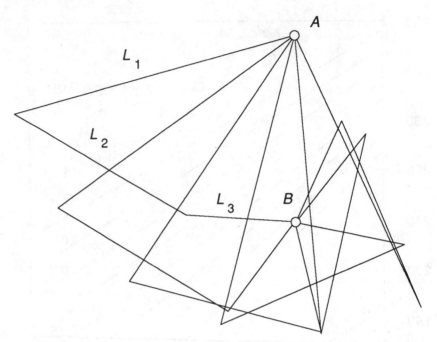

Figure 5.24 Motion of an oscillating linkage that produces large output angular motions.

Comments

The charts are based on the use of an input link length that maximizes the output angular range that may be obtained for each given combination of values of μ_{min} and L_3/L_4. Therefore, for either a given input link angular range or a given output link-length ratio, the output link angular range found by using Fig. 5.19 is the maximum range that can be obtained without exceeding that deviation of the transmission angle from 90° which was indicated on that figure.

It will be noted that when the counterclockwise motion of the output link is limited by L_2 becoming parallel to L_1, the ratio of the angular velocity of L_3 to the angular velocity of L_1 becomes zero at that limit. This position of the example linkage is seen at the right-hand side of Fig. 5.24. This produces an approximate "dwell" in the motion of output link L_3. Most of the linkages designed by using the charts in this paper will have such a dwell.

In some applications this dwell may be highly desirable in order to allow time for some machine function to occur. At and near the dwell position, the mechanical advantage in terms of output torque relative to input torque is extremely large. This, too, may be desirable in many applications in which large output forces or torques are required. For example, both the timing feature and the torque feature might be advantageous if the output link is required to perform a lamination function involving heating and pressure.

In cases in which it is not desirable to have a dwell at one particular end of the range of operation of the linkage, the linkage configuration can be arranged to place the dwell at the other end of the operating range. In cases in which a dwell at neither end is tolerable, a linkage design may be chosen that has more range than needed, and the resulting linkage can then be used over less than its full range.

5.7 Velocity Analysis of Inverted Crank-Slider Mechanisms

The velocity analysis of an inverted crank-slider linkage can be performed either graphically or analytically in a manner that is basically similar to the methods described in Sec. 3.9 for crank-slider linkages. However, because in an inverted crank-slider linkage the slide or slot is moving, it becomes necessary to resolve the slider's velocity into a component parallel to the slide and another component perpendicular to the slide. The first of these components is called the *slip velocity,* and the second of these components is called the *velocity of transmission.* The use and significance of these components are seen most easily in a graphical velocity analysis.

Graphical velocity analysis of inverted crank-slider linkages

Figure 5.25a depicts a slider body L_2 that is sliding in a slide or slot s in a body L_3. Both bodies L_2 and L_3 are moving in such a manner that a point B_2 on body L_2 has a velocity of $\mathbf{V}_{B_2}$ and such that a point B_3, which is coincident with point B_2 at this particular instant but which is on body L_3, has a velocity of $\mathbf{V}_{B_3}$. The point B_2 can have a velocity that differs from that of point B_3 *in the direction of the slide* because L_2 can slide relative to L_3. However, for B_2 to have a velocity component that is different in *the direction perpendicular* to the slide from that of point B_3, the body L_2 would have to escape from the slide in body L_3.

We may therefore draw the velocity vectors of these two points as shown in Fig. 5.25a. In this figure it is seen that the tips of the velocity vectors of the two points lie on a line abc that is parallel to slide s. Note that the difference between these two velocities is given by the vector $\mathbf{V}_{B_s}$ that extends from point b to point c, and this difference is the *slip velocity,* which of course is parallel to slide s. The velocity of each of points B_2 and B_3 will have a component perpendicular to the

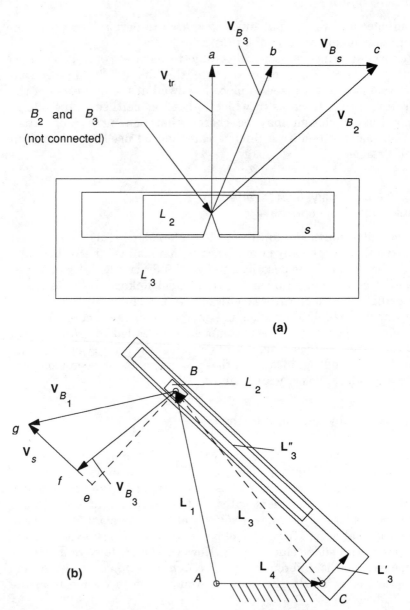

Figure 5.25 Slip velocity and velocity of transmission in inverted crank-slider linkages.

slide that will be a vector from point B to point a. This vector component is the velocity of transmission and is labeled $\mathbf{V}^{tr}$. This is the velocity that body L_2 *transmits* to body L_3, or vice versa.

Figure 5.25b shows an inverted crank-slider linkage of the type shown in Figs. 5.1b and 5.2a, except that the linkage in Fig. 5.25b has been rotated counterclockwise from that in Fig. 5.1b and has the axis of its slide offset by a distance L'_3 from the pivot C. The configurations in Figs. 5.1b and 5.2a are seen to be just special cases of a configuration such as in Fig. 5.25b but in which L'_3 is zero. The linkage in Fig. 5.25b is indicated to consist of the ground, represented by vector $\mathbf{L}_4$, a link represented by vector $\mathbf{L}_1$ from pivot A to pivot B, a slider represented by the block labeled L_2, and the slotted link, which is represented by the dashed vector $\mathbf{L}_3 = \mathbf{L}'_3 + \mathbf{L}''_3$.

The velocity of point B_1 at pivot B on link L_1 is perpendicular to L_1 and has a magnitude of $L_1\omega_1$ and is shown perpendicular to $\mathbf{L}_1$ as $\mathbf{V}_{B_1}$ in Fig. 5.25b. The velocity $\mathbf{V}_{B_3}$ of a point B_3 that is coincident with point B_1 but which is attached to the slotted link L_3 must be perpendicular to the *line from the pivot O to that point* B_3. The magnitude of this velocity will be $\omega_3 OB_3 = \omega_3 L_3$. This velocity vector is shown perpendicular to the dashed vector $\mathbf{L}_3$ in the figure, and it is to be remembered that just as in Fig. 5.25a, the tips of $\mathbf{V}_{B_1}$ and $\mathbf{V}_{B_3}$ must both lie on the same *line efg that is parallel to the slide*. Then, because the directions of both these velocity vectors are known, if the magnitude of one is known, the magnitude of the other can be determined. Thus, if the angular velocity of link L_1 is known, the angular velocity of link L_3 can be determined, and vice versa.

A vector from point B to point e will be perpendicular to the slide and will represent the velocity of transmission because it is the component of the velocities of points B_1 and B_3 that is perpendicular to the slide. The vector $\mathbf{V}^s$ is the difference between the velocities of points B_1 and B_3, so it is the slip velocity.

An example velocity analysis of an inversion of the crank-slider mechanism is included in Sec. 5.9.

Another crank-slider inversion (the one shown in Fig. 5.1c) is redrawn in Fig. 5.26 in a "vertically flipped" position, with line AB horizontal and with the addition of an offset $\mathbf{L}'_2$ of the slot in link L_2. The configuration in Fig. 5.1c is seen to be just a special case of a configuration such as shown in Fig. 5.26 but in which there is no offset of the slide.

The linkage in Fig. 5.26 is indicated to consist of the ground, represented by vector $\mathbf{L}_4$, a link L_1 from ground pivot A to pivot C, a slider represented by the block labeled L_3, and the slotted link labeled L_2, which is represented by the vector sum $\mathbf{L}'_2 + \mathbf{L}''_2$. This configuration differs from that in Fig. 5.25b because although its crank L_1 can rotate through 360°, the angle of slide link L_2 can only oscillate up and down. By having its sliding block pivoted to ground rather than having its slotted link pivoted to ground, point B_3 at pivot point B and attached to block L_3 must have zero velocity, so its velocity of transmission also must be zero. The velocity of a point B_2 on link L_2 but coincident with point B_3 therefore must have no velocity component perpendicular to the direction of the slot. That is, the velocity $\mathbf{V}_{B_2}$ of point B_2 on slotted

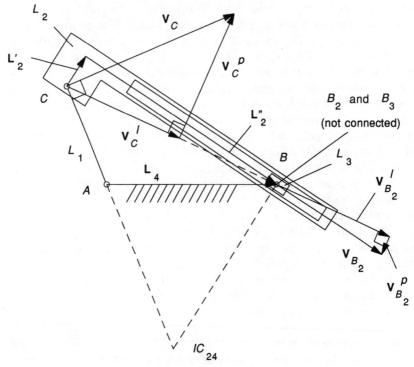

Figure 5.26 Graphical velocity analysis of an inverted crank-slider linkage.

member L_2 must consist entirely of slip velocity and must be parallel to the vector $\mathbf{L}''_2$. However, as discussed in Secs. 2.5 and 3.9, because points A and B_2 are attached to the same rigid body, the longitudinal velocity $\mathbf{V}^l_{B_2}$ (which is the component of this velocity that is parallel to the line CB) must be equal to the longitudinal velocity $\mathbf{V}^l_C$ of point C (which is of course also parallel to CB).

Then, if the angular velocity ω_1 of link L_1 is known, the velocity $\mathbf{V}_C$ of point C is perpendicular to $\mathbf{L}_1$ and its magnitude is $\omega_1 L_1$, and a vector representing this velocity can be drawn to some convenient scale, as shown in Fig. 5.26. This vector can then be projected perpendicularly onto line CB to obtain the longitudinal component $\mathbf{V}^l_C$ as shown. Because the longitudinal component $\mathbf{V}^l_{B_2}$ of the velocity of point B_2 must be equal to the longitudinal component $\mathbf{V}^l_C$ of the velocity of C, this component $\mathbf{V}^l_{B_2}$ can be laid out on an extension of line CB as shown. The *total* velocity $\mathbf{V}_{B_2}$ of point B_2 on link L_2 must be *parallel to the slide,* as discussed above, and thus can be laid out as shown in Fig. 5.26.

Note that the perpendicular components of the velocities of points C and B_2 are merely the components perpendicular to line CB of the velocity vectors. These components are shown as $\mathbf{V}_C^p$ and $\mathbf{V}_{B_2}^p$, respectively. The angular velocity ω_2 of the body L_2 is given by the magnitude of the difference in these two perpendicular component vectors divided by the length of line segment CB. That is,

$$\omega_2 = \left(\frac{V_C^p + V_{B2}^p}{L_2} \right) \tag{5.11}$$

Because the *direction* of the velocity of each of two points (i.e., C and B_2) on link L_2 is known, the location of the instant center of link L_2 can be found by drawing a line through each of the two points and perpendicular to the velocity of that point. These lines are shown dashed. The instant center is at the intersection of these two lines. In Fig. 5.26 this instant center is indicated at IC_{24}. Then, considering the link L_2 to be instantaneously rotating about that center and using a velocity-distribution triangle, velocities of all points on body L_2 can be related to each other and to the angular velocity of that body. Thus, if the angular velocity of either L_1 or L_2 is given, or if the velocity of any point on the linkage is given, all other velocities can be found.

Comments on the inversions
of the crank-slider linkage

In the crank-slider linkage inversions in Fig. 5.1a and d, neither the slider nor the slot rotates. The motion of the configuration in Fig. 5.1a was analyzed in Chap. 3. The motion of the configuration in Fig. 5.1d is exactly the same as that of the configuration in Fig. 5.1a except that the slider remains stationary and the other links translate and rotate relative to it. It can be analyzed as though it were the same as the configuration in Fig. 5.1a, and then the negatives of the displacement, velocity, and acceleration of the slider can be added to each of the other links to give the final answers. The inversions shown in Figs. 5.25b and 5.26 are the only two of the inversions from Fig. 5.1 in which the slider and slot rotate.

It will now be shown that many apparently different arrangements of slide and slider give the same motions, are kinematically the same, and can be analyzed in similar manners. To see this, consider Fig. 5.27, where Fig. 5.26 is redrawn at reduced scale as Fig. 5.27a.

Compare the dashed rectangle $BQCP$ in Fig. 5.27b with the dashed rectangle $BQCP$ in Fig. 5.27a. If the slide offsets in the two figures are equal, the widths QB of the two rectangles are equal. If the lengths of links L_1 and L_4 are equal in the two figures, then for a given angular

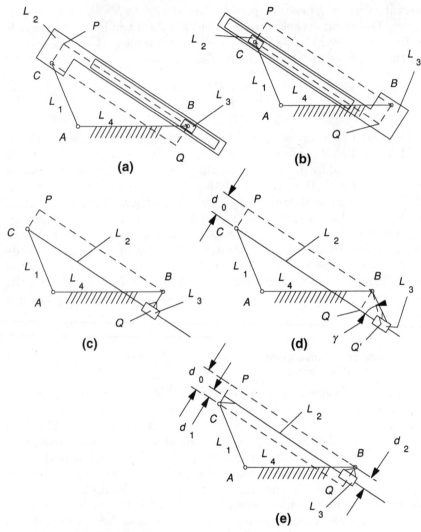

Figure 5.27 Effects of offsets of the slide in an inverted crank-slider linkage.

orientation of L_1, the diagonals CB of the two rectangles will be equal, so the lengths BP of the two rectangles will be equal. It is thus seen that for given motion of link L_1, the translational and rotational displacements, velocities, and accelerations of L_2 will be the same in the two configurations. Also, for the same conditions, the rotational displacements, velocities, and accelerations of L_3 will be the same in the two configurations. It can then be concluded that the slider and slide (or slot) are interchangeable in these configurations, as long as the offset is made the same.

Analysis of the configuration in Fig. 5.27b is probably most appropriately performed like that shown in Fig. 5.25b because of the similarities of the two structures.

Carrying the preceding reasoning further, the diagrams in Fig. 5.27c and d can be seen to give the same motions as does the configuration in Fig. 5.27a, as long as the corresponding link lengths and offset dimensions are kept the same in the various configurations. In some texts, the slider is shown and dimensioned in terms such as length BQ' and angle γ in Fig. 5.27d. Such dimensions are easily converted to the offset d_0 shown. In most cases in actual practice, the physical parts will be designed and dimensioned in terms of the offset (such as d_0) of a pivot from the axis of a slide or slider rather than in terms of γ and BQ'.

Figure 5.27e shows a configuration with an offset of d_1 in the slide and an offset of d_2 in the slider. The motions of L_2 and L_3 in this configuration will be identical to those of the corresponding links in the other configuration in Fig. 5.27 if the total offset $d_1 + d_2$ is equal to offset d_0 in those other configurations and the other corresponding link lengths are equal.

This interchangeability pertains to the other inversions also. This gives engineers considerable flexibility in design because they can synthesize inverted crank-slider mechanisms as easily visualized schematic configurations and then implement them in any of many equivalent physical configurations.

Analytical velocity analysis of an inverted crank-slider linkage

Analytical velocity analysis also produces terms that are identifiable as slip velocity and transmission velocity terms. As in Chaps. 3 and 4, analytical velocity analysis starts with the vector loop closure equation in complex exponential form and differentiates it. For the linkage in Fig. 5.25b, the vector loop closure equation is

$$\mathbf{L}_1 - \mathbf{L}_3 - \mathbf{L}_4 = 0 \tag{5.12}$$

where $\mathbf{L}_3 = \mathbf{L}_3' + \mathbf{L}_3''$. Then, writing Eq. (5.12) in terms of complex exponentials,

$$L_1 e^{j\theta_1} - L_3' e^{j(\theta_3 - \pi/2)} - L_3'' e^{j\theta_3} - L_4 e^{j\theta_4} = 0 \tag{5.13}$$

Differentiating with respect to time, remembering that $e^{j(\pi/2)} = j$, $\dot{\theta} = \omega$, and L_3'' is variable, and transposing terms to the right-hand side of the equation, we get

$$\omega_1 L_1 e^{j(\theta_1 + \pi/2)} = \omega_3 L_3' e^{j\theta_3} + \omega_3 L_3'' e^{j(\theta_3 + \pi/2)} + \dot{L}_3'' e^{j\theta_3} \tag{5.14}$$

The term on the left-hand side of the equation can be seen to be the velocity $\mathbf{V}_{B_1}$ of point B_1 as caused by rotation of link L_1. It is perpendicular to L_1, and its magnitude is $\omega_1 L_1$. The first two terms on the right-hand side represent the velocity $\mathbf{V}_{B_3}$ of point B_3 that is *attached to rigid, rotating link* $\mathbf{L}_3$ and is therefore perpendicular to $\mathbf{L}_3 = \mathbf{L}_3' + \mathbf{L}_3''$. The third term on the right represents the slip velocity resulting from the rate of change of L_3'', and it is therefore parallel to the vector $\mathbf{L}_3''$.

Comparing Eq. (5.14) with the vectors in Fig. 5.25b shows that this equation expresses the relationship $\mathbf{V}_{B_1} = \mathbf{V}_{B_3} + \mathbf{V}^s$ represented by the vector triangle *Bfg*.

Note in addition that the second term on the right-hand side of Eq. (5.14) represents the velocity of transmission $\mathbf{V}_{B_3}^{tr}$ of point B_3 and is the component of $\mathbf{V}_{B_3}$ that is perpendicular to $\mathbf{L}_3''$ just as shown dashed in Fig. 5.25b.

The vector Eq. (5.14) is equivalent to two linear algebraic equations in the three variables ω_1, ω_3, and $\dot{L}_3''$, so if a value for any one of these three variables is known, the equations can be solved for the remaining two values, just as was done in Chaps. 3 and 4.

A similar procedure can be used for the analysis of any inverted crank-slider linkage, starting by writing the vector loop closure equation for the particular inversion of interest.

> Remember that regardless of the presence or absence of offset, slip velocity is parallel to the slide and the velocity of transmission is perpendicular to the slide.

5.8 Acceleration Analysis of Inverted Crank-Slider Mechanisms

Acceleration analysis of inverted crank-slider linkages is basically the same as the acceleration analyses described in Chaps. 3 and 4 but with some additional terms. These additional terms arise because the position, velocity, and acceleration of a point on the slider must be expressed in terms of a vector that not only rotates but also changes in length. The result is that acceleration analysis of inverted crank-slider linkages involves not only the tangential and normal components of acceleration of points, as in previously described acceleration analyses, but also includes Coriolis acceleration components. These Coriolis components are easily calculated and are merely additional vectors that must be included in the vector polygons in graphical analysis or in the equations in analytical acceleration analysis. The procedure for doing this is shown in the example in Sec. 5.9.

Graphical acceleration analysis
of inverted crank-slider linkages

A rough visualization of the source of the Coriolis acceleration term in slider motion can be obtained from examination of Fig. 5.28. In Fig. 5.28a, a slider to which is attached a point P is shown sliding with a constant magnitude *slip* velocity $\mathbf{V}^s$ in a slide in a body B that is rotating with a constant counterclockwise angular velocity ω. The position of point P is described by the vector $\mathbf{R}$ from the grounded pivot. The point P will have a velocity of transmission of $\mathbf{V}^{tr}$ that is perpendicular to the slide and which has a magnitude of $V^{tr} = \omega R$.

A very short time later the system in Fig. 5.28a has rotated slightly counterclockwise to the position shown in Fig. 5.28b, and the slider point P has moved slightly upward to position P'. (These displace-

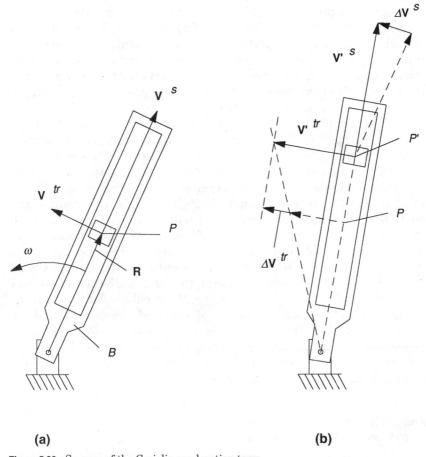

(a)   (b)

Figure 5.28 Sources of the Coriolis acceleration term.

ments are exaggerated here for better visibility.) It can be seen that the slip velocity has changed from its previous direction (shown dashed) by a vector amount ΔV^s. For a very small time interval between the conditions in Fig. 5.27a and b, it is seen that this change vector will be almost perpendicular to the slide. Also, for such a small time interval, the magnitude of this change is seen to be approximately equal to the angular change in direction of V^s times the magnitude of V^s. This angular change is, of course, proportional to the angular velocity ω of the slide. Therefore, the magnitude of ΔV^s is seen to be proportional to $\omega V^s = \omega \dot{R}$.

Note also that the point P has moved from its original position in the slide at P to P', which is at a greater distance from the ground pivot. Then the dashed velocity-distribution triangle in Fig. 5.28b shows that the velocity of transmission has increased to V'^{tr} from the previous vector value V^{tr} (which is shown dashed) by an amount whose component perpendicular to the slide is ΔV^{tr}. It can be seen that the magnitude of this component of change is proportional to the magnitude of V^{tr}, which, in turn, is proportional to ω. The magnitude of the change is also proportional to $\dot{R}$ because $\dot{R} \Delta t$ is the amount by which the point P will move upward in a given small time interval Δt. Thus we see that the magnitude of this component of change is proportional to $\omega \dot{R}$, just as was the change in the slip velocity.

There is therefore an acceleration (rate of change of velocity) component called the *Coriolis acceleration component* that is perpendicular to the slide and which consists of a rate of change of the slip velocity vector plus a rate of change of the transmission velocity. Each of these components of rate of change is perpendicular to the slide, and each of their magnitudes is proportional to $\omega \dot{R}$. In the subsequent subsection on analytical acceleration analysis it will be shown more rigorously that the magnitudes of these rates of change are equal to each other and that the magnitude of the total Coriolis acceleration component is given by $A^c = 2\omega \dot{R}$. This magnitude is easily calculated from the results of a velocity analysis. In general, the *total* acceleration of point P also will include the usual tangential and normal acceleration terms.

Graphical acceleration analysis of inverted crank-slider mechanisms is therefore performed using acceleration vector addition in polygons in the same manner as demonstrated in Chaps. 3 and 4 but with the added inclusion of the Coriolis terms wherever a sliding joint occurs. Such an analysis is shown in Sec. 5.9.

Analytical acceleration analysis of inverted crank-slider linkages

Expressions for use in analytical acceleration analysis are obtained by differentiating the velocity equation obtained by differentiating the vector loop closure equation for the mechanism being analyzed. To illustrate this process, we will derive expressions for the inverted crank-slider linkage configuration in Fig. 5.25b.

The vector loop closure equation for this configuration is given as Eq. (5.12) in the preceding section, and the velocity equation that results from differentiating that loop closure equation is Eq. (5.14). The velocity equation is rewritten here as

$$\omega_1 L_1 e^{j(\theta_1 + \pi/2)} = \omega_3 \left(L_3' e^{j\theta_3} + L_3'' e^{j(\theta_3 + \pi/2)} \right) + \dot{L}_3'' e^{j\theta_3} \qquad (5.15)$$

Now, noting that θ_1, θ_3, ω_1, ω_3, and L_3'' are all variable, Eq. (5.15) can be differentiated with respect to time to give

$$\alpha_1 L_1 e^{j(\theta_1 + \pi/2)} - \omega_1^2 L_1 e^{j\theta_1} = \alpha_3 \left(L_3' e^{j\theta_3} + L_3'' e^{j(\theta_3 + \pi/2)} \right)$$

$$- \omega_3^2 \left(L_3' e^{j(\theta_3 - \pi/2)} + L_3'' e^{j\theta_3} \right) + \ddot{L}_3'' e^{j\theta_3} + 2\omega_3 \dot{L}_3'' e^{j(\theta_3 + \pi/2)} \qquad (5.16)$$

where $\omega = \dot{\theta}$ and $\alpha = \dot{\omega}$, and α is angular acceleration.

As discussed previously, by using Euler's formula, this equation in complex variables can be written as two equations in real variables. It can be seen that once position and velocity analyses have been completed, the only variables in these equations whose values are unknown are α_1, α_3, and $\ddot{L}_3$, and these two real equations are linear in these three variables. Therefore, if a value is known for any one of these three variables, these two equations can be solved for the values of the other two.

The two terms on the left-hand side of Eq. (5.16) will be recognized as the tangential and normal acceleration components of point B_1 on link L_1, and their sum is the total acceleration of that point.

In the term containing α_3 on the right-hand side, the quantity in parentheses can be seen to represent just the vector $\mathbf{L}_3$ rotated 90° counterclockwise. This term containing α_3, then, represents the tangential acceleration component of point B_3 attached to link L_3. In the term containing ω_3^2 on the right-hand side, the quantity in parentheses can be seen to represent just the vector $\mathbf{L}_3$. This term containing ω_3, then, represents the normal acceleration component of point B_3 attached to link L_3. Then, because link L_3 is a rigid body rotating about fixed point O, the sum of these two terms represents the total acceleration of point B_3 that is attached to that body.

Then we can write Eq. (5.16) as

$$\mathbf{A}_{B_1}^{t} + \mathbf{A}_{B_1}^{n} = \mathbf{A}_{B_3}^{t} + \mathbf{A}_{B_3}^{n} + \mathbf{A}_{B_1 B_3} \qquad (5.17)$$

which is an acceleration-difference equation in which the third term on the right-hand side is the acceleration of point B_1 relative to point B_3, and this relative acceleration consists of the last two terms on the right-hand side of Eq. (5.16). The first of these last two terms on the right-hand side of Eq. (5.16) is seen to represent a vector in a direction parallel to the slide or slot in link L_3. Its magnitude is equal to the second rate of change $\ddot{L}_3$ of the location along the slide of point B_1. Point B_1 is attached to link L_1, so this term is the *slip acceleration* of point B_1 relative to point B_3 on the slide. The last term on the right-hand side of Eq. (5.16) is perpendicular to the slide, and it is the Coriolis term that was discussed qualitatively at the beginning of this section.

Then the third term on the right-hand side of Eq. (5.17) can be written as the last two terms from Eq. (5.16), so Eq. (5.17) becomes

$$\mathbf{A}_{B_1}^t + \mathbf{A}_{B_1}^n = \mathbf{A}_{B_3}^t + \mathbf{A}_{B_3}^n + \mathbf{A}_{B_1B_3}^s + \mathbf{A}_{B_1B_3}^c \qquad (5.18)$$

where the last two terms are, respectively, the slip acceleration and the Coriolis acceleration terms. They appear in more detail as the last two terms on the right-hand side of Eq. (5.16).

By examining the last term on the right-hand side of Eq. (5.16) it is seen that this Coriolis acceleration vector is rotated 90° counterclockwise from the vector $\mathbf{L}_3''$ (and $\mathbf{L}_3''$ is parallel to the slide). Then, if the scalar product $\omega_3 \dot{L}_3$ is positive, the Coriolis acceleration vector is of magnitude $\omega_3 \dot{L}_3$ and is directed 90° counterclockwise from the position vector $\mathbf{L}_3$. If the scalar product $\omega_3 \dot{L}_3$ is negative, the Coriolis acceleration vector is also of magnitude $\omega_3 \dot{L}_3$ but is directed 90° clockwise from the position vector $\mathbf{L}_3$.

Both sides of Eq. (5.18) express the acceleration of the same point B_1. The left-hand side involves only tangential and normal acceleration components because the point's position, velocity, and acceleration are therein expressed in terms of the vector $\mathbf{L}_1$, which, although it rotates, is of constant length. On the right-hand side, however, the position, velocity, and acceleration of the same point are expressed in terms of vectors $\mathbf{L}_3'$ and $\mathbf{L}_3''$, which not only rotate, but vector $\mathbf{L}_3''$ also varies in length, so the slip and Coriolis terms must be included.

5.9 Analysis of a Geneva Mechanism

Figure 5.29 depicts a mechanism that is known as a *Geneva mechanism*. It is used to convert continuous rotary motion of an input shaft to intermittent unidirectional rotation of an output shaft. It consists of two rigid bodies that are labeled L_1 and L_3 in Fig. 5.29. The left-hand body is the input link, and it is pivoted to ground at point A. It consists of a disk to which are attached a pin at point B and a cam that is shaped like an almost-full moon as shown. The right-hand body L_3 is the output link, and it is a slotted disk that is pivoted to ground at point C. In the mechanism in this example, the input link carries

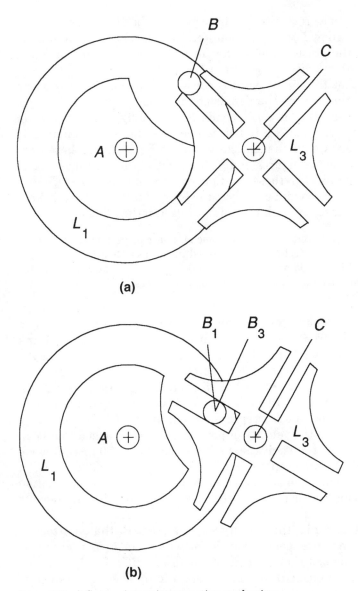

Figure 5.29 A Geneva intermittent-motion mechanism.

one pin and the output link contains four slots. Other combinations of numbers of pins and slots are also possible.

For this example, the input link L_1 is assumed to be rotating clockwise at 10 rad/s. In Fig. 5.29a the pin at B is seen to be just entering a slot in output link L_3, and the cam that is attached to link L_1 and which has until now restrained L_3 from rotating has just reached a position that will allow L_3 to rotate counterclockwise. Further rotation of L_1 will cause the pin to slide further into the slot and thus to rotate link L_3 counterclockwise, as shown in Fig. 5.29b.

By noting that the pin at point B constitutes a slider in a rotating slot, it is seen that this mechanism is an inverted crank-slider linkage. Further, it is seen that it corresponds to the configuration shown in Fig. 5.27b but with no offset. It also can be seen that the configuration in Fig. 5.27b also would be identical to that in Fig. 5.25b if there were no offsets in these two figures and if the ratios of the length of link L_1 to the length of link L_4 in these two figures were the same. The Geneva mechanism therefore can be analyzed using the techniques that were discussed in connection with the configuration in Fig. 5.25b.

Indeed, any inverted crank-slider mechanism can be analyzed using a combination of the techniques discussed in connection with the configurations in Figs. 5.25b and 5.26. Care must be taken to keep track of the signs of the variables and the names of the vectors used to represent the link positions.

Consider the Geneva mechanism to be in the position shown in Fig. 5.29b. A line from point A to point C represents the ground link L_4, which is given as 1.414 in long and is oriented as drawn as line AC in Fig. 5.30a. A line from point A to point B represents the input link L_1, which is 1.0 in long, is inclined at 15°, is rotating clockwise at a constant angular velocity of 10 rad/s, and is drawn in Fig. 5.30a as line AB. In the figure the line BC represents output link L_3.

The magnitude of the velocity of point B_1 that is attached to link L_1 at the center of the pin is given by $\omega_1 L_1 = (10)(1.0) = 10$ in/s. This velocity of point B_1 is shown as the vector $\mathbf{V}_{B_1}$ in Fig. 5.30a. It is, of course, perpendicular to link L_1 and is in the direction of rotation of that link. This vector is drawn to a scale such that 1 in represents 5 in/s.

Just as in Fig. 5.25a, the velocity $\mathbf{V}_{B_3}$ of a point B_3 that is attached to link L_3 but instantaneously coincident with point B_1 is drawn perpendicular to the slot (i.e., perpendicular to line BC). Its tip must lie on a line r (shown dashed) that is through the tip of $\mathbf{V}_{B_1}$ and which is parallel to the slot, as shown in Fig. 5.30a. Measurement of the length of vector $\mathbf{V}_{B_3}$ gives a length of 1.41 in, so the magnitude of the velocity of point B_3 is $V_{B_3} = (1.41 \text{ in})(5 \text{ in/s per inch}) = 7.05$ in/s. Similarly, the velocity of slip is $V_B^s = (1.41)(5) = 7.05$ in/s toward pivot C.

The angular velocity ω_3 of output link L_3 is found by dividing V_{B_3} by the length from pivot C to point B_1. This length is measured to be 0.52 in on the scale drawing in Fig. 5.30a. Thus $\omega_3 = 7.05/0.52 = 13.6$ rad/s counterclockwise.

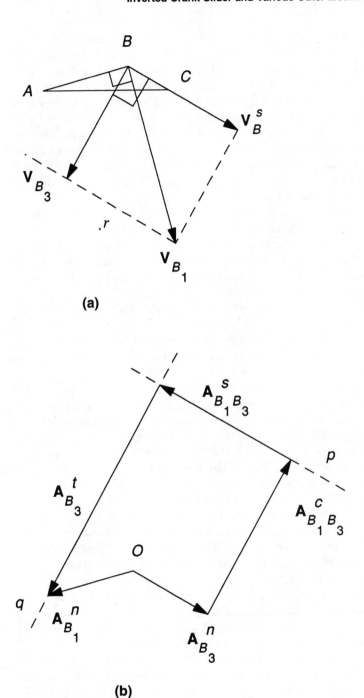

(a)

(b)

Figure 5.30 Graphical velocity and acceleration analysis of a Geneva intermittent-motion mechanism.

There is now enough information to proceed with the acceleration analysis. This analysis is based on Eq. (5.18), which in effect says that the vectors on the right-hand side of the equation must add up to give the same total vector as that obtained by adding the vectors on the left-hand side. The tangential acceleration $\mathbf{A}_{B_1}^t$ of point B_1 on link L_1 is zero because the input link is rotating at constant angular velocity. Therefore, the left-hand side of the equation consists of only the normal acceleration vector $\mathbf{A}_{B_1}^n$. This vector is directed from point B_1 toward pivot A, and its magnitude is given by $A_{B_1}^n = \omega_1^2 L_1 = (10)^2(1.0) = 100$ in/s^2. This vector is drawn from an origin O in Fig. 5.29b to scale such that 1 in represents 100 in/s^2.

All the vector terms on the right-hand side of Eq. (5.18) must add up to equal the vector just drawn. That is, together with that vector, they must form a closed vector polygon. The value of the first term on the right-hand side is not known, so proceed to the second term, i.e., the normal acceleration term. The normal acceleration of point B_3 is a vector directed from point B_3 toward pivot C, and it has a magnitude given by $A_{B_3}^n = \omega_3^2 L_3 = (13.6)^2(0.52) = 96.2$ in/s^2. This vector is drawn as $\mathbf{A}_{B_3}^n$ from the origin O in Fig. 5.30b.

The value of the slip acceleration $\mathbf{A}_{B_1B_3}^s$ is not known, so proceed to the final term on the right-hand side, i.e., the Coriolis term $\mathbf{A}_{B_1B_3}^c$. As can be seen from Eq. (5.16), the magnitude of this term is given by $A_{B_1B_3}^c = 2\omega_3\dot{L}_3 = 2(13.6)(7.05) = 191.8$ in/s^2. As shown previously, this Coriolis acceleration vector component is perpendicular to the slot (i.e., perpendicular to line BC in Fig. 5.30a). Referring to the discussion pertaining to Fig. 5.28, it can be seen that *if* the slider were traveling *away from* the point about which the slot is rotating, the Coriolis component of acceleration would be in the direction that the rotation of the slot is moving the slider. That is, it would be in the direction of the transmission velocity. However, in the present Geneva mechanism example, the slider is moving *toward* the point about which the slot is rotating, so the Coriolis component of acceleration is in the direction *opposite* to the transmission velocity. Therefore, in the present example the Coriolis term $\mathbf{A}_{B_1B_3}^c$ is directed upward and to the right and is drawn starting at the tip of the vector $\mathbf{A}_{B_3}^n$, as shown in Fig. 5.30b.

It remains, then, to determine the values of the vectors $\mathbf{A}_{B_3}^t$ and $\mathbf{A}_{B_1B_3}^s$. Although the magnitudes of these vectors are not known, their *directions are known*. The slip vector $\mathbf{A}_{B_1B_3}^s$ must be parallel to the slot, so the dashed line p may be drawn through the tip of the vector $\mathbf{A}_{B_1B_3}^c$ as shown, and the slip acceleration vector must lie along that line. The tangential acceleration vector $\mathbf{A}_{B_3}^t$ must be perpendicular to the slot, and it must complete the vector polygon, so the dashed line q may be drawn through the tip of the vector $\mathbf{A}_{B_1}^n$. The tangential acceleration vector $\mathbf{A}_{B_3}^t$ must lie along this line. Thus it can be seen that if the vectors $\mathbf{A}_{B_3}^t$ and $\mathbf{A}_{B_1B_3}^s$ are drawn as shown, these vectors have the required directions, and they close the polygon.

Measuring the vectors in Fig. 5.30*b* and noting that 1 in represents 100 in/s^2, it is found that $\mathbf{A}_{B_1B_3}^s = 167$ in/s^2 *away* from pivot C, and $\mathbf{A}_{B_3}^t$ $= 262$ in/s^2 downward and to the left. The angular acceleration of link L_3 is obtained by dividing $A_{B_3}^t$ by the length BC, so $\omega_3 = (262)/(0.52) = 504$ rad/s^2 in a counterclockwise direction.

By examining Fig. 5.30*b* it can be seen that the Coriolis acceleration vector of length 192 in/s^2 accounts for almost three-quarters of the 262 in/s^2 of the tangential acceleration $A_{B_3}^t$ and therefore accounts for almost three-quarters of the angular acceleration of the output link. *In inverted crank-slider linkages, Coriolis terms can, as in this example, account for a very large portion of the resulting accelerations.*

Output links of Geneva mechanisms can experience large accelerations, and if these output links must drive appreciable inertias, the forces involved and the resulting vibrations can be quite severe.

Chapter

6

The General Four-Bar Linkage

6.1 Introduction

This chapter describes four-bar linkage configurations that use several different combinations of sliding joints and pin joints. Such linkages have characteristics that can provide many useful applications. Also, the relationships between these linkages and their characteristics furnish useful *stimuli for creative mechanism design.*

Because the sliding joints may be located in various positions in the linkage, methods for solving the position-closure equation are classified in four cases, and these methods are presented in this chapter. Although the techniques discussed in previous chapters are still applicable to velocity and acceleration analysis, the additional terms discussed in Chap. 5 are now included, and the analysis techniques are extended accordingly.

6.2 Combinations of Joint Types and Link Types in Four-Bar Linkages

The discussion in this section consists of a rudimentary form of *type synthesis.* Type synthesis involves consideration of the topology ("connectedness") required of possible linkages in order to perform a desired function. Type synthesis also involves *number synthesis,* which relates the numbers of links, joints, loops, and degrees of freedom required to perform the desired function. Type and number syntheses are very helpful guides to creative design of linkages. Such syntheses are discussed in more detail in Chap. 7.

As seen in Chap. 1, a planar linkage can consist of links of any of several types joined together by any of three types of joints. The types of links available are classified according to the number of joints by which they are attached to other links, and such link types are shown in Fig. 1.21. The types of joints available are shown in Fig. 1.20. Note that the first two types of joints (revolute and slider) possess only one degree of freedom, whereas the third type (sliding pin) possesses two degrees of freedom.

This chapter is concerned only with linkages that contain exactly four links and in which the vectors representing the links form a single closed loop. Chapter 4 showed that such a linkage can consist of four binary links joined together by four revolute joints. If ternary links (containing three joint connection points) are used in a four-bar linkage, there must be either two or four of them, and such use will result in *two* or *three* link vector closed loops, respectively. Not only will multiple loops result, but it will be necessary to use sliding-pin joints in such linkages if one or more degrees of freedom are to be provided (i.e., if the linkages are to be movable). An example of a four-bar linkage that contains two ternary links is shown in Fig. 6.1a, where links L_1 and L_3 are binary links and links L_2 and L_4 are ternary links. That is, link L_2 is attached to other links by joints at B, C, and D, while ground link L_4 is connected to other links by joints at A, D, and E. The joints A, B, and C are revolute joints, and the joints D and E are sliding-pin joints.

The two vector loops involved in this mechanism can be drawn by first drawing vectors that represent the links in some manner that is convenient for specifying the positions of those links and joints. The vectors shown in Fig. 6.1b are a convenient set for specifying the positions of the links and joints. Then, polygons can be formed of these vectors. For this linkage configuration, the two such polygons (loops) *could* be *ABCDFA* and *FDCEGF*.

Applying Gruebler's formula [Eq. (1.2)] to the configuration in Fig. 6.1a, we see that $F = 3(L - 1) - 2J_1 - J_2 = 3(4 - 1) - 2(3) - 2 = 1$, so this linkage possesses one degree of freedom.

Not only is the linkage in Fig. 6.1a a two-loop linkage, but if the sliding-pin joints are replaced by pivoted sliding blocks or by infinitely long revolute-jointed links, the linkage will be found to be a Stephenson six-bar linkage, which is discussed in Chap. 7. It is not possible to use ternary or higher-order links in a single-loop *four*-bar linkage.

From the foregoing it can be seen that by limiting the discussion in this chapter to single-loop four-bar linkages, the discussion is limited to linkages consisting of four binary links and four joints, i.e., to linkages with $L = 4$ and $J_1 + J_2 = 4$ or $J_2 = 4 - J_1$. Then, for these linkages, Gruebler's formula can be written

$$F = 3(4 - 1) - 2J_1 - (4 - J_1)$$

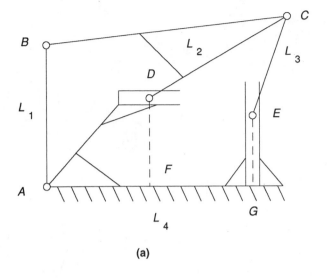

(a)

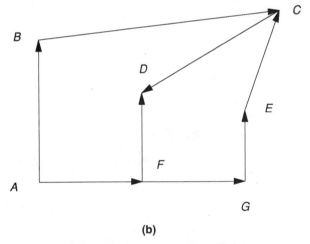

(b)

Figure 6.1 A four-bar linkage using ternary links.

from which we can write

$$J_1 = 5 - F \tag{6.1}$$

Then, from Eq. (6.1) and the fact that as noted above $J_2 = 4 - J_1$, we may write

$$J_2 = F - 1 \tag{6.2}$$

From Eqs. (6.1) and (6.2) it may be seen that if the linkage is to have one degree of freedom, it must contain four single-degree-of-freedom

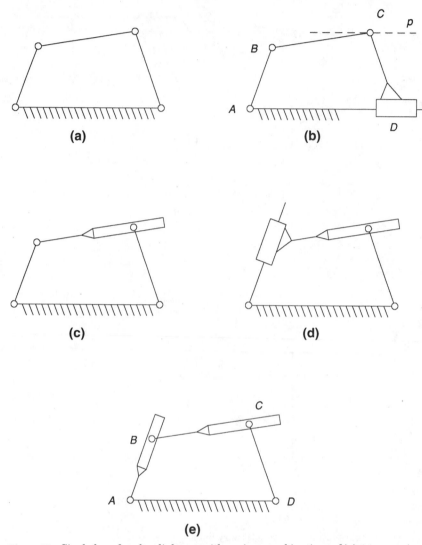

Figure 6.2 Single-loop four-bar linkages with various combinations of joint types.

joints (i.e., $J_1 = 4$) and no two-degree-of-freedom joints (i.e., $J_2 = 0$). Examples of configurations having two possible combinations of single-degree-of-freedom joints are shown in Fig. 6.2a and b. More of these single-degree-of-freedom configurations will be discussed in Sec. 6.3.

From these same equations it can be seen that if the linkage is to have two degrees of freedom, it will have three single-degree-of-freedom joints and one two-degree-of-freedom joint. Example configurations of this type are shown in Fig. 6.2c and d. Such two-degree-of-freedom configurations will be discussed in Sec. 6.4.

If the single-loop four-bar linkage is to have more than two degrees of freedom, it must contain two or more sliding-pin joints, as indicated by Eqs. (6.1) and (6.2). Figure 6.2e shows a possible configuration having three degrees of freedom and therefore containing two sliding-pin joints. Being a three-degree-of-freedom linkage, it will require three actuators (drives) if it is to be controlled. An input shaft could be placed at each of pivots A and D to provide rotational actuation at those pivots. In actual practice, providing the third required actuation would require modifying one of the sliding-pin joints by the addition of a fifth link. For example, if a translational actuation were provided at joint B, that sliding-pin joint would be replaced by a block (a fifth link) that is pivoted to the link BC and which slides in a slide attached to link AB. That block would constitute a piston in a hydraulic or pneumatic cylinder or would constitute an armature of a linear electromagnetic actuator.

It is thus seen that single-loop four-bar linkages having more than two degrees of freedom are best replaced by linkages having more than four bars. Such linkages are covered in Chap. 7. *This chapter will therefore consider only single-loop linkages with one or two degrees of freedom.*

6.3 Four-Bar Linkages with One Degree of Freedom

As discussed in the preceding section, all joints in a single-loop, single-degree-of-freedom four-bar linkage must be single-degree-of-freedom joints. The two types of joints that possess one degree of freedom are the pin joint and the sliding joint. Figure 6.2a shows a linkage in which all the joints are pin joints. The synthesis and analysis of this type of linkage are covered in Chap. 4.

Four-bar linkages with one sliding joint

In the linkage shown in Fig. 6.2b, one of the four joints is a sliding joint, whereas the remaining three joints are pin joints. The link CD is constrained to slide horizontally without rotating, so all points on that link (including point C) must move horizontally along lines parallel to AD. Thus point C must travel back and forth along the dashed line p. Comparing Fig. 6.2b with Fig. 3.9 shows that the linkages in these two figures experience the same sort of motion, except that the offsets of the slider paths relative to the crank pivots at A and O are in opposite directions. The linkage in Fig. 6.2b is thus seen to be an offset crank-slider linkage. The synthesis and analysis of such linkages are covered in Chap. 3, and the synthesis and analysis of their inversions are covered in Chap. 5.

Some four-bar linkages with two sliding joints

In each of the linkages shown in Fig. 6.3, two of the joints are sliding joints and the remaining two are pin joints. In Fig. 6.3a the link L_4 is grounded, and it can be seen that as a consequence, the only link that can rotate is link L_1. As the crank L_1 rotates, the link L_2 slides up and down on link L_3, and as L_2 is moved right and left by the crank L_1, L_3 also moves right and left as it slides on link L_4. Because L_2 is a nonrotating rigid body and point A on it moves along a circular path, all points on L_2 must follow identical circular paths. The path followed by point C on L_2 is shown dashed and centered at point e, where radius eC is equal to and parallel to AB. When C on L_2 is at point f, link L_3 will be at its rightmost position (shown dashed). When C on L_2 is at point g, link L_3 will be at its leftmost position. This mechanism is a scotch yoke, similar to that described in Secs. 3.2 and 3.3. The translational motion of link L_3 will be a sinusoidal function of the angular position of L_1. The similarity can be seen if the angles δ and γ in Fig. 6.3a are set to 90° and if the offset distance d (from B to h) of the slide on link L_2 in that figure is set to zero. The result of these settings is shown in Fig. 6.3b, where it will be noted that the length of link L_2 is zero, so points B and C coincide. Then visualize the motions of the linkages in Figs. 3.1a and 6.3b.

Consider next a kinematic inversion of the linkage in Fig. 6.3a that would result from grounding link L_2 instead of link L_4. Such an inversion is shown in Fig. 6.3c. Links L_2, L_3, and L_4 are still restrained from rotating. By following reasoning such as used above, it can be seen that this inversion is also a scotch yoke. However, in this case, as link L_1 rotates, link L_3 is caused to slide sinusiodally up and down relative to grounded link L_2 rather than sliding back and forth on link L_4, as was the case in Fig. 6.3a and b.

Scotch yokes are usually made with their slides at right angles to each other. However, if space or other restrictions dictate, then, as in Fig. 6.3a, the slides can be inclined differently, and an offset can be used.

Figure 6.4 shows a third kinematic inversion of the linkage shown in Fig. 6.3. In Fig. 6.4a link L_1 (the link to which are attached both pin joints) is grounded instead of link L_2 or link L_4. Because the joints at B and C prevent relative rotation between the links joined by these joints, links L_2, L_3, and L_4 must rotate in unison, although sliding can occur between them. Thus, if an input shaft at pivot B causes link L_2 to rotate (as indicated by ω_1 in the figure), an output shaft that is

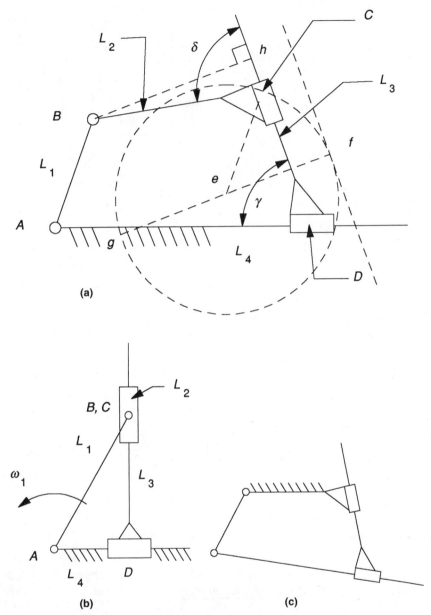

Figure 6.3 Four-bar linkages with two sliding joints and two pin joints.

placed at pivot A and which is parallel to but not coincident with the input shaft will be caused by link L_4 to rotate in exactly the same manner. When used this way, this linkage is an *Oldham coupling,* and usually the length of link L_2 is made zero and the slides are made perpendicular to each other, as shown in Fig. 6.4*b*. The physical

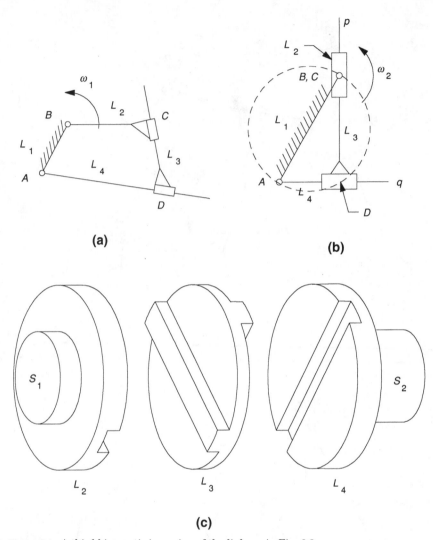

Figure 6.4 A third kinematic inversion of the linkage in Fig. 6.3.

implementation of this linkage usually appears as shown exploded in Fig. 6.4c, where the links are labeled to correspond with those in Fig. 6.4b, the input shaft is S_1, and the output shaft is S_2. These two shafts are parallel but are usually not coaxial. The sliding joints between the three links shown are the mating grooves and raised ridges on those links. Notice that in both Figs. 6.4b and 6.4c the link L_3 is connected to links L_2 and L_4 by sliding joints that are at right angles to each other.

Referring again to Fig. 6.4b, consider the line p that is drawn through pivot B and along the slide in the pivoted block L_2, and consider the line q that is drawn through pivot A and along the slide in link L_3. These lines will always intersect at some point D that is attached to link L_3. If the slides are made perpendicular to each other (as is usually the case in an Oldham coupling), these lines also will remain at right angles to each other. Then, because these lines pass through points A and B, their intersection (which is point D where the slides cross) will always lie on a circle whose diameter is line AB. Such a circle is shown dashed in the figure.

If the input shaft at pivot B in the linkage in Fig. 6.4b starts at an angle such that line p is parallel to line AB, point D will lie at point A. As the input shaft rotates line p counterclockwise, point D attached to link L_3 will follow a semicircular path to the right and upward (shown dashed) until, when line p is perpendicular to line AB, point D becomes coincident with point B. As the input shaft continues to rotate line p counterclockwise, point D attached to link L_3 will follow a semicircular path to the left and downward (shown dashed) until, when line p is again parallel to line AB, point D again becomes coincident with point A. Thus, during 180° of rotation of the input shaft, point D on link L_3 where the slides cross will follow a path that is a complete circle (shown dashed). It can be seen, then, that as the input shaft rotates through a full 360°, the point on link L_3 where the slides cross travels *twice* around a circular path whose diameter is the line connecting the input and output pivots.

When the sliders in an Oldham coupling are replaced by extremely long pivoted links (see Procedure 8.1), the linkage becomes a drag-link mechanism. It is not surprising, then, that the travel of coupler point D twice around a circle for each rotation of the input shaft in the foregoing discussion bears some similarity to the *double-looped coupler* point paths shown in Fig. 8.24a and b for a drag-link mechanism.

Even if the slides are not perpendicular to each other in the linkage just discussed, the point of intersection of the axes of the slides will still trace a circular path. Relationships among the parameters associated with an arc of such a circle are indicated in Fig. 6.5a. Because the angle α between the sliders on links L_2 and L_3 is constrained to be constant, point D (the intersection of the axes of the slides) must lie on a circle through points A and B, so the angle α is an inscribed angle in that circle. The circle will have some radius r, and its center will be at some point p as indicated. From high school geometry we know that the central angle (shown as 2α) is twice the inscribed angle α. Then, from examination of the two equal right triangles AeP and BeP, we may write

$$\sin(\pi - \alpha) = \frac{L_1}{2r} \qquad (6.3)$$

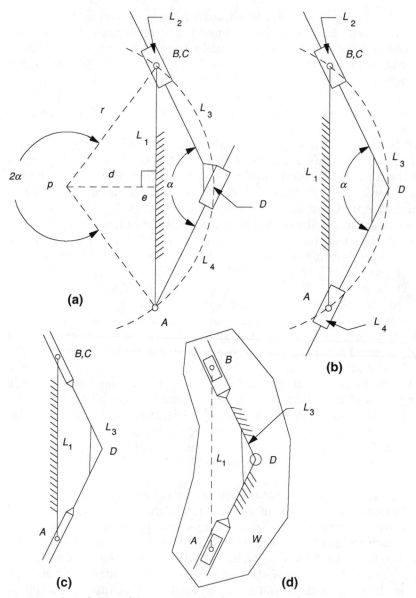

Figure 6.5 A mechanism for generating circular arcs of very large radius.

and
$$d = \frac{L_1}{2} \tan(\alpha - \pi/2) \tag{6.4}$$

where L_1 is the distance AB and d is the distance from line AB to the center of the circle traced by point D. Note that if $\alpha < \pi/2$, then d is negative, indicating that the center p is on the same side of line AB as point D.

This mechanism can be used to draw arcs of very large radius or arcs whose centers are otherwise not conveniently accessible. For example, if an arc of radius 60 in is to be drawn and points A and B on that arc are 20 in apart, Eq. (6.3) can be used to find that $\alpha = 170.4°$. Using this setting and attaching the stylus at point D, the mechanism could be used to draw the required arc without having access to the center. Equation (6.4) indicates that the center would be 59.16 in from chord AB.

For actual implementation of a drawing mechanism such as just discussed, the sliding joint at D in Fig. 6.5a could be replaced by interchanging its slide and slider and moving the resulting sliding joint along the slide axis to coincide with point A to give the configuration of Fig. 6.5b. Then the slides and sliders in Fig. 6.5b could be replaced by sliding pins, as shown in Fig. 6.5c.

If it is desired to machine a part to have a cut, slot, or track that is of large radius, the mechanism could take the form indicated in Fig. 6.5d. In this figure the tool (a milling cutter, nibbler, band saw blade, etc.) would be fixed to ground at point D, and the work piece W would be moved relative to it. Blocks would be pivoted *to the work piece* at points A and B on the desired arc, and the blocks would slide in slots that are attached to the base of the machine doing the cutting. These slots would be inclined to each other at the desired angle α, and the centerlines of the slots would intersect at the center of the tool. This is obviously an inversion of the linkage in Fig. 6.5b, where now link L_3 becomes the grounded machine and link L_1 becomes the moving work piece. (Of course, the pivoted blocks could be replaced by sliding pins as in Fig. 6.5c.)

Caution: Although point D in the preceding linkages is caused to travel along a circular path, the center of the circle is *not an instant center for any of the links.* None of the links rotates about that point, so the mechanism cannot be used (by itself) to furnish a virtual pivot for a body. A *virtual pivot* would be a point neither contained in nor attached to the body but about which the body would rotate. The mechanism could, however, be used in machining a curved track that could provide such a virtual pivot.

The remaining inversion of the four-bar linkage having two sliding joints is shown schematically in Fig. 6.6. In this inversion the link L_3 is grounded so that the only link that can rotate is link L_1. Links L_2 and L_4 merely slide relative to link L_3 without rotating. The motions of points A and B are therefore constrained to be along the straight lines indicated dashed as p and q, respectively, as link AB rotates.

The most common manifestation of this inversion is in the *elliptical trammel*, which is shown in Fig. 6.6b and c. In this configuration the length of link L_2 has been reduced to zero, and the slides have been

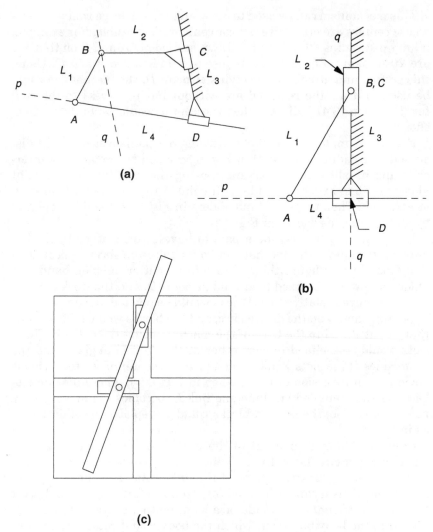

Figure 6.6 A fourth kinematic inversion of the linkage in Fig. 6.3.

made perpendicular to each other. In this form, all points along the centerline of link L_1 trace elliptical paths, and the mechanism is often used to draw ellipses. It also has been used in machining elliptical forms. Figure 6.6c shows the familiar toy form of this mechanism, whose action is fascinating to observe.

In a four-bar linkage with two sliding joints, each sliding joint is associated with a variable-length link, so there are two such variable-length links in such a linkage. In the linkages just discussed, the two variable-length links were connected to each other by a sliding joint. (Links L_3 and L_4 were connected to each other by a sliding joint at D.) Another form of the four-bar linkage containing two sliding joints is indicated in Fig. 6.7a. In this linkage, the two variable-length links are L_1 and L_4, and they are connected to each other by the revolute joint at point A. A third form of four-bar linkage with two sliding joints is indicated in Fig. 6.7b. In this configuration the variable-length links L_2 and L_4 are not contiguous; i.e., they are not connected directly to each other. Numerous other configurations of four-bar linkages with two sliding joints can be derived by offsetting the slide axes from the pivots and by interchanging slides and sliders.

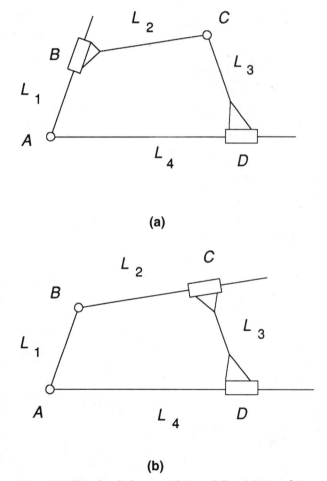

(a)

(b)

Figure 6.7 Four-bar linkages with two sliding joints and two pin joints but in which the two variable-length links are *not* connected to each other by a sliding joint.

Four-bar linkages with three or four sliding joints

Figure 6.8*a* shows a four-bar linkage with three sliding joints. Although Gruebler's formula indicates that this linkage has one degree of freedom, none of the links can be rotated. The freedom consists of the ability of link L_2 to slide up and down. For example, L_2 could be slid up to a position at which points B and C would be at points B' and C'.

Figure 6.8*b* shows a four-bar linkage with four sliding joints. No rotation is possible in this linkage either. Although Gruebler's formula

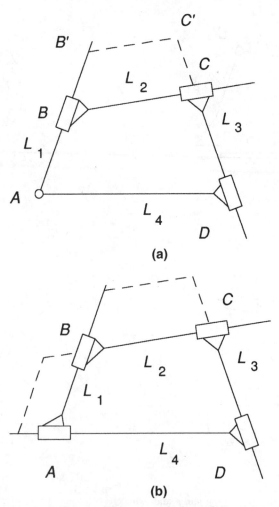

(a)

(b)

Figure 6.8 Four-bar linkages with three sliding joints and with four sliding joints.

indicates that this linkage has one degree of freedom, the linkage cannot be assembled unless the sum of the angles between the slides on adjacent links is sufficiently accurately made equal to 360°. If this requirement *is* satisfied, it will be found that the linkage has *two practical* degrees of freedom. That is, it will be noted that link L_2 can be slid up and down *and* link L_1 can be slid left and right to positions such as shown dashed, and that these motions can be performed independently or simultaneously. This is another example of the reason that Gruebler's formula, although very useful, must be used with caution. The nature of each computed degree of freedom should be visualized.

Unless *both* the two practical degrees of freedom provided by the linkage having four sliding joints are required, it would be preferable to use the configuration in Fig. 6.8a and avoid the alignment-accuracy requirements associated with that in Fig. 6.8b.

6.4 Four-Bar Linkages with Two Degrees of Freedom

As pointed out in Sec. 6.2, a single-loop, four-bar linkage with two degrees of freedom must have three single-degree-of-freedom joints and one two-degree-of-freedom joint. Each of the three single-degree-of-freedom joints can be either a revolute joint or a sliding joint. The two-degree-of-freedom joint must be a sliding-pin joint. Some examples of linkages of this sort are shown schematically in Fig. 6.9.

A most likely use for linkages of this type is in adjustable mechanisms, where the sliding-pin joint is used as an adjustable pivot point. It will be seen that if the motion of link L_1 in Fig. 6.9a and b is considered the input motion and link L_3 is held stationary, the coupler link L_2 will slide back and forth and rock about the now stationary pivot at C. Moving the sliding pin at C to a new position by rotating L_3 about its pivot at D will change the motion of coupler L_2. A detailed discussion of adjustable linkages and the additive and multiplicative combining of inputs is contained in Secs. 8.2 and 8.3.

As an illustration of one way in which a two-degree-of-freedom, single-loop, four-bar linkage has been used, compare Fig. 6.9c with Fig. 8.7. The link L_1 in Fig. 6.9c represents the shuttle of Fig. 8.7, and that shuttle slides horizontally on the ground link L_4. Pivoted to L_1 at point P is link L_2, which has, at its far end A, a sliding pin that slides in a slide (slot) in link L_3. This link L_2 represents the corresponding link from point P to point A in Fig. 8.7, but in Fig. 6.9c the pivoted sliding block at point A has been replaced by the equivalent sliding-pin joint. By changing the inclination of the slotted link L_3 in each figure, the amount and direction of the rocking of link L_2

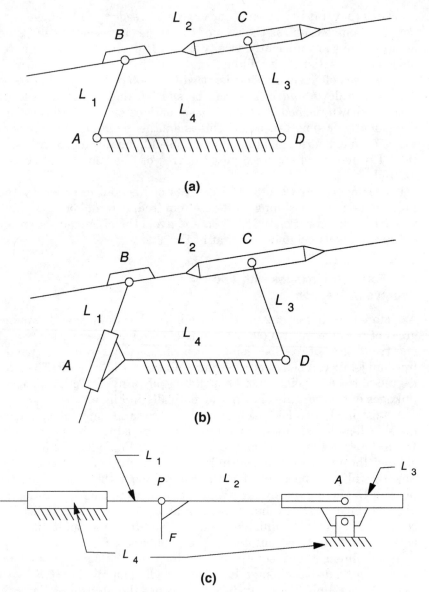

Figure 6.9 Examples of four-bar linkages that have two degrees of freedom.

as it slides back and forth can be changed. The discussion of Example 8.4 in Sec. 8.3 describes how this linkage is used in a cam-driven mechanism to adjust the stroke of the shuttle. In Fig. 6.9 the cam follower's input motion would be a horizontal displacement at point F.

It is often useful initially to conceive of a two-degree-of-freedom linkage as having a sliding-pin joint as in Fig. 6.9c. Then, after working out details, the sliding-pin joint can be replaced by a pivoted sliding block, as in Fig. 8.7, or by a pivoted long binary link.

By comparing Figs. 6.9b and 6.9c it can be seen that the same combination of joints is used in both configurations. It also will be seen that the order in which these joints are distributed around the loop is the same, except that in Fig. 6.9c the sliding pin is connected to link L_2 and the slide (slot) is connected to link L_3, whereas in Fig. 6.9b exactly the reverse is true.

Obviously, the joint combinations can be arranged in many ways, and the angles and offsets of the slots can be chosen to have many different values, resulting in a rich variety of possible two-degree-of-freedom, single-loop, four-bar linkages. If the sliding-pin joints are replaced by sliding blocks, these linkages become five-bar linkages, which are discussed in Chap. 7.

6.5 Position Analysis of the General Four-Bar Linkage

Computerized analysis or simulation usually will be employed to determine the positions of links and points in linkages, particularly if more than one position of the linkage is to be investigated. However, a visualization and an increased understanding of the interactions of the parts are often provided by a graphical analysis of a representative condition. Also, such graphical analysis is often required for making CAD drawings of mechanisms.

When investigating the positions of links in a linkage, it is useful to represent each link by a vector. Such a vector is considered to extend between two specified points on the link and thus to be attached to the link and therefore to translate and rotate with it. Figure 6.10a schematically depicts a *pin-jointed* four-bar linkage in which the links are represented by the four vectors $\mathbf{L}_1$, $\mathbf{L}_2$, $\mathbf{L}_3$, and $\mathbf{L}_4$. Each of these vectors extends from one of the pivots on its respective link to the other such pivot. For example, $\mathbf{L}_1$ extends from a point A_1 on link L_1 at the pivot at point A to a point B_1 on link L_1 at the pivot at point B. The vector $\mathbf{L}_2$ extends from a point B_2 on link L_2 at the pivot on point B to a point C_2 on link L_2 at the pivot at point B. Because the pivots are rigidly attached to the links, the points A_1 and B_1 are firmly attached to link L_1 and the points A_2 and B_2 are firmly attached to link L_2. Similarly, the points A_4 and D_4 are firmly attached to link L_4 and the points D_3 and B_3 are firmly attached to link L_3. The point A_1 coincides with point A_4, B_1 coincides with point B_2, etc.

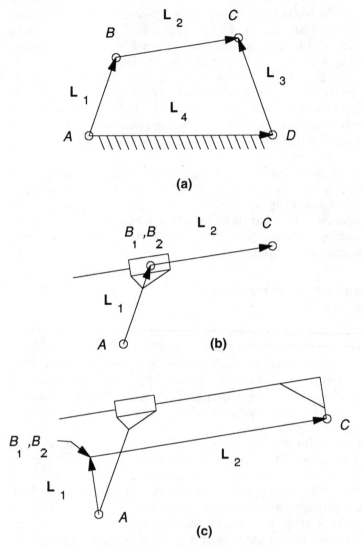

Figure 6.10 Vector representation of links in four-bar linkages.

The sense of each of these vectors (i.e., which end has the arrow-head) is arbitrary. However, it usually is convenient to place the tail of the arrow at the end where the link is pivoted to ground if the link contains such a pivot (e.g., links L_1 and L_3).

In a linkage containing only pin joints, such as in Fig. 6.10a, each and every one of the representative vectors will be of constant length. If, however, the linkage contains one or more sliding joints, one or more of the representative vectors will have a variable length. Consider the pair of links depicted in Fig. 6.10b. The link L_1 is represented by the vector $\mathbf{L}_1$ that extends from point A_1 on link L_1 to point B_1 on that same link. Neither of these two points moves *relative to* link L_1. The link L_2 is represented by the vector $\mathbf{L}_2$ that extends from point B_2 on link L_2 to point C on link L_2. Because, in this case shown, the joint at point C is a pin joint, point C is firmly attached to link L_2 and therefore will move in unison with L_2. However, although point B_2 is considered firmly attached to L_2 *at the instant for which analysis is being performed*, it is considered to have a different location along that slide relative to link L_2 at each different instant. As a result, the representative vector $\mathbf{L}_2$ (which extends from B_2 to C) will have a variable length.

When drawing a representative vector for a link containing a sliding joint and a pin joint, one end of the vector should lie at the pin joint, and the vector should be made parallel to the sliding action, as in Fig. 6.10b. Figure 6.10c shows an example of an *offset* sliding-jointed link pair and the most easily used vector representation. It can be seen that drawing the vector $\mathbf{L}_2$ this way affects the choice of location of points B_2 and B_1 and therefore affects the way in which the vector $\mathbf{L}_1$ must be drawn.

Recall from previous chapters that analysis of link positions is based on solving the vector loop closure equation either graphically or analytically. The vector equation is equivalent to two algebraic equations, and, therefore, it can be solved if and only if two of the variable quantities are unknown. Because of the possible presence of combinations of both pin joints and sliding joints in the same linkage, each of these two variables can be either a link angle or a link length. Chace has classified these combinations into four cases.[1] Each of these four cases is discussed in one of the following four subsections.

Case 1:   The angle and length of a single link (vector) are unknown

Figure 6.11a depicts an inverted crank-slider linkage in which the length of link L_2 between points B and C is unknown and the angle θ_2 of that link is also unknown. The unknown length and angle of this link L_2 are quite easily found by simple graphical construction by laying out the known links L_1 and L_4 to scale and connecting their tips.

---

[1]Chace, M. A., Development and Application of Vector Mathematics for Kinematic Analysis of Three-Dimensional Mechanisms, Ph.D. thesis, The University of Michigan, Ann Arbor, 1964.

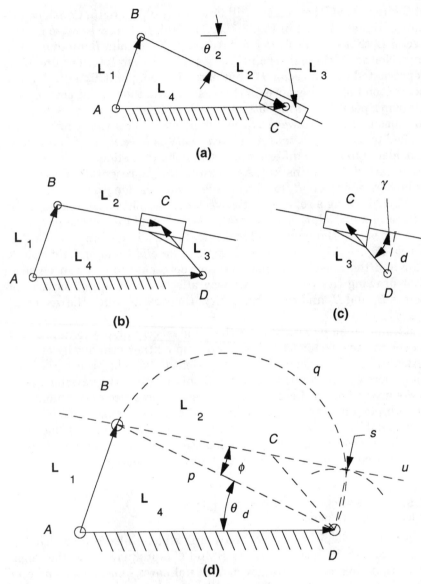

Figure 6.11 Position analysis of a Case 1 four-bar linkage.

A variation of this inverted crank-slider linkage, but with an offset, is shown in Fig. 6.11b. The solution of the vector loop equation is not quite as direct in this case. However, as shown in Fig. 6.11c, the offset distance d of the pivot at D from the slide (slot) in link L_3 can be computed from the link length L_3 and the fixed angle γ as $d = L_3 \sin \gamma$.

Then the graphical construction becomes quite direct, as described in the following and as shown in Fig. 6.11d.

In Fig. 6.11d the vectors L_1 and L_4 have been drawn to scale, and the dashed line p has been drawn connecting their tips (points B and D). Then a semicircle q is constructed, using that line p as a diameter. Using point D at the end of L_4 as a center, an arc of radius d is drawn to intersect semicircle q at point s. The dashed line u through points B and s is the line along which link L_2 must lie. (Note that a line from D to point s will be perpendicular to line u and will have a length d.) The location of point C can then be determined by constructing a line through point D that is inclined to the line from D to s by an angle of magnitude $90 - \gamma$.

This graphical construction provides an excellent hint for performing an *analytical* position analysis of this linkage. Note that triangle BsD is a right triangle, so

$$(Bs)^2 = (BD)^2 - (Ds)^2 \tag{6.5}$$

where Ds is the perpendicular offset d *of* the slide on L_2, and the length BD is found by solving the triangle ABD using the law of cosines, as discussed in Sec. 3.8 along with Fig. 3.15. Then length L_2 from point B to point C (Fig. 6.10b) is found as

$$L_2 = (Bs) - d \cot \gamma \quad \text{(see Fig. 6.10c and d)} \tag{6.6}$$

The angle θ_d is found in the solution of triangle ABD using the law of cosines as above and as in Sec. 3.8. Then the angular orientation θ_2 of link L_2 can be found from

$$\theta_2 = \theta_d - \phi \quad \text{where } \sin \phi = d/(Bs) \tag{6.7}$$

The analytical position analysis of this linkage also can be performed starting with the vector loop closure equation as in previous chapters. That equation is then broken into its two equivalent scalar equations. These equations contain terms in which the unknowns L_2 and θ_2 appear in expressions such as L_2, $\sin \theta_2$, and $\cos \theta_2$. By devious appropriate use of a trigonometric identity, these otherwise apparently intractable equations can be reduced to forms equivalent to Eqs. (6.5), (6.6), and (6.7) and the relations from which the triangle ABD can be solved. It would seem easier just to use these equations and relations as described above rather than bother to solve the vector loop closure equation.

Case 2:   The unknowns are associated
with two separate vectors

When the unknowns are associated with two separate vectors, the unknowns can consist of three combinations of angles and magnitudes (length). These combinations are covered in the following three subcases.

Case 2a: The lengths of two links (vectors) are unknown

Figure 6.12a shows a linkage in which two of the joints are sliding joints. Therefore, the lengths of two of the links are variable. *However, it should be noted that if link* L_1 *is the input link, the angles of all the links are known.* If this linkage is driven by rotating link L_1 while link L_4 is grounded, then the lengths of links L_2 and L_3 will be unknown. Because the angle α is fixed, the angle of link L_3 will be known. Then, because the angle β also is fixed, the angle of link L_2 also will be known. Thus the angular orientations of all the links are known. To find the unknown lengths of links L_2 and L_3, lay out from an origin A the completely known link vectors $\mathbf{L}_1$ and $\mathbf{L}_4$ to scale, as shown in Fig. 6.12b. Then draw a line r through the tip of vector $\mathbf{L}_4$ at an angle

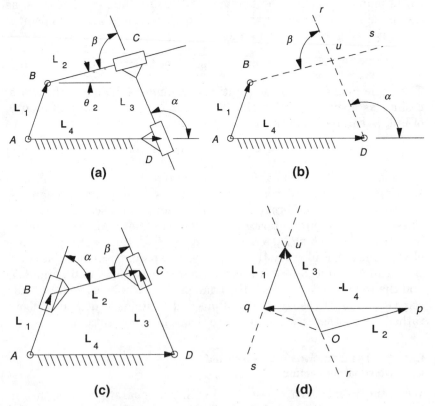

Figure 6.12 Position analysis of a Case 2a four-bar linkage.

equal to α. Then draw a line s through the tip of vector $\mathbf{L}_1$ at an angle equal to θ_2, where

$$\theta_2 = \alpha + \beta - 180°$$

These lines intersect at point u. Then length Bu is the length of link L_2, and length Du is the length of link L_3.

Analytical position analysis of this configuration can be performed by straightforward use of the vector loop closure equation, which, when reduced to its equivalent scalar equations, consists of two linear simultaneous equations in the unknown lengths. Analytical solution for the position of the linkage in Fig. 6.12a also can be performed by first solving the triangle ABD in Fig. 6.12b using the law of cosines. Then, knowing length BD, angle ADB, and the given angles α and β, one side and all angles of triangle BDu are known, and thus triangle BDu can be solved using the law of sines.

The links with the unknown lengths are contiguous with each other in Fig. 6.12a. That is, they share a common joint at C. In Fig. 6.12c, however, if the *angle* of link L_1 is considered to be the input, the links with the unknown lengths are L_1 and L_3, and they are separated from each other by links L_2 and L_4. They are *not* contiguous with each other. It is seen that if the angular position of link L_1 relative to link L_4 is known, the angles of all other links are known.

The values of the unknown lengths can be found by a method that is similar to the foregoing but in which the *order* of the links is re-arranged. To illustrate this procedure, consider the vector loop closure equation, which for this linkage is seen to be

$$\mathbf{L}_1 + \mathbf{L}_2 = \mathbf{L}_4 + \mathbf{L}_3$$

The completely known vectors are $\mathbf{L}_2$ and $\mathbf{L}_4$. Transpose terms in this equation to place the completely known terms on the left-hand side:

$$\mathbf{L}_2 - \mathbf{L}_4 = \mathbf{L}_3 - \mathbf{L}_1 \tag{6.8}$$

Then, starting at an origin O, draw the left-hand side of the equation to scale, as shown in Fig. 6.12d, where following the vectors $\mathbf{L}_2$ and $-\mathbf{L}_4$ from tail to head produces a path from the origin O to point p and then to point q. Then a vector from point O to point q is the sum vector $\mathbf{L}_2 - \mathbf{L}_4$, which, according to Eq. (6.8), must be equal to $\mathbf{L}_3 - \mathbf{L}_1$.

Next, construct a line r through the origin O and at the known orientation of link L_3. Then construct a line s through point q and at the known orientation of link L_1. These two lines (shown dashed) intersect at a point u. Tracing a path from the origin O to point u and then to point q must be equivalent to following the vectors $\mathbf{L}_3$ and $-\mathbf{L}_1$. These vectors are labeled in Fig. 6.12d, and their lengths and angles can be measured from the drawing.

The procedure represented by Fig. 6.12d has made use of the commutativity of vector addition. That is, it used the property that a group of vectors may be added in any order without affecting the resulting sum. As always, the signs and senses of the vectors must be carefully kept track of. This commutativity property is useful in solving many vector problems.

Analytical position analysis of this configuration can be performed by straightforward use of the vector loop closure equation, which, when reduced to its equivalent scalar equations, consists of two linear simultaneous equations in the unknown lengths. Analytical solution for the position of the linkage in Fig. 6.12c also can be performed by first solving the triangle Opq in Fig. 6.12d using the law of cosines. Then, knowing angle Oqp and the angles of all the links, all angles of triangle Ouq can be found, and triangle Ouq can be solved using the law of sines.

Case 2b: The length of one vector and the angle of another vector are unknown

Figure 6.13a shows a linkage in which link L_1 is assumed to be the input link, the angle θ_2 of link L_2 is unknown, and the length of link L_4 is unknown. Graphical determination of the values of these unknown quantities can be performed simply, as shown in Fig. 6.13b. First, the known vectors $\mathbf{L}_0$ and $\mathbf{L}_1$ are plotted tail to head from an origin O. Then, because the angle of link L_4 is known, a dashed line p can be drawn through point O at that known angle. Then an arc q with a radius equal to the known length of link L_2 is drawn with its center at the head of vector $\mathbf{L}_1$. The arc q intersects the line p at points s and s' as shown. The length Os or Os' is the sought length of link L_4, and the angle θ_2 can be measured as shown. There are obviously two solutions to this analysis. The reader must choose the one that is consistent with the mechanism design.

Figure 6.13c depicts a linkage in which link L_1 is assumed to be the input link and the variable-length link L_4 is assumed to be fixed in direction. The unknown quantities in this case are the angle θ_2 of link L_2 and the length of link L_4. The links to which these unknowns pertain are not contiguous. We therefore make use of the commutativity of addition of vectors, which corresponds to writing the vector loop closure equation in the form

$$\mathbf{L}_1 - \mathbf{L}_3 = \mathbf{L}_4 - \mathbf{L}_2 \qquad (6.9)$$

That is, as shown in Fig. 6.13d, we plot the vectors $\mathbf{L}_1$ and $-\mathbf{L}_3$ tail to head from an origin O. Their sum, which is given by the left-hand side

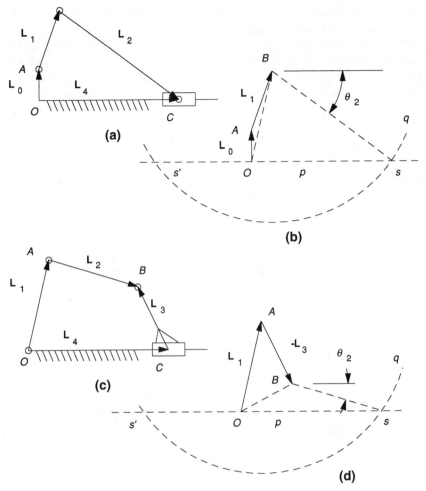

Figure 6.13 Position analysis of a case 2b four-bar linkage.

of the preceding equation, is represented by a dashed vector from O to B in the figure. Then, through the origin O a dashed line p is drawn at the known angle of link L_4. Using point B as a center, draw an arc q that has a radius equal to the known length of link L_2. This arc intersects line p at points s and s'. The length Os or Os' is the sought length of link L_4, and the angle θ_2 can be measured as shown. Again, there are obviously two solutions to this analysis. The reader must choose the one that is consistent with the mechanism design.

Notice that the motion of point B in Fig. 6.13c must be along a horizontal line parallel to $\mathbf{L}_4$. This motion is that of an offset crank-slider mechanism such as shown in Fig. 3.14a, so it also can be analyzed by the methods of Sec. 3.8. Because of the manner in which the mechanism is designed and dimensioned, however, it may be more convenient to use the method just described.

The analytical solution of this case 2b can consist of first solving the triangle OAB in Fig. 6.13b or d using the law of cosines because an angle and two side lengths are known. Then, all the angles in triangle OBs are easily computed, and triangle OBs can be solved using the law of sines.

Case 2c: The angles of two vectors are unknown

In Figure 6.14a, the linkage shown is a pin-jointed four-bar linkage. If the input link is link L_1, then the angles of links L_2 and L_3 are un-

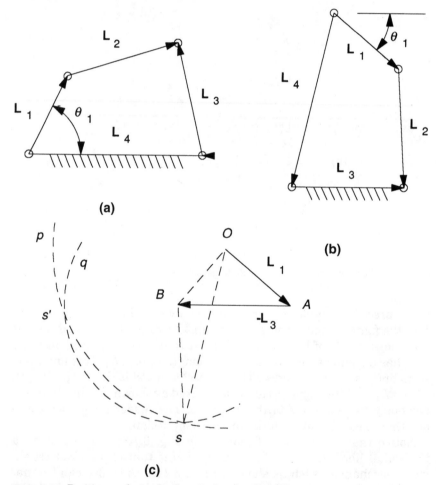

(a)

(b)

(c)

Figure 6.14 Position analysis of a Case 2c four-bar linkage.

known. The values of these link angles can be found by the method described in Sec. 4.8.

In the very unusual case shown in Fig. 6.14b, link L_1 is the input link and, therefore, its angle is known. Because link L_3 is grounded, its angle is also known. The angles of links L_2 and L_4 are unknown, and these links are not contiguous with each other. Again, as in previous cases of noncontiguous links, the problem is solved by using the commutativity of vector addition as embodied in Eq. (6.9) and as shown in Fig. 6.14c. In this figure, in accordance with the left-hand side of the equation, the vectors $\mathbf{L}_1$ and $-\mathbf{L}_3$ are drawn tail to head from an origin O. Then, in accordance with the right-hand side of the equation, a circular arc p of radius equal to the length of link L_4 is drawn with the origin O as its center. Another circular arc q is drawn with the tip of the vector $-\mathbf{L}_3$ as its center and with a radius equal to the length of link L_2. These arcs intersect at the points s and s'. It can be seen that if a vector from point O to point s is chosen as the link vector $\mathbf{L}_4$, and if a vector from point B to point s is chosen as the link vector $\mathbf{L}_2$, then these vectors will satisfy Eq. (6.9). It is also seen that these vectors also will match the corresponding vectors in Fig. 6.14b. Notice also that if a vector from point O to point s' is chosen as the link vector $\mathbf{L}_4$, and if a vector from point B to point s' is chosen as the link vector $\mathbf{L}_2$, then these vectors also would satisfy Eq. (6.9). The reader must choose the solution that is consistent with the mechanism being designed or analyzed.

Analytical solution for the link positions proceeds as in the previous cases by solving the triangle AOB in Fig. 6.14c using the law of cosines. In this Case 2c, the triangle OBs also must be solved using the law of cosines because, although lengths of all three of its sides are known, none of its angles is known.

6.6 Velocity Analysis of the General Four-Bar Linkage

At the beginning of Sec. 6.5 it was shown that if a four-bar linkage contains only pin joints, its link positions can be represented by four vectors whose lengths are constant, although their angles vary. The velocity analysis of such linkages is covered in Sec. 4.8. Section 6.5 also showed that when a four-bar linkage contains one or more sliding joints, the length of one or more of the representative link vectors will vary. Therefore, velocity analysis of four-bar linkages containing sliding joints must include the effects of slip velocity, which were introduced in Sec. 5.7.

Figure 6.15 is a redrawing of Fig. 6.10b, and it shows a pair of links that are connected to each other by a sliding joint. Link L_2 is represented by the vector $\mathbf{L}_2$, which ends at the pivot joint at point C on link L_2 and

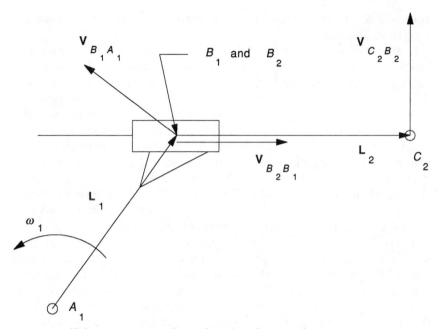

Figure 6.15 Sliding-joint vector relationships for velocity analysis.

which is drawn parallel to the sliding action that takes place between links L_1 and L_2. Link L_1 is then represented by the vector $\mathbf{L}_1$, which starts at the pivot joint at point A_1 on link L_1 and extends to a point B_1 on link L_1, where B_1 coincides with a point B_2 at the end of vector $\mathbf{L}_2$.

We may write the velocity-difference equation relating the velocity of point B_1 to that of point A_1 as

$$\mathbf{V}_{B_1} = \mathbf{V}_{A_1} + \mathbf{V}_{B_1 A_1} \tag{6.10}$$

where $\mathbf{V}_{B_1 A_1}$ is the velocity of point B_1 relative to point A_1. Because link L_1 and therefore its vector $\mathbf{L}_1$ are of constant length, the velocity of point B_1 relative to point A_1 is perpendicular to the vector $\mathbf{L}_1$, as shown as vector $\mathbf{V}_{B_1 A_1}$ in Fig. 6.15, and it has a magnitude of

$$V_{B_1 A_1} = \omega_1 L_1 \tag{6.11}$$

Because slippage occurs at the joint at B, point B_2 will have a velocity $\mathbf{V}_{B_2 B_1}$ relative to point B_1. This is, of course, the slip velocity, and it is parallel to the slide, as shown in Fig. 6.15. Then we may extend the velocity-difference relationship of Eq. (6.10) to relate the velocity of point B_2 to that of point A_1 by writing

$$\mathbf{V}_{B_2} = \mathbf{V}_{B_1} + \mathbf{V}_{B_2 B_1} = \mathbf{V}_{A_1} + \mathbf{V}_{B_1 A_1} + \mathbf{V}_{B_2 B_1} \tag{6.12}$$

Point B_2 is coincident with point B_1 at the instant considered but is attached to link L_2, so points B_2 and C_2 are both points attached to the same rigid body L_2. Therefore, the velocity $\mathbf{V}_{C_2B_2}$ of point C_2 relative to point B_2 is perpendicular to $\mathbf{L}_2$, as shown in Fig. 6.15, and it has a magnitude of $V_{C_2B_2} = \omega_2 L_2$.

Now, extending the velocity-difference relationship of Eq. (6.12) to relate the velocity of point C_2 to that of point A_1, we may write

$$\mathbf{V}_{C_2} = \mathbf{V}_{B_2} + \mathbf{V}_{C_2B_2} = \mathbf{V}_{A_1} + \mathbf{V}_{B_1A_1} + \mathbf{V}_{B_2B_1} + \mathbf{V}_{C_2B_2} \qquad (6.13)$$

By noting the progressive development of relationships from Eq. (6.10) to Eq. (6.13), it can be seen that the velocities of points along a chain of links can be related to each other by proceeding from point to point, writing the velocity-difference equation for each pair of points along the chain. If the chain forms a closed loop, the velocity of a point on the linkage can be *related to itself* by relating the velocities of successive points around the loop to each other by means of the velocity differences. That is, the sum of all the velocity differences around the loop will be zero. *This amounts to velocity vector loop closure.* The principle of velocity vector loop closure results in a vector equation that can be used to solve for two unknown velocity vector component magnitudes. Because all the vector directions are known from a position analysis, the corresponding scalar equations are linear in the unknown vector component magnitudes. Consequently, the equation(s) can be solved simply, both analytically and graphically.

Alternatively, the velocity differences can be summed from some chosen point to another point partway around the loop, and then also summed between the same two points by going around the loop in the opposite direction. The two sums must be equal. Such a procedure is used in Example 6.1.

Example 6.1: Graphical Position and Velocity Analysis of a Four-Bar Linkage Having One Sliding Joint

Figure 6.16a depicts a four-bar linkage that contains revolute joints at points A, C, and D and one sliding joint at point B. This linkage is of a type discussed in Sec. 6.5 as Case 2b. The slide at point B is inclined to line L_1 at an angle of $45°$. The link lengths are $AD = 3.5$ units, $AB = 1.2$ units, and $CD = 1.625$ units. The condition to be analyzed is that for which $\theta_1 = 55°$ and $\omega = -10$ rad/s. That is, link L_1 is at an angle of $55°$ from link L_4, and L_1 is rotating clockwise at 10 rad/s.

The first step in analysis is position analysis. First, the links whose positions are completely known or specified are laid out graphically. Thus in Fig. 6.16a, L_4 is drawn as a 3.5-in horizontal line AD, and L_1 is drawn as a line 1.2 in long from the left-hand end of L_4 upward at an angle of $55°$. It is known that link L_2 is inclined at $45°$ from L_1, but its length is unknown. It is therefore indicated as a dashed line p through point B at an angle of $45°$ from L_1 (at an angle of $\theta_2 = 10°$ from the horizontal), and its length is yet to

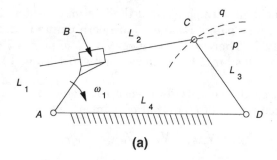

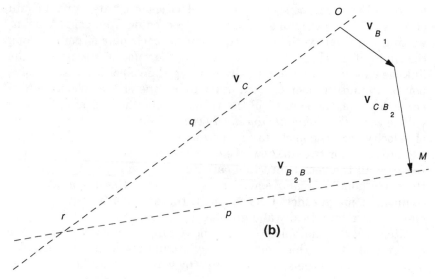

Figure 6.16 Graphic position and velocity analysis of a four-bar linkage having one sliding joint.

be determined. The length of L_3 is known but its direction is unknown, so an arc q of radius 1.625 in is drawn with its center at point D. This arc q intersects line p at point C, which is the location of the revolute joint connecting L_2 to L_3. Measurements of this drawing give values of $L_2 = 1.88$ in and $\theta_3 = 126.2°$ for the unknown quantities. (Note that the arc q can intersect line p at another point to the right of point C, so another solution also is possible.)

The velocity analysis will be performed by graphically computing the velocity of point C by two independent means and equating the results. The first computation will use an equation such as Eq. (6.13), which, because the velocity of grounded point A is zero, becomes

$$\mathbf{V}_C = \mathbf{V}_{B_1} + \mathbf{V}_{B_2 B_1} + \mathbf{V}_{CB_2} \tag{6.14}$$

The graphical construction corresponding to this equation is shown in Fig. 6.16b. First, the vector velocity $\mathbf{V}_{B_1}$ of point B_1 on link L_1 is drawn. The velocity of point B_1 is perpendicular to link L_1, and because that link is rotating clockwise, it is downward and to the right. Its magnitude is $V_{B_1} = \omega_1 L_1 = (10)(1.2)$ units/s. In Fig. 6.16b this vector is drawn to a scale, where 1 in represents 10 units/s, so the vector is drawn from the origin O as 1.2 in long and perpendicular to L_1.

The vector $V_{B_2 B_1}$ is unknown, so construction proceeds to the vector V_{CB_2}. This vector represents the velocity of point C relative to point B_2, and because both these points are attached to the same body (link L_2), this vector is downward and perpendicular to L_2 and has a magnitude of $V_{CB_2} = \omega_1 L_2 = (10)(1.88)$ units/s. This vector is drawn to scale as 1.88 in long in Fig. 6.16b, with its tail at the tip of vector V_{B_1}. The remaining vector $V_{B_2 B_1}$ on the right-hand side of Eq. (6.14) should be attached to the tip of the vector V_{CB_2}. However, although it is known that this vector $V_{B_2 B_1}$ (the slip velocity) must be parallel to L_2, its magnitude is unknown. Therefore, it will be represented initially as the dashed line p parallel to L_2.

The second computation notes that $V_C = V_{CD}$ because point D has a velocity of zero. Because both points C and D are attached to the same body (link L_3), the velocity of point C is perpendicular to link L_3 and has an unknown magnitude of $V_C = \omega_3 L_3$. This vector is therefore represented initially by the dashed line q drawn through the origin O in Fig. 6.16b.

The lines p and q intersect at a point r. Then it can be seen that if a vector from the origin O to point r is considered to be V_C and a vector from the tip of vector V_{CB_2} to point r is considered to be $V_{B_2 B_1}$, the two graphical vector sums (corresponding to the two paths from O to r) give the same value for the velocity vector of point C.

Measuring the length of $V_{B_2 B_1}$ (from point M to point r) in Fig. 6.16b and scaling the result gives $V_{B_2 B_1} = 63.8$ units/s. Thus it is seen that point B_2 on link L_2 is sliding leftward relative to point B_1 on link L_1 with a velocity of 63.8 units/s.

Measuring the length of V_C (from point O to point r) in Fig. 6.16b and scaling the result gives $V_C = 61.8$ units/s leftward. Because $V_C = \omega_3 L_3 = \omega_3 (1.625)$, computation gives $\omega_3 = 38.0$ rad/s. The angular velocity of L_3 is counterclockwise (positive) because the velocity of point C is leftward.

6.7 Acceleration Analysis of the General Four-Bar Linkage

Acceleration analysis of a linkage, particularly a linkage with sliding joints, usually will be performed with the aid of a computer. However, when some undesirable and/or unexpected acceleration is indicated by such analysis, graphical analysis often can provide visual insight into the sources and cures for such acceleration components. Acceleration analysis is performed using acceleration-difference relationships between successive points on the linkage and progressing around the linkage in a manner similar to that used for velocity analysis. As in the case of velocity analysis, when a sliding joint is encountered, care must be taken to define a point on each of the two bodies (links) involved at that joint. Thus, for the pair of links shown in Fig. 6.15, we may write the acceleration of point C_2 relative to point A_1 as

$$\mathbf{A}_{C_2A_1} = \mathbf{A}_{B_1A_1} + \mathbf{A}_{B_2B_1} + \mathbf{A}_{C_2B_2} \qquad (6.15)$$

Because points B_1 and A_1 are both on the same rigid body, the acceleration $\mathbf{A}_{B_1A_1}$ of point B_1 relative to point A_1 will consist of a normal (or centripetal) component parallel to L_1 and a tangential component perpendicular to L_1. Thus,

$$\mathbf{A}_{B_1A_1} = \mathbf{A}^{\,n}_{B_1A_1} + \mathbf{A}^{\,t}_{B_1A_1} \qquad (6.16)$$

where the magnitudes of the vectors on the right-hand side of this equation are given by $A^{\,n}_{B_1A_1} = \omega_1^2 L_1$ and $A^{\,t}_{B_1A_1} = \alpha_1 L_1$ (see Sec. 2.7).

Points B_1 and B_2 are on links that slide relative to each other, so as shown in Secs. 5.8 and 5.9, the acceleration of point B_2 relative to point B_1 not only will consist of the normal and tangential acceleration components but also will include a Coriolis component and a slip component. Thus,

$$\mathbf{A}_{B_2B_1} = \mathbf{A}^{\,n}_{B_2B_1} + \mathbf{A}^{\,t}_{B_2B_1} + \mathbf{A}^{\,c}_{B_2B_1} + \mathbf{A}^{\,s}_{B_2B_1} \qquad (6.17)$$

where the magnitudes and directions of the normal and tangential components are computed in the same manner as above. By choosing the points B_1 and B_2 to be coincident at the instant being analyzed, the distance between these two points is made zero, and, therefore, the first two terms on the right-hand side of Eq. (6.17) vanish. The slip component will be parallel to the direction of the slide, and the Coriolis component will be perpendicular to the direction of the slide. The slip component is the rate at which the slip velocity is changing, and the magnitude of the Coriolis component is given by

$$A^{\,c}_{B_2B_1} = 2\omega_2 V^{\,s}_{B_2B_1} \qquad \text{(see Sec. 5.8)}$$

Because points C_2 and B_2 are attached to the same rigid body, the last term on the right-hand side of Eq. (6.15) will consist of two components in a manner similar to that expressed by Eq. (6.16). Thus it is seen that acceleration analysis of general four-bar linkages can be performed by summing the acceleration differences between the joints, just as in the case of pin-jointed four-bar linkages in Sec. 4.9, but that when a sliding joint is encountered, the acceleration difference between the two points on the two connected links at that joint must be summed with the other acceleration differences between the joints. Such an analysis is demonstrated in Example 6.2.

Example 6.2: Graphical Acceleration Analysis of a Four-Bar Linkage Containing One Sliding Joint The linkage in Fig. 6.16a will be analyzed to determine the accelerations experienced by links L_2 and L_3 for the conditions of Example 6.1. For this analysis, the link L_1 will be assumed to be accelerating clockwise at 1000 rad/s^2 in addition to having a clockwise angular velocity of 10 rad/s. The analysis will use the technique of expressing the acceleration of point C by two independent means, in a manner similar to that used for velocity analysis in Example 6.1.

Because of the similarity between Figs. 6.15 and 6.16a, the acceleration of point C can be computed from an equation of the form of Eq. (6.15). Points A, D, and C are common to both the links that are joined at each of these points, so subscripts are not necessary for their symbols. The pivots at points A and D are grounded, so the accelerations of these points are zero. Then, when applied to this example, Eq. (6.15) becomes

$$\mathbf{A}_C = \mathbf{A}_{B_1} + \mathbf{A}_{B_2 B_1} + \mathbf{A}_{CB_2} \qquad (6.18)$$

The graphical analysis then starts by graphically summing the vectors on the right-hand side of this equation. The first term on the right-hand side of this equation consists of two components, so for this example Eq. (6.16) can be written as

$$\mathbf{A}_{B_1} = \mathbf{A}_{B_1}^{n} + \mathbf{A}_{B_1}^{t} \qquad (6.19)$$

so the two vectors on the right-hand side of Eq. (6.19) will be drawn in Fig. 6.17 starting from an origin O.

The normal component $\mathbf{A}_{B_1}^{n}$ is parallel to link L_1 and is directed from the point B_1 toward point A. Its magnitude is given by $A_{B_1}^{n} = \omega_1^2 L_1 = (10)^2(1.2) = 120$ units/s. Using a scale such that 1 in corresponds to 400 units/s^2, this vector is plotted 0.3 in downward and to the left from O in Fig. 6.17. The tangential component $\mathbf{A}_{B_1}^{t}$ is perpendicular to L_1, and because the angular acceleration of L_1 is clockwise, it is directed downward and to the right. Its magnitude is given by $A_{B_1}^{t} = \alpha_1 L_1 = (1000)(1.2) = 1200$ units/s^2. Then $\mathbf{A}_{B_1}^{t}$ is drawn from the tip of $\mathbf{A}_{B_1}^{n}$, with a length of 3.0 in as shown.

The second term on the right-hand side of Eq. (6.18) consists of four components, as indicated by Eq. (6.17). Because points B_1 and B_2 have been chosen to be coincident at the instant being analyzed, the first two terms on the right-hand side of Eq. (6.17) are zero, so that equation becomes

$$\mathbf{A}_{B_2 B_1} = \mathbf{A}_{B_2 B_1}^{c} + \mathbf{A}_{B_2 B_1}^{s} \qquad (6.20)$$

The first vector on the right-hand side of Eq. (6.20) is the Coriolis component, and it is perpendicular to the slide (i.e., perpendicular to L_2). Its magnitude is given by $A_{B_2 B_1}^{c} = 2\omega_1 V_{B_2 B_1} = 2(10)(63.8) = 1275.6$ units/s^2 (see Sec. 5.8). This vector is drawn from the tip of $\mathbf{A}_{B_1}^{t}$, as shown in Fig. 6.17. It is directed upward because B_2 is moving leftward away from B_1 while the slide is rotating with a clockwise angular velocity. The second vector on the right-hand side of Eq. (6.20) is the slip acceleration of point B_2 relative to B_1.

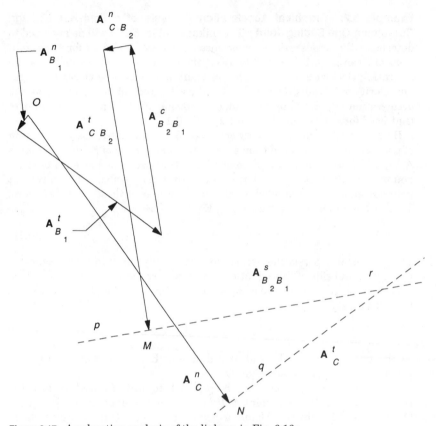

Figure 6.17   Acceleration analysis of the linkage in Fig. 6.16a.

Although it is known that this vector is parallel to the slide, its magnitude cannot be calculated at this time. It will be considered after all the calculatable vectors are drawn.

The last vector on the right-hand side of Eq. (6.18) relates the acceleration of two points that are both attached to L_2, so that vector consists of two components:

$$\mathbf{A}_{CB_2} = \mathbf{A}_{CB_2}^{n} + \mathbf{A}_{CB_2}^{t} \tag{6.21}$$

The vectors on the right-hand side of this equation are parallel and perpendicular, respectively, to L_2. Their magnitudes are $A_{CB_2}^{n} = \omega_2^2 L_2 = (10)^2(1.88)$ = 188 units/s² and $A_{CB_2}^{t} = \alpha_1 L_2 = (1000)(1.88) = 1880$ units/s², respectively. These two vectors are then drawn in Fig. 6.17 as shown.

The remaining vector component from Eq. (6.20) must now be accounted for. As mentioned previously, this component $\mathbf{A}_{B_2 B_1}^{s}$ is parallel to L_2, and its magnitude is unknown. Therefore, it is represented temporarily by a dashed line p through the tip of $\mathbf{A}_{CB_2}^{t}$ (point M) and parallel to L_2, as shown in Fig. 6.17.

All the vectors in Eq. (6.18) have now been accounted for. Next, draw another set of vectors which, when summed, also give the acceleration $\mathbf{A}_C$ of point C. It can be seen in Fig. 6.16a that because point D is stationary, the acceleration of point C is given by

$$\mathbf{A}_C = \mathbf{A}_C^n + \mathbf{A}_C^t \qquad (6.22)$$

where the vectors on the right-hand side of this equation are parallel and perpendicular, respectively, to L_3. Their magnitudes are $A_C^n = \omega_3^2 L_3 = (38.0)^2(1.625) = 2347$ units/s^2 and $A_C^t = \alpha_3 L_3$, respectively. The first of these two vectors is then drawn in Fig. 6.17 from the origin O as shown. The direction of the second of these two vectors is known, but its magnitude is not known. Therefore, it is represented temporarily by a dashed line q through the tip of $\mathbf{A}_C^n$ (point N) and perpendicular to L_3 as shown.

The two dashed lines p and q intersect at a point r. If a vector $\mathbf{A}_{B_2B_1}^s$ is drawn from point M at the tip of vector $\mathbf{A}_{CB_2}^t$ to point r and a vector $\mathbf{A}_C^t$ is drawn from point N at the tip of $\mathbf{A}_C^n$ to point r, then it is seen that the sum of vectors found by following vectors from tail to tip from point O to point r is the same regardless of which of the two paths in Fig. 6.17 is used. Both paths give a vector sum that is equal to $\mathbf{A}_C$.

Measuring vector lengths in Fig. 6.17 and recalling that 1 in represents 400 units/s^2 gives a value of $A_{B_2B_1}^s = (4.03)(400) = 1610$ units/s^2 and a value of $A_C^t = (3.20)(400) = 1280$ units/s^2. Thus we find that point B_2 is accelerating (in slip) rightward away from point B_1 at a rate of 1610 units/s^2. Also, because the magnitude of the tangential acceleration of point C is given by $A_C^t = \alpha_3 L_3 = \alpha_3(1.625) = 1280$ units/s^2, α_3 is computed to be 789 rad/s^2. This angular acceleration is clockwise because the tangential acceleration $\mathbf{A}_C^t$ of point C relative to point D is rightward.

Chapter 7

Multiloop Linkages and Other Linkages with More than Four Links

7.1 Introduction

Creative and optimal design of a mechanism is most likely to be achieved when engineers performing the design have many and varied tools and concepts at their disposal. The range of mechanism concepts available to engineers is not limited to linkages consisting of only four links such as presented in Chap. 6, nor is it limited to the many other variations of such four-bar linkages. Therefore, this chapter extends discussion to linkages having more than four links and includes procedures for synthesis and analysis of multiloop linkages. Such procedures include the use of *number synthesis*.

A linkage with more than four links may be either a single-loop linkage with more than one degree of freedom or a multiloop linkage. In either case, the synthesis and analysis techniques discussed in previous chapters are still applicable to the synthesis and analysis of *each individual loop*. In some cases the loops may be treated individually in sequence. In other cases the loops must be treated simultaneously. This chapter describes the techniques and equations used in such syntheses and analyses, and introduces computer techniques that are needed for the extensive computations required.

Because of the extensive computation involved, analysis of all but the simplest of multiloop linkages, as well as analysis for more than a very few positions of the linkages, ordinarily will be done by means of computer software. Therefore, the material in Secs. 7.4 and 7.5 will be of interest primarily to readers who may wish to write their own software for such analyses.

7.2 Synthesis of Multiloop Linkages

Often, a multiloop linkage or multi-degree-of-freedom linkage can be developed quite naturally by combining individual simpler loops and their features. Examples of such approaches are described in the first three subsections of this section. Occasionally, however, when developing a mechanism configuration, it is desired to make an extensive search of a large variety of link and joint combinations, hoping to find combinations that might be capable of providing the desired mechanism function. The most promising of these combinations are then subjected to more detailed analysis in order to choose an optimal candidate for detailed design. In such a search, the technique of *number synthesis,* which is described in the last subsection of this section, is a powerful tool.

Linkages with more than four links, considered as extensions from four-bar linkages

Often a basic motion that has *many or most* of the characteristics desired for a given application can be provided by a simple four-bar linkage. Then, by displacing one of the grounded pivots, the original motion can be altered to more closely match the desired motion. An example of such an approach is shown in Fig. 7.1. In this figure the

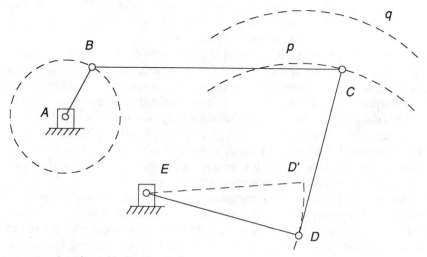

Figure 7.1 An adjustable four-bar linkage.

link *DE* is initially considered fixed in a position as shown. Thus, in effect, the pivot at *D* becomes a grounded pivot, and the links *AD*, *AB*, *BC*, and *CD* constitute a four-bar crank-rocker linkage. As the crank *AB* rotates through 360°, the rocker *CD* rocks back and forth, and point *C* travels back and forth along the dashed arc *p*.

By adjusting the angle of link *DE* to a new value such that pivot *D* is moved to location *D'*, the path along which pivot *C* travels back and forth is shifted upward to arc *q*. This is seen to be an adjustable four-bar linkage in which the location of pivot *D* is adjustable. It is, of course, really a two-degree-of-freedom five-bar linkage, but when considered as a four-bar linkage, it is easily visualized, synthesized, and analyzed. The link *DE* may itself be an output link of another linkage, so the total combination of links could constitute at least a seven-bar linkage.

Adjustable linkages are discussed in further detail in Sec. 8.3. In each of the examples in that section, adjustment is accomplished by altering the location of a pivot on a link. The motion of the adjustable mechanism can be easily visualized and synthesized with that pivot considered fixed. Adjusting the pivot location involves using additional links whose kinematics can be approximately treated separately from the kinematics of the linkage motion being adjusted. Of course, because approximating assumptions are made in this approach, the resulting total linkage kinematics should be analyzed or simulated to see whether the desired action has been provided. Usually such analysis or simulation will be done on a computer.

Another opportunity for extending a four-bar linkage by adding more links to provide an additional output occurs when some additional motion must be provided and that motion must be synchronized with the motion of the original four-bar linkage. In such a case, the motion of some point in the original four-bar linkage often can be "borrowed" and used as an input to move the added links. For example, in the mechanism shown in Fig. 4.28 and described in Procedure 4.9, the approximately elliptical path of point *C* in the figure can be used as an input motion to another set of links that will provide a function in addition to that of the original "walking beam" motion. Because the body *M* in the figure translates but does not rotate, all points on it follow identical paths. Figure 7.2 shows the motion of a point *D* on *M* being used as an input to links *DE* and EO_3F, which are used to raise the trailing edge of each transported part after its last advance to the left and during the return stroke of the walking beam.[1] When the pivot at

[1]This walking-beam system as actually implemented is required to perform *several* functions. By adding further links to perform these functions, the system became a nineteen-bar, two-degree-of-freedom linkage. Because all the links are tightly interconnected, the synchronism between the functions is always precisely maintained without need for adjustment.

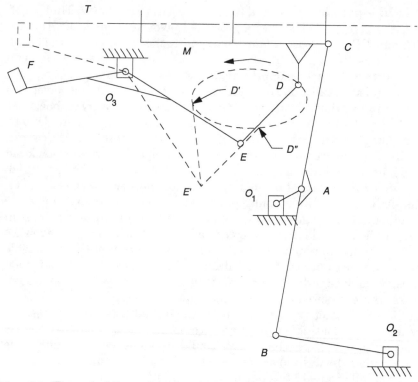

Figure 7.2 "Borrowing" the motion of a point in a walking-beam linkage to drive the motion of additional links.

D has progressed to position D', the links DE and EO_3F will be in the positions shown dashed. When the mechanism is in that position, point F on link EO_3F will be level with the table T. As the pivot D moves from point D' to point D'', point F on link EO_3F will rise above table level and then descend to that level. Thus the part that has been pushed by the walking beam into a position between positioning guides is lifted clear of the guides and can be transported perpendicular to the plane of this diagram before the next part is advanced to the left.

Linkages that produce parallel motion by using cognates and parallelogram loops

In the walking-beam system shown in Figs. 4.28 and 7.2 it was necessary to prevent rotation of body M that carries the fingers that push the parts which are to be transported. That is, *parallel motion* was necessary. As described in Procedure 4.9 and shown in Fig. 4.28, this was accomplished by adding portions of a cognate linkage. The result was a six-bar linkage that developed naturally from the four-bar linkage that was synthesized to produce the desired path of point C.

Section 8.7 describes two alternative methods for providing parallel motion. These methods are illustrated in Fig. 8.26. If a four-bar linkage were used to move the configuration shown in Fig. 8.26a around some orbital path, the total linkage would be a ten-bar linkage. The configuration in Fig. 8.26b is an eight-bar linkage. Both configurations result from adding links to a basic four-bar linkage that was designed to give the desired orbital motion of a point.

Synthesis of individual loops in multiloop linkages

An excellent example of the separation of the synthesis of a six-bar linkage into the synthesis of a four-bar linkage and of an additional pair of links is described in the paper by B. C. Vierstra and A. G. Erdman, entitled, "Redesign of a New Instant Camera Mirror Positioning Mechanism," in *Proceedings of the Fourth National Applied Mechanisms and Robotics Conference, Cincinnati, Ohio,* December 10–13, 1995.[2] This paper describes the designing of a linkage that would coordinate the folding and unfolding motions of a mirror body and a shutter housing in a folding camera. The resulting system is depicted schematically in Fig. 7.3a. The basic action required of this system consists of folding the mirror link L_2 down into a position parallel to the camera body link L_1 and between that camera body and the shutter housing link L_5, which is simultaneously folded down to a position parallel to L_1. Because of constraints on where pivots and links could be located and the space in which they could move, a six-bar, two-loop linkage of the general type shown was chosen. It then remained to choose the dimensions of the links such that the camera would fold and unfold without undue forces and without collisions between the mirror link and the shutter housing link.

The synthesis procedure that was followed started by assuming link L_5 to be stationary, as shown in Fig. 7.3b, while folding link L_1 upward and simultaneously folding link L_2 counterclockwise into the space between L_5 and L_1. That is, *kinematic inversion* was used in the synthesis. If link L_1 were stationary, point B on link L_2 would move along a circular arc such as p that is centered at pivot C. However, holding L_5 stationary and rotating L_1 and L_2, point B would follow a path such as that from location B through points B' and B'' to point B'''. The shape

---

[2]Contact Prof. A. G. Erdman, Mechanical Engineering Department, University of Minnesota, Minneapolis, regarding availability of this paper.

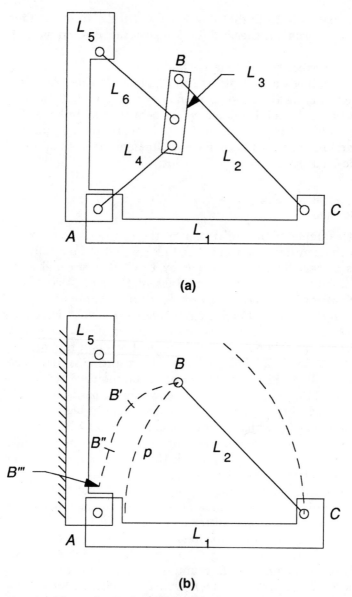

(a)

(b)

Figure 7.3 The synthesis of a camera mirror positioning mechanism.

of this latter path will depend on the relative rates of rotation of links L_1 and L_2 but must not be allowed to collide with the body of link L_5. Notice that this path of point B when referred to Fig. 7.3a is a path of a point on coupler L_3 of the four-bar linkage consisting of links L_4, L_3, L_6, and the grounded link L_5. Vierstra then chose as precision points positions B and B''' and two points along a promising coupler point path that satisfied the preceding collision constraint.

The computerized four-bar linkage synthesis package called LINCAGES 4® was then used to synthesize a four-bar linkage that would produce the assumed coupler point path. By iteratively adjusting the locations of the intermediate precision points on the trial path and synthesizing the corresponding linkages, a four-bar configuration was found that possessed link lengths, pivot locations, and link motions that satisfied all the camera design constraints. This is a very brief description of the very thorough synthesis and analysis that was performed. Nonetheless, it shows how a challenging task of synthesizing a six-bar, two-loop linkage was broken down into a manageable four-bar synthesis problem.

Other instances in which the synthesis of a multiloop linkage can be broken down into the synthesis of individual loops include the synthesis of dwell mechanisms and quick-return mechanisms. In the dwell mechanisms described in Sec. 8.8, an output rocker link is caused to have a displacement that dwells for some portion of its motion. This is done by connecting it through a second link to a pivot on a link that follows an appropriate path. This latter pair of links is thus assumed to be connected to a four-bar linkage that can be synthesized separately to produce the desired coupler point path by methods described in Chap. 8.

The quick-return mechanisms described in Secs. 5.3, 5.4, and 5.5 are usually connected to some additional links, the timing of whose motion is to have prescribed characteristics. Thus these mechanisms are also usually parts of multiloop mechanisms, and the quick-return-generating portions of these mechanisms can be synthesized separately.

An extreme application of dwell and quick-return mechanisms in a multiloop linkage is contained in U.S. patent no. 3,964,523. In the mechanism described in this patent, a dwell is produced in the manner described in Sec. 8.8 and shown in Fig. 8.27a. The coupler point path used to drive the output link, as shown in Fig. 8.27a, was generated by a four-bar crank-rocker linkage whose input crank was in turn driven by the output of a drag-link mechanism. The quick-return feature of the drag-link mechanism caused the duration of the dwell produced by the remainder of the linkage to be "streched out" so that the overall mechanism produced an approximate dwell with a duration of about 230° of rotation of the input to the drag-link mechanism. The resulting mechanism constituted an eight-bar, three-loop linkage. The drag-link loop, the coupler path-generating loop, and the final pair of links that provide the output rocker motion could each be synthesized individually and in sequence.

Number synthesis

The foregoing synthesis procedures consist of extending the features of a four-bar linkage or of combining the features of two or more four-bar linkages. This usually involves preconceived ideas of how the final mechanism will be configured. There are times when designers would like be more creative and therefore would like to make more comprehensive searches of possible combinations and numbers of links and joints. Such a search can make use of the procedure of determining the number and type of links and joints that will be required in a linkage to perform a required function, and this procedure is known as *number synthesis*. For a complete design, this number synthesis must be followed by a synthesis of the *dimensions* of the mechanism.

Number synthesis starts with knowledge of the number of degrees of freedom that must be provided, i.e., the number of input motions or forces that will be used to produce the desired output or outputs. As shown in Sec. 1.12, Greubler's formula relates the number of degrees of freedom to the numbers of links and joints. This formula is repeated here:

$$F = 3(L - 1) - 2J_1 - J_2 = 3(L - 1) - 2(J_1 + J_2) + J_2 \qquad (7.1)$$

where F is the number of degrees of freedom in the mechanism, L is the total number of links in the mechanism, and J_n is the number of joints each of which possesses n degrees of freedom.

As discussed in Sec. 1.12 and indicated in Fig. 1.21, each link in a linkage can be joined to other links by two, three, four, or more joints. *Half* of each joint is associated with each of the two links that are joined by that joint. Therefore, we can express the total number of joints as

$$J_1 + J_2 = 0.5(2B + 3T + 4Q + \cdots) \qquad (7.2)$$

where B = number of *binary* links (links that are joined to other links by two joints)

T = number of *ternary* links (links that are joined to other links by three joints)

Q = number of *quaternary* links (links that are joined to other links by four joints)

etc.

In most mechanisms, the links are joined to each other in such a manner that they form one or more loops. For example, consider Fig. 7.4a. In this figure eight links are joined together by eight joints to form a loop. It is seen that regardless of the form of the links, the num-

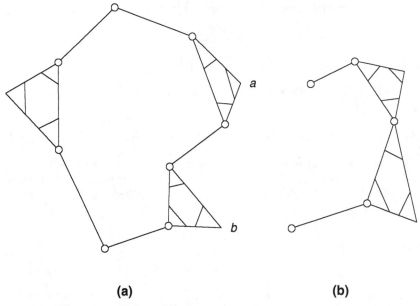

(a)   **(b)**

Figure 7.4 The joining of links to form loops.

ber of links in a single loop equals the number of joints *in that loop.*
Then we may write for that single loop

$$J_{\text{single}} = L_{\text{single}} \qquad (7.3)$$

where J_{single} and L_{single} are the numbers of joints and links in that single loop.

If the linkage consists of more than one loop, we speak of it as consisting of a number N of independent loops. These loops are not independent in the sense that they move independently of each other. Rather, a loop is considered independent of other loops if it contains one or more links that are not contained in any of the other loops. The linkage in Fig. 7.4a could be made into a linkage consisting of two independent loops by joining the chain in Fig. 7.4b to it at points a and b. The number of joints in the *chain* in Fig. 7.4b is one more than the number of links in that chain. Thus it can be seen that each time an independent loop is added to the mechanism, the number of joints increases by one more than does the number of links. Then we may write

$$N_{\text{added}} = J_{\text{added}} - L_{\text{added}} \qquad (7.4)$$

where N_{added} = number of independent loops added
  J_{added} = number of joints added
  L_{added} = number of links added to the original loop

Next, we may write the total number of joints and links as

$$J_{\text{total}} = J_{\text{single}} + J_{\text{added}} \qquad (7.5)$$

and

$$L_{\text{total}} = L_{\text{single}} + L_{\text{added}} \qquad (7.6)$$

Subtracting Eq. (7.6) from Eq. (7.5) and noting the equality in Eqs. (7.3) and (7.4) gives

$$J_{\text{total}} - L_{\text{total}} = J_{\text{added}} - L_{\text{added}} = N_{\text{added}} \qquad (7.7)$$

The total number of independent loops N is one more than N_{added} because of the existence of the original loop. The total number of joints is given by $J_{\text{total}} = J_1 + J_2$, and the total number of links is denoted by L. Then we may write

$$N = J_1 + J_2 - L + 1 \qquad (7.8a)$$

or

$$J_1 + J_2 = L + (N - 1) \qquad (7.8b)$$

From Eq. (7.8b) the expression $L + (N - 1)$ can be substituted for $J_1 + J_2$ in the right-hand side of Eq. (7.1), which, after a little rearranging, gives

$$F = L - 1 - 2N + J_2 \qquad (7.9)$$

Number synthesis can then be based on the use of Eqs. (7.9), (7.8b), and (7.2).

Procedure 7.1:   Number Synthesis

1. Decide how many input motions, forces, or torques are to be used to drive or actuate the mechanism that is to be devised. That is, how many inputs must the action of the mechanism depend on? This number will be F, the required number of degrees of freedom.

2. Decide how many sliding-pin joints, if any, will be tolerated in this mechanism. If the mechanism is to operate at high speed and/or with large forces, and if long life is to be required, such sliding-pin joints probably should be avoided. The number of such joints to be allowed will then be called J_2.

3. Choose a number N of independent loops of links in the linkage. Initially, try $N = 1$.

4. Use Eq. (7.9) to calculate L, the required number of links.

5. Use Eq. (7.8b) to calculate $J_1 + J_2$, the required total number of joints. From this total, subtract the assumed number J_2 from step 2 to get the value for J_1.

6. Using the value of L calculated in step 4, choose a combination of numbers B of binary, T of ternary, Q of quartenary, etc., links such that $B + T +$

$Q + \cdots = L$ and such that Eq. (7.2) gives the same value for $J_1 + J_2$ as calculated in step 5.

7. Sketch a schematic diagram of a linkage that conforms to the values of L, J_1, J_2, and N just calculated. Use symbols such as shown in Figs. 1.20 and 1.21.

8. Choose another combination of numbers B of binary, T of ternary, Q of quartenary, etc. links as in step 6, and repeat steps 6, 7, and 8 until the possible combinations have been exhausted.

9. Increase N, the number of loops, by 1, and repeat steps 4 through 9 until you feel that you have enough linkage configuration candidates for a creative exploration of the design possibilities.

10. The resulting collection of schematic diagrams tells how many links can be used and which links each link must be connected to. However, they do not indicate the size and shape of any link nor *where the joints are located on any link*. The final linkage that is designed may look very different from its corresponding schematic diagram as drawn in step 7. Exploration of size and shape requires techniques such as those discussed earlier in this chapter. Some iteration between these techniques and number synthesis can lead to insights that greatly aid in the search for an optimal design.

Comments and suggestions concerning number synthesis

The foregoing procedure will always indicate that a single-loop mechanism consisting of only binary links will provide the required number of degrees of freedom. Then why should the engineer bother to look for other combinations, as suggested in steps 8 and 9? The answer lies in the fact that these other combinations involve greater numbers of links and different kinds of links and thereby can provide a greater variety of *output* motions or forces. Readers will find that as they explore the approximate dimensional synthesis of a few of the configurations suggested by number synthesis, they will develop increasing facility in evaluating the suitability of subsequent configurations.

It also should be noted that the number of combinations possible for small numbers of degrees of freedom and small numbers of loops is quite limited, so the search process need not be extensive. This is shown in the following simple example.

Example 7.1: Number Synthesis of a Single-Degree-of-Freedom Linkage

For the number synthesis of a single-degree-of-freedom linkage that uses *no* two-degree-of-freedom joints (i.e., $J_2 = 0$), Eq. (7.9) becomes

$$1 = L - 1 - 2N + 0 \qquad \text{or} \qquad L = 2 + 2N \qquad (7.10)$$

Initially choosing $N = 1$ results in a single-loop linkage having four links (i.e., $L = 4$). From Eq. (7.8b), $J_1 + J_2 = J_1 = 4$. In Eq. (7.2) we can choose the four links to be binary links ($B = 4$) so that $J_1 + J_2 = J_1 = 4$, just as computed in Eq. (7.8b). No other combination of link types will cause Eq. (7.2) to give the same number of joints as computed in Eq. (7.8b). This linkage is a pin-jointed four-bar linkage.

Next, choosing $N = 2$ to give a two-loop linkage, when substituted into Eq. (7.10), gives $L = 6$. Then Eq. (7.8b) gives $J_1 + J_2 = J_1 = 7$. Two ternary links ($T = 2$) and four binary links ($B = 4$), when entered into Eq. (7.2), also give $J_1 + J_2 = J_1 = 7$, so this is a possible combination of links and joints for this example. The two useful arrangements of this combination of links are shown in Fig. 7.5. In Fig. 7.5a the two ternary links are joined together by the joint at A, and this configuration is called a *Watt six-bar linkage*. It can be used either with one of the binary links grounded (and often referred to as a *Watt I six-bar linkage*) or with one of the ternary links grounded (and often referred to as a *Watt II six-bar linkage*).

In the configuration shown in Fig. 7.5b, the two ternary links are *not* joined directly together. This type of linkage is known as a *Stephenson six-*

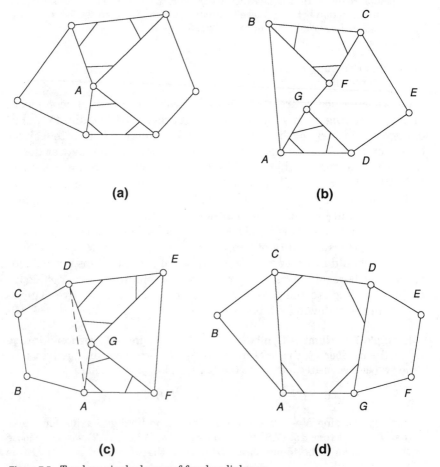

(a)   (b)

(c)   (d)

Figure 7.5   Two-loop single-degree-of-freedom linkages.

bar linkage. Binary links *AB* and *FG* are each joined to both ternary links. If the linkage is used with either of these binary links grounded, the linkage is often referred to as a *Stephenson I six-bar linkage.* Binary links *CE* and *DE* are each joined to one ternary link and to one binary link. If the linkage is used with either of these binary links grounded, the linkage is often referred to as a *Stephenson II six-bar linkage.* If the linkage is used with either of the ternary links grounded, the linkage is often referred to as a *Stephenson III six-bar linkage.*

It is found that the only six-bar single-degree-of-freedom linkages without sliding-pin joints are the Watt six-bar linkage and the Stephenson six-bar linkage. Although the joints are shown as pin (revolute) joints, any or all of them could be single-degree-of-freedom sliding joints (not sliding-*pin* joints) and the same freedom would exist.

Figure 7.5c indicates another possible arrangement of the six bars. Note that the two ternary links, when joined together by joint *G* and by binary link *EF*, form a single rigid body that could be represented by a link *AD*, which is shown dashed. This configuration is thus no more than the equivalent of a pin-jointed four-bar linkage. If a quaternary link is chosen as one of the six links and entered into Eq. (7.2), it is found that the equations are all satisfied. However, when the schematic diagram of such a combination is drawn, it appears as in Fig. 7.5d. There it can be seen that the quaternary link *ACDG* and binary links *AB* and *BC* are interconnected in such a manner as to constitute a single rigid body, so this total configuration also is equivalent to a pin-jointed four-bar linkage. Analogous arguments hold even if some of the joints are sliding joints.

Readers are encouraged to explore the nature of linkages that result from increasing the number *N* of loops in this example to three. Such linkages will have eight links and ten joints.

7.3 Position Analysis of Linkages with Many Links

The analysis of linkages containing more than four links can become arduous, and consequently, it is usually performed with the aid of a computer. Large computer-aided design (CAD) packages usually will assemble a collection of links when told which links are to be joined to which other links. Such an assembly drawing indicates the positions of the links and thus provides a position analysis. Large CAD packages also frequently include an added kinematic-analysis package that can provide velocity and acceleration information. Dynamic simulation packages also will assemble links to form linkages. In addition, these simulation packages can provide velocity, acceleration, force, and torque information. Some inconvenience can occur in the use of these dynamics packages because the vibration frequencies of the links as simulated can cause initial transients in the simulation results and/or can require short integration intervals and long computation times.

The algorithms for position, velocity, and acceleration analysis of a linkage consisting of many links are basically quite simple, and adventurous readers may wish to write a program using such algorithms. All the position, velocity, acceleration, and coupler point path data in this book were produced using a limited form of such a program that was written by the author. The algorithms for this program are described in papers by Eckhardt,[3] and a brief summary of these algorithms is given in the remainder of this chapter.

A linkage with many links can consist of a single loop of links that has several degrees of freedom, or it may consist of two or more loops of links. Inasmuch as a multiloop linkage can be considered to consist of a collection of interconnected individual loops, the analysis of such individual loops will be considered first.

Link positions in an individual link loop

As in the case of four-bar linkages, the position analysis of more complex linkages is facilitated by considering the links to be represented by vectors. Each such representative vector extends from one joint on a link to another joint on the same link. In analysis of four-bar linkages it was found convenient to represent links that are joined to ground by vectors whose tails were located at the respective grounded joints. When tracing around such a four-bar vector loop from joint to joint, the path traced sometimes proceeded in the direction indicated by a vector and sometimes in the opposite direction from that indicated by a vector. In a linkage with many links, it is often more convenient to choose vectors that are all directed such that tracing around the loop follows the direction of each and every vector. Such a choice is shown in Fig. 7.6a, where tracing around the loop in a clockwise direction follows each successive vector from tail to head.

**Graphical position analysis
of an individual loop**

As can be seen in Fig. 7.6a, if the loop is to be closed, the sum of all the vectors must be zero. That is, for this case

$$\sum_{i=1}^{7} \mathbf{L}_i = 0 \qquad (7.11)$$

[3]Eckhardt, H. D., The use of a closed form algorithm in planar multi-loop linkage analysis, 1984, ASME paper 84-DET-136; H. D. Eckhardt, Speed enhancement of multi-loop linkage analysis by the use of a simple closed form algorithm, in *Proceedings of the Ninth Applied Mechanics Conference, Kansas City, MO*, October 1985, paper no. III-B-VI. (Contact the author at 27 Laurel Drive, Lincoln, MA 01773 concerning availability of these papers.)

As was shown in analysis discussions in previous chapters, a vector loop closure equation such as this can be broken down into two scalar equations that can be solved if they contain two and only two unknown variables. Section 6.5 showed that these unknowns fall into four cases. In Case 1, the length and angle of a single vector are unknown, and the vector equation can be solved simply by graphically summing the

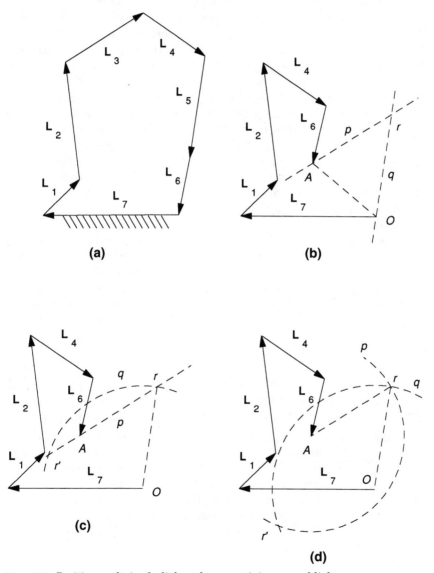

Figure 7.6   Position analysis of a linkage loop containing several links.

known vectors and noting that the vector required to complete a closed loop is the unknown vector.

Figure 7.6b illustrates case 2a, in which the *lengths* of vectors $\mathbf{L}_3$ and $\mathbf{L}_5$ are unknown. As shown, the known vectors $\mathbf{L}_7$, $\mathbf{L}_1$, $\mathbf{L}_2$, $\mathbf{L}_4$, and $\mathbf{L}_6$ are drawn tail to head from an origin O, forming a chain from that origin to point A. A vector from point A to O represents the required sum $\mathbf{L}_3 + \mathbf{L}_5$ of the vectors of unknown length. As in Sec. 6.5, a line p parallel to the known direction of $\mathbf{L}_3$ is drawn through point A, and a line q parallel to the known direction of $\mathbf{L}_5$ is drawn through point O. These lines intersect at point r. It is then seen that a vector $\mathbf{L}_3$ from A to r and a vector $\mathbf{L}_5$ from r to O would have the specified respective directions and would satisfy the vector loop closure equation (Eq. 7.11).

Figure 7.6c illustrates the solution for case 2b, where the length of $\mathbf{L}_3$ is unknown and the angle of $\mathbf{L}_5$ is unknown. In a manner analogous to that in Sec. 6.5, a line p parallel to the direction of $\mathbf{L}_3$ is drawn through point A and an arc q of radius L_5 is drawn centered at point O to produce intersections at r and r'. Again, it is then seen that a vector $\mathbf{L}_3$ from A to r would have the specified direction and a vector $\mathbf{L}_5$ from r to O would have the specified length, and together they would satisfy the vector loop closure equation [Eq. (7.11)]. Note that there is also an alternative solution, as indicated by the intersection r'.

Figure 7.6d illustrates the solution for case 2c, where the angles of $\mathbf{L}_3$ and $\mathbf{L}_5$ are unknown. In a manner analogous to that used in Sec. 6.5, arc p and q of radii equal to lengths L_3 and L_5 are drawn centered at points A and O, respectively. These arcs intersect at points r and r'. It is again seen that a vector $\mathbf{L}_3$ from A to r and a vector $\mathbf{L}_5$ from r to O would have the specified lengths, and together they would satisfy the vector loop closure equation [Eq. (7.11)]. Note that there is also an alternative solution, as indicated by the intersection at r'.

Analytical (computer-aided) position analysis of an individual loop

Just as in the case of graphical analysis, analytical position analysis of a linkage loop is based on a vector loop closure equation such as Eq. (7.11). The summation implied in this equation involves repeated operations that are easily performed by subroutines that are called repeatedly in a computer program or by stored "macros" on a pocket calculator. The algorithms used in these subroutines are based on equations that are easily derived from the vector loop closure equation by using the complex number notation for vectors. Then that equation becomes

$$\sum \mathbf{L}_i = \sum L_i e^{j\theta_i} = 0 \tag{7.12}$$

The summation includes one or two vectors for which length and/or angle are unknown. Therefore, define the sum of the vectors for which lengths and angles are completely known as a vector denoted by $\mathbf{L}_s$ (known as the *known-sum vector*), where

$$\mathbf{L}_s = L_s e^{j\theta_s} = \overset{*}{\underset{i}{\sum}} L_i e^{j\theta_i} \qquad (7.13)$$

and where the asterisk above the summation sign denotes that the sum does *not* include vectors for which the angles and/or lengths are unknown. In the subsequent discussion, the length and angle and x and y components of this known-sum vector will always be assumed to have been computed as a first step in the computation. The sum of *all* vectors as indicated in Eq. (7.12) must include the one or two vectors that each has an unknown angle and/or length. Therefore, Eq. (7.12) may be written

$$\sum \mathbf{L}_i = L_s e^{j\theta_s} + L_m e^{j\theta_m} + L_n e^{j\theta_n} = 0 \qquad (7.14)$$

where the terms containing subscripts m and n denote vectors whose length and/or angle are unknown and must be solved for. If the linkage configuration corresponds to Case 1 described in the foregoing graphical analysis (only one vector contains the unknowns), Eq. (7.14) will not contain the term containing the subscript n. If the linkage configuration corresponds to Case 2 described in the foregoing graphical analysis, Eq. (7.14) will contain both the term containing subscript m and that containing n. If values for the variables having unknown values are entered by the user as U (for "unknown"), a computer program to perform the linkage analysis can be programmed to recognize which terms cannot be summed to form the known-sum vector $\mathbf{L}_s$ but must later be solved for. Analytical solutions for link positions must be applicable to all four of the cases treated in the graphical analyses just described.

For *Case 1,* the length and angle of one vector are unknown, so only one vector will be excluded from the summation. Therefore, the vector that has subscript n will be omitted from Eq. (7.14). Then the equation can be written

$$L_s e^{j\theta_s} + L_m e^{j\theta_m} = 0 \qquad (7.15)$$

Using Euler's formula, this complex variable equation can be written as two real equations:

$$L_s \cos \theta_s = \overset{*}{\underset{i}{\sum}} L_i \cos \theta_i = -L_m \cos \theta_m \qquad (7.16)$$

and

$$L_s \sin \theta_s = \overset{*}{\underset{i}{\sum}} L_i \sin \theta_i = -L_m \sin \theta_m \qquad (7.17)$$

The summations in these equations are the sums of known values. The terms on the right-hand sides are, respectively, negatives of the x and y components of the unknown link vector. The length and angle of that link vector are then easily computed.

For *Case 2*, the algorithm is also based on Eq. (7.14). However, in all subcases of Case 2, the two unknown quantities are contained in two vectors. Then, using Euler's formula, Eq. (7.14) can be written as the two real equations:

$$L_s \cos \theta_s = \sum_i^* L_i \cos \theta_i = -L_m \cos \theta_m - L_n \cos \theta_n \qquad (7.18)$$

and

$$L_s \sin \theta_s = \sum_i^* L_i \sin \theta_i = -L_m \sin \theta_m - L_n \sin \theta_n \qquad (7.19)$$

For *Case 2a*, the *lengths* L_m and L_n of the two vectors are unknown, so once the known vectors are summed as indicated on the left-hand sides of these equations, Eqs. (7.18) and (7.19) are linear equations in those two variables L_m and L_n, so they are readily solved for these unknown values.

For *Case 2b*, the length of one vector (L_m, for example) and the angle of another vector (θ_n, for example) are unknown. For this case, it is convenient to multiply each complex term in Eq. (7.14) by $e^{-j\theta_m}$, which then gives

$$L_s e^{j(\theta_s - \theta_m)} + L_m + L_n e^{j(\theta_n - \theta_m)} = 0 \qquad (7.20)$$

The two real equations corresponding to this equation are

$$L_s \cos(\theta_s - \theta_m) = -L_m - L_n \cos(\theta_n - \theta_m) \qquad (7.21)$$

and

$$L_s \sin(\theta_s - \theta_m) = -L_n \sin(\theta_n - \theta_m) \qquad (7.22)$$

The only unknown in Eq. (7.22) is θ_n, so it is easily solved for by using only that equation. Notice that because two values of $(\theta_n - \theta_m)$ give the same value for $\sin(\theta_n - \theta_m)$, there are two solutions to this equation. The computer program can be instructed to choose the appropriate solution by having the user enter an *estimate* of the correct value. Then, knowing θ_n, the value of L_m is easily computed from Eq. (7.21).

For *Case 2c*, the *angles* of the two vectors that are subscripted m and n are unknown. For this case it is convenient to multiply each complex term in Eq. (7.14) by $e^{-j\theta_s}$, which, after rearranging terms, gives

$$-L_s = L_m e^{j(\theta_m - \theta_s)} + L_n e^{j(\theta_n - \theta_s)} \qquad (7.23)$$

This equation corresponds to the two real equations

$$-L_s = L_m \cos(\theta_m - \theta_s) + L_n \cos(\theta_n - \theta_s) \qquad (7.24)$$

and $$0 = L_m \sin(\theta_m - \theta_s) + L_n \sin(\theta_n - \theta_s) \qquad (7.25)$$

Transpose the first term on the right-hand sides of Eqs. (7.24) and (7.25) to the left-hand sides, square both sides of the resulting equations, and add these squared equations. After simplifying and rearranging terms, the result is

$$L_s^2 + L_m^2 + 2L_mL_s \cos(\theta_m - \theta_s) = L_n^2 \qquad (7.26)$$

Then transpose the last term on the right-hand sides of Eqs. (7.24) and (7.25) to the left-hand sides, square both sides of the resulting equations, and add these squared equations. After simplifying and rearranging terms, the result is

$$L_s^2 + L_n^2 + 2L_nL_s \cos(\theta_n - \theta_s) = L_m^2 \qquad (7.27)$$

Each of Eqs. (7.26) and (7.27) will be recognized as the trigonometric law of cosines. Because all the lengths in these equations are known, the angles $(\theta_m - \theta_s)$ and $(\theta_n - \theta_s)$ can be computed easily.

Note, however, that there are two solutions to each of these equations. To see how to specify the proper ones to use in computing θ_m and θ_n, refer to Fig. 7.7. In Fig. 7.7a the known-sum vector $\mathbf{L}_s$ is shown. The vector $\mathbf{L}_m$ is then drawn from the tip of $\mathbf{L}_s$, and the vec-

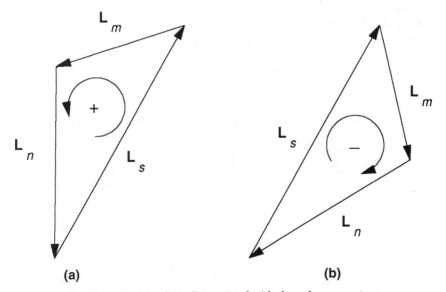

(a)   **(b)**

Figure 7.7 Specifying the "circulation" associated with the unknown vectors.

tor $\mathbf{L}_n$ is drawn from the tip of $\mathbf{L}_m$ to complete a closed triangle. Tracing around this triangle in the direction of the arrows follows a path that is counterclockwise, and we will refer to this as a *positive circulation*. If the vectors $\mathbf{L}_m$ and $\mathbf{L}_n$ had been drawn in the same order but in the orientations shown in Fig. 7.7b, the circulation would have been negative. These two triangles correspond to the two solutions to the loop closure equation for this case. Engineers performing an analysis will have beforehand knowledge of the rough orientation of the unknown vectors $\mathbf{L}_m$ and $\mathbf{L}_n$ well enough to make a rough sketch and determine which sign of this circulation is appropriate for the linkage being analyzed. The sign of this circulation cannot change without having the linkage pass through a condition of *lockup* or indeterminacy.

A position-analysis computer program can then, when it senses the existence of case 2c, ask the user for the sign of the circulation. Then the sign that should be assigned to $(\theta_m - \theta_s)$ is the same as the sign of that circulation, and the sign that should be assigned to $(\theta_n - \theta_s)$ is the opposite of the sign of that circulation.

Then, because the value θ_s of the angle of the known-sum vector has been computed previously, the values of θ_m and θ_n can be computed from $(\theta_m - \theta_s)$ and $(\theta_n - \theta_s)$.

7.4 Position Analysis of Multiloop Linkages

Because of the extensive computations involved, analysis of all but the simplest of multiloop linkages and analysis for more than a very few positions of the linkages ordinarily will be done by means of computer software. Therefore, the material in Secs. 7.4 and 7.5 will be of interest primarily to readers who may wish to write their own software for such analysis.

Multiloop linkages, as the name implies, can be represented by a number of closed vector loops. For example, the Stephenson III linkage in Fig. 7.8a can be represented by the two vector loops in Fig. 7.8b and c. The vector loop in Fig. 7.8b will be referred to as loop number 1, and that in Fig. 7.8c will be referred to as loop number 2. The first subscript on each vector indicates the loop number, and the second subscript indicates the number of the vector in the loop.

If this linkage is to be driven by the motion of the link represented by vector $\mathbf{L}_{11}$, and if link ADG in Fig. 7.8a is grounded, the angles of vectors $\mathbf{L}_{11}$ and $\mathbf{L}_{14}$ will be known, and the angles of the two vectors $\mathbf{L}_{12}$ and $\mathbf{L}_{13}$ can be found by means already described. Note that links CD and BCE in Fig. 7.8a are each represented by the two vectors ($\mathbf{L}_{13}$ and $\mathbf{L}_{22}$ in the case of link CD and $\mathbf{L}_{12}$ and $\mathbf{L}_{23}$ in the case of link BCE). It is seen that $\theta_{22} = \theta_{13} + 180°$ and that, similarly, $\theta_{23} = \theta_{12} + \alpha$. Then, be-

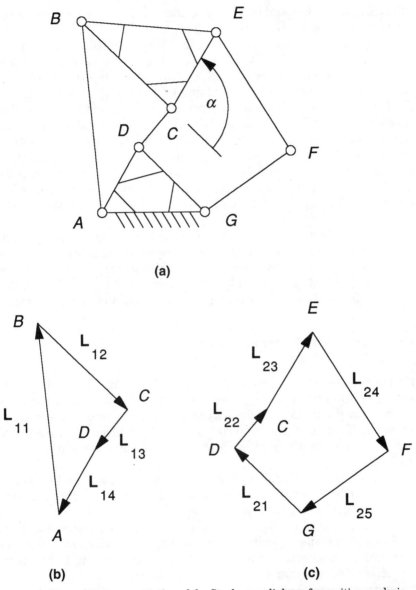

(a)

(b)   (c)

Figure 7.8 Vector loop representation of the Stephenson linkage for position analysis.

cause the link represented by vector L_{21} is grounded, θ_{21} is known, and the only unknowns in loop number 2 are θ_{24} and θ_{25}; values for these variables can be solved for by means that were discussed in Sec. 7.3.

It is thus seen that if the input to the linkage in Fig. 7.8a is applied to link AB, the positions of the linkage can be analyzed by sequentially analyzing the individual loops. The algorithm for doing so is described in the second reference in footnote 3.

If the input to the mechanism in Fig. 7.8a is applied to a link in loop number 2, say, link FG, it is seen that the only link angles in that loop which are known are θ_{21} and θ_{25}. The remaining three unknown link angles cannot be solved for using only the loop closure equation for loop number 2. Because this linkage is a one-degree-of-freedom linkage, its position analysis *can* be performed by *simultaneously* solving the scalar equations corresponding to the loop closure equations for the two loops. These scalar equations are nonlinear, so iterative procedures are required for their solution. An algorithm that uses individual loop-analysis algorithms such as described in Sec. 7.3 is described in the second reference in footnote 3. That algorithm is also iterative.

The mechanism indicated in Fig. 7.8a uses only revolute joints, and the algorithms described in the references in footnote 3 also consider only revolute joints. However, those algorithms are easily augmented by the incorporation of the algorithms described in Sec. 7.3 so that they can be used to analyze linkages that also include sliding joints.

7.5 Velocity and Acceleration Analyses of Multiloop Linkages

Because of the extensive computations involved, analysis of all but the simplest of multiloop linkages and analysis for more than a very few positions of the linkages ordinarily will be done by means of computer software. Therefore, the material in Sec. 7.5 will be of interest primarily to readers who may wish to write their own software for such analysis.

After the position analysis has been performed, all the link-vector lengths and link-vector angles are known. The equations that are written for use in velocity analysis are then found to be constant-coefficient linear equations in the velocities. After the velocity analysis has been performed, all the velocities are known. The equations that are written for use in acceleration analysis are then found to be constant-coefficient linear equations in the accelerations. All these linear velocity equations can then be solved simultaneously by matrix computations, regardless of the number of link loops involved and regardless of the nature and number of inputs. The same is true for the acceleration equations.

Multiloop velocity analysis

The velocity-analysis equations, and thus the algorithms that may be used in either manual or computerized analysis, are most easily derived by differentiating the vector loop closure equation [Eq. (7.12)] with respect to time. The result for loop number i is

$$\frac{d}{dt}\left(\sum_k L_{ik} e^{j\theta_{ik}}\right) = \sum_k \dot{L}_{ik} e^{j\theta_{ik}} + \sum_k j\dot{\theta}_{ik} L_{ik} e^{j\theta_{ik}} = 0 \qquad (7.28)$$

The variables in this equation have been given two subscripts to accommodate the existence of a number of loops as well as a number of links in each loop. The first subscript denotes the number of the loop, and the second denotes the number of the link in the loop. Noting that $\dot{\theta}_i = \omega_i$, Eq. (7.28) becomes

$$\sum_k \dot{L}_{ik} e^{j\theta_{ik}} + \sum \omega_{ik} L_{ik} e^{j(\theta_{ik} + \pi/2)} = 0 \qquad (7.29)$$

This equation is equivalent to the following two equations in real variables:

$$\sum_k \dot{L}_{ik} \cos\theta_{ik} - \sum_k \omega_{ik} L_{ik} \sin\theta_{ik} = 0 \qquad (7.30)$$

and

$$\sum_k \dot{L}_{ik} \sin\theta_{ik} + \sum_k \omega_{ik} L_{ik} \cos\theta_{ik} = 0 \qquad (7.31)$$

All the lengths L_{ik} and all the angles θ_{ik} in these equations are known from the position analysis. A pair of equations such as Eqs. (7.30) and (7.31) pertain to each loop in the linkage. The variables $\dot{L}_{ik}$ and ω_{ik} in these equations can be related to each other not only by these equations but also by formulas that the user enters into the computer (or computation) in order to interconnect and constrain the linkage. These formulas can take the forms of Eqs. (7.32) and (7.33):

$$L_{ik} = L_{pq} + F_{ik} \qquad (7.32)$$

where F_{ik} is the length that link L_{ik} would have if it were not variable (in which case L_{pq} would be omitted from the formula) and which is an additive constant length that relates the length of link L_{ik} to the length of link L_{pq} if both L_{ik} and L_{pq} vary in length in unison.

$$\theta_{ik} = G_{ik}\theta_{pq} + H_{ik} \qquad (7.33)$$

where G_{ik} is a multiplying or proportionality constant and H_{ik} is an additive angle, both of which relate the angle θ_{ik} to the angle θ_{pq}. For example, as discussed in Sec. 7.4 and shown in Fig. 7.8, $G_{22} = 1.0, p = 1$, $q = 3$, and $H_{22} = 180$ because $\theta_{22} = \theta_{13} + 180°$.

In cases where θ_{ik} is constant, the first term on the right-hand side of Eq. (7.33) is omitted. In cases where θ_{ik} is "driven," the θ_{pq} in the first term on the right-hand side of Eq. (7.33) is replaced by a drive angle

D_{pq}. In cases where θ_{ik} is an unknown that must be solved for, this is indicated by entering the symbol U as the entire right-hand side of Eq. (7.33).

Then each and every θ_{ik} can be expressed as a constant, as a drive angle, as an unknown angle, or as another link angle times a multiplying constant. In this last case, the "other" angle can, in turn, be expressed as a constant, as a drive angle, as an unknown angle, or as another link angle times a multiplying constant. Thus, eventually, each and every angle can be expressed as a constant, as a drive angle, or as an unknown angle times a multiplying constant.

Differentiating Eq. (7.33) with respect to time gives

$$\omega_{ik} = G_{ik}\omega_{pq} \tag{7.34}$$

By following the same argument used relative to Eq. (7.33), it is found that eventually each and every angular velocity is either zero, can be expressed as a drive rate, or can be expressed as a multiplying constant times an unknown angular velocity that must be solved for. Similarly, by considering Eq. (7.32), it is found that eventually each and every slip velocity $\dot{L}_{ik}$ is either zero, can be expressed as a drive rate, or can be expressed as a multiplying constant times an unknown slip velocity that must be solved for.

Then the simultaneous Eqs. (7.30) and (7.31) can be written with the sums of all the unknown terms on the left-hand side and all the known terms on the right-hand side. In matrix form, this set of equations becomes

$$[\mathbf{P}] \times [\mathbf{U}_V] = [\mathbf{Q}_V] \tag{7.35}$$

where $\mathbf{P}$ is a matrix consisting of two rows per linkage loop. Each element of this matrix will either be zero or will contain a combination of one or more of the following: a constant G_{pq}, a sine or cosine of a link angle, or a link length. $\mathbf{U}_V$ is a column vector, each of whose elements is an unknown angular velocity or slip velocity, and $\mathbf{Q}_V$ is a column vector of elements, each of which is a number which is the sum of known terms on the right-hand side of the appropriate rearranged Eqs. (7.30) and (7.31) for the corresponding row in $\mathbf{P}$. The unknown angular and slip velocities in $\mathbf{U}$ are then computed from

$$[\mathbf{U}_V] = [\mathbf{P}]^{-1} \times [\mathbf{Q}_V] \tag{7.36}$$

Multiloop acceleration analysis

The acceleration-analysis equations, and thus the algorithms that may be used in either manual or computerized analysis, are most easily derived by differentiating the vector velocity loop equation (Eq. 7.28) with respect to time. The result for loop number i is

$$\frac{d}{dt}\left(\sum_k \dot{L}_{ik}e^{j\theta_{ik}} + \sum_k j\dot{\theta}_{ik}L_{ik}e^{j\theta_{ik}}\right) = \sum_k \ddot{L}_{ik}e^{j\theta_{ik}} + \sum_k jL_{ik}\alpha_{ik}e^{j\theta_{ik}}$$

$$- \sum_k L_{ik}\omega_{ik}^2 e^{j\theta_{ik}} + \sum_k j2\omega_{ik}\dot{L}_{ik}e^{j\theta_{ik}} = 0 \quad (7.37)$$

This equation is equivalent to two equations in real variables:

$$\sum_k \ddot{L}_{ik}\cos\theta_{ik} - \sum_k L_{ik}\alpha_{ik}\sin\theta_{ik} - \sum_k L_{ik}\omega_{ik}^2\cos\theta_{ik}$$

$$- \sum_k 2\omega_{ik}\dot{L}_{ik}\sin\theta_{ik} = 0 \quad (7.38)$$

and

$$\sum_k \ddot{L}_{ik}\sin\theta_{ik} + \sum_k L_{ik}\alpha_{ik}\cos\theta_{ik} - \sum_k L_{ik}\omega_{ik}^2\sin\theta_{ik}$$

$$+ \sum_k 2\omega_{ik}\dot{L}_{ik}\cos\theta_{ik} = 0. \quad (7.39)$$

Previously performed position and velocity analyses will have determined values for all the symbols in these equations except for slip accelerations $\ddot{L}_{ik}$ and the angular accelerations α_{ik}. In a manner analogous to that used in the case of velocity analysis, it can be shown that each of these variables is either zero, can be expressed as a drive slip acceleration or drive angular acceleration, or can be expressed as an unknown slip acceleration or a multiplying constant times an unknown angular acceleration. When all the terms containing unknown values are transposed to the left-hand sides of these equations and the remaining terms are placed on the right-hand sides of the equations, the set of simultaneous equations for all the loops can be written in matrix form as

$$[\mathbf{P}] \times [\mathbf{U}_A] = [\mathbf{Q}_A] \quad (7.40)$$

where $\mathbf{P}$ is the same coefficient matrix as that which appeared in Eqs. (7.35) and (7.36), $\mathbf{U}_A$ is a column vector each of whose elements is an unknown angular acceleration or slip acceleration, and $\mathbf{Q}_A$ is a column vector of elements each of which is a number that is the sum of known terms on the right-hand side of the appropriate rearranged Eq. (7.38) or (7.39) for the corresponding row in $\mathbf{P}$.

The unknown slip and angular accelerations can then be solved for by using

$$[\mathbf{U}_A] = [\mathbf{P}]^{-1} \times [\mathbf{Q}_A] \quad (7.41)$$

Comments on multiloop velocity and acceleration analyses

Calculating the values for the unknown velocities and accelerations using Eqs. (7.36) and (7.41) implies considerable computation. However, many of these computations are quite repetitive so that they can be performed by repeatedly using short sections of computer code. Also, by suitably grouping the terms in $\mathbf{Q}_V$ and in $\mathbf{Q}_A$, terms using similar coefficients can be computed in sequence, and the need to reevaluate trigonometric functions can be minimized.

The analysis procedures just described are essentially the same as those described in the references in footnote 2 but with some terms added to account for the addition of the ability to analyze linkages with sliding joints. The fact that analyses such as these consider the links to be represented only by vectors allows the analyses to be used very early in the design synthesis procedure when little or nothing is known about the size or shape of the parts of the mechanism being synthesized.

Chapter

8

Qualitative Approaches to Linkage Synthesis

8.1 Introduction

Relatively complicated motions of linkages often may be viewed as combinations of simpler, more easily understood motions. Mechanisms that produce these simpler motions can then be synthesized and combined to produce the more complicated motions. This chapter explores by example means by which simply produced displacements and path curvatures may be added, subtracted, multiplied, magnified, diminished, and otherwise combined and distorted to produce a desired mechanism action. Also illustrated are some uses of quadrature motions in path generation, in timing, and in phase shifting. Although these techniques are largely qualitative, the quantitative aspects thereof can be used to develop concepts and initial designs that can be further refined with the aid of computer methods.

Illustrations are given of the use of these qualitative techniques in the synthesis of straight-line mechanisms, conveying mechanisms, other specialized path-generation mechanisms, and adjustable mechanisms.

8.2 Adding, Subtracting, Multiplying, and Dividing Functions of Linkages

It is no longer practical to use mechanical motions for actual computing. Computing is now much more easily and accurately performed electrically. However, the principles that were used formerly in the design of mechanical analog computers are often very useful in the design of mechanisms in general. Principles of computing mechanisms were described and analyzed in great detail by Svoboda.[1] In this book, Svoboda concentrated on means for performing computations accurately. This section will deal with mechanism features which, although they perform only *approximate* arithmetic functions, are very effective in producing desired mechanism operations. Although this section and Sec. 8.3 present some specific examples of applications of adding and multiplying linkages, it is important to realize that the concepts of adding and multiplying are very broadly applicable to the synthesis and visualization of linkage motions and particularly to the visualization of means for tailoring such motions to particular uses.

Combining motions by addition (and subtraction)

The output displacement of one linkage can be added to (or subtracted from) the output displacement of another linkage to produce a more complicated motion or to allow one linkage to adjust the output of the other. Such combinations of motions often can be used instead of clutches and brakes when a machine motion must be intermittently interrupted or disabled. They also can be used for making vernier adjustments to motions. Examples of uses of motion addition are given in Examples 8.1 and 8.2 and in Sec. 8.3.

Means for performing the addition (or subtraction) of displacements can be derived from the geometry shown in Fig. 8.1. In this figure a link L, to which are attached the three points A, B, and C, is shown located such that point A is a horizontal distance x_A from a reference axis Y. The link is inclined by an angle θ from the Y axis, and as a result, points B and C are at horizontal distances x_B and x_C, respectively, from the Y axis.

For purposes of visualizing this adding action, consider link L to be nearly vertical (i.e., θ is close to zero), consider its length to be about 10 in, consider the distance AC to be 5 in, and consider the distance BC to be 5 in. Then, if point A were displaced rightward by a distance $\Delta x_A = 1$ in while point B was held fixed, point C would be displaced rightward by a distance $\Delta x_C = 0.5$ in. Similarly, if point B were displaced rightward by a distance $\Delta x_B = 1$ in while point A was held fixed, point C would be displaced rightward by a distance $\Delta x_C = 0.5$ in also. If points A and B were both displaced rightward by a distance $\Delta x_A = \Delta x_B = 1$ in, point C would be displaced rightward by a distance $\Delta x_C = 1.0$ in. In each of these displacements, the horizontal displacement of point C is seen to be the average of the horizontal displacements of

---

[1]Svoboda, A., *Computing Mechanisms and Linkages*, Dover Publications, New York, 1965. This book originally was published by McGraw-Hill Book Company in 1948. It was Vol. 27 of the Massachusetts Institute of Technology Radiation Laboratory Series.

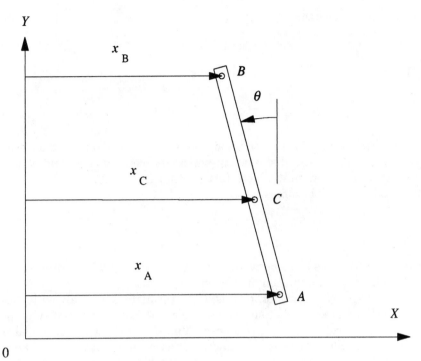

Figure 8.1 Adding of displacements by means of a link.

points A and B. In other words, the link adds the horizontal displacements of points A and B according to

$$\Delta x_C = 0.5(\Delta x_A + \Delta x_B) \qquad (8.1)$$

A somewhat more rigorous and general treatment of this action can be obtained by writing from Fig. 8.1

$$x_A = x_C + d_1 \sin \theta \qquad (8.2a)$$

$$x_B = x_C - d_2 \sin \theta \qquad (8.2b)$$

where d_1 is the distance AC and d_2 is the distance BC. Multiplying Eq. (8.2b) by d_1/d_2 gives

$$\left(\frac{d_1}{d_2}\right) x_B = \left(\frac{d_1}{d_2}\right) x_C - d_1 \sin \theta \qquad (8.3)$$

Adding Eqs. (8.2a) and (8.3) gives

$$x_A + \left(\frac{d_1}{d_2}\right) x_B = \left(1 + \frac{d_1}{d_2}\right) x_C \qquad (8.4)$$

which can be written

$$k_A x_A + k_B x_B = x_C \tag{8.5}$$

where

$$k_A = \frac{d_2}{d_1 + d_2} \quad \text{and} \quad k_B = \frac{d_1}{d_1 + d_2} \tag{8.6}$$

Note that Eq. (8.5) represents simple linear adding of the distances x_A and x_B to give the distance x_C, where the constants k_A and k_B are merely scaling constants. Note further that if $d_1 = d_2$, then $k_A = k_B = 0.5$, and Eq. (8.5) becomes

$$\frac{x_A + x_B}{2} = x_C \tag{8.7}$$

Equation (8.7) also can be written as $x_A = 2x_C - x_B$, which represents a combination of displacements by subtraction.

The foregoing discussion of Eqs. (8.2) through (8.7) dealt with distances that were added or subtracted. Because Eqs. (8.5) and (8.7) represent linear relationships, readers will note that these equations also apply to *changes* in those distances. That is, if x_A, x_B, and x_C were replaced by *displacements* Δx_A, Δx_B, and Δx_C rather than absolute distances, the equations would still apply. Equation (8.7) would then become Eq. (8.1). Subsequent discussions will deal with *displacements* from arbitrary reference positions.

The book by Svoboda referred to previously describes linkages that perform the *exact* summations indicated by Eqs. (8.5) and (8.7). These linkages use slides to cause the displacements that are to be added to be straight-line displacements. In most machine applications, it is not necessary that the addition be exact. Figure 8.2 shows a link combination that can be used to combine displacements x_A and x_B to give an approximate-sum displacement x_C or x_E. In this figure, length $FC = CG = DC = HC$. If points A, B, and E are constrained to move in such a manner that links AG, BF, and HE each remain more or less horizontal, then we may write

$$x_C \approx \frac{x_A + x_B}{2} \tag{8.8}$$

and

$$x_C \approx \frac{0 + x_E}{2} \tag{8.9}$$

From Eqs. (8.8) and (8.9) we see that

$$x_E \approx x_A + x_B \tag{8.10}$$

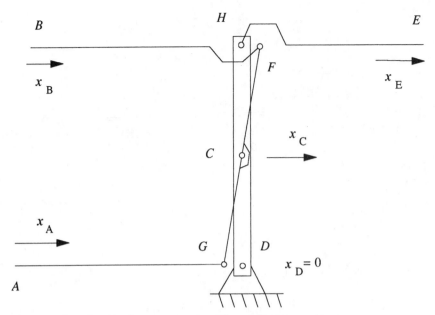

Figure 8.2   A mechanism for approximate adding of displacements.

Equation (8.10) also can be written as

$$x_A \approx x_E - x_B \qquad \text{or} \qquad x_B \approx x_E - x_A \qquad (8.11)$$

showing that this link combination also can be used for subtraction.

The lengths FC, CG, DC, and HC in Fig. 8.2 can be varied independently of each other to provide various scalings of the displacements as combined in the summation. That is, the configuration discussed in connection with Eqs. (8.8), (8.9), (8.10), and (8.11) is a special but very useful adder configuration.

Example 8.1: Using an Adding Linkage to Disengage a Feeding Finger in a Progressive Transport Mechanism (Walking Beam)   An application of the adder configuration shown in Fig. 8.2 is shown in Fig. 8.3. This figure shows a progressive transport (or walking-beam) mechanism similar to that shown in Fig. 4.28. For clarity, the cognate portion of the linkage that is used to keep beam HF_1 horizontal is not shown in Fig. 8.3. Like the mechanism in Fig. 4.28, the purpose of the mechanism in Fig. 8.3 is to move small parts to the right, one by one. In Fig. 8.3, however, finger F_1 descends from *above* to a position behind part M to be moved from position P_1, and then pushes it to the right before ascending for the return motion to the left for the start of the next cycle. Finger F_1 pushes one part to the right from position P_1 into the accumulation pit at position P_2 during each rotation of the input shaft at pivot point S. (During the return motion of finger F_1, another mechanism that is not shown places another new part at position P_1.) As

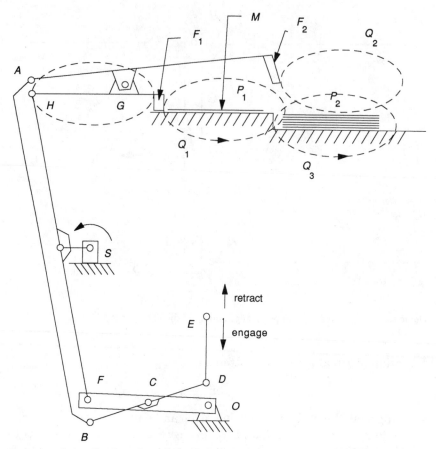

Figure 8.3 An application of an adding linkage to a walking-beam mechanism.

finger F_1 travels along orbital path Q_1 once per machine cycle, finger F_2 travels in unison with it along a similar orbital path Q_2, never making contact with any part. One part is added to the pile of parts in the accumulation pit at position P_2 during each rotation of the shaft at S. As soon as a desired number of parts has accumulated in the pit (as indicated by some measuring or counting instrument), orbital path Q_2 is lowered to position Q_3 for one rotation of the shaft at S. During this rotation, then, finger F_2 pushes the accumulated pile of parts to the right into a package which is then carried away.

The switching of the orbital path of finger F_2 between positions Q_2 and Q_3 is accomplished through the use of rocker AGF_2, coupler AB, and an adding link combination involving points B, C, D, E, F, and O. Note the similarity between the adding link combination at the bottom of Fig. 8.3 and the mechanism in Fig. 8.2. From this similarity it can be seen that vertical displacements of points B, D, E, and F are related by

$$y_B \approx y_F - y_D \approx y_F - y_E \tag{8.12}$$

If y_E is held constant at a value such that pivots D and O coincide, points B and F will coincide and will move together. Then, because link HF_1 is prevented from rotating by the cognate linkage portion that is not shown, and link AG is pivoted to HF_1 at G, and $AG = HG$, point A will remain at an almost constant distance almost directly above point H. Point A will follow a path that is almost identical to that followed by point H but slightly above it. Therefore, finger F_2 will follow a path Q_3 that is almost identical to path Q_1 of finger F_1 but to the right of it and lower than it. Thus, in this condition, finger F_2 will push the accumulated parts rightward in unison with the action of finger F_1 as F_1 pushes parts rightward.

Equation (8.12) shows that raising point E lowers point B. Lowering point B lowers point A without appreciably altering the shape of its orbital path. Lowering the orbital path of point A raises the orbital path of finger F_2. Raising point E by a constant amount therefore adds a constant upward displacement to each point on the orbital path of finger F_2. Thus, by using an actuator to raise point E, finger F_2 may be caused to remain above the parts so that it will not move them. In this manner, parts are allowed to accumulate at position P_2 until the desired number of parts has accumulated, and then the actuator is caused to lower point E for one rotation of shaft S so that finger F_2 will push the accumulated parts to the right.

An actuator such as a spring-loaded electric solenoid, a pneumatic actuator, or a stepping motor can be used to position point E, and that actuator can be attached to the base of the machine.

In an alternative implementation of this mechanism, link DE and the actuator at point E were omitted, and link BC was pivoted to ground at point O and was spring-loaded downward against a stop on link FCO so as to press pivot B into coincidence with pivot F. Then a latch was placed such that, when actuated, it would extend beneath pivot B and prevent pivot B from descending below its uppermost position. By actuating this latch as pivot B passes near its uppermost position, the pivot is caused to remain in the elevated position while points F and H descend. Then, because pivot A remains high as point H descends, finger F_2 is forced downward into a position to push parts rightward until the latch is released (deactuated).

The mechanism shown in Fig. 8.3 would cause finger F_2 to follow a path whose shape was almost identical to that of the path followed by finger F_1 but which was displaced upward or downward by the actuator at point E. However, in the alternative implementation, the *shape* of the path of finger F_2 would be *altered* when the latch was actuated. This altered path *shape* would cause the finger to engage the parts and push them rightward.

Using an adding mechanism such as described in this example eliminates the need for a separately driven mechanism for the intermittently used finger and/or eliminates the need for a clutch to disengage the driving of that finger mechanism. The use of a clutch introduces the problem of reestablishing synchronism when reengaging the clutch. The mechanism described here automatically maintains synchronism. Further examples are given in Secs. 8.3, 8.4, and 8.5.

Using a bell crank in an adding mechanism

Figure 8.4a shows a mechanism by means of which the horizontal displacement of point G can be made to be the approximate sum of the horizontal displacement of point B plus the vertical displacement of point A. If point B is displaced horizontally while holding point A fixed, the parallelogram linkage $BDEF$ prevents the bell crank FBC (a bent ternary link) from rotating, so the horizontal displacement of point C (and G) will be equal to that of point B. If, in addition, point A is displaced vertically, the parallelogram will cause the bell crank to rotate, and the horizontal displacement of points C and G will be given by

$$x_G \approx x_C \approx x_B + y_A(d_2/d_1) \tag{8.13}$$

Example 8.2: Use of a Bell-Crank Addition to Modify a Motion Figure 8.4b shows an application of additive combination of motions by means of a bell crank. This mechanism carries a printing head at point C, and this head is used to print on the sides of parts that are placed successively as shown. Point B is held fixed by holding point G fixed by means of an actuator. In an actual implementation of this configuration, the actuator was a pneumatic cylinder. The bell crank is caused to rock back and forth about pivot B by crank DE and coupler EF, and thus the print head at point C prints on parts as they are placed successively in the position shown. Crank DE rotates at constant angular velocity, but it is occasionally necessary to prevent the print head from moving to the print position because a defective

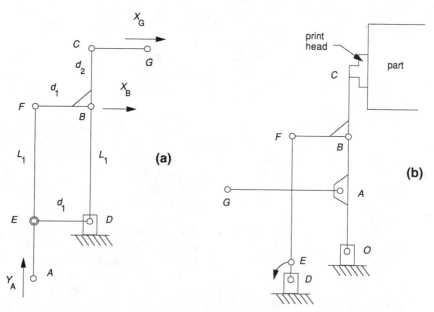

Figure 8.4 Use of a bell crank in an adding linkage.

part is present or a part is positioned improperly. The print head is re-strained from moving to the print position by moving pivot point B leftward through the use of an actuator at point G.

Because the purpose of combining motions in the mechanism in Fig. 8.4b is simply to enable or disable engagement of the print head with the part being printed on, the addition function can be extremely approximate. Therefore, although vestiges of the parallelogram in Fig. 8.4a can still be seen in Fig. 8.4b, the printing rocking and hold-off functions of the mechanism can be made adequately independent of each other even though links BO and EF are unequal in length and do not remain parallel. The principle of approximate additive combination of displacements is present, however.

It may be noted that the use of the bell-crank link is topologically the same as that of link FCG in Fig. 8.2 except that the bell crank has been bent and is actuated from a different direction. Also note that the bell crank need not be bent at exactly 90°.

Combining motions by multiplication (and division)

It is sometimes desired to modify a motion by magnifying or diminishing part or all of the motion pattern involved rather than by simply adding some displacement to it. An application of this type could occur in a scanning mechanism in which the angular excursion of the mirror on an output member of the mechanism must be adjusted to accommodate different sizes of images that are to be scanned. Means for multiplying the original motion by some other displacement are useful for such applications.

The influences that the lengths between the pivots on the links in Fig. 8.1 have on the constants k_A and k_B in Eq. (8.5) suggest that if these lengths could be varied as functions of input displacements, linkages could be used to multiply as well as to add. To see how this can be accomplished, consider Fig. 8.5a. In this figure a link is pivoted to ground at point O and is inclined from a vertical reference line through O by an angle θ. The distances from this reference line to points A and B are, respectively,

$$x_A = d \sin \theta \quad \text{and} \quad x_B = L \sin \theta$$

from which we see that

$$x_A = x_B(d/L)$$

If d were made proportional to some input displacement, x_A would become proportional to x_B times that input displacement. Figure 8.5b shows a method for an approximate multiplication of this sort. In this figure the pivot A of Fig. 8.5a is allowed to slide up and down link OB

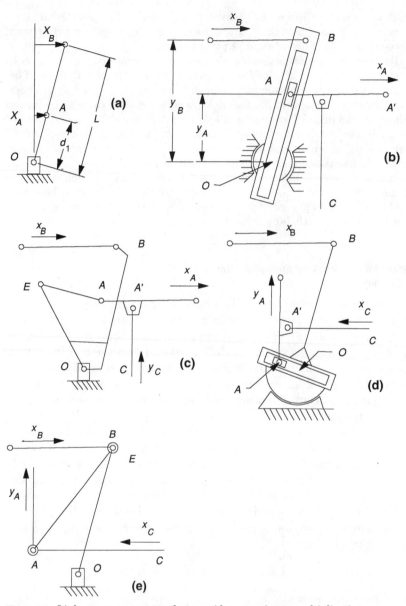

Figure 8.5 Linkage arrangements that provide approximate multiplication.

so that d can be varied by raising or lowering point C. The horizontal displacements of points A and B from a vertical reference line in Fig. 8.4b are given by

$$x_A = y_A \tan \theta \quad \text{and} \quad x_B = y_B \tan \theta$$

from which

$$x_A = x_B \left(\frac{y_A}{y_B} \right) = \frac{x_B y_A}{L \cos \theta}$$

where L is distance OB. If the angle θ is kept small, $\cos \theta \approx 1$, and we may write the approximate relationship

$$x_A \approx \frac{x_B y_A}{L} \qquad (8.14)$$

Again, as in the discussions of adding mechanisms, these discussions will turn to consideration of displacements rather than distances.

Section 1.5 of the previously cited book by Svoboda describes multiplying mechanisms that perform multiplication without making the approximations just described. However, Svoboda's mechanisms make extensive use of sliders. Because we are usually interested only in combining displacements in the manner of an approximate multiplication, it is possible to use mechanisms that use primarily pin joints or only pin joints. Indeed, the sliding pin joint in Fig. 8.5b can be replaced by link and pin joints at the expense of another small deviation from exact multiplication. This modified mechanism is shown in Fig. 8.5c. In this figure the pivot at point A is at the end of a link AE that is pivoted to link EOB at point E so that pivot A can travel relative to link EOB along an arc centered at E. The location of pivot E and the length of link AE are chosen such that pivot A travels relative to link EOB along an arc which, over the expected range of variation of y_C, is at all points close to the line connecting pivots O and B. That is, the motion of pivot A relative to link EOB is made *similar* to that which it would experience if it were in a slot in link OB. The proportions shown in Fig. 8.5c would restrict the usable range of y_C. To provide a greater range for y_C, the length of AE could be increased by moving pivot E leftward.

Comments and suggestions

When conceiving and synthesizing approximate computing linkages such as those discussed in this chapter, it is helpful to synthesize them initially using sliders because the motions of these configurations are usually simpler to visualize. Then, once the desired operation has been achieved conceptually, the slides can be converted to their pin-jointed approximate equivalents. The method for making such conversions is given in Procedure 8.1.

Now note that the slot in link *OB* in Fig. 8.5*b* need not be aligned right along the line *OB* but could be at an angle to it, as shown in Fig. 8.5*d*. In this particular example the slot is perpendicular to line *OB*. Then, if the slide were replaced by a pivoted link, the group of link segments *OE*, *EA*, *AA'*, and *A'C* in Fig. 8.5*c* also would be rotated and flipped as a group until pivot *E* lies on link *OB*, as shown in Fig. 8.5*e*. The resulting configuration will be seen to perform an approximate multiplication in which $y_A \approx (x_B x_C)/L$ because the distance *OA* is approximately proportional to the horizontal displacement of point *C*, as can be seen by comparing the position of point *A* in Fig. 8.5*d* and *e*.

The slot could be oriented on link *OB* at any angle depending on the direction of motion desired for the output, and the corresponding pivoted link that could be used to replace the slide would be rotated accordingly.

Procedure 8.1: Converting a Sliding Joint to a Pivoted Link   To convert a sliding joint to a pivoted link, make the following constructions *on the body that contains the slide* (slot). In Fig. 8.5*b* and *d*, that body is link *OB*.

1. Determine the expected extreme positions of the slider's pivot (*A* in the previous discussion) relative to the slide.
2. Draw a line connecting these extreme positions of the sliding pivot.
3. Construct the perpendicular bisector of the line drawn in step 2.
4. Place a pivot (*E* in the previous discussion) on the body that carries the slide such that the pivot lies on the bisector drawn in step 3 at some distance from the slide.
5. Pivot a link (*EA* in Fig. 8.5*c*) at the pivot chosen in step 4, and make that link of such a length that its other end traces an arc which, between the expected extremes of slider travel, is close to the slide.

8.3   Adjustable Linkages

Many reasons are given for providing adjustability in machines and products. Readers should examine these reasons carefully when considering incorporating adjustability in any design. Then, because any adjustable mechanism is susceptible to *misadjustment,* readers should search for means for eliminating the need for adjustment. Some common reasons for providing adjustments and some ways for avoiding problems associated with those adjustments are as follows:

1. The machine or product must be adjustable because it must process or act on raw materials that vary unpredictably in pertinent characteristics. Try to make the machine action or process insensitive to any such variability. For example, a toaster must act on bread that varies from very moist and soft to very dry and firm. This variability

can be accommodated by incorporating a sensor that detects the "brownness" of the toast and ejects it when the desired condition is reached. This, of course, constitutes automatic adjustment rather than eliminating adjustment. Another example is the processing of web material that varies in stretchability as a function of humidity. This can result in registration problems if long lengths of the material occur between stations at which the web is handled. Obviously, minimizing such lengths will minimize these problems. Also, processing the material in the form of short, discrete pieces can reduce this problem. In any case, readers should search for types of processes that are insensitive to variability in the raw material.

2. The machine must be adjustable because it is to be applied to a variety of tasks. For example, the machine must be able to make long widgets as well as short widgets. If time is available between performance of the different types of tasks, perhaps the machine can be provided with quickly and easily changed parts that are suitable for the different tasks. If adjustability *must* be provided, it should not be continuously variable but should consist of "switching" between discrete parameter values or discrete mechanism component positions.

3. The machine must be adjustable because design parameters cannot be determined precisely in advance due to the complexity of the required operation. This reason has become increasingly less valid as computerized analysis and simulation have become ubiquitous. Such computerized aids should be used in the synthesis process to determine precise design dimensions for the machine. These analyses and simulations may have to be supplemented by predesign experimentation on some components in cases where complex flow and/or deformation phenomena are involved in the process. If adjustments *must* be provided, calibration marks on the adjustment "knobs" also should be provided so that the adjustments, if disturbed, can be returned to settings that are found to be appropriate. Also, setup manuals should be provided to describe adjustment procedures. Once proper adjustment values have been found, they should be locked in position.

Despite careful efforts during design, it occasionally is necessary to design an adjustable mechanism. The mechanisms described in Examples 8.1 and 8.2 are adjustable in the sense that by changing the position of a part of the mechanism, changes in the nature of the output motion of that mechanism are produced. These adjustments are not continuous but are in the nature of switching between two types of operation.

Often it is desired to provide continuously variable adjustment. Such adjustment can be provided by using the same principles of

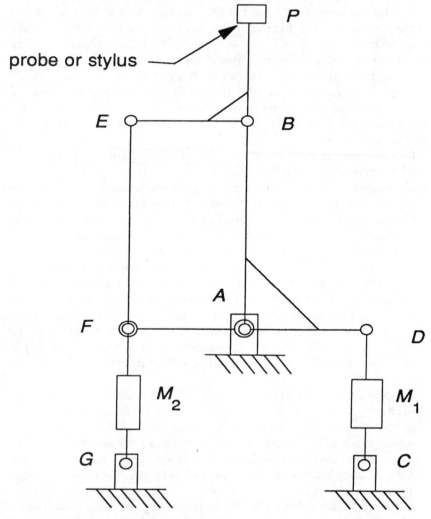

mechanical addition and multiplication described in Sec. 8.2. Examples 8.3, 8.4, and 8.5 illustrate such applications, but it should be remembered that these are only three illustrations; the concepts of mechanical addition and multiplication have applications to many other tasks.

Example 8.3: Use of Mechanical Adding to Provide Rapid Slewing plus Micropositioning Figure 8.6 schematically depicts a mechanism that might be used to position a stylus or probe that must be positioned very rapidly (slewed) to widely varied positions but also must be positioned very

Figure 8.6 A mechanism that provides rapid slewing plus micropositioning.

precisely at those positions. The system uses an actuator M_1 to provide the large travel and high velocity required for initial positioning (slewing). This actuator moves bell crank DAB, which, in conjunction with parallelogram linkage $ABEF$, moves the probe at point P in unison with point B. The actuator M_1 might be a stepper motor that will move the probe rapidly over a wide range of positions but which would be limited in the resolution with which it could be controlled.

The actuator M_2 could be a device such as a piezoelectric or magnetostrictive actuator. Such devices are capable of extremely small displacements and fine resolution, but their output velocities and ranges of motion are quite limited. By adding the displacements produced by these two actuators, using a mechanism such as shown in Fig. 8.6, the required positioning functions can be provided. It will be noted that the configuration in Fig. 8.6 is essentially the same as that in Fig. 8.4

Example 8.4: Use of Mechanical Adding to Provide Adjustment of the Endpoints of the Motion of a Reciprocating Shuttle Mechanism   A reciprocating shuttle mechanism for sequentially moving parts (or a web of material) from left to right is indicated schematically in Fig. 8.7a. (The somewhat awkward schematic arrangement in this figure is chosen in order to show clearly the kinematic components of the system. The actual implementation used a barrel cam with conjugate surfaces and a spring-loaded second follower and was somewhat more compact.) The shuttle is caused to slide back and forth horizontally by means of a cam C. A follower F is connected to the shuttle by link (bell crank) FPA, which is in turn pivoted to the shuttle body at point P. The end A of link FPA is pivoted at point A to a slider L that slides back and forth in the slide S between points A and B as the shuttle reciprocates. With slide S in the position shown, the slider moves horizontally, keeping arm PA horizontal as the shuttle reciprocates. The shuttle therefore moves in unison with cam follower F, and the stroke of the shuttle (distance from one extreme of its motion to the other) is equal to the rise and fall distances of the cam.

In Fig. 8.7b the slide S has been rotated clockwise about point A so that point B has been raised slightly. Because the slider position is at the same location in both Fig. 8.7a and b, follower F is in the same position relative to the shuttle in both figures, and the shuttle is therefore in the same position at the right-hand end of its stroke. However, as follower F and the shuttle move toward the left-hand end of the stroke, arm PA rotates counterclockwise in Fig. 8.7b as the slider rises slightly from position A to position B, and therefore, point P moves leftward relative to follower F. The result is that when the follower reaches its leftmost position (as determined by the low dwell portion of cam C), point P and the shuttle will be further leftward in Fig. 8.7b than they were in Fig. 8.7a.

Thus it is seen that by raising end B of slide S without raising point A of the slide, the length of the stroke of the shuttle has been increased by moving the left-hand end of that stroke leftward without altering the rightmost end position of that stroke. By using a large ratio of a length of arm PA to length of arm PF, the follower can be moved horizontally through very small

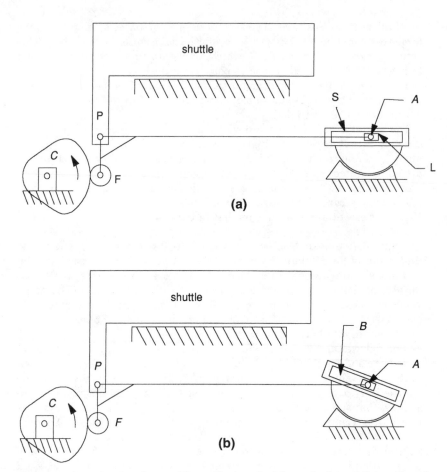

Figure 8.7 A mechanism for providing adjustment of the endpoints of the motion of a shuttle.

distances relative to the shuttle, and thus very fine vernier adjustments can be made to stroke length. It also should be noted that if point A of the slide were raised or lowered without raising or lowering point B, the position of the right-hand end of the shuttle stroke could be adjusted without affecting the left-hand end. Therefore, this sort of mechanism could be used to adjust each end of the stroke independently. This adjustment scheme is in use as part of an automatic registration control system.

The slide S could be replaced by a pivoted link, as described in Procedure 8.1, if space restrictions permit.

Example 8.5: Use of a Multiplication Mechanism to Adjust the Horizontal Dimension of an Oval Orbital Motion A portion of the mechanism that was used in the walking-beam system used to illustrate Procedure 4.9 and which was shown in Fig. 4.25 is shown in Fig. 8.8a. The purpose of this portion of the mechanism is to cause point C to orbit along the oval path that is shown dashed between locations C and C'. By considering the similar triangles BAO_1 and BCC'', it can be seen that the end positions C and C' of the orbit are a horizontal distance apart from $CC' = 2rL/d$, where $r = O_1A$, $L = BC$, and $d = BA$. The vertical dimension of the oval is seen to be $DD' = 2r$ and is therefore independent of the lengths L and d. The horizontal travel of point C of the walking beam therefore can be varied by varying the ratio L/d without affecting the vertical motion or the timing of the motion perceptibly.

The ratio L/d can be varied by varying the distance AB. Such a variation changes both L and d but in so doing changes the ratio. Figure 8.8*b* schematically shows a scheme for varying the distance AB by means of a *ground-mounted actuator*. Compare the positions of pivots A, B, and C in Fig. 8.8*a* and Fig. 8.8*b*. By raising pivot H in Fig. 8.8*b*, which is attached to the actuator, pivot D has been raised relative to link AC, and thus pivot B has been caused to slide upward in slot S in link AC, thereby increasing the ratio L/d. The horizontal travel of point C is approximately equal to the horizontal travel of point A *multiplied by* a factor that is controlled by the displacement of the actuator at pivot H.

Note that pivot E does *not* slide in slide S. It is in a plane that is closer to the viewer than the plane of that slide. Recall also that the slider and

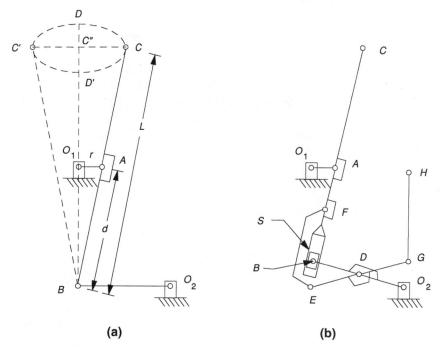

(a)   (b)

Figure 8.8   Use of a multiplication mechanism to adjust the horizontal dimension of an oval orbital motion.

slide S can be replaced by a suitable pivoted link in accordance with Procedure 8.1.

8.4 Combining and Distorting Curvatures and Curve Shapes

The curves that are followed by points on all links *that are pivoted to ground* are circles or arcs of circles. This is not true of curves that are followed by points on links such as couplers that are *not* pivoted to ground, and Sec. 8.6 discusses the myriad shapes that such coupler point paths can have. This section introduces some of the simpler forms of these coupler point curves, describes how the concepts of addition and multiplication can be used to visualize the effects that linkage dimensions have on the curve shapes, and how these concepts can be used in synthesizing mechanisms to generate these curves. Although some quantitative relationships are discussed, the important aspects of these concepts lie in their use in *qualitatively visualizing* the effects that mechanism features (e.g., dimensions, etc.) have on the characteristics of the coupler curves generated.

Multiplication (magnification) of the size of a curve

Figure 8.9 depicts a crank-slider mechanism and the coupler point curves that are followed by coupler points A, C, D, E, and F. Point A, being also on crank OA, which is pivoted to ground, follows a circular path of radius $r = OA$. If crank OA is considered to be rotating counterclockwise, point A will have a leftward horizontal component of velocity when in the position shown. It can be seen that in this position the horizontal component of the velocity of each of points C, D, E, and F on the coupler also will be leftward. Thus it can be seen that the

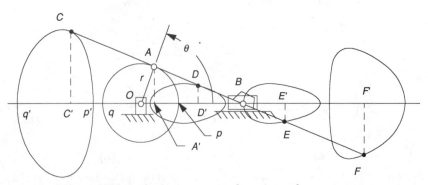

Figure 8.9 Effects of linkage dimensions on coupler point path curvatures.

points on the left side of pivot B will follow counterclockwise paths, just as point A does. However, points E and F follow their paths in a clockwise direction.

The circular path followed by point A intersects the horizontal line through points O and B at points p and q, which are separated by the distance of 2 times OA, and these intersections occur when θ is $0°$ and $180°$. At these same values of θ, coupler points C, D, E, and F also lie on the horizontal line through points O and B, so their coupler point curves intersect the horizontal line when $\theta = 0°$ and $\theta = 180°$. Because the distances along the coupler between points A, C, D, E, and F are fixed, it can be seen that for *each* of these coupler points, the distance between points at which its path curve intersects the horizontal line is 2 times OA. That is, each of the curves is of the same horizontal width, which is 2 times OA.

The vertical distance between point C and the horizontal line through point O is indicated as dashed line CC'. By similar triangles, it can be seen that length $CC' = AA'(BC/BA)$. The vertical distance from the horizontal line to point C is therefore equal to the distance from that horizontal line to point A *multiplied by* the ratio BC/BA. This same multiplication factor pertains to all points on the path of point C, so it can be seen that because the vertical height of the closed path of point A is $2(OA)$, the vertical height of the closed path of point C is $2(OA)(BC/BA)$. The path of point C is very roughly similar to that of point A except that although its horizontal width is the same as that of point A, its vertical height is equal to that of point A *multiplied by* BC/BA. Its height has been *magnified* relative to that of point A.

The same sort of relationship also pertains to each of the other coupler point curves shown. Note that because $BD/BA < 1.0$, the vertical height of the path of point D has been multiplied by a number less than unity, so the vertical height of its curve is diminished relative to that of point A and so its curve is flattened. The corresponding effects are seen on the curves for points E and F. The multiplication effect has applied to the vertical dimensions of all four of the curves. Their horizontal dimensions (widths) are unchanged relative to that of the path of point A.

Multiplication of the curvature of a curve

It can be seen that the path of point C is curved much more sharply at its top than is the top of the path of point A. A measure of the "sharpness" of this curving is known as the *curvature* of the curve. The curvature of a curve at any point can be defined as being equal to 1.0 divided by the radius of curvature of that curve at that point. The radius of curvature is the radius of the circular arc that best fits the curve near that point. For example, if we denote the curvature at point A in Fig. 8.9 by the symbol K_A, we may write $K_A = 1/\rho_A$, where $\rho_A = AO$ is the radius of curvature of the path of point A.

It can be shown that in Fig. 8.9 the curvature K_C of the path of C at *its uppermost point* is equal to the curvature K_A of the path of A multiplied by BC/BA. That is,

$$K_C = \frac{1}{\rho_C} = K_A \frac{BC}{BA} = \frac{1}{AO} \frac{BC}{BA} \qquad (8.15)$$

The curvature at the top of the path of point A has been *multiplied by* a constant to give that of the path of C.

A similar observation can be made that the curvature of the path of A has been multiplied by a constant (BD/BA) that is less than 1.0 to give the smaller curvature at the top of the path of point D. Thus,

$$K_D = \frac{1}{\rho_D} = \frac{1}{AO} \frac{BD}{BA} \qquad (8.16)$$

This concept of multiplying curvatures by constants can be used as a qualitative and rough quantitative guide in the development of mechanisms for producing curve segments that have desired curvatures. In actual implementation, the sliding joint at point B can be replaced by a pivoted rocker, as described in Procedure 8.1.

It also can be seen that the sides (near the horizontal line OB) of the path of point C are much less sharply curved than the corresponding portions of the path of point A. The ratio of the curvature of the path of C at the points p' and q' to the curvature of the path of A is a more complicated function than that described above. If, however, length AO is small compared with length AB so that link CAB never departs greatly from horizontal, and if length AC is not large compared with length AB, then to a very rough first approximation the curvature of the path of point A will be multiplied by a factor of approximately $(AB/BC)^2$ to give the curvature of the path of C at points p' and q'. [If AO is not small compared with length AB, then Eqs. (8.22) and (8.24) apply.] Thus, for example,

$$K_{p'} = \frac{1}{\rho_{p'}} = \frac{1}{AO} \left(\frac{AB}{BC} \right)^2 \qquad (8.17)$$

Note also that the curvature of the path of D at points p'' and q'' is greater than that of the path of A because the approximate curvature multiplying factor $(AB/BD)^2$ is greater than 1.0.

A distinction can be made between the curvature at points such as C and at points such as p' and q'. The former will be called *longitudi-*

nal curvatures because the curve at that point is more or less longitudinal or parallel to the coupler. The latter will be called *perpendicular curvatures* because the curve at that point is perpendicular to the coupler.

Adding curvatures

As mentioned in the preceding subsection, the slider and slide at B in the mechanism discussed in connection with Fig. 8.9 can be replaced by a rocker that is pivoted to ground. If that rocker link is long relative to the horizontal movement of the pivot by which it is connected to the coupler, the motion of the linkage will be very similar to that which would result if a slider were used. This is so because the rocker would oscillate through only a small angular excursion, and thus the arc described by its pivot to the coupler would be approximately a straight line. If, however, the rocker is not long, the pivot corresponding to slider pivot B in Fig. 8.9 would experience appreciable vertical motion. This vertical motion would approximately add to the vertical motion produced at point C by the motion of point A (with scaling factors depending on the lengths between A, B, and C). In effect, at the top of the path of C, the curvatures of the paths of A and B would be added together approximately to give the curvature of that portion of the path of C.

As an aid to visualizing this approximate adding of curvatures, consider the linkages in Fig. 8.10. Figure 8.10a is similar to the portion of Fig. 8.9 that involves pivots O, A, B, and C except that in Fig. 8.10a the slide in which the slider at B slides has been raised so that it is in line with pivots A and C. We will consider only the motion of the mechanism in which point C is near the top of its path. Just as in the case of Fig. 8.9, the curvature of the path (shown as a dashed curve) of point C at the top of its travel will be approximately the curvature of the path of pivot A multiplied by CB/AB. That is,

$$K_C = \frac{1}{\rho_C} = \frac{1}{AO_1}\left(\frac{CB}{AB}\right) \tag{8.18}$$

Note that the denominator of this multiplying factor is the distance between the slider and the point whose path curvature is to be multiplied, and the numerator is the distance between the slide r and the point whose path is to result from that multiplication.

Next, consider the mechanism in Fig. 8.10b. In this mechanism the slider is at A, between points C and B. Using the curvature approximate multiplying factor that is analogous to that in the preceding

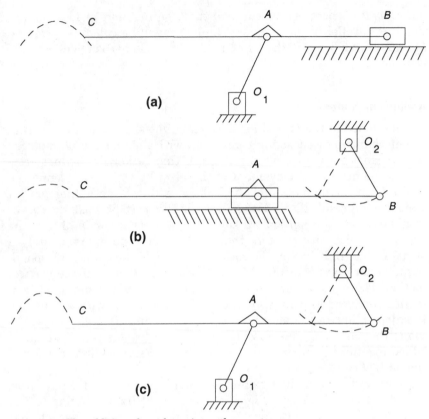

Figure 8.10 The addition of coupler point path curvatures.

paragraph, the curvature of the path of point B (shown as a dashed curve) will be multiplied by CA/AB to obtain the curvature of the path of C at the top of its path. That is,

$$K_C = \frac{1}{\rho_C} = \frac{1}{BO_2}\left(\frac{CA}{AB}\right) \qquad (8.19)$$

Now define the curvature of a curve segment that points upward at its ends as a *positive* curvature and of one that points downward at its ends as *negative* curvature. Therefore, both path curvatures shown in Fig. 8.10a are negative. In Fig. 8.10b the curvature of the path of pivot B is positive, but because points B and C are on opposite sides of the slider at B, the curvature multiplying factor is negative, so the curvature of the path of C is negative, as was the curvature at C in Fig. 8.10a.

Now consider the configuration in Fig. 8.10c. In this mechanism neither point A nor point B is on a slider. Both are on links that are pivoted to ground. These two links are the same as the respective links in Fig. 8.10a and b. The resulting curvature of the path of C is approximately the sum of the curvatures of the paths of C in Fig. 8.10a and Fig. 8.10b. That is,

$$K_C = \frac{1}{\rho_C} = \frac{1}{AO_1}\left(\frac{CB}{AB}\right) + \frac{1}{BO_2}\left(\frac{CA}{AB}\right) \tag{8.20}$$

This addition of *curvatures* is analogous to the addition of *displacements* that was described in connection with the adding links in Figs. 8.1, 8.2, and 8.3 in Sec. 8.2. Readers should compare this addition of curvatures according to Eq. (8.20) with the addition of displacements in Fig. 8.1.

Example 8.6: Generating a Coupler Curve by Using Motion-Adding and -Multiplying Concepts

The principles described in this example are useful in conceiving the type and general arrangement of a mechanism for use in generating a coupler curve segment of a prescribed type. However, the *quantitative* aspects of the task are more appropriately performed by using a technique similar to that described in Procedures 4.7 and 4.8. Both the conceptual and the quantitative phases of synthesis are illustrated in the solution of the following problem.

Problem: It is desired to cause a point P in a mechanism to follow a path that is roughly semicircular with a radius of 0.5 in about a given fixed point O_3 (shown as a dashed semicircle in Fig. 8.11a). The mechanism is to be located off the end of the diameter of the approximate semicircle, as shown in Fig. 8.11a. The input shaft for this mechanism is to be located about 5 in from the center of the semicircle, and that shaft is to rotate back and forth through a total angle of 60° while causing point P to travel back and forth along the semicircular path.

In searching for a mechanism type that would satisfy the requirements stated in this problem, it is noted that the mechanism in Fig. 8.9 produces a sharply curved output motion of point F during excursions of the input rotation of link OA near $\theta = -90°$. Let us see whether some variant of this mechanism can be used with some added curvature, in the manner described in the preceding subsection, to solve this problem.

Figure 8.11b shows the required semicircular path (dashed) and an input rocker O_1A whose grounded pivot O_1 (the input shaft) is about 5 in to the right of the center of the semicircle. The total horizontal excursion of output point P as it travels along the required curve from location P to P' is 1 in, so the length of the input rocker is chosen to be 1 in, so that the 60° required motion of the input shaft will cause pivot A to also have a total horizontal travel of 1 in as it goes from A to A', as shown.

Draw a coupler connecting pivot A at location A with output point P, as shown in Fig. 8.11b. As drawn, this coupler is 5 in long. Locate a pivot B at an initial trial position along coupler AP as shown. As drawn, pivot B is 3 in from point P. Tentatively consider pivot B to be connected to a slider that slides horizontally. If such a slider is used, the resulting path of point P will

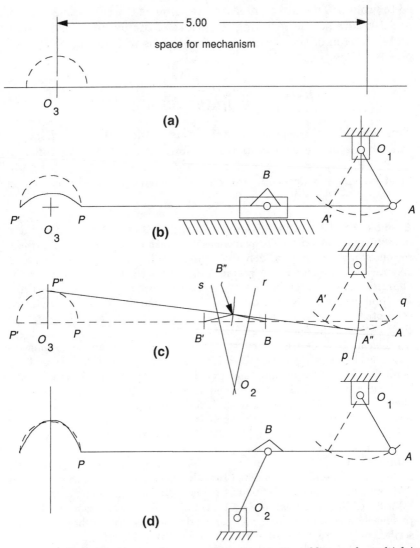

Figure 8.11 Generating a coupler curve by using motion-adding and -multiplying concepts.

be solid, as shown. Note that as this input rocker rotates back and forth through 60°, the path that P follows is not nearly as curved as is the required semicircle. Obviously, some curvature must be added to the solid curve if it is to be made to approximate the required semicircular curve. As discussed in connection with Fig. 8.10, this can be done by using a rocker instead of a slider at pivot B.

To determine the nature of this rocker, a procedure similar to Procedure 4.7 will be used. Figure 8.11c shows coupler AP drawn in positions AP, $A'P'$, and $A''P''$ such that the points P, P', and P'' are on the required semicircular output curve. Point P'' is chosen to be midway along the required curve, and point A'' is determined by constructing an arc p of radius equal to the coupler length (i.e., AP) centered at point P''. Where this arc intersects arc q of radius AO_1 centered at point O_1 is the location of point A''.

Next, draw pivot B at a distance 3 in from point P on coupler AP, draw pivot B' at a distance 3 in from point P' on the coupler in position $A'P'$, and draw pivot B'' at a distance 3 in from point P'' on the coupler in position $A''P''$. The rocker that must be substituted for the tentative slider at pivot B must cause pivot B to follow an arc that passes through points B, B', and B''. The center of this arc (and thus the grounded pivot of the rocker) will lie at the intersection of the perpendicular bisectors of line segments BB'' and $B''B'$. These perpendicular bisectors are drawn in Fig. 8.11c as line segments r and s, respectively, and their intersection is indicated as O_2.

The resulting linkage is shown in Fig. 8.11d, where the resulting path followed by point P is shown as the solid curve at the left. This path can be compared with the required semicircle, which is shown as a dashed curve. The degree of approximation is easily seen. This mechanism configuration is a *first estimate* of the mechanism to be used. If a more convenient physical arrangement of mechanism parts is desired, the dimensions of the mechanism can be varied gradually from these first estimates, and new resulting coupler curves can be produced by computerized analysis or simulation in a search for an improved configuration.

If rocker length O_2B for this first approximation had been computed by solving Eq. (8.20) for a rocker length that would give a path curvature of 1.0/0.5 in at point P'' in Fig. 8.11, the result would have been $O_2B = 5.0$ in. Although such a rocker would have produced the desired path *curvature* at its topmost point, the height of the curve would not have been as great as obtained in the preceding first approximation. More iterations would have been required.

Example 8.7: Generating and Distorting a Coupler Curve by Using Motion-Adding and -Multiplying Concepts   *Problem:* It is desired to cause a roller to apply paint to the vertical (receiving) surface of each of a series of parts as they are sequentially placed at a station on an assembly line. The roller must roll down along that receiving surface on the part while applying relatively uniform pressure to the part. After traversing the surface, it must move horizontally away from the surface and return to a position from which to start its action on the next part to be presented. Figure 8.12a indicates a part and the sort of path (dashed) that the roller must follow.

The more or less oval path indicated in Fig. 8.12a suggests that a path such as that followed by point E in Fig. 8.9 might be used as a first approximation in the synthesis process, provided that the mechanism of Fig. 8.9 is rotated 90°. Figure 8.12b shows such a mechanism applied to this problem. The link lengths used are $AO = 1.5$ in, $AB = 8.0$ in, and $BC = 4.0$ in. The resulting oval path is 3.0 in high and 1.5 in wide. Unfortunately, the left-hand side of this oval is convex rather than being a straight line as desired, so the roller would have to penetrate the part or

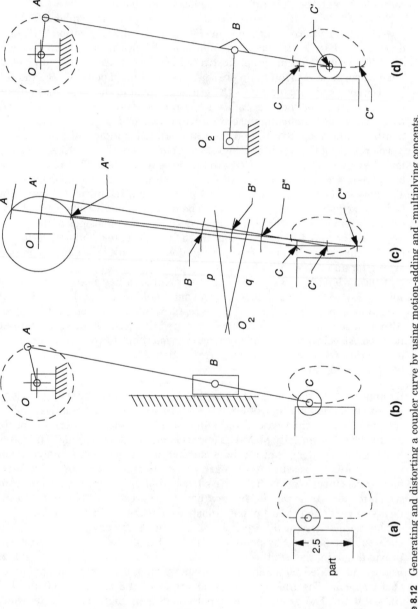

Figure 8.12 Generating and distorting a coupler curve by using motion-adding and -multiplying concepts.

it would not contact the part over its total height. This side of the oval path could be made approximately straight by adding a suitable curvature to it in a direction such as to cancel the unwanted curvature. As shown previously (Figs. 8.10 and 8.11), this can be done by substituting a rocker for the slider at B.

The dimensions and location of the rocker will be determined by the same process as used in Example 8.6 (i.e., using Procedure 4.7 or 4.8). In Fig. 8.12c, most of the left-hand side of the oval has been tentatively replaced by a straight line, and three points C, C', and C'' have been chosen on that line. (Only a part of the side of the oval has been replaced by a straight line, because if too much is replaced, unsatisfactory results can occur.) The 12-in-long coupler AC is drawn in the three positions that are determined by locations of C, C', and C'' and by the circular path that pivot A must follow. Points B, B', and B'' are drawn on the respective coupler positions as shown. Then, as in Example 8.6, the perpendicular bisectors p and q of BB' and $B'B''$, respectively, are drawn, and pivot O_2 is located at their intersection.

The final mechanism and roller path are shown schematically in Fig. 8.12d. Although the path passes through the three positions C, C', and C'', it is not precisely straight between these points. However, a very small amount of compliance in the roller would accommodate this inaccuracy. The very close approximation of a straight line is adumbrative of phenomena discussed in Sec. 8.5.

Note that the substitution of a rocker for the slider has bent the entire symmetrical oval in Fig. 8.12b into the unsymmetrical shape in Fig. 8.12d.

Comments and suggestions

In each of the foregoing discussions, the two points whose path curvatures were to be added and the coupler point generating the output curve were all three on the same straight line on the coupler. Although the effects become a bit more complicated if the three points are not on the same straight line on the coupler, it is not necessary that they be located this way. Synthesis can start by assuming them to be on a single straight line, and then, by iteration, readers can progressively move one of the points from that line and recompute the resulting motion. With a bit of trial and error, a more desirable mechanism might be obtained. More discussion of various types of coupler point paths is contained in Sec. 8.6.

In Examples 8.6 and 8.7, the grounded pivot location was determined on the basis of the location of three positions of the coupler and its point B. Sometimes it is possible to obtain a better fit to a desired curve by using more than three positions. This will result in indicating several *different, conflicting* locations for the grounded pivot of the rocker. The *average* of these locations may give a better fit to the desired curve.

8.5   Straight-Line Mechanisms

It is often desired to cause some portion of the path of some point in a mechanism to be a straight-line segment. Of course, it is possible to attach the point to a slider that is constrained to move in a straight line. There are now many good slider and slide components available commercially. Many of these products use rolling elements (balls or rollers) so that the friction involved is minimal. Also, many of these products are very high precision units. Nonetheless, if the required motion can be an *approximate* straight-line motion, and if the compactness that can be provided by using a slide and slider is not required by space limitations, it is often preferable to use a pin-jointed mechanism to provide this motion. Use of such a pin-jointed mechanism can simplify lubrication and bearing protection problems.

In Example 8.7 and Fig. 8.12 it was shown that a pin-jointed linkage can be used to provide a coupler point path, a portion of which approximates a straight line. Over the last 200 years, engineers have searched for mechanisms that can generate *exact* straight-line motions. This search has uncovered several types of mechanism that can generate approximate straight-line motions and has uncovered some types that can generate exact straight lines. These exact straight-line mechanisms will be discussed first.

Exact straight-line mechanisms

In France in 1873, Charles-Nicolas Peaucellier described his invention of what is considered to have been the first discovered exact straight-line-generating planar pin-jointed linkage. This mechanism is known as the *Peaucellier straight-line linkage* and is shown schematically in Fig. 8.13a. The link proportions are $AC = AD = BC = BD$ and $O_2C = O_2D$ and $O_1A = O_1O_2$. As pivot A travels along a portion of the circular path shown dashed, pivot B travels along an exact straight-line path that is perpendicular to a line through pivots O_1 and O_2, as shown at the right.

While the Peaucellier linkage consists of eight links, another linkage, the *Hart straight-line linkage,* consists of only six links and also can generate an exact straight-line path. The Hart straight-line linkage is shown in Fig. 8.13b. The proportions of the links are $ED = CF$ and $EC = DF$. With these proportions, a line CD will be parallel to a line EF. Pivots O_2 and A and point B are located on a line that is parallel to these two lines, so $EO_2/EC = EA/ED = FB/FC$. If pivot A is connected by link AO_1 to grounded pivot O_1 such that $O_1A = O_1O_2$, then as pivot A travels along a portion of the circular path shown, point B will travel along a straight-line path that is perpendicular to a line through pivots O_1 and O_2, as shown dashed at the right.

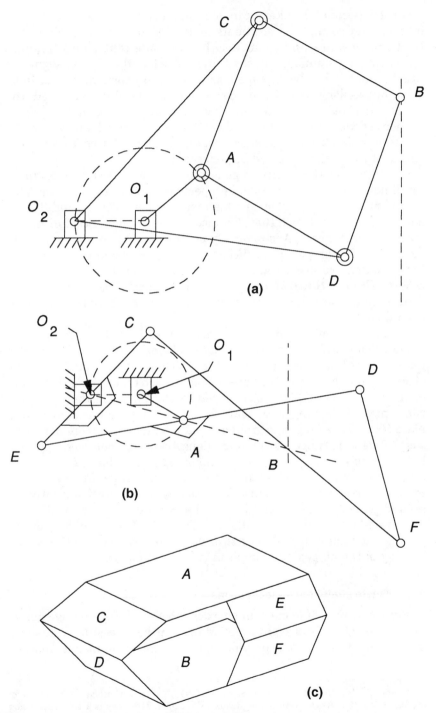

Figure 8.13 Exact straight-line mechanisms.

Obviously, no real linkage can generate a straight line extending to infinity, and so the line segments generated by the Peaucellier and Hart linkages are limited in length. Also note that point A cannot travel entirely around the dashed circle. The lengths of these segments will be limited by the linkage size and the proportions of the links used, but as long as those proportions specified above are used, the paths generated will be straight-line segments. For example, the proportion between links $O_2 C$ and AC is *not* specified above for the Peaucellier linkage, so this proportion could be varied to vary the length of the path that could be generated.

In addition to the ability to generate straight-line segments, these two linkages have the additional property that they are *inversors*. This means that they move in such a manner that points O_2, A, and B lie in a straight line, and such that distance $O_2 A$ times distance $O_2 B$ is constant; i.e., distance $O_2 A$ is inversely proportional to distance $O_2 B$, and vice versa. The properties of these two linkages, including this inverting property, are discussed in considerable detail in a book entitled *How to Draw a Straight Line,* by A. B. Kempe.[2] That book also describes many variations of these linkages and other linkages derived from them.

Another exact straight-line mechanism, which unlike the Peaucellier and Hart linkages is *not a planar linkage,* is the Sarrus linkage, which is shown in Fig. 8.13c. As shown, this linkage consists of six rectangular plates that are hinged to each other along their common edges. That is, plate A is hinged to plate C along their common edge, plate C is hinged to plate D along their common edge, and plate D in turn is hinged to grounded plate B along their common edge. Also, plate A is hinged to plate E along their common edge, plate E is hinged to plate F along their common edge, and plate F in turn is hinged to grounded plate B along their common edge. Plate A is able to move vertically in straight-line motion without rotation relative to ground plate B. This spatial linkage has zero degrees of freedom, as computed by Gruebler's formula (see Sec. 1.7). Therefore, in order to provide the required single *practical* degree of freedom and thus avoid binding in the hinges, the hinges must be aligned very accurately.

Approximate straight-line mechanisms

The *exact* straight-line mechanisms described in the preceding subsection consisted of six or eight links. Many four-bar pin-jointed linkages have been devised that provide *approximate* straight-line motion.

---

[2]Kempe, A. B., *How to Draw a Straight Line,* National Council of Teachers of Mathematics, 1906 Association Drive, Reston, VA, 22091. This book is a 1977 reprinting of a book that was originally published by Macmillan in London in 1877.

Some of the more common of these linkages are shown in Fig. 8.14. Variations and generalizations of some of these linkages will be discussed later in this subsection.

The linkage shown in Fig. 8.14a is known as the *Watt straight-line mechanism*. The midpoint C of link AB will travel along a path that is approximately a horizontal straight line for appreciable displacements from the position shown.

Figure 8.14b shows a linkage known as an *isosceles* or *Scott Russel* or *Freemantle straight-line linkage*. The link-length proportions are OA = AB = AC. This is an *exact* straight-line mechanism, but it uses a

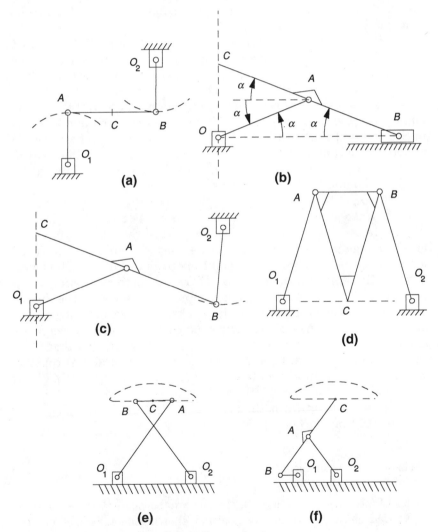

Figure 8.14 Some approximate straight-line mechanisms.

straight-line slider at point B. Its operation is best visualized by noting that triangle OAB remains an isosceles triangle with a horizontal base as the slider at B is displaced along the slide. Also, by noting the equality of the four angles labeled α, it can be seen that the isosceles triangle OAC remains isosceles with a vertical base OC as slider B is displaced. Therefore, point C remains on the vertical dashed line shown.

Because the slider at B in the linkage in Fig. 8.14b experiences only small displacements as point C moves over appreciable distances along its straight-line path, the slider at B can be replaced by a pivoted link without introducing much inaccuracy into the straight-line generation function. Such a modification is shown in Fig. 8.14c, and the resulting linkage is known as a *grasshopper* or *Evans straight-line mechanism*.

The *Roberts straight-line linkage* is shown in Fig. 8.14d. Point C traces an approximately straight horizontal line as it is moved back and forth between pivot points O_1 and O_2. The proportions of this linkage are $O_1A = AC = CB = BO_2$ and $O_1O_2 = 2AB$.

Figure 8.14e shows the *Tchebycheff* (sometimes *Chebychev*) *approximate straight-line linkage*. This linkage was discussed and analyzed in Sec. 2.5 and Example 2.1, where it was shown that the motion of midpoint C of link AB is along an approximate straight line over an appreciable portion of its path. The proportions of this linkage are $O_1O_2 = 2AB$ and $O_1A = O_2B = 2.5AB$.

The linkage shown in Fig. 8.14f, known as a *Hoekens* or *Tchebycheff lambda linkage,* is a cognate of the Tchebycheff linkage shown in Fig. 8.14e. As discussed in Sec. 4.7, a linkage that is a cognate of a given linkage possesses a coupler point that will trace a path that is identical to the path traced by a point in the given linkage. Thus point C in the Hoekens linkage in Fig. 8.14f traces a path containing an approximately straight-line segment, just as does point C in the linkage in Fig. 8.14e. The proportions of the Hoekens linkage are $O_1O_2 = 2AB$ and $AB = AC = O_2A = 2.5AB$. This linkage has the advantages that it can be driven by a continuously rotating shaft at O_1 and that if the shaft rotates at constant speed, point C will travel at almost constant velocity along the approximate straight-line portion of its path. Figure 8.14e and Fig. 8.14f are drawn to the same scale, and point C in each case traces an identical path. For this to be true, O_1B in Fig. 8.14f must be 0.5 times AB in Fig. 8.14e.

Devising other approximate straight-line mechanisms

In the Watt linkage of Fig. 8.14a, it can be seen that as link AO is rotated clockwise from the position shown, the travel of pivot A

along the dashed arc causes pivot A to move downward as well as rightward. Simultaneously, pivot B moves rightward through a horizontal distance that is approximately equal to the horizontal displacement of pivot A. Because pivot B also travels along an arc, it rises as it moves rightward. As pivot A moves downward, pivot B moves upward, and the midpoint C between them, although traveling rightward with the two pivots will have a vertical component of displacement that is approximately the average of the vertical displacements of the two pivots (see Sec. 8.2). That is, point C will have very little vertical displacement. Its path will be approximately a horizontal straight line.

Generalizing the Watt linkage

The straight-line-generating phenomenon in the Watt linkage can be viewed as either the adding of vertical displacements of pivots A and B, as discussed in Sec. 8.2 (see Fig. 8.1), or as the adding of the curvatures of their paths, as discussed in Sec. 8.4 (see Figs. 8.9 and 8.10). In the configuration in Fig. 8.14a, point C was the midpoint of coupler AB, and links AO_1 and O_2B were of equal length. However, it was shown in Secs. 8.2 and 8.4 that the dimensions associated with the location of the coupler point relative to the coupler pivots affect the factors by which the perpendicular[3] displacements or longitudinal path curvatures of those pivots are multiplied before they are added (combined) to give the resulting coupler point perpendicular displacement or longitudinal path curvature. If unequal lengths of the links O_1A and O_2B were used, point C could be placed away from the midpoint of link AB to compensate for this inequality in the curvatures of the paths of A and B and thus cause C to trace an approximate straight line. The proportions for such a configuration, shown in Fig. 8.15a, can be obtained by using Eq. (8.20) with $K_C = 0$ and noting that the curvatures of the paths of pivots A and B are of opposite sign. This gives the result

$$\frac{AC}{BC} = \frac{O_2B}{O_1A} \tag{8.21}$$

---

[3]The definitions used here are as follows: A *perpendicular displacement* is a displacement of a point perpendicular to the line connecting the coupler pivots, and a *longitudinal displacement* is a displacement in a direction parallel to that line. A *perpendicular curvature* is a curvature of a curve segment whose tangent is approximately perpendicular to the line connecting the coupler pivots. A *longitudinal curvature* is a curvature of a curve segment whose tangent is parallel to the line connecting the coupler pivots. For an analogy, see Sec. 2.5 for definitions of perpendicular and longitudinal components of velocity.

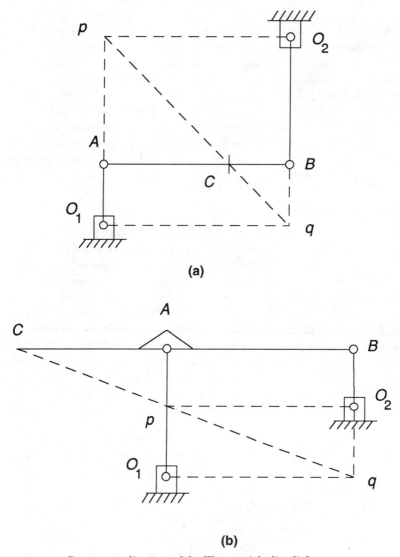

(a)

(b)

Figure 8.15   Some generalizations of the Watt straight-line linkage.

Procedure 8.2: Synthesis of a Generalized Watt Straight-Line Linkage   A
simple graphical construction that can be used to generate the proportions
for a linkage that will produce approximately straight-line motion is also
indicated by dashed lines on Fig. 8.15a. The dashed construction consists
of a rectangle, two of whose sides are perpendicular to the coupler and
which pass through the coupler pivots A and B and are therefore coincident
with links O_1A and O_2B. The other two sides of the rectangle are parallel
to AB. The grounded pivots O_1 and O_2 lie at diagonally opposite corners of

the rectangle, and the straight-line-generating coupler point C lies at the intersection of the coupler line AB and the diagonal pq that *does not* contain the grounded pivots. The validity of this construction is seen easily by noting that triangle ApC is similar to triangle BCq and that $Bq = OA_1$ and $Ap = BO_2$.

This rectangle is drawn easily if the grounded pivot locations O_1 and O_2 are known. Then the diagonal pq is drawn to locate point C. If coupler point C is known, draw lines perpendicular to AB at points A and B and then draw an arbitrary diagonal line through point C. This diagonal line will intersect the perpendicular lines through A and B at points p and q. The required grounded pivots O_1 and O_2 are then located at the corners of the rectangle that *do not* contain the points p and q.

A further generalization of the Watt linkage recognizes that coupler point C that generates the straight line need not lie *between* pivots A and B. A configuration in which point C lies on an extension of coupler line AB is shown in Fig. 8.15b. This will be recognized as the same sort of configuration as that used in Example 8.7 rotated 90° (and left-right reversed). This mechanism could be synthesized by the same method used in that example. It also should be noted that Eq. (8.20) can be used for this configuration too. Substituting $K_C = 0$ into that equation and noting that the curvatures of paths of pivots A and B are both of the same sign but that they subtract for the arrangement in Fig. 8.15b gives the required proportions as the same as given by Eq. (8.21).

Procedure 8.3: Synthesis of an Alternate Form of the Generalized Watt Straight-Line Linkage   A simple graphical construction that can be used to generate the proportions for an alternate form of the Watt linkage is indicated in Fig. 8.15b. This construction consists of a rectangle, two of whose sides are perpendicular to the coupler and which pass through coupler pivots A and B and thus coincide with links O_1A and O_2B. The other two sides of the rectangle are parallel to AB. This rectangle is shown dashed in Fig. 8.15b. The grounded pivots O_1 and O_2 lie at diagonally opposite corners of the rectangle, and coupler point C that generates an approximate straight line lies at the intersection of coupler line AB and an extension of the rectangle's diagonal pq that *does not* contain the grounded pivots. The validity of this construction is easily seen by noting that triangle ApC is similar to triangle BCq and that $Bq = OA_1$ and $Ap = BO_2$ so Eq. (8.21) is satisfied.

The dashed rectangle is drawn easily if the grounded pivot locations O_1 and O_2 are known. That is, draw lines parallel to AB through O_1 and O_2, and complete the rectangle by extending link AO_1 or BO_2 if necessary. Then the diagonal pq is drawn to locate point C on coupler line AB. If coupler point C is known, draw lines perpendicular to AB at points A and B and then draw an arbitrary diagonal line through point C. This diagonal line will intersect the perpendicular lines through A and B at points p and q. Complete the rectangle by drawing lines parallel to AB through p and q. The required grounded pivots O_1 and O_2 are then located at the corners of the rectangle that *do not* contain the points p and q.

Generalizing the Evans linkage

The linkage configuration in Fig. 8.16a is similar to that shown in Fig. 8.9, and differs from the isosceles linkage in Fig. 8.14b only in the proportions of its link lengths. It can be shown that the curvature of the path of point C at position p in Fig. 8.16a and at position p' in Fig. 8.9 (the perpendicular curvature) is given by

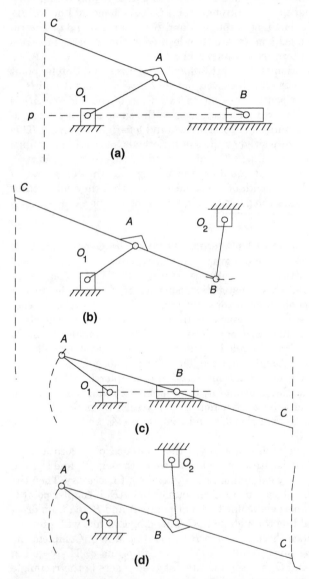

Figure 8.16 Some generalizations of the Evans straight-line linkage.

$$K_C = \frac{1}{\rho_C} = \frac{(AC)(AO_1) - (AB)^2}{(AO_1)(BC)^2} \qquad (8.22)$$

This equation could be used to choose linkage proportions that would cause the path of point C to have any of a wide range of curvatures at those points (p and p' in the respective figures). By setting $K_C = 0$, Eq. (8.22) shows that the linkage will produce a zero curvature at those points if the linkage proportions are such that

$$(AC)(AO_1) = (AB)^2 \qquad \text{or} \qquad \frac{(AC)}{(AB)} = \frac{(AB)}{(AO_1)} \qquad (8.23)$$

Equation (8.23) therefore gives the proportions for an Evans-type straight-line linkage. These proportions are consistent with the proportions in the isosceles and Evans linkages where $AC = AO_1 = AB$. However, whereas the isosceles linkage generates a perfect straight line (which, of course, has zero curvature at all points), the proportions just discussed give zero path curvature *only* near those points p and p' on the line through pivot O_1 and B in the linkages in Figs. 8.9 and 8.16a. Nonetheless, such proportions provide a very good first approximation of a linkage design that will give an approximate straight-line coupler point path.

Example 8.8: A Modified Evans linkage To illustrate the use of Eq. (8.23), the linkage shown in Fig. 8.16b has been synthesized using the proportions $AO_1 = 1.0$ units, $AB = 1.5$ units, and $AC = 2.25$ units. Then the slider at pivot B was replaced by a pivoted link $BO_2 = 1.0$ units in the manner used to convert the isosceles linkage to an Evans linkage in Fig. 8.14. It can be seen that this linkage provides a good approximation to a straight-line coupler point path, with only slight deviations at the ends of the path segment shown.

Equation (8.22) pertains to the condition in Fig. 8.9 when $\theta = 0°$. As was noted in discussing that figure, there is also a region of low coupler point path curvature at position q' that corresponds to the condition when $\theta = 180°$. The path curvature at that point is given by

$$K_C = \frac{1}{\rho_C} = \frac{(AC)(AO_1) + (AB)^2}{(AO_1)(BC)^2} \qquad (8.24)$$

To make the curvature zero, it would be necessary to make

$$(AC)(AO_1) = -(AB)^2 \qquad \text{or} \qquad \frac{(AC)}{(AB)} = -\frac{(AB)}{(AO_1)} \qquad (8.25)$$

Equation (8.25) can be satisfied if $AC < 0$, where a negative value for AC means that point C is to the right of point A rather than to the left as shown. Figure 8.16c shows a configuration with such a rightward placement of point C. The linkage proportions are $AO_1 = 1$ unit, $AB = 2$ units, and $AC = -4$ units.

Figure 8.16d shows the result of replacing the slider in Fig. 8.16c with a pivoted link BO_2, thereby giving another generalization of the Evans linkage. It can be seen that this type of linkage produces a straight-line segment of quite limited length. Although there may be some instances in which the configurations of Fig. 8.16c and Fig. 8.16d may be preferred, in general, they occupy more space for a given length of straight-line segment generated than do the configurations in Fig. 8.16a and Fig. 8.16b.

Comments and suggestions for devising general linkages for generating straight-line segments

The foregoing straight-line linkages had rather specialized geometries in the sense that the coupler points that were used to generate the straight lines were located on the line connecting the coupler pivots. In applications in which such restricted coupler point placement is undesirable, recourse may be taken to Procedures 4.7 and 4.8. To use these procedures, specify the three desired coupler point positions P_1, P_2, and P_3 in the procedures to be on the desired output straight-line path segment. Care must be taken in using these procedures because although the coupler point in the synthesized linkage will pass through the three points, it may deviate appreciably from a straight line *between* the points. Trials and error may be required.

In devising a new straight-line mechanism, it is sometimes helpful to start with a mechanism that is known to generate a good approximation of the desired straight line and then to explore modifications to that linkage. This can minimize the path deviation from a straight line *between* the specified path points. For example, any of the linkages in Fig. 8.14 could be chosen, and the position of the coupler at three locations of the coupler point along the straight-line portion of its path would be noted. Then the links that are pivoted to ground would be discarded, and Procedure 4.4 or 4.5 would be used to determine the links between the coupler and ground. Two examples of the use of this method are shown in Fig. 8.17.

Modifying the isosceles straight-line linkage

Figure 8.17a shows a mechanism described by Chironis[4] as an invention that was patented by James A. Daniels, Jr. This mechanism can be developed by starting with the isosceles mechanism of Fig. 8.14b.

---

[4]Chironis, N. P., *Mechanisms and Mechanical Devices Sourcebook*, McGraw-Hill, New York, 1991, pp. 110–111.

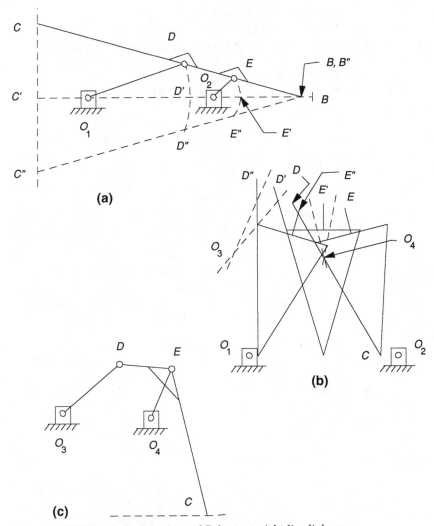

Figure 8.17 Modifying the isosceles and Roberts straight-line linkages.

The coupler BC of the mechanism in Fig. 8.14b is shown in three positions (BC, $B'C'$, and $B''C'''$) in Fig. 8.17a. Then, in accordance with Procedure 4.4, coupler pivots D and E are chosen at convenient locations on the coupler, and their locations at the three coupler positions are plotted as shown. The perpendicular bisectors of DD' and $D'D''$ locate the ground pivot O_1, and the perpendicular bisectors of EE' and $E'E''$ locate the ground pivot O_2. This method allows considerable freedom in the choice of pivot locations. However, a transmission angle problem occurs if coupler CDE must pass near or through position

$C'D'E'$. The coupler portion BE can be discarded if only the vertical straight line shown is to be generated. Note also that the coupler position $C''D''E''$ could have been chosen to be *above* position $C'D'E'$ rather than below as shown, and the analogous geometric construction could be used to locate the ground pivot points.

For this type of synthesis or linkage modification, line CC''' need not be perpendicular to line $C'B$.

Example 8.9: Modifying the Roberts Straight-Line Linkage Figure 8.17b and c shows the result of using this method for modifying the Roberts straight-line mechanism of Fig. 8.14d. In Fig. 8.17b the coupler ABC of Fig. 8.14d is shown as an unlabeled isosceles triangle in each of its three possible positions. Moving pivots D and E are then placed at convenient locations on that coupler. For this example, the locations are chosen such that pivot D is along an extension of the left-hand side of the triangle. The pivot E is located above the midpoint of the short side of the triangle. The successive positions of these pivots are then shown as D, D' D'', E, E', and E''. Then, in accordance with Procedure 4.4, the perpendicular bisectors of DD' and $D'D''$ (shown dashed) locate the ground pivot O_3 and the perpendicular bisectors of E_1E_2 and E_2E_3 (shown dashed) locate the ground pivot O_4. Figure 8.17c shows the resulting linkage and the path followed by its point C.

Instead of choosing locations for the moving pivots D and E in two examples in Fig. 8.17, the ground pivots O_1 and O_2 (or O_3 and O_4) could have been chosen, and Procedure 4.5 could have been used to locate the moving pivots. In all cases in which the commonly used straight-line mechanisms are modified, the resulting coupler point paths should be checked to see whether the desired degree of approximation to a straight line has been achieved.

8.6 Types of Coupler Point Paths

In the design of a mechanism, it is often necessary to cause some point in that mechanism to follow some noncircular path. This path may be required in order to position a product at a sequence of prescribed locations, or it may be required as an input to another portion of the mechanism. This section and the next discuss such applications. It will be seen that paths of points on the couplers of four-bar linkages are useful. This section describes the wide variety of path shapes that can be generated, and suggests how to find points on a coupler that provide the various types of curves.

The coupler of a four-bar linkage can be considered to be subjected to rotation about a point on that coupler (such as either of its pivots) plus translation of that point. This combined motion can cause other points that are attached to the coupler (coupler points) to travel along paths (coupler point paths) that have a wide variety of shapes. Some

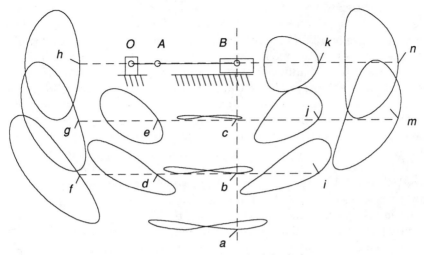

Figure 8.18 Coupler point paths for an in-line crank-slider linkage.

examples of the more simple path shapes were discussed in Secs. 8.4 and 8.5. An excellent compilation of thousands of varied coupler point paths and the crank-rocker linkage that produces each is given in Hrones and Nelson.[5] This book can be searched for the coupler point path shape desired, and the linkage proportions and coupler point location that will give that path can be determined from the dimensions given.

The purpose of this section is to provide guidance in searching an atlas such as Hrones and Nelson or, if such an atlas is not available, to provide guidance in using kinematics software in a trial-and-error search. The motion phenomena that affect the coupler curve shapes are described, and a systematic arrangement of the locations of points on the coupler versus the type of path generated by each is presented.

Coupler point paths for an in-line crank-slider linkage

Figure 8.18 depicts an in-line crank-slider linkage that consists of a crank OA that is pivoted to ground at point O and a coupler AB that is pivoted to the crank at A and to the slider at B. Attached to that coupler and therefore *moving with it* is a grid that is shown dashed. This grid consists of a line $abcB$ that is perpendicular to the "axis" AB of the coupler and consists of a set of lines *fdbi, gecjm,* and *hABkn* that are

---

[5]Hrones, J. A., and Nelson, G. L., *Analysis of the Four-Bar Linkage,* MIT Press and Wiley, New York, 1951 (currently out of print).

perpendicular to line $abcB$. Coupler points indicated by the lowercase letters a through n are placed on the grid lines, and as the coupler moves, these points follow the paths shown as curves connected to the points. The small tic marks on these curves and on similar curves in subsequent figures indicate the positions of the coupler points when the crank is at zero angle. It will be noted that the paths followed by points h, A, k, and n on the uppermost line are essentially the same curves as shown in Fig. 8.9. If the crank OA is considered to be rotating counterclockwise, it will be found that all points that are on the same side of line $abcB$ as pivot A that connects the coupler to the crank will travel around their paths in a counterclockwise direction, while the points on the opposite side of that line will travel along their paths in a clockwise direction. The points that lie exactly on the line $abcB$ will follow paths that are "figure-eights." These points will travel around the loops of the figure-eights that are closest to pivot A in a counterclockwise direction and around the loops that are farthest from pivot A in a clockwise direction.

Only curves for points below the coupler axis AB are shown. The curves for points above that axis are mirror images of those shown. Note the gradual change in curve shape as we choose different positions for the point relative to the coupler. As we choose a coupler point farther and farther from pivots A and B, the path followed by that point appears to be stretched out more and more along an arc that is centered in the general vicinity of the pivots. This effect can be understood by noting that the coupler motion can be considered to consist of a translational motion of pivot B plus a rotational motion about pivot B. The farther a coupler point is from pivot B, the greater is the influence of that rotation on the motion of that point, until the rotational contribution to the point's motion greatly dominates the translational contribution.

In order to gain some insight into the reasons for the shapes of the coupler curves, refer to Fig. 8.19. In Fig. 8.19a the linkage is shown in the position where the crank angle is zero, just as shown in Fig. 8.18. The coupler points b, d, and i (corresponding to the points with the same labels in Fig. 8.18) are shown rigidly connected to coupler link AB. It can be seen that because crank pivot A is moving upward, the coupler is rotating clockwise. Therefore, point d will have a velocity that is upward and to the left, as indicated. In Fig. 8.19b the linkage is shown with crank OA having been rotated counterclockwise to an angle of 90°. In this condition the coupler will have zero angular velocity and the velocity of point d will be horizontal, just as is that of pivot A. By comparing the motions implied for point d in Fig. 8.19 with the shape of the path as indicated for point d in Fig. 8.18, it can be seen that point d travels along its path in a counterclockwise direction.

Similar comparisons of the motions implied for point i in Fig. 8.19 with the path shown for that point in Fig. 8.18 will show that point i travels around its path in a clockwise direction.

The motion of point b is perhaps a bit more subtle. Note that in both Fig. 8.19a and Fig. 8.19b the velocity of point b is horizontal and leftward. However, point b is displaced vertically upward in Fig. 8.19b relative to its position in Fig. 8.19a. This sequence corresponds to the leftward S-shaped portion of the curve from b in Fig. 8.18. In Fig. 8.19c the velocity of point b is horizontal and rightward, but point b has been displaced downward from its position in Fig. 8.19b. This portion of the sequence corresponds to the left-hand end portion of the curve in Fig. 8.18. It can thus be seen that the coupler point travels around the left-hand loop of the figure-eight in a counterclockwise direction. By

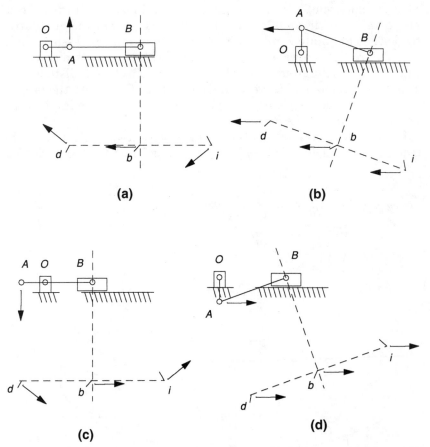

(a) (b)

(c) (d)

Figure 8.19 Direction of motion of various coupler points during a cycle of input link motion.

continuing the comparisons of Fig. 8.19c, d, and a with the remainder
of the path of point b in Fig. 8.18, it can be seen that the point travels
around the right-hand loop of the figure-eight in a clockwise direction.

Alternatively, it can be noted that point b in Fig. 8.19 passes through
two vertical high points (at Fig. 8.19b and d) and two vertical low
points (at Fig. 8.19a and c) during each complete rotation crank OA.
However, in each complete rotation of the crank, point b passes
through only one leftward extreme position and one rightward extreme position. The path can then be viewed as a Lissajous figure with
a vertical frequency of twice its horizontal frequency, and such a figure
is generally a figure-eight figure.

Coupler point paths for an offset crank-slider linkage

Slightly more complicated relationships pertain to the natures of coupler point paths in an offset crank-slider linkage, but the qualitative
tendencies are quite similar to those for the in-line crank-slider linkage. Figure 8.20 shows an offset crank-slider linkage with a crank OA
that is pivoted to ground at point O that is offset by a distance y =
1 unit from the line of travel of the slider at B. Otherwise the proportions of this linkage are the same as those of the linkage in Fig. 8.18;
i.e., OA = 1 unit and AB = 3 units.

In this linkage the transmission angle (see subsection entitled "The
transmission angle in a crank-slider linkage" in Sec. 3.6) varies be-

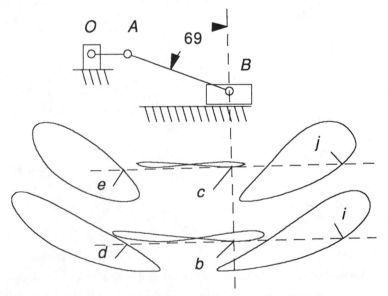

Figure 8.20 Coupler point paths for an offset crank-slider linkage.

tween extreme values of about 48° and 90°. The grid line bcB is chosen to be inclined to the coupler axis AB at an angle of 69°, which is the average of the extreme values of the transmission angle. Call this line the *coupler grid reference line*. (Although this line appears to be nearly vertical with the crank in the position shown, this is merely coincidental.) By choosing such an inclination of the grid reference line, as the linkage passes through its entire cycle of motion, the extremes of the angular position of that line lie at equal angles on either side of the vertical, just as did the grid reference line bB shown in Fig. 8.19b and d. Therefore, if the coupler grid is based on a grid reference line that is inclined in this manner, the discussion relative to Figs. 8.18 and 8.19 is also applicable to the offset crank-slider linkage. Because of the similarity of curves for the two types of linkages, only a few curves are shown in Fig. 8.20. Some of the "up-down" symmetry of the curve set for the in-line linkage is lost in the offset linkage.

**Coupler point paths for
a crank-rocker linkage**

Figure 8.21 shows a crank-rocker linkage with a crank length $O_1A = 1$ unit, a coupler length $AB = 3$ units, a rocker length $BO_2 = 2.5$ units,

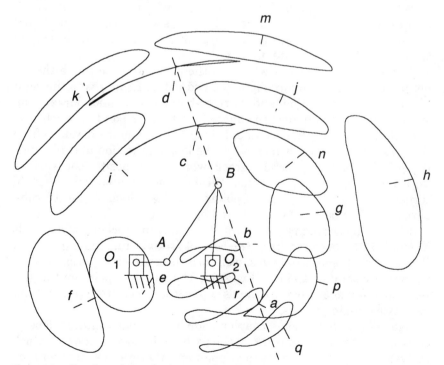

Figure 8.21 Coupler point paths for a crank-rocker linkage.

and a ground link length $O_1O_2 = 2.5$ units. The transmission angle for this linkage varies between extreme values of 30° and 78° (see subsection entitled "Transmission angles in pin-jointed four-bar linkages" in Sec. 4.3 and also Fig. 4.3b and c). The grid reference line $abBcd$ is chosen to be inclined at 54° from the coupler axis line AB as shown. This angle is the average of those extreme values of transmission angle, so as the linkage passes through its entire range of positions, this grid reference line will attain an extreme position of 24° on each side *of the output rocker* BO_2. Notice that these grid reference line extreme positions are symmetrical relative to the rocker, which is a moving link; they are not symmetrical relative to vertical, as was the case in the crank-slider linkages. Nonetheless, most of the general observations that were made for crank-slider linkages relative to the types of curves generated by coupler points at various locations on the coupler also can be applied to crank-rocker linkages.

Notice that coupler points a and b in Fig. 8.21 are located on the grid reference line and that they follow figure-eight–shaped paths which, although they are not exactly symmetrical, each have two loops that are roughly comparable in size. It is also found that as in the case of crank-slider linkages, the points such as e, f, k, and i that are on the same side of the grid reference line as the crank-to-coupler pivot A travel around their paths in a counterclockwise direction if the crank rotates counterclockwise, whereas the points such as g, h, j, m, n, and p that are on the opposite side of the grid reference line travel around their paths in the opposite direction.

Points q and r, which are near to the grid reference line although not on it, also follow figure-eight–shaped paths. Point q is to the right of the grid reference line, and its right-hand loop is much larger than its left-hand loop, whereas the opposite is true for point r, which is to the left of the grid reference line. It is found that as the distance from the grid reference line is increased, the loop on the side toward which the point is moved becomes larger, while the opposite loop shrinks until it becomes a cusp and then becomes merely a sharply rounded corner, as seen in the paths of points p and g, which should be compared with the paths of points a and b, respectively.

Because the curved path of pivot B in the crank-rocker linkage adds a complication not present in the case of straight-line motion of pivot B in the crank-slider linkages, well-formed figure-eight paths are not followed by points c and d in Fig. 8.21. The path of a point on the grid reference line above point B tends to enclose very little area and can have two crossing points.

Figure 8.22 shows some coupler point paths that contain cusps. A cusp appears as a sharp point on the path, and it is a position at which the velocity vector of that coupler point exactly reverses direction; the velocity vector is exactly zero for an instant. Cusps occur on the paths of coupler points that are located on the moving centrode of the coupler relative to ground (see Sec. 2.5). Such points on the coupler can be found by kinematic inversion (considering the coupler to be grounded while the other links move relative to it). In Fig. 8.23a, the polygon O_1ABO_2 represents the linkage with the same labels in Fig. 8.22. In Fig. 8.23 the coupler AB is considered grounded, and the crank O_1A is chosen to be at some convenient angle to it. (In this instance the angle corresponds to an input crank angle of 150° relative to the ground link O_1O_2.) The positions of links O_1O_2 and O_2B are drawn. The two links that are pivoted to the coupler (O_1A and O_2B) are then extended until they intersect at point b as shown. Point b is a point attached to the coupler and on the moving centrode. This coupler point is shown in the position for a zero crank angle as point b in Fig. 8.22,

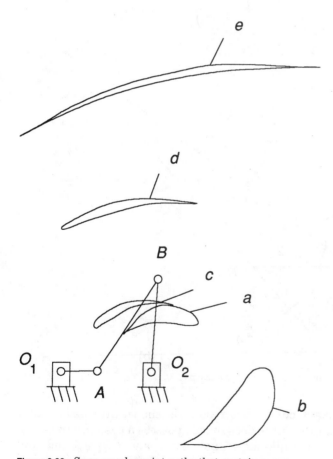

Figure 8.22 Some coupler point paths that contain cusps.

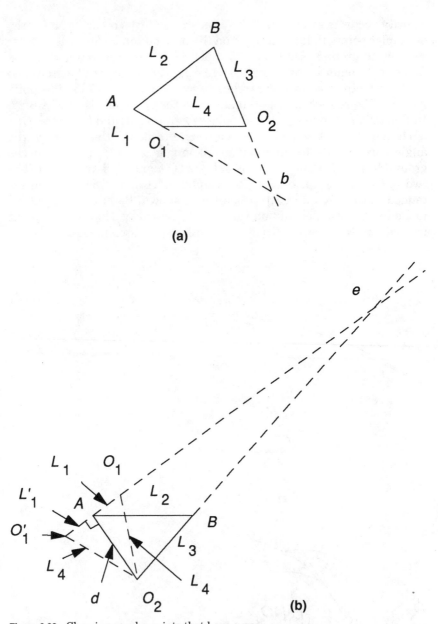

Figure 8.23 Choosing coupler points that have cusps.

and the coupler point curve that results is shown attached to that point; it is seen that the curve does indeed possess a cusp. By the same process with different crank angles, points a, c, and d were found, and their associated curves are shown in Fig. 8.22; it is seen that these curves also possess cusps.

It is possible to find one coupler point whose path will have two cusps, although the location of this point may be quite inconvenient. Determining the location of this point involves kinematic inversion, and the construction is shown in Fig. 8.23b. The coupler AB is shown forming a triangle in which the two other sides are rocker O_2B and a length of d, where the length d is obtained from the expression $d^2 = (O_1O_2)^2 - (O_1A)^2$. Then at point A a line (shown dashed) is constructed perpendicular to the side AO_2, and this line is extended until it intersects an extension of side O_2B of the triangle. The point of intersection is the desired point attached to the coupler, and it is shown as point e in Fig. 8.23b. This point is a coupler point that will follow a path that has two cusps. The validity of this construction can be seen by noting that the link O_1A can be positioned as shown in position O_1A and also in the position $O_1'A$ as shown dashed. Both these positions represent possible positions of the linkage in this inversion. In both these positions the link O_1A, when extended, intersects the link O_2B at the same point e, so the moving centrode passes through point e twice.

Point e is attached to coupler AB and moves with it, generating the path at e in Fig. 8.22. It is seen that as is often the case, the curve encloses very little area; it almost retraces its path on the return stroke.

Comments and suggestions on crank-rocker coupler point paths

The properties of a crank-rocker coupler point path that are usually of interest are (1) the possible existence of approximately straight-line segments of the path, (2) the sharpness of curvature of the path (whether approximately constant or varying appropriately), and (3) the direction of travel of the point around the path. The use of straight-line portions of coupler paths was discussed earlier in this chapter. The use of approximately constantly curved portions of coupler point paths is discussed in Sec. 8.8.

It will be noted that points that are near the coupler's pivots tend to describe paths that are convex at all points on the path. The convex curvature on the side of the path nearest to coupler axis AB tends to decrease (become more straight) as the point is moved away from that axis until it becomes more or less straight and then becomes concave.

The unique features of figure-eight–shaped paths are that (1) the coupler point passes through the same point twice during each cycle of the input crank (although the direction of passage is different in the two passes) and (2) the coupler point travels around the two loops in opposite directions.

The unique feature of a cusp on a path is that the x and y components of the coupler point's velocity both go to zero simultaneously at the cusp point. The coupler is rotating about that coupler point for that instant. Before using a coupler point path because of the properties of its one or two cusps, readers should consider whether a point on the rocker would provide the needed motion characteristics. The velocity vector of a rocker point also goes to zero and reverses at each end of its travel, but the path of a point on a rocker encloses zero area, unlike the path of a coupler point.

Coupler point paths for a Grashof-type double-rocker linkage

A Grashof-type double-rocker linkage is a pin-jointed four-bar linkage in which the only link that is capable of rotating through 360° relative to ground is the coupler, and the coupler is also the shortest link in such a linkage. The other two movable links in this linkage can only rock back and forth through limited angles. An example in which the path of a coupler point in a Grashof-type double-rocker linkage is useful is shown in Fig. 8.14e. This linkage is the Tchebycheff straight-line mechanism.

Grashof-type double-rocker linkages are seldom operated throughout their full possible ranges of operation because the coupler is not pivoted to ground, and, therefore, there is difficulty in driving the coupler in continuous rotation. To provide a continuous rotation to the coupler, additional components must be connected to the coupler, such as a flexible shaft, gearing, a chain and sprockets, or a belt and pulleys. Fortunately, any coupler point path that can be produced by a Grashof-type double-rocker linkage also can be provided by a crank-rocker linkage that is a cognate of the double-rocker linkage. The method for finding such a cognate linkage is given in Procedure 4.9. An example of the substitution of such a cognate linkage for a Grashof double-rocker linkage is shown in Fig. 8.14f, which is a crank-rocker cognate of the double-rocker linkage in Fig. 8.14e. The crank-rocker linkage in Fig. 8.14f obviously can be driven quite conveniently by a continuously rotating shaft at point O_1.

Coupler point paths for a drag-link mechanism (double-crank linkage)

Examples of characteristics that are unique to coupler point paths in drag-link mechanisms are indicated in Fig. 8.24. In each part of this figure the input crank O_1A is assumed to be rotating counterclockwise, and the linkage is shown in the position for zero input crank angle. A point p is shown in that zero crank position on the path that it follows as the crank is rotated. Parts a, b, and c of Fig. 8.24 are drawn separately because even though point p is located quite differently relative

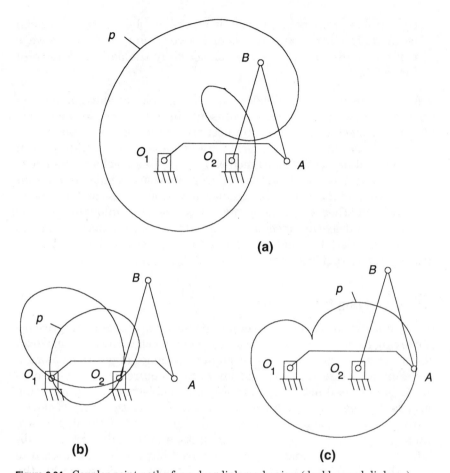

(a)

(b)   (c)

Figure 8.24 Coupler point paths for a drag-link mechanism (double-crank linkage).

to the coupler in each of these parts, the paths depicted overlap greatly and would be difficult to distinguish if they were all in one figure.

Unique features that these paths possess are as follows:

1. In all cases the coupler point travels along its path in the same direction as the rotation of the input crank, regardless of the location of the coupler point on the coupler.

2. Each path can consist of one or two loops, and if a path consists of two loops, either the loops will overlap as in Fig. 8.24*b* or one loop will be inside the other as in Fig. 8.24*a*. This is so because the coupler point always travels around all loops in the same direction as the rotation of the input crank. Note that the curve in Fig. 8.24*b* contains three crossing points.

3. By suitable choice of point location on the coupler, an inner loop can be made to shrink until it becomes a cusp, and that cusp always points *inward* toward the area enclosed by the path, as shown in Fig. 8.24c.

Although coupler point paths in a drag-link mechanism possess unique characteristics, their use presents inconvenience in the means for making mechanical output connections to the chosen coupler point. The complexity of the mechanical arrangement required can be seen either by making a cardboard model of a drag-link mechanism such as shown in Fig. 4.14d or by drawing a front view and a side view of that mechanism and then visualizing how the various links can be made to pass by each other and pass by the various pivots. Unfortunately, the cognates of a drag-link mechanism are also drag-link mechanisms, so if path characteristics that are unique to drag-link coupler point paths are absolutely required, only a drag-link linkage can be used.

8.7 Conveying Mechanisms

A very effective application of coupler point paths occurs in progressive conveying mechanisms. The purpose of such mechanisms is to move an object or objects progressively through a sequence of positions along a table or guideway. The general operation required of these mechanisms is indicated in Fig. 8.25. In Fig. 8.25a objects are shown resting on a table together with a frame that has fingers that project downward between the objects. The arrow a indicates that the fingers are lowered until they are between the objects, and then the arrow b indicates that the frame and fingers are moved horizontally rightward to push the objects to the right. At the rightmost end of its stroke, the frame and its fingers are raised clear of the objects, as indicated by arrow c, and are then moved leftward, as indicated by arrow d, to a position from which the sequence is repeated. It can be seen that this operation moves the objects progressively rightward through a sequence of positions. In a manufacturing process, a different operation might be performed on each object at each position during the time when the frame and fingers were in their return strokes (arrow d).

Figure 8.25b shows that a similar action could be accomplished by a set of fingers that extend upward through a slot in the table, and Fig. 8.25c shows that such an action could be used with platforms on the fingers to lift the objects off the table surface each time they are moved horizontally.

This periodic sequence of horizontal and vertical displacements of the frame and fingers is often called a *box motion*. It can be provided by a pair of cam and follower systems, one of which moves the frame and fingers vertically and the other of which moves the frame and

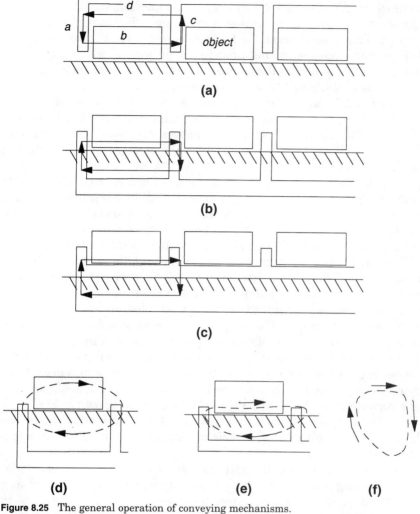

Figure 8.25 The general operation of conveying mechanisms.

fingers horizontally. There are several commercially available self-contained units or *cam boxes* that can provide this action. These units work well at low to moderate speeds and with low to moderate masses. However, instead of such a unit, it is often possible to use a four-bar linkage coupler point path in a mechanism that is simpler and which operates more smoothly. The following paragraphs describe such applications.

Figure 8.25*d* shows an arrangement in which the frame and fingers are caused to move along an oval path rather than the box path shown in Fig. 8.25*b*. This oval path is the same coupler point path as used in the conveying system shown in Figs. 4.28 and 8.3 and which was used to illustrate Procedure 4.9. Not only is this mechanism simpler than a cam-driven system, its operation is much smoother than that of a cam system. This is so because the horizontal and vertical accelerations of the primary masses in this linkage-driven system are almost sinusoidal, and therefore, as discussed in Chap. 10, the tendencies to cause vibrations in the machine are minimized.

Figure 8.25e and f shows examples of other types of coupler point paths that might be used in conveying mechanisms. The curve in Fig. 8.25e could be the curve followed by point f or point i in Fig. 8.21. These curves have relatively straight motions during pushing of an object. The curve in Fig. 8.25f is the curve followed by point g in Fig. 8.21. This latter curve has been rotated 180°, and it drops rather abruptly and deeply at both ends of a relatively straight horizontal travel. Such abrupt changes in motion direction can be advantageous in some applications. The corners on the curve may be "sharpened" by suitably choosing the coupler point location.

It usually is necessary to cause the frame and fingers to remain horizontal as they are raised, lowered, and moved horizontally. Of course, this can be done by using sliding guides. If the frame is long, however, it usually is more practical to use some parallelogram mechanism such as shown in Fig. 8.26a. This mechanism can be added to a box motion or any of the other types of conveying motions shown in this section.

If a four-bar coupler-point path-generating mechanism is used in the conveying system, the frame and fingers can be made to remain horizontal as it moves by two other methods. In the first method, portions of a cognate of the four-bar linkage can be used, as shown in Figs. 4.27 and 4.28 and as described in Procedure 4.9. This method adds two links and three pivots to the basic driving linkage, resulting in a six-bar linkage.

In the second method, the output rocker and a portion of the coupler of the driving four-bar linkage can be duplicated and connected to a second point on the finger frame. Such an arrangement is shown in Fig. 8.26b. The four-bar linkage driving the conveyor in this figure is linkage O_1ABO_2, with coupler point C pivoted to the finger frame. This is the same linkage as shown in Fig. 8.14f, so the fingers travel at almost constant velocity as they push the objects. Output rocker BO_2 is duplicated as DO_3, and a *portion* of coupler ABC is duplicated as DE. By using a link such as FG to form a parallelogram, links DO_3 and DE are caused to move exactly the same as links FO_2 and BC so that point E follows a path that is identical to that followed by point C and the finger frame remains horizontal. This second method adds four links and six pivots to the driving four-bar linkage, resulting in an eight-bar linkage.

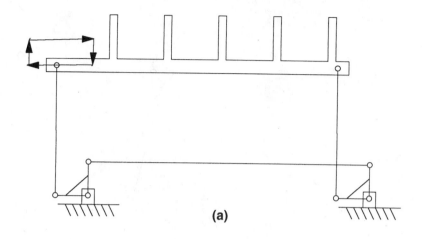

(a)

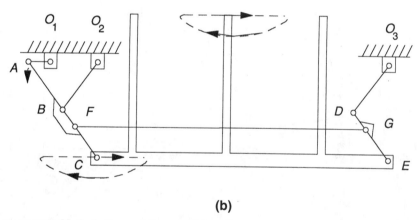

(b)

Figure 8.26 Methods for providing parallel motion.

8.8 Dwell Mechanisms

Coupler point paths can be used to provide mechanisms whose motions contain appreciable periods in which the output velocity is near zero. That is, they can be used to provide outputs with "dwells." An illustration of such a use of a coupler point path is given in Fig. 8.27a. In this figure the dashed curve is the path followed by point i in the mechanism in Fig. 8.21. (The actual path-generating mechanism is not shown in Fig. 8.27a.) It can be seen that the portion of the curve that lies between points p and q in Fig. 8.27 has approximately a constant radius of curvature. Although the four-bar linkage that produces this

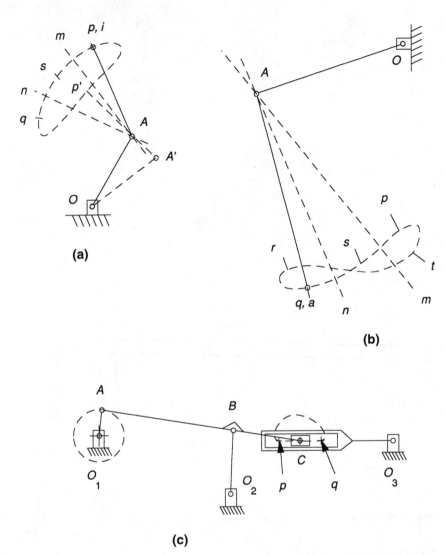

Figure 8.27 The use of coupler point paths in providing dwells.

curve is not shown in Fig. 8.27, a link iA is shown pivoted to coupler point i when point i is at location p, and the other end A of the link is shown at the center of a circular arc that closely fits the coupler point curve. The center of the arc is at the intersection of lines m and n, which are perpendicular bisectors of lines (not drawn) connecting points p and s and points s and q, respectively. The link iA is pivoted to a rocker AO that is pivoted to ground at point O. The resulting combination of the path-generating or driving four-bar linkage plus the two links shown in Fig. 8.27a is a six-bar linkage. During the portion of the driving four-bar linkage's motion that moves point i between positions p and q on the path, link iA will merely pivot about point A (approximately), so point A will not move and link OA will not rotate (it will dwell). During the remainder of the cycle of the motion of the four-bar linkage, link AO will rock back and forth. The duration of the dwell will equal the time it takes coupler point i to travel between points p and q, which in this case corresponds to 120° of input crank rotation in the linkage in Fig. 8.21. The total angular motion of output rocker link AO is determined by the width sp' of the coupler point path, the length AO of that link, and the location of pivot O.

The coupler point path to be used in a dwell mechanism can be chosen from an atlas of such curves, it can be synthesized by techniques described earlier in this chapter, or it can be found by trial and error using a computer program and by using the discussion in Sec. 8.6 as a guideline.

Figure 8.27b shows the figure-eight–shaped curve that is followed by coupler point a in Fig. 8.21. A portion of the curve that has an almost constant radius of curvature is seen to exist between points p and q in Fig. 8.27b. Coupler point a is shown as having moved along the curve to location q, and a link aA is shown pivoted to the coupler of the driving four-bar linkage (not shown) at that coupler point a when it has traveled to location q. Pivot A is located at the center of the circular arc that best fits this portion of the curve between points p and q. As in Fig. 8.27a, link aA is pivoted to a rocker AO that is pivoted to ground at point O. Also, as in Fig. 8.27a, the rocker will dwell as the coupler point a moves between points p and q. The dwell produced by this six-bar linkage persists during approximately 90° of rotation of the crank in the path-generating four-bar portion of the linkage. However, when point a moves from location q to r, the rocker will rotate clockwise from the dwell position; it will then rotate counterclockwise until the coupler point approaches a location such as indicated at t, at which time the output rocker AO will be counterclockwise of the dwell position. It is thus seen that the dwell in Fig. 8.27b occurs between the extreme positions of the output rocker rather than at either of the extremes, as was the case in Fig. 8.27a.

The dashed lines m and n in Fig. 8.27b are the perpendicular bisectors of line segments ps and sq, respectively, and the intersection of lines m and n indicates the location of pivot A during the dwell.

Figure 8.27c shows a dwell mechanism that uses a coupler point path that has an approximate straight-line section. The driving four-bar linkage O_1ABO_2 is the same linkage synthesized in Fig. 8.12. A rocker with a slide is pivoted at O_3 such that both the slide and pivot O_3 are located in line with the straight-line portion of the path of coupler point C. It can be seen that as the coupler point C travels along the straight-line portion of its path, the slider at point C will slide in the rocker without rotating that rocker. That is, the rocker will dwell during that time. For the particular link proportions in this example, the dwell will persist during about $110°$ of rotation of input crank O_1A. When coupler point C is in other portions of its curve, the rocker will be rotated clockwise from its dwell position.

8.9 Combining Motions in x and y Directions

Useful motions can be produced by combining an output motion of one linkage or mechanism with that of a second mechanism, particularly when these output motions are not parallel to each other. For example, if in Fig. 8.28 a horizontal reciprocating motion x_A of point A is the output of one linkage (not shown) and a vertical reciprocating motion y_B of point B is the output of a second linkage (not shown), then pivot C will experience horizontal and vertical reciprocating motion components that are, respectively, approximately equal to those two motions (if links AC and BC are long relative to the magnitudes of the displacements so that links AC and BC do not rotate appreciably).

If, then, the two motions x_A and y_B are identical except that x_A is a horizontal reciprocation and y_B is a vertical reciprocation, and if these motions are synchronized as indicated in Fig. 8.28b, then the y displacement of pivot C at any time will be equal to its x displacement at that time, and the motion of pivot C will be approximately a straight line inclined at $45°$, as shown in Fig. 8.28a. By varying the magnitudes of the x and y motions independently, the combined motion of point C can be varied in both length and direction.

Consider shifting the entire upper curve in Fig. 8.28b to the right relative to the lower curve so that the y_B displacement goes from negative to positive (i.e., moves vertically upward) before the x_A displacement goes from negative to positive (i.e., goes from left to right). Such a timing is shown in Fig. 8.28c. The resulting motion of point C will be an approximate box motion such as shown in Fig. 8.28d and discussed in Sec. 8.7.

Figure 8.28e schematically shows a five-bar linkage that is frequently used in the cloth feed mechanism of sewing machines. The inputs to this linkage are the rocking of links O_1A and O_2B, which causes pivots A and B to oscillate horizontally and vertically, respectively. Links O_1A and O_2B are driven by cam followers in a timing relationship very much like that shown in Fig. 8.28c, although the cams do not produce sharp corners on the curves. The result is that as in the

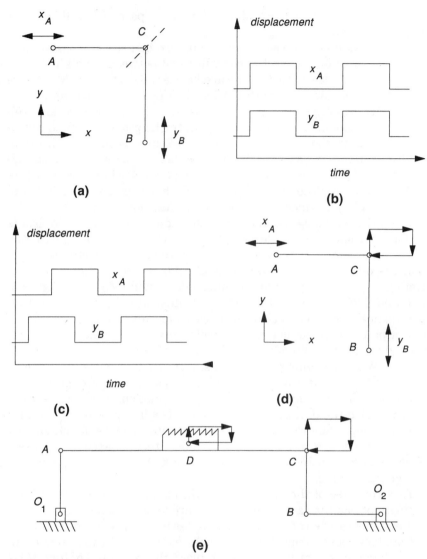

Figure 8.28 Combining motions in x and y directions.

case in Fig. 8.28d, point C in Fig. 8.28e follows a box-type path as shown. (Note that the combination of links AC and BC is the same in Fig. 8.28d and Fig. 8.28e except for their proportions.) Point D will have essentially the same horizontal displacement as points A and C, but because point D is about halfway between pivots A and C, point D will have about half as much vertical displacement as point C, as explained in Sec. 8.2. Thus point D will follow a box-type path, just as will point C, but the vertical dimension of the path of D will be about half that of point C.

The serrated edge above point D will rise beneath the sewing machine's presser foot, push the cloth forward between each stitch, and then drop below the table for a return before the next stitch. Means are provided in the mechanism that drives link O_1A to vary the amplitude of the rocking of that link and to reverse the phase of that rocking relative to the motion of rocker O_2B so that the length of the stitch can be varied and so that the cloth can be fed in the reverse direction. Multiplication such as described in Sec. 8.2 can be used for such adjustments. Also, in order to save space, in actual sewing machines, link BC is usually replaced by a sliding block that is pivoted at point B to the link O_2B and which slides in a horizontal slot in link AC.

From the foregoing it can be seen that if a point in a mechanism is to follow a path that encloses a nonzero amount of area (i.e., if the point is not simply to retrace the same path back and forth), the components of its motion in two perpendicular directions must differ in timing or phase rather than in amplitude. It will be seen in Figs. 8.3, 8.9, and 8.25 and in the example that illustrates Procedure 4.9 that the vertical displacements in these cases are about one-quarter cycle out of phase with the horizontal displacements. Periodic actions that are a quarter-cycle out of phase with each other are said to be in *quadrature*. When designing a mechanism for feeding parts or material and/or intermittent processing of such fed parts, it is frequently helpful to search a mechanism for quadrature motions that can be combined to give the desired timing between feeding, processing, and return actions. Quadrature motion components can be found in any motion in which a point traces a planar path that encloses an area. This, of course, includes all points on cranks because such points follow circular paths.

As a converse of the preceding, it will be noted that if two reciprocating motions that are in quadrature are available, those *reciprocating* motions can be combined in a manner to produce a *rotary* motion.

Consider now a mechanism such as shown schematically in Fig. 8.28e but in which the serrated block that is on top of link AC is constrained to move with link AC in the x and y directions but that is caused to slide in a direction z that is perpendicular to the plane of the page by some additional mechanism. Such an arrangement is shown schematically in front and top views in Fig. 8.29. In the front view, point D is indicated to be following a box-type path, just as in Fig. 8.28e. In the top view, the serrated block s is shown connected to a link by a spherical joint at point e so that by moving that link in the z direction, the block can be caused to move in the z direction relative to link AC. If this z displacement is synchronized with the x displacement of link AC in the same manner as shown in Fig. 8.28b,

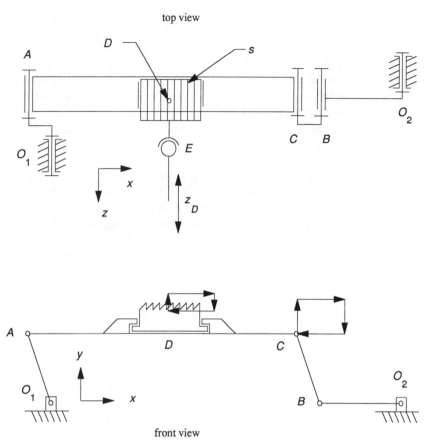

Figure 8.29   Combining motions in x, y, and z directions.

the top view of the path followed by point D will look like that in
Fig. 8.28a. That is, in the xz plane the motion of point D will be back
and forth along an approximate straight line whose direction and
length can be varied by varying the magnitudes of the x and z displacements applied, just as described in connection with Fig. 8.28b
and c. This means that the rectangular path shown in the front view
can be tilted out of the xy plane about the y axis by any desired angle.
With such an arrangement, the cloth can be fed in any direction in
the xz plane.

Lissajous figures and geared
five-bar linkages

In the foregoing, the input motions in the two perpendicular directions
were shown as being square-wave functions of time, so a point whose
displacement components were approximately equal to those input

motions traced a rectangular path. In the example mechanisms shown in Fig. 8.3 and in the example that illustrates Procedure 4.9, the horizontal and vertical input motions were approximately sinusoidal functions of time or of input shaft angle, and the path of the point that combined those input motions was approximately elliptical. This was the result of using input motions that were of the same frequency and were about 90° out of phase with each other. By using input motions that differ from each other in frequency and/or phase, a variety of path shapes can be produced. These paths tend to resemble what are known as *Lissajous figures.*

A Lissajous figure is generated by a point whose x displacement is sinusoidal, such as $x = a \sin \theta$, and whose y displacement is sinusoidal such as $y = b \sin n(\theta + \phi)$. The constants a and b merely determine the x and y size of the curve. An example of a Lissajous figure in which $a = 2b$, $n = 3$, and $\phi = 0$ is shown in Fig. 8.30a. It will be noted that there are three points at which the curve is a maximum in the y direction and

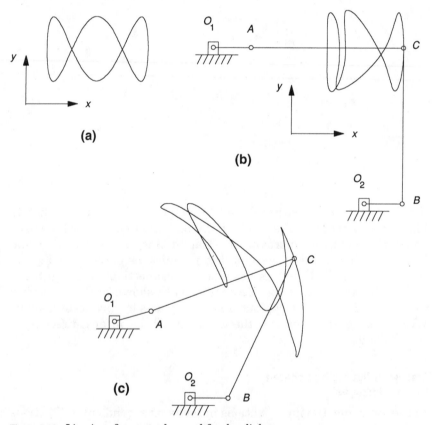

Figure 8.30 Lissajous figures and geared five-bar linkages.

three points at which it is at a minimum in the y direction, for a total of six points of extreme y values. It also will be noted that there is one point of maximum x value and one point of minimum x value, for a total of two points of extreme x values. The ratio of these total numbers of extreme points is $6/2 = 3 = n$. The constant n always equals the ratio of the numbers of extreme points in the curve. The effect of varying ϕ is to shift the locations of these extreme value points right and left in the case of the y extremes and up and down in the case of the x extremes.

As shown in Fig. 8.28e, a five-bar linkage can be used to combine motions that take place in two different directions. Figure 8.30b shows a geared five-bar linkage that is driven in such a manner that the angular velocity of input crank O_2B is three times the angular velocity of input crank O_1A so that O_2B rotates through three revolutions, while O_1A rotates through only one. Because links AC and BC are long, the motion of point A imparts primarily sinusoidal motion in the x direction to point C, while the motion of point B imparts primarily sinusoidal motion in the y direction to point C. The result is a path for point C that generally resembles the Lissajous figure of Fig. 8.30a. Because the frequency of the y component of motion of point B is three times that of the x component of the motion of point A, the ratio of the numbers of extreme value points on the curve is $6/2 = 3$, just as in the curve of Fig. 8.30a.

In Fig. 8.30b the linkage proportions were chosen so that links AC and BC do not rotate appreciably (their motions are primarily translations) and so that they remain approximately perpendicular to each other. Under such conditions, the path of point C will resemble a Lissajous figure relatively closely. As conditions deviate from these, the curve becomes progressively more distorted. A relatively severely distorted curve is shown as generated by the linkage in Fig. 8.30c. Note, however, that even in this case three upper extremes, three lower extremes, one rightward extreme, and one leftward extreme are discernible, as would be expected because the crank O_2B is rotating at three times the rate at which O_1A is rotating. If the ratio of the input angular velocities is not the ratio of relatively small integers, the resulting curve can be extremely complicated.

8.10   Combining Motions that Are Parallel to Each Other (Phase Shifting)

Section 8.2 has described the combining of motions by addition and multiplication, but it did not consider the timing of those motions. It is sometimes useful to combine motions that differ in frequency and/or phase to provide adjustable phase or some such timing characteristic in the combined output.

To see how two motions that have the same frequency but are 90° out of phase with each other (i.e., are in quadrature) can be combined to produce a combined output motion having another chosen phase relationship to the inputs, consider the trigonometric identity

$$A \sin \theta + B \cos \theta = C \sin(\theta + \phi) \qquad (8.26)$$

where

$$C^2 = A^2 + B^2 \quad \text{and} \quad \tan \phi = B/A \qquad (8.27)$$

Figure 8.31 shows a five-bar mechanism that could be used to implement Eq. (8.26). In this figure the rocker O_1D would be driven (by means not shown) such that its angle α would be given approximately by $\alpha = k \sin \theta$, and the rocker O_2E would be driven similarly such that its angle β would be given approximately by $\beta = k \cos \theta$, where θ is the angular position of an input shaft (not shown) and k would be chosen so that the two rockers would oscillate through small angles. Because the angles α and β remain small, the displacements of points D, E, and F will be approximately vertical. The displacement of point D will be approximately vertical and of magnitude $a\alpha$, and the displacement of points E and F will be approximately vertical and of magnitude $b\beta$, where a and b are the lengths of the respective rockers and α and β are in radians. From Sec. 8.2 we recall that the vertical displacement of point G will be approximately half the sum of the displacements of points D and F. Then this vertical displacement of the point G can be written approximately as

$$y_G = 0.5(a\alpha + b\beta) = 0.5(ak \sin \theta + bk \cos \theta) \qquad (8.28)$$

It can be seen that the right-hand side of Eq. (8.28) becomes the same as the left-hand side of Eq. (8.26) if $A = 0.5ak$ and $B = 0.5bk$. Thus it

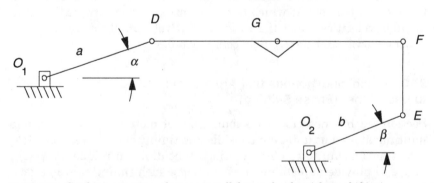

Figure 8.31 Combining motions that are parallel to each other (phase shifting).

is seen that the vertical motion of point G is a sinusoidal function of the input shaft angle but with a phase angle ϕ that is determined by suitable choice of the rocker lengths a and b.

This method for producing an output motion that has a prescribed phase relationship to the rotation of an input shaft can be compared with a simple method in which the output is produced by using a four-bar linkage that is driven by an eccentric or crank that is attached to the input shaft at an angle that corresponds to the prescribed phase angle. If a single such output is required, the eccentric or crank system is probably simpler and cheaper. However, if a large number of such outputs are required for the same input shaft, using a mechanism such as shown in Fig. 8.31 for each output might be preferable. This is so because this latter system can use flexural pivots such as elastomeric bushings for all the pivots at points O_1, O_2, D, E, and F for each of the outputs, whereas a full bearing would have to be used in each of the eccentrics or cranks. Also, the rockers such as O_1D and O_2E for each output could be placed on shafts that are common to all other outputs. A large number of individually phased outputs is required in a machine that is required to perform a large number of operations in a closely spaced time sequence along a line of locations. Such a sequence resembles a peristaltic action and has been used in "wave shedding" textile weaving machines and in machines for progressively shaping stationary long metal strips.

8.11 Frequency-Doubling Mechanisms

It is occasionally desired to perform a given function twice during each cycle of the operation of a particular machine. Of course, an auxiliary shaft that turns two revolutions during each machine cycle can be used to drive this function, or a two-lobed cam can be used. However, it often is convenient to use a linkage motion for such a drive.

Figure 8.32 shows a six-bar linkage that produces a twice-per-cycle output motion. The input to this linkage is a continuous rotation of the input crank O_1A about a shaft at O_1 at a rate of one revolution per machine cycle. As this crank rotates, rocker O_2B rocks back and forth between positions O_2B' and O_2B'', which are shown dashed. At each of these positions, rocker link O_3C will be at a clockwise or downward position. When rocker O_2B is in a position such that O_2, B, and C all lie in a straight line, rocker link O_3C will be at its most counterclockwise or uppermost position, as shown by solid lines. It thus can be seen that output rocker link O_3C will be low twice during each $360°$ of rotation of input crank O_1A, and it also will be high twice during each $360°$ of rotation of input crank O_1A. That is, there will be two cycles of oscillation of rocker O_3C during each cycle of rotation of the input crank.

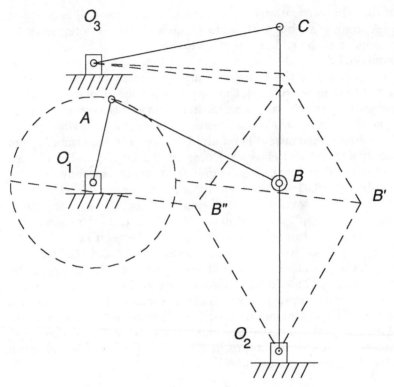

Figure 8.32   A frequency-doubling mechanism.

8.12 Summary

This chapter has presented several concepts that can be used as qualitative guides in selecting types of mechanisms that are useful in producing various functions. The qualitative concepts covered include the use of mechanical addition and multiplication: (1) to combine and alter the displacements and curved paths produced by linkages and (2) to provide adjustable linkages. Methods for producing desired path curvatures and straight lines were discussed. The effects that the location of a point attached to a coupler has on the type of path it follows were described, and the uses of coupler point paths in dwell mechanisms and conveying mechanisms were presented.

Once a mechanism has been chosen using these concepts as guides, this initial choice can then be explored quantitatively using various manual and computerized synthesis and analysis methods. Progressive refinement of the initial mechanism concept will then provide a final design.

Dyad Synthesis and Computer-Aided Synthesis of Linkages

9.1 Introduction

The very powerful technique of dyad synthesis is described in this chapter so that readers can develop an understanding of the theory on which much computerized synthesis software is based. Because of the extensive computation involved in the use of dyad synthesis, and because of the multiplicity of possible solutions which it can produce for a given problem, *manual* dyad synthesis is arduous and therefore is not discussed in this book. However, the multiplicity of solutions makes it important that users of dyad synthesis software understand the implications of the results. The vast majority of the solutions that can be provided by these syntheses are far from optimal, and engineers must vary several parameters at their disposal in a search for a satisfactory answer. To aid readers in such searches, the sources and the implications of the multiplicity of these solutions are discussed.

9.2 Dyad Synthesis Compared with Graphical Synthesis

Procedures 4.4 and 4.5 in Chap. 4 are graphical procedures by means of which engineers can synthesize four-bar linkages that will cause a body to pass through a prescribed sequence of positions. In the descriptions of these motion synthesis procedures and in the discussions associated with these descriptions, two important phenomena were noted.

First, it was noted that it is possible to choose arbitrarily varied locations for some of the pivots. It also was noted that the locations chosen for the pivots affect the resulting locations of other pivots, the resulting motions of the links, the resulting variations in the transmissions angle, and other important features of the linkage performance. Figures 4.17 through 4.19 and their associated discussion illustrate this fact. In Figs. 4.17 and 4.18 the locations of the pivots on the coupler were chosen arbitrarily. In Fig. 4.19 the locations of the grounded pivots were chosen arbitrarily. Usually the locations *that can be chosen arbitrarily* for pivots are not strictly dictated by the overall mechanism design task. Rather, it is usually only necessary that these pivots be located in certain prescribed areas. Therefore, in the synthesis process, it becomes necessary to explore several possible locations for these pivots in a search for an optimal mechanism design. Such an exploration by graphical means can become laborious.

Second, it was noted in Figs. 4.17 and 4.18 that the location chosen for each pivot on the coupler affects only the resulting synthesized location of the grounded pivot that was connected to the same link, as was the chosen coupler pivot. In Fig. 4.19 the location chosen for each grounded pivot affects only the resulting synthesized location of the coupler pivot that was connected to the same link, as was the chosen grounded pivot. In the comments following Procedure 4.4 it was pointed out that pivots A and O_1 in Fig. 4.17 constitute such an interdependent pair and that pivots B and O_2 constitute the other such pair. These pairs of pivot locations are called *Burmester point pairs,* and they can be treated as independent pairs and the locations in each pair can be synthesized independently of locations in the other pair.

Dyad synthesis is an analytical synthesis technique devised by Professor G. N. Sandor that takes advantage of this second phenomenon. It is the basis of the computer software KINSYN developed by Professor Kaufman,[1] and it is the basis of the commercially available computer software LINCAGES that is described in a book by Professors Sandor and Erdman.[2] Although dyad synthesis is an analytical technique and therefore can provide extremely precise results, searching for optimal mechanism configurations by exploring the possible free choices with dyad synthesis manual calculations can become arduous. Also, being analytical, dyad synthesis tends to be less directly visual and intuitive than graphical techniques. Fortunately, the above-mentioned software is quite graphic and interactive and greatly en-

[1]Kaufman, R. E., Mechanism design by computer, *Machine Design* (October 1978), pp. 94–100.

[2]Sandor, G. N., and Erdman, A. G., *Advanced Mechanism Design: Analysis and Synthesis,* vol. 2, Prentice-Hall, Englewood Cliffs, NJ, 1984, pp. 177–188.

hances the search process. Fortunately also, the intuitive aspects of a synthesis task often can be enhanced by preceding the use of the software by brief preliminary exploration using graphical synthesis.

The nature of dyad synthesis provides the identification of those mechanism variables for which values can be arbitrarily (or freely) chosen so that the remaining unknowns can be solved for. Subsequent sections of this chapter present a tabulation of these variables and describe the theory behind the synthesis technique so that readers will have insight into the power and limitations of dyad synthesis.

9.3 Development of the Dyad Synthesis Relationships

Figure 9.1 indicates a rigid body L that is to be the coupler in a four-bar linkage to be synthesized. Attached to this coupler is a pivot at

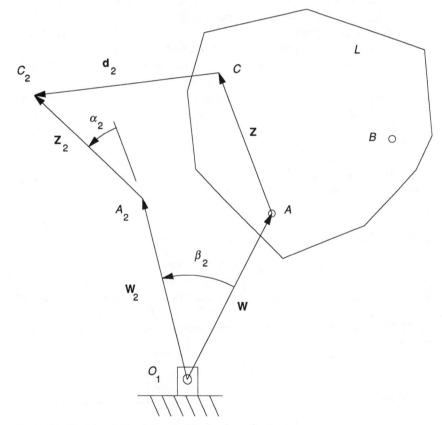

Figure 9.1   Vector relationships used in dyad synthesis.

point A, and another pivot would be attached at point B. The pivot at point A is also attached to a link that is pivoted to ground at point O_1 and which is represented by the vector $\mathbf{W}$. Points A and O_1 thus constitute a Burmester point pair which, as discussed previously, can therefore be considered independently of a Burmester point pair involving point B, so the motion of point B will be ignored for now. Also attached to body L is a so-called *tracer point* C. The position of point C relative to the pivot at point A is represented by the vector $\mathbf{Z}$, which is of course attached to the body L. The pair of vectors $\mathbf{W}$ and $\mathbf{Z}$ is called a *dyad*. The purpose of the synthesis procedure described in the following paragraphs is to find values for these vectors $\mathbf{W}$ and $\mathbf{Z}$ such that the links represented by them will provide some prescribed motion of the body L.

If the body L is moved from the initial position shown in Fig. 9.1 to some subsequent position, point A will move to some new position indicated by A_2, and tracer point C will move to a new position indicated by C_2. The vectors in the dyad will assume new positions indicated by $\mathbf{W}_2$ and $\mathbf{Z}_2$. The displacement of point C from its initial position to its new location is represented by the vector $\mathbf{d}_2$, as shown. During this displacement, the body will in general rotate through an angle α_2, so the vector $\mathbf{Z}$ will have rotated from its initial position by that same angle, as indicated in Fig. 9.1. The point O_1, being the location of a grounded pivot, will remain fixed, and the vector $\mathbf{W}$ will have rotated through an angle β_2. The lengths of vectors $\mathbf{W}$ and $\mathbf{Z}$ are assumed to be constant. Therefore, as the linkage that is to be synthesized moves, the pivot at point A will be constrained to move around the fixed pivot at O_1 along a path that is a circle or a circular arc. The point A is therefore called a *circling point* or, more briefly, a *circle point*. The point O_1 is the center of the circular path of point A, so point O_1 is called a *center point*.

The vectors $\mathbf{W}_2$, $\mathbf{Z}_2$, $\mathbf{d}_2$, $\mathbf{Z}$, and $\mathbf{W}$ are seen to constitute a closed vector loop. Then, summing these vectors around the loop gives

$$\mathbf{W}_2 + \mathbf{Z}_2 - \mathbf{d}_2 - \mathbf{Z} - \mathbf{W} = 0 \tag{9.1}$$

It is helpful to express some of these vectors in the equivalent complex variable form that was introduced in Chap. 2. Thus we may write vector $\mathbf{W}$ in terms of its magnitude or length W and its angular orientation θ_W as

$$\mathbf{W} = We^{j\theta_W} \tag{9.2}$$

Similarly, we may write

$$\mathbf{W}_2 = We^{j(\theta_W + \beta_2)} = e^{j\beta_2}We^{j\theta_W} = e^{j\beta_2}\mathbf{W} \tag{9.3}$$

In like manner, we may write

$$\mathbf{Z}_2 = Ze^{j(\theta_Z + \alpha_2)} = e^{j\alpha_2}Ze^{j\theta_Z} = e^{j\alpha_2}\mathbf{Z} \qquad (9.4)$$

Substituting Eqs. (9.3) and (9.4) into Eq. (9.1) and grouping and rearranging terms gives

$$\mathbf{W}(e^{j\beta_2} - 1) + \mathbf{Z}(e^{j\alpha_2} - 1) = \mathbf{d}_2 \qquad (9.5)$$

Equation (9.5) is known as the *standard-form equation for dyad synthesis*.

Each vector in Eq. (9.5) represents two variables, i.e., either an angle and a length or two Cartesian components. Thus the three vectors and the angle values α_2 and β_2 together represent eight variables. If the displacement of body L from its first position to its second position shown in Fig. 9.1 is prescribed (as would be the case in a motion synthesis task), the values represented by $\mathbf{d}_2$ and α_2 will be known, and values for the five variables represented by $\mathbf{W}$, $\mathbf{Z}$, and β_2 will be unknown. Then, because the vector equation [Eq. (9.5)] represents two scalar equations, only two unknown values can be solved for. It is therefore possible and necessary to choose values for three of the variables in order to allow solution for the remaining two variables. These value choices are known as *free choices*. Thus, for motion synthesis for which two coupler positions are prescribed, there are three free choices for variable values. These choices can be values for any combination of the variables represented by $\mathbf{W}$, $\mathbf{Z}$, and β_2.

Once Eq. (9.5) has been solved, the vectors $\mathbf{W}$ and $\mathbf{Z}$ are known, and the location of the pivot points A and O_1 *relative to the initial or reference position* of tracer point C can be calculated. Notice that if, for example, a value for β_2 were chosen as one of the free choices, an infinite number of different values could be chosen for that variable. A different solution for Eq. (9.5) would result for each such value chosen, so there would an infinite number of solutions corresponding to that free choice. A similar situation exists for each of the other two free choices available in the foregoing discussion, so altogether there will be a threefold infinity of solutions possible for a given two-position motion synthesis task. This offers the engineer a wide variety of possible solutions to choose from. However, it also presents a challenge to search this threefold infinity of free choices in an attempt to find a best solution.

Dyad synthesis for more than two positions

Notice that the angles α_2 and β_2 and the vector $\mathbf{d}_2$ do not represent absolute measurements from some fixed reference system. Rather, they are measures of the displacement of body L and the dyad from their first or reference positions. Then, if the synthesis were to be performed for three positions of body L, an additional equation such as Eq. (9.5) could be written relating that third position to the initial position. The result would be

$$\mathbf{W}(e^{j\beta_3} - 1) + \mathbf{Z}(e^{j\alpha_3} - 1) = \mathbf{d}_3 \qquad (9.6)$$

where β_3, α_3, and $\mathbf{d}_3$ are displacements that are measured *from the initial positions* of $\mathbf{W}$, $\mathbf{Z}$, and C, respectively, to their third positions. Thus, for the motion synthesis of a mechanism for three positions, one more vector equation (two more scalar equations) would be added to Eq. (9.5), and one more unknown (i.e., β_3) would be added to the previous list of unknowns. Indeed, for each additional body position prescribed, two more scalar equations and one more unknown would be added. The added vector equations would have the same form as Eq. (9.5) and could be expressed in the general form

$$\mathbf{W}(e^{j\beta_i} - 1) + \mathbf{Z}(e^{j\alpha_i} - 1) = \mathbf{d}_i \qquad (9.7)$$

where the subscript i refers to the ith displacement.

We may therefore construct Table 9.1, which is a table of numbers of prescribed body positions, number of scalar equations available, total number of variables other than those for which values are prescribed, number of free choices, and number of solutions associated with these free choices.

Note in this table that by making the number of free choices of values (for unknowns) indicated in the fourth column, the number of unknowns that must be solved for is reduced to a number equal to the

TABLE 9.1 Relationships between numbers of prescribed positions, numbers of equations, numbers of variables, and numbers of solutions in motion synthesis problems.

No. of prescribed positions	No. of scalar equations	Total no. of unknown variables	No of free choices	No. of unknowns remaining	No. of solutions
2	2	5 $(\mathbf{W}, \mathbf{Z}, \beta_2)$	3	2	∞^3
3	4	6 $(\mathbf{W}, \mathbf{Z}, \beta_2, \beta_3)$	2	4	∞^2
4	6	7 $(\mathbf{W}, \mathbf{Z}, \beta_2, \beta_3, \beta_4)$	1	6	∞
5	8	8 $(\mathbf{W}, \mathbf{Z}, \beta_2, \beta_3, \beta_4, \beta_5)$	0	8	Finite

number of scalar equations available. For two-position motion synthesis, the three free choices allow a threefold infinity (indicated by the notation ∞^3) of possible solutions to the synthesis problem. For three-position motion synthesis, the two free choices allow a twofold infinity of possible solutions to the synthesis problem. Exploration of the varieties of solutions for these two cases can become laborious, but the equations are amenable to linear solutions. For more than three prescribed positions, the equations become nonlinear. The previously mentioned LINCAGES software solves the equations for the case of four prescribed positions. It does so by first determining the range of all values for β_2 that are consistent with the given set of prescribed displacements and by then solving for $\mathbf{W}$ and $\mathbf{Z}$ for each of these possible values of β_2 in that sequence. As noted previously, once $\mathbf{W}$ and $\mathbf{Z}$ are known for each value of β_2, the position of the moving pivot (circle point) and the grounded pivot (center point) relative to the initial position of body L can be computed. The software plots these two positions for each assumed value of β_2. This produces curves of all the possible positions of the *initial positions* of the pivots relative to the initial position of point C. These curves are known as *Burmester curves,* one of which is the *circle-point curve* and the other of which is the *center-point curve,* corresponding, respectively, to the locus of possible positions of the moving pivot and the locus of possible positions of the grounded pivot.

In using this software, the user enters data defining the four positions of the tracer point and the successive angular displacements (α_i) of body L from its initial position. The software plots the circle-point curve and the center-point curve. The user then chooses a point on either of these curves, and the software indicates the corresponding point on the other curve. One of these points is a possible center point, and the other is the corresponding circle point (according to which curve each point lies on). A choice of one such Burmester pair defines the locations of the pivots associated with one of the links that connects the coupler to ground. Locations of the other pair of pivots in the four-bar linkage (such as those at point B and some point O_2 on the ground) are found simply by choosing another pair of points on the same curves.

Readers who wish to develop algorithms for solving the dyad synthesis standard form equations will find detailed discussions of such solution procedures in the reference given in footnote 2.

9.4 Dyad Synthesis for Motion, Path, and Function Synthesis

Section 9.3 developed the dyad synthesis relationships on the basis of an illustration of *motion synthesis.* Dyad synthesis is also very effective in path synthesis and function synthesis. It will be shown that these latter forms of synthesis are merely variations of motion synthesis.

Dyad synthesis for path generation

In Sec. 9.3 the locations of a succession of positions of tracer point C were prescribed by the succession of vectors $\mathbf{d}_2$, $\mathbf{d}_3$, etc. This succession of point positions obviously defines a path that point C is to follow, so by merely arbitrarily specifying values for α_2, α_3, etc. and performing motion synthesis, a path synthesis is implicitly performed. This is somewhat wasteful, however, because the arbitrary choices that would be assigned to α_2, α_3, etc. can be put to better use rather than just being ignored as not being part of path prescription.

Note the similarity of the form of the term containing $\mathbf{W}$ to the form of the term containing $\mathbf{Z}$ in Eq. (9.7). If values were prescribed for β_2, β_3, etc. instead of for α_2, α_3, etc., the effect would be that of prescribing a coordination between the successive values of $\mathbf{d}_2$, $\mathbf{d}_3$, etc. and the successive values of β_2, β_3, etc. (see Fig. 9.1). This constitutes prescribing a path generation with a prescribed timing of the positions along the path (or prescribed coordination of those positions along the path with the angular position of the link represented by the vector $\mathbf{W}$). The *form* of Eq. (9.7) would not be changed, so the procedure for the solution would be the same as described above. Thus the LINCAGES software also can be used for *path synthesis with prescribed timing*.

Because the form of the term containing $\mathbf{W}$ is identical to the form of the term containing $\mathbf{Z}$ in the set of equations represented by Eq. (9.7), prescribing a set of values for β_2, β_3, etc. instead of that same set of values for α_2, α_3, etc. would, for a given set of values of $\mathbf{d}_2$, $\mathbf{d}_3$, etc., result in a solution in which the resulting values of $\mathbf{W}$ and $\mathbf{Z}$ would be interchanged. This effect is illustrated in Fig. 9.2. In this figure assume that the vectors $\mathbf{W}$ and $\mathbf{Z}$ represent the solutions to a synthesis problem in which values were prescribed for $\mathbf{d}_2$, $\mathbf{d}_3$, etc. and for α_2, α_3, etc. If the same set of values that was prescribed for α_2, α_3, etc. was *instead* prescribed for β_2, β_3, etc., the solution to the set of dyad synthesis equations would give the vectors $\mathbf{W}'$ and $\mathbf{Z}'$ shown. The vectors $\mathbf{W}'$ and $\mathbf{Z}'$ represent the linkage dimensions required for path generation with the prescribed timing represented by the sequence of angular displacement values β_2, β_3, etc.

Compare the parallelogram formed by the vectors $\mathbf{W}$, $\mathbf{Z}$, $\mathbf{W}'$, and $\mathbf{Z}'$ in Fig. 9.2 (i.e., parallelogram O_1ACA') with parallelogram O_1ACG in Fig. 4.26a. Because in Fig. 4.26a the link O_1G must remain parallel to

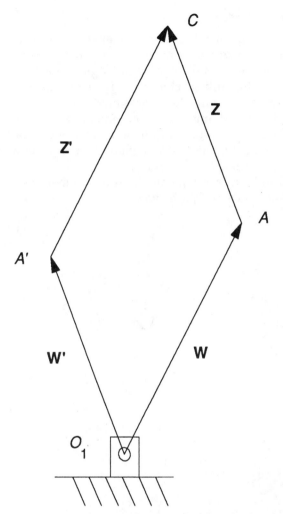

Figure 9.2 The vector equivalence between dyad motion synthesis and dyad path synthesis with prescribed timing.

line AC, the link O_1G must rotate in exactly the same manner as does the coupler ABC. Thus it is seen that if the linkage O_1ABO_2 is synthesized to provide motion generation, the cognate linkage O_1GFH whose coupler point (tracer point) C will follow the same path as point C on linkage O_1ABO_2 will provide path generation with prescribed timing. It is thus seen that a linkage that provides path generation with prescribed timing is a cognate of a linkage that provides motion generation. This can provide an alternative procedure for path synthesis with prescribed timing.

Dyad synthesis for function generation

In a function-generating linkage, an output link is required to perform a sequence of angular displacements that possess a one-to-one correspondence to a prescribed sequence of displacements of an input link. Consider the function-generating linkage shown in Fig. 9.3a. The output link O_2C experiences the angular displacements δ_2, δ_3, and δ_4 in response to some corresponding input angular displacements β_2, β_3, and β_4 (not shown) of input link O_1A. As the output link experiences the prescribed angular displacements, the pivot at point C is displaced along a circular arc to the positions C_2, C_3, and C_4.

By considering point C to be a tracer point, the function synthesis task can be treated as a special case of path synthesis with prescribed

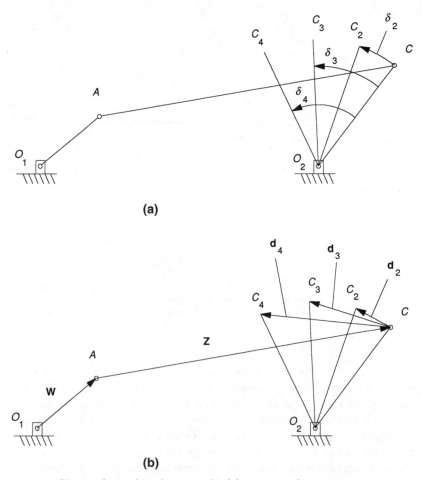

(a)

(b)

Figure 9.3   Vector relationships for use in dyad function synthesis.

timing. For this purpose, successive displacements $\mathbf{d}_2$, $\mathbf{d}_3$, and $\mathbf{d}_4$ can be prescribed for point C, as shown in Fig. 9.3b, and corresponding successive angular displacements β_2, β_3, and β_4 can be prescribed for link O_1A. The vectors $\mathbf{W}$ and $\mathbf{Z}$ then represent the links, as indicated in Fig. 9.3b. The remainder of the process is identical to that of path synthesis with prescribed timing and consists of finding values for the vectors $\mathbf{W}$ and $\mathbf{Z}$.

If the function-generating linkage produced by the foregoing process is too large or too small, it may be shrunk or magnified, and the angular *displacement* relationships between input and output will not be affected. It also may be rotated to any orientation, and the angular *displacement* relationships between input and output will not be affected. Such options are obviously not available for motion synthesis or for path synthesis.

9.5 Comments and Suggestions for Dyad Synthesis and Computer-Aided Synthesis

As pointed out in preceding sections, dyad synthesis and computer-aided synthesis usually produce a manifold infinity of answers to a synthesis problem. Most of these answers are highly unsatisfactory, but interactive software such as LINCAGES can aid in the search for satisfactory answers. In addition, the nature of the answers available and the difficulty of searching for satisfactory answers among those available depend greatly on the actions prescribed for the linkage to be synthesized. It is therefore advisable to carefully consider how much is *really required* of the linkage performance. For example, it is generally more difficult to search for and find a satisfactory linkage that will perform a four-position motion generation than it is for a two-position motion generation. Therefore, it would be wise to ask whether a two-position motion generation would be adequate for the job at hand.

For the same reason, it often is very helpful to perform quick preliminary explorations using only two- or three-position motion, path, or function synthesis before proceeding to computer-aided synthesis for four positions. The preliminary explorations can provide starting points for the search conducted on a computer. Graphical synthesis techniques such as described in Chap. 4 can be very appropriate for these preliminary trials because of the intuitive nature of the graphical constructions. CAD systems can provide rapid, flexible means for preforming these constructions.

The linkage that is synthesized should be analyzed to see whether it meets additional requirements such as Grashof's condition, suitability of transmission-angle variations, and the ability to pass through the required sequence of output positions in the desired order and without disassembling the linkage between positions. The LINCAGES software provides auxiliary displays that greatly aid in checking to see whether such additional requirements have been satisfied.

Chapter

10

Cams

10.1 Introduction

Very frequently a mechanism must be designed such that its output displacement will be a given function of its input displacement. In previous chapters function synthesis was used to design linkages for such purposes, and it was found that the range of functions that can be generated by such linkages is somewhat limited. A much broader range of functions may be provided by cam systems.

This chapter presents the basic concepts and parameters associated with the design of cams. It derives, describes, and compares the most common standard cam motion programs. Shops that produce a lot of cams generally have stored programs by means of which they can automatically machine cams on the basis of customer-specified follower motion and type of standard motion program desired. In cases where such stored programs are not available, machining coordinates may have to be computed using the cam motion program equations and follower relationships given in this chapter. A summary of the general cam design procedure is presented in Sec. 10.8.

The "smoothness" of operation of cam systems is discussed, and the effects that the choice of the standard cam motion programs and the choice of dwell duration have on this smoothness are shown. The effects that cam size and pressure angle have on cam follower forces, camshaft torque fluctuations, and system vibration are described. Procedures for designing cam systems are presented based on the preceding considerations.

10.2 Terminology, the Principles of Cam Operation, and Types of Cam Systems

The basic concept of cam operation can be illustrated by referring to Fig. 10.1. In this figure a body C that has some prescribed shape is

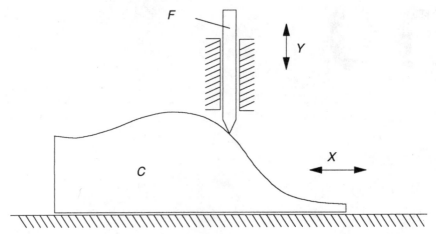

Figure 10.1 The basic concept of cam operation.

constrained such that it can only slide horizontally in the $\pm X$ direction. A second body F is constrained such that it can only slide vertically in the $\pm Y$ direction as shown. If the body F is held in contact with the upper, shaped surface of body C as body C is slid horizontally rightward or leftward, body F will be caused to move up or down. The resulting motion of F will be a function of the motion of C, and that function will be determined by the shape that is prescribed for C. The shaped body is called a *cam,* and the body F is an example of a *follower* because it is caused to "follow" the surface of the cam.

It can be seen that by suitably shaping the cam surface, the function relating the motion of the follower to the motion of the cam can be caused to be any of a wide variety of functions. Indeed, when engineers choose to use a cam system for an application, they often assume that any arbitrarily chosen function can be produced by such a system. However, as will be shown in subsequent sections, there are practical considerations that can limit the types of motion that can be produced.

Types of cams

A cam of the general type indicated in Fig. 10.1 is often called a *wedge cam* or a *sliding cam.* In some applications of this type of cam the mechanism is kinematically inverted so that the cam is stationary, and the slide in which the follower slides up and down is caused to move horizontally so that the follower is caused to move up and down as it slips horizontally along the cam surface.

Other commonly used types of cams are shown in Fig. 10.2. In Fig. 10.2a the cam takes the form of a disk of prescribed shape, and it is mounted on a shaft so that it rotates rather than translating. The

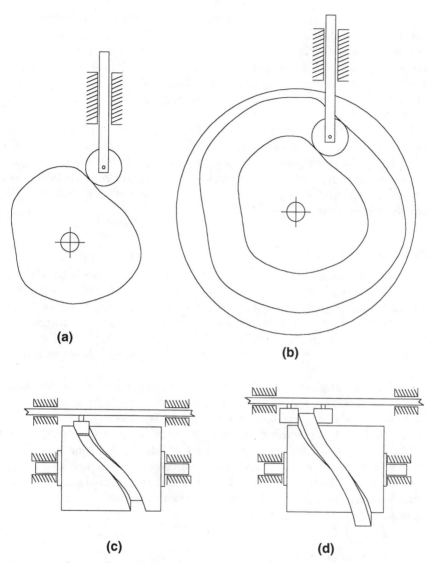

(a)

(b)

(c)   **(d)**

Figure 10.2 Commonly used types of cams.

follower shown rises and falls as the cam rotates. This type of cam is often referred to as a *plate cam*. The follower in Fig. 10.2*b* possesses a roller on its tip, and that roller is "captured" in a groove or track that is machined in the face of the disk so that as the disk rotates, the follower is forced up and down in a prescribed manner. This is referred to as a *face cam* or a *track cam*. The groove in the cam shown in Fig. 10.2*c* is machined into the cylindrical surface of the cam, which rotates about a horizontal axis that is parallel to the page. As the cam rotates, the follower that possesses a roller captured in the groove is caused to slide rightward and leftward. A similar arrangement is shown in Fig. 10.2d. In this arrangement, the groove in Fig. 10.2c is replaced by a raised ridge, and the follower uses two rollers, one on each side of the ridge. The cams in parts c and d of Fig. 10.2 are known as *barrel cams.*

In addition to the disk and cylinder shapes illustrated above, cams can assume many other configurations, including conical and spherical shapes. However, because plate cams are probably the most commonly encountered type, and because their analysis and synthesis are most directly and easily treated, the remainder of this chapter will concentrate on plate cams.

Types of followers

Some of the types of follower configurations most commonly used with plate cams are shown in Fig. 10.3. The follower in Fig. 10.3a is a *sliding* or *translating follower,* which uses a roller at its tip to contact the cam. Such rollers are commercially available as units with antifriction bearings and which are specifically designed for this use. A roller-type follower, of course, has the advantages of low friction, low wear, and long life.

The translating *pointed follower* shown in Fig. 10.3b is simpler than the roller follower, but unless it is used with very light loads and/or low speeds, it is subject to rapid wear. The translating *flat-faced follower* in Fig. 10.3c somewhat reduces these wear problems. Figure 10.3d and e shows follower systems in which the follower motion is rotary instead of translational. Such follower systems are known as *oscillating followers* or *swing followers.* Most automotive engine valve cam systems use one or another of the systems shown in Fig. 10.3c, d, or e.

Maintaining contact between follower and cam

In order to perform its intended function, a follower must remain in contact with the cam surface. In cams of the types shown in Fig. 10.2b, c, and d, this is accomplished by confining the follower between two shaped cam surfaces. Another way of providing confinement of the follower system is shown in Fig. 10.4a. This system uses two cams that are rigidly attached to each other and two followers that are rigidly attached to each other. These two cams are known as *conjugate cams,* and their shapes, although different from each other, must be carefully

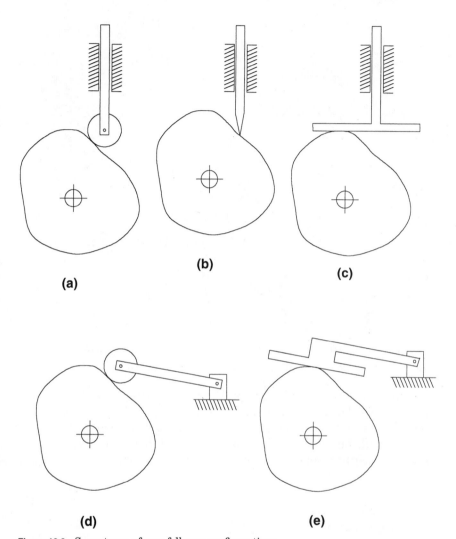

Figure 10.3 Some types of cam follower configurations.

coordinated with each other. Indeed, the shapes of the two surfaces used in each of the cams shown in Fig. 10.2b, c, and d and Fig. 10.4a must be very accurately coordinated with each other if the follower system is to be moved without excessive backlash and impacts, while at the same time avoiding undue stresses in the follower system or on the cam surfaces.

The most commonly used cam systems have only a single cam surface such as is shown in Fig. 10.4b. If the motions involved are slow enough, gravity can be used to keep the follower in contact with the

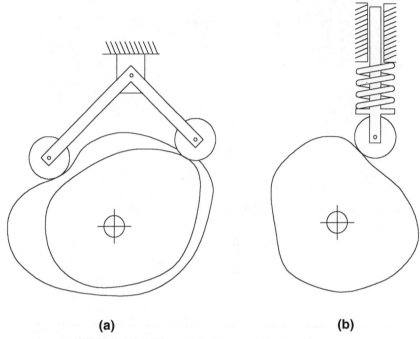

(a)   **(b)**

Figure 10.4   Means for maintaining contact between follower and cam.

cam surface. However, in most applications a spring is used, as shown in Fig. 10.4b. The spring is usually a coil spring, although other types such as pneumatic cylinders sometimes can be used. The spring force must be sufficient to maintain contact despite forces caused by friction, inertia, and other phenomena in the mechanism being driven by the follower.

Cam systems compared with linkage systems

Many actions required of a machine can be performed either by a cam system or by a linkage system. When choosing which type of system to use, engineers should carefully compare the advantages and disadvantages of both types. The following are lists of some of those advantages and disadvantages.

Advantages of cam systems

1. Cam systems can provide a wider range of functional relationships between input and output than can linkage systems. In particular, cams can produce true dwells (see Sec. 10.3), whereas linkages can only produce approximate dwells (see Sec. 8.8). Cams also can produce

constant-velocity motions over a specified displacement. Before choosing a cam system for a given application, the engineer should make sure that the special capability of a cam is really needed.

2. Designing a cam system to provide a prescribed function is generally more direct and intuitive than designing a corresponding linkage system. This has become less important as computer-aided linkage synthesis software has continued to be developed.

3. For a given function, a cam system almost always occupies less space than does a linkage system.

Disadvantages of cam systems

1. The cam follower always contacts the cam surface at a single point or along a single line. Therefore, the forces between cam and follower are highly concentrated rather than being distributed over an area, such as would be the case in a revolute or slider joint in a linkage. As a consequence, for wear and/or vibration reasons, more careful attention must be paid in cam systems to surface hardness, surface finish, and lubrication. Whereas revolute joints in linkages can consist of enclosed bearings, the exposed nature of cam surfaces can cause problems of keeping those surfaces clean and lubricated, sometimes requiring the enclosure of the entire cam and follower system.

2. Means must be provided for keeping the follower in contact with the cam surface. This usually involves the use of a spring or of two carefully coordinated cam surfaces. In either case, the tolerances and/or compliances in the follower system must be tailored carefully to accommodate and oppose the forces that are reflected to the follower by the moving output machine parts and which could cause separation of the follower from the cam surface(s).

3. At high speeds, the tendency of a follower to separate from the cam surface can result in impacts between follower and cam and/or can require very high spring loads to prevent separation. These impacts and loads are concentrated at a single point or line of contact between follower and cam, as mentioned before.

4. The complicated shape of cam surfaces can require more expensive machining than is required in the manufacture of linkage systems. Note that follower systems often involve some links, so that the use of a cam does not always completely replace a linkage.

10.3 Timing, Displacement, Velocity, Acceleration, and Jerk Diagrams

In most applications of cams, the output motion of the follower system must occur in a certain time relationship with other motions in the

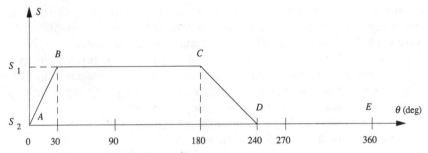

Figure 10.5 An example of a timing diagram.

machine or product being designed. The cam design process therefore often starts with the drawing or sketching of a *timing diagram* such as shown in Fig. 10.5. The purpose of such diagrams is to show when in the cycle of a machine's operation certain events must occur. Because the shaft on which a plate cam is attached will be the main shaft driving the machine or will be synchronized to that main shaft, the horizontal axis of this diagram usually represents camshaft angle as shown. The vertical axis represents the output action of the follower system.

The example diagram in Fig. 10.5 indicates that the follower system output must be at some position (represented by S_1 on the vertical axis) while the shaft angle is between 30° and 180°, and that it must be at another position (represented by S_2 on the vertical axis) while the shaft angle is between 240° and 360°. To indicate these requirements, the horizontal lines BC and DE, respectively, are drawn. The follower system must move from position S_2 to position S_1 as the shaft angle goes from 0° to 30°, so arbitrarily a straight line AB is drawn to indicate that transition. Similarly, the follower system must move from position S_1 to position S_2 as the shaft angle goes from 180° to 240°, so arbitrarily a straight line CD is drawn to indicate that transition.

The horizontal-line portions of a diagram such as Fig. 10.5 represent portions of the cam action during which the follower system remains at rest, and these portions are called *dwells*. A sloping line segment such as AB or CD represents a portion of a cam action during which the follower is moving from one position to another, and such a portion is called either a *rise* or a *fall,* usually depending on whether the follower is moving, respectively, away from or toward the center of the cam.

If the machine were to operate only at extremely low speeds, perhaps a cam would be made that would produce a follower system motion exactly like that indicated in Fig. 10.5. Notice, however, that the follower system would be at a constant position for shaft angles from 30° to 180° (corresponding to the horizontal line from point B to point

C) but would move at constant velocity for shaft angles from 180° to 240° (because of the constant slope of the straight line from point C to point D). Therefore, the follower system *velocity* must change from zero to some constant nonzero value instantaneously at point C, resulting in infinite acceleration. Similar infinite acceleration requirements occur at points A, B, and D. Consequently, the line segments AB and CD are not intended to represent exact motions produced by the cam. A timing diagram such as Fig. 10.5 is intended only to indicate the timing of position requirements such as represented by line segments BC and DE.

As described in Sec. 10.4, cams usually are designed to produce motions that would correspond to smooth curves between points A and B and between points C and D. An example *displacement diagram* for such a cam is shown in Fig. 10.6a. It is seen that the sloping straight-line segments in Fig. 10.5 have been replaced by smooth curves and that the sharp corners at points A, B, C, and D have been eliminated.

Because cam systems usually are parts of dynamic systems, the follower system velocities and accelerations are of concern to engineers. The curves in Fig. 10.6b and c show the follower velocity and acceleration variations, respectively, associated with a cam having a displacement diagram such as that in Fig. 10.6a. Notice that while the displacement curve is relatively smooth, the velocity curve is less smooth and the acceleration curve is even less smooth. That is, these curves of the higher derivatives of displacement are progressively less smooth in the sense that they tend to possess more maxima and minima, more points of inflexion, more reversals of sign, and more sharp corners.

In mechanisms that involve appreciable mass, the accelerations of the follower system are important because the cam must apply sufficient force to the follower to produce those accelerations in the follower system mass. Because these forces not only produce stresses in the mechanism parts but also can cause machine vibration and noise, any lack of smoothness in the acceleration curve should be of concern to mechanism designers. If the forces in a machine vary slowly and smoothly, little shaking of the machine parts or of the machine mount will be perceived, even if the forces are large. If, however, the acceleration-induced forces change suddenly and rapidly, the machine can experience objectionable vibration, can produce undesirable noise, and/or can be considered to be a "rough-running machine." An often-used indicator of acceleration curve "roughness" is the rate of change (time derivative) of acceleration. This rate of change is called *jerk,* and a plot of the jerk that is associated with the curve of Fig. 10.6c is given as Fig. 10.6d.

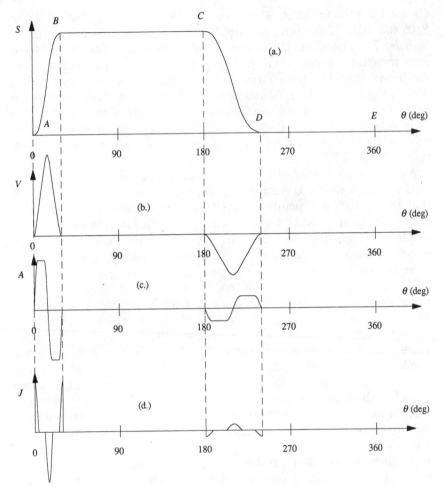

Figure 10.6 Displacement, velocity, acceleration, and jerk diagrams (SVAJ diagrams).

It is seen that this curve is even less smooth than is the acceleration curve, and that it vividly indicates points at which the acceleration curve is "rough." The concept of roughness is discussed more extensively in Sec. 10.4.

Notice that the rise occurs during only 30° of cam rotation, whereas the fall occurs during 60° (twice the duration). As a consequence, the velocity, acceleration, and jerk are much smaller during the rise than during the fall. *Short rises and falls produce roughness.*

Plots such as those in Fig. 10.6 are often referred to as a cam's *SVAJ diagrams* (because of the variable labels on their vertical axes). When designing a cam system, all or most of the SVAJ diagrams for that cam should be plotted and examined.

10.4   Smoothness of Operation and Some Standard Cam Motion Programs

Principal objectives in cam design

Of course, the main purpose for which a cam system is designed is to provide some output motion or displacement that approximates as closely as possible a prescribed function of some input motion or displacement. Usually, it is only important that the output correspond in a prescribed manner to the input over a *limited portion of the input and output ranges*. Therefore, engineers have some freedom in choosing the correspondence between input and output over the remainder of those ranges. Care should be exercised to be sure that only the input-output correspondences that are absolutely necessary for satisfactory operation of the machine are prescribed, thereby maximizing the engineer's freedom in the cam system design.

The second objective in cam design is to provide the preceding input-output correspondence in a manner that does not result in undue forces in the machine, undue vibration, or undue noise. This objective could be expressed as the objective of providing smooth operation. The terms *smooth, rough, smoothness,* and *roughness* tend to be qualitative terms, and as discussed briefly at the end of Sec. 10.3, they can be thought of in terms of steepness of curves in one or more of the SVAJ diagrams and in terms of sharpness of corners in those curves. The remainder of this section will discuss means for minimizing such roughness features. Another indicator of smoothness is discussed in Sec. 10.5.

The quest for smoothness of operation has resulted in the development of several standard types of cam motion programs. These programs are derived and described in this section, and their relative advantages and disadvantages are compared in the discussion associated with Table 10.1 later in this section.

One of the most commonly encountered types of timing diagrams is exemplified by that in Fig. 10.5. The corresponding cam contains a "high" dwell period and a "low" dwell period, separated from each other by a rise and a fall. Because no motion (and hence no roughness) can occur during a dwell, only the rise portion of cam motion programs will be discussed in this section. Note that a fall can be just like a rise except for its direction of motion.

Although this chapter discusses outputs as motions of the follower, it should be noted that some output parts of the machine that contain important masses and/or which experience important forces may be connected to the follower by a mechanism that causes the motions of those parts to differ appreciably from the follower motion. In such cases it may be more appropriate to design for smooth operation of those parts rather than for follower motion. Mechanical advantages may become important in such cases. In any event, a cam system must be designed as part of an overall system.

The harmonic cam program (low-speed cams)

As mentioned earlier, a cam could be designed to produce a plot of follower motion versus cam motion (a displacement diagram) that would look just like Fig. 10.5. If the cam were to be moved slowly and/or only occasionally as a means of adjustment, perhaps such a cam would be satisfactory.

Generally, however, it is desirable to eliminate the sharp corners represented at points A, B, C, and D in Fig. 10.5. This can be done quite directly by replacing the straight line segments AB and CD with cosine curves. The resulting rise and fall curves are known as *harmonic cam motion programs*. Such a curve for the cam rise AB would appear as the rise displacement diagram shown in Fig. 10.7a. The velocity, acceleration, and jerk diagrams associated with this rise motion program are given as Fig. 10.7b, c, and d.

It can be seen that the curves in the displacement S and velocity v diagrams are smooth throughout the rise, including their transitions to the dwells at each end. Therefore, any forces in the follower system that are proportional to or dependent on follower displacement (such as spring forces) or which are proportional to or dependent on follower velocity (such as friction or viscosity forces) will vary smoothly.

However, the curve in the acceleration a diagram has a discontinuity at each end of the rise, where there is an instantaneous change from zero acceleration during the dwells to a nonzero acceleration in the rise. This is reflected in the infinite values of jerk indicated in the curve in the jerk j diagram. Therefore, any forces in the follower system that are proportional to or dependent on follower acceleration (such as inertial reaction forces) will change abruptly at those instants. Thus a cam system using a simple harmonic rise or fall motion program and having appreciable mass in the follower system will exhibit considerable roughness if operated at any but very low speeds.

The analytical expressions that correspond to the preceding displacement, velocity, and acceleration diagrams can be written as

$$S = \frac{h}{2}\left[1 - \cos\left(\pi\frac{\theta}{\beta}\right)\right] \tag{10.1}$$

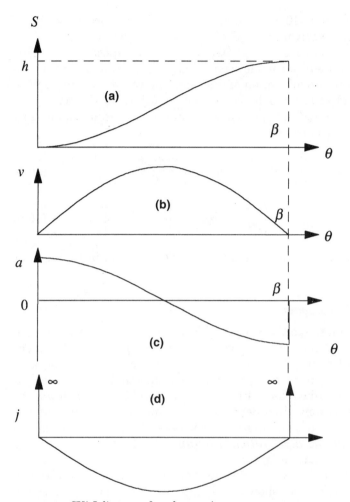

Figure 10.7 SVAJ diagrams for a harmonic cam program.

$$v = \frac{dS}{d\theta} = \frac{h}{2}\frac{\pi}{\beta}\sin\left(\pi\frac{\theta}{\beta}\right) \tag{10.2}$$

$$a = \frac{d^2S}{d\theta^2} = \frac{h}{2}\frac{\pi^2}{\beta^2}\cos\left(\pi\frac{\theta}{\beta}\right) \tag{10.3}$$

where, as shown in Fig. 10.7, h is the height of the rise, θ is the cam rotation measured from the beginning of the rise, and β is the duration of the rise measured in terms of cam rotation.

Notice that in Eqs. (10.2) and (10.3) the velocity and acceleration are given in terms of derivatives of S with respect to cam angle θ in radians rather than in terms of derivatives of S with respect to time. This provides a convenient means for comparing cam motion programs, independent of the rate of motion of the cam. The values of v and a from Eqs. (10.2) and (10.3) can be converted to values for velocity V in inches per second and acceleration A in inches per second squared by the following relationships:

$$V = \frac{dS}{dt} = v\left(\frac{d\theta}{dt}\right) = v\omega = v\left[\frac{2\pi(\text{rpm})}{60}\right] \tag{10.4}$$

$$A = \frac{d^2S}{dt^2} = a\left(\frac{d\theta}{dt}\right)^2 = a\omega^2 = a\left[\frac{2\pi(\text{rpm})}{60}\right]^2 \tag{10.5}$$

where, of course, ω is the angular velocity of the cam in radians per second.

The cycloidal cam program

It has been remarked in the foregoing that the progressively higher derivatives of the follower displacement become progressively more rough. Then it was noted that the displacement and velocity diagrams for the harmonic cam motion program were smooth, but the acceleration diagram contained discontinuities that would be unacceptable for many cam applications. In order to develop a motion program that will be suitable for operation at appreciable speeds, then, let us start by devising an *acceptable acceleration diagram* and deriving the velocity and displacement diagrams from it.

Let us choose a sine-wave curve for the acceleration diagram for a cam motion program as shown in Fig. 10.8c. The equation for this curve can be written as

$$a = a_{\max} \sin\left(2\pi\frac{\theta}{\beta}\right) \tag{10.6}$$

This equation can then be integrated with respect to θ to give

$$v = -\left(\frac{\beta}{2\pi}\right)a_{\max}\cos\left(2\pi\frac{\theta}{\beta}\right) + k_v \tag{10.7}$$

Then this equation too can be integrated with respect to θ to give

$$S = \left(\frac{\beta^2}{4\pi^2}\right)a_{\max}\sin\left(2\pi\frac{\theta}{\beta}\right) + k_v\theta + k_S \tag{10.8}$$

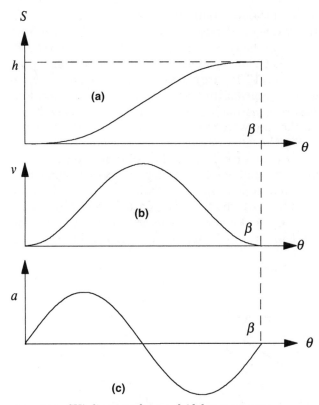

Figure 10.8 SVA diagrams for a cycloidal cam program.

The values of the coefficients $a_{\max}$, k_v, and k_S can then be found by noting that the boundary conditions are $v = S = 0$ when $\theta = 0$ and $v = 0$ and $S = h$ when $\theta = \beta$. Substituting these values into Eqs. (10.6), (10.7), and (10.8) and solving the resulting equations gives

$$k_S = 0 \qquad k_v = \frac{h}{\beta} \qquad \text{and} \qquad a_{\max} = 2\pi \frac{h}{\beta^2} \qquad (10.9)$$

Substituting these values into Eqs. (10.6), (10.7), and (10.8) gives

$$S = h\left[\frac{\theta}{\beta} - \frac{1}{2\pi}\sin\left(2\pi\frac{\theta}{\beta}\right)\right] \qquad (10.10)$$

$$v = \frac{h}{\beta}\left[1 - \cos\left(2\pi\frac{\theta}{\beta}\right)\right] \qquad (10.11)$$

$$a = 2\pi\frac{h}{\beta^2}\sin\left(2\pi\frac{\theta}{\beta}\right) \qquad (10.12)$$

Again, as in the case of the harmonic cam motion program, the velocity and acceleration values *with respect to time* can be obtained from these expressions by using the relationships in Eqs. (10.4) and (10.5).

The displacement, velocity, and acceleration diagrams corresponding to Eqs. (10.10), (10.11), and (10.12) are shown in Fig. 10.8*a*, *b*, and *c*, respectively. A cam motion program that corresponds to these equations is known as a *cycloidal cam motion program.* This appellation is the result of the fact that if the cam were to rotate at constant angular velocity, the displacement *S* of the follower during the rise would be the same as the *horizontal displacement component* of a point on the circumference of circle of radius $h/2\pi$ that is rolling at constant velocity on a horizontal line starting with that point in contact with the line and ending with that point in contact with the line. The path followed by such a point is called a *cycloid.* Cycloidal motion is described in the subsection on centrodes in Sec. 2.5 and is illustrated in Figs. 2.15 and 2.16.

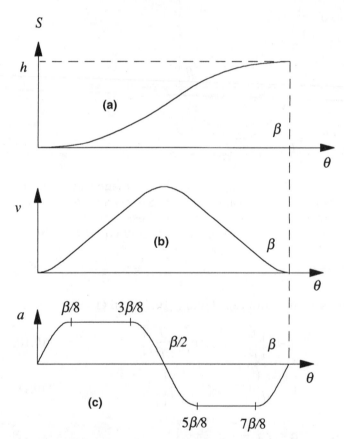

Figure 10.9 SVA diagrams for a modified trapezoid cam program.

The modified trapezoid cam program

It will be noticed that in Fig. 10.8c the acceleration builds up gradually to a momentary peak value and then gradually decreases to zero before going to a negative peak value. It would appear that equivalent velocity and displacement effects could be obtained with a *lower maximum* acceleration if the acceleration were built up more quickly and then if the maximum value were held for a longer time before rapidly decreasing the acceleration to zero. Such an acceleration diagram is shown in Fig. 10.9c.

The portion of the curve from $\theta = 0$ to $\theta = \beta/2$ in Fig. 10.9c can be considered to look somewhat like a trapezoid in which the sloping sides have been curved. The remainder of the curve also looks somewhat like a trapezoid below the axis. In this acceleration diagram, the curved portions are quarter-cycles of sine waves. A cam motion program that has an acceleration diagram of this form is known as a *modified trapezoid acceleration motion program* or simply a *modified trapezoid cam program* (often referred to as *mod trap*).

As in the case of the cycloidal motion program, the velocity and displacement diagrams can be derived by integrating this acceleration curve with respect to θ and then applying the boundary conditions to the resulting equations to determine values for the coefficients in the equations. The results are shown in Fig. 10.9b and a, respectively. The formulas for these curves are as follows:

For $0 \le \theta \le \beta/8$:

$$S = h\left[0.3889845\frac{\theta}{\beta} - 0.0309544 \sin 4\pi\left(\frac{\theta}{\beta}\right)\right]$$

$$v = 0.3889845\frac{h}{\beta}\left[1 - \cos 4\pi\left(\frac{\theta}{\beta}\right)\right] \quad (10.13a)$$

$$a = 4.8881238\frac{h}{\beta^2}\sin 4\pi\left(\frac{\theta}{\beta}\right)$$

For $\beta/8 \le \theta \le 3\beta/8$:

$$S = h\left[2.4440619\left(\frac{\theta}{\beta}\right)^2 - 0.2220309\left(\frac{\theta}{\beta}\right) + 0.0072341\right]$$

$$v = \frac{h}{\beta}\left[4.8881238\left(\frac{\theta}{\beta}\right) - 0.2220309\right] \quad (10.13b)$$

$$a = 4.8881238\left(\frac{h}{\beta^2}\right)$$

For $3\beta/8 \le \theta \le 5\beta/8$:

$$S = h\left[1.6110155\left(\frac{\theta}{\beta}\right) - 0.0309544 \sin 4\pi\left(\frac{\theta}{\beta} - \frac{1}{4}\right) - 0.3055077\right]$$

$$v = \frac{h}{\beta}\left[1.6110155 - 0.3889845 \cos 4\pi\left(\frac{\theta}{\beta} - \frac{1}{4}\right)\right] \qquad (10.13c)$$

$$a = 4.8881238\left(\frac{h}{\beta^2}\right)\sin 4\pi\left(\frac{\theta}{\beta} - \frac{1}{4}\right)$$

For $5\beta/8 \le \theta \le 7\beta/8$:

$$S = h\left[4.6660928\left(\frac{\theta}{\beta}\right) - 2.4440619\left(\frac{\theta}{\beta}\right)^2 - 1.2292649\right]$$

$$v = \frac{h}{\beta}\left[4.6660928 - 4.8881238\left(\frac{\theta}{\beta}\right)\right] \qquad (10.13d)$$

$$a = -4.8881238\left(\frac{h}{\beta^2}\right)$$

For $7\beta/8 \le \theta \le \beta$:

$$S = h\left[0.6110155 + 0.3889845\left(\frac{\theta}{\beta}\right) + 0.0309544 \sin 4\pi\left(\frac{\theta}{\beta} - \frac{3}{4}\right)\right]$$

$$v = 0.3889845\frac{h}{\beta}\left[1 + \cos 4\pi\left(\frac{\theta}{\beta} - \frac{3}{4}\right)\right] \qquad (10.13e)$$

$$a = -4.8881238\left(\frac{h}{\beta^2}\right)\sin 4\pi\left(\frac{\theta}{\beta} - \frac{3}{4}\right)$$

The velocities and accelerations *with respect to time* can be computed using Eqs. (10.4) and (10.5).

The skewed modified trapezoid cam program

The modified trapezoid cam motion program just described produces accelerations that are of equal magnitude in both positive and negative directions, as can be seen in Fig. 10.9c. In some cam applications it is important that the magnitude of the acceleration be limited more in one direction than in the other. For example, if the cam follower is to raise a platform from rest at some lower position to rest at a higher position in order to elevate some product that is sitting on it, the deceleration of the platform as it approaches the upper rest position should not exceed the acceleration of gravity. Otherwise, the product would fly up off the platform and then fall back onto it at the top of the motion. A so-called *skewed modified trapezoid cam acceleration motion program* such as indicated in Fig. 10.10 could be used in such an application. Using such a motion program, the upward acceleration would, as shown, be greater than the downward acceleration, but the areas under the two "trapezoids" would be equal so that the velocity would be zero at the beginning and end of the rise.

Derivation of relationships corresponding to Eq. (10.13a) through Eq. (10.13e) for such a skewed motion program would be done in the same manner as was used for Eqs. (10.13). That is, the acceleration for each of the five segments of the motion program shown would be integrated in sequence to give the equations for the velocity and displacement programs (taking care to determine expressions for the constants of integration for each segment). The parameters h (total rise height), β, and $R = a_2/a_1$ would be chosen as independent variables for which values would be prescribed by the user. Using these variables and the fact that velocity and acceleration must both be zero at the start and end of the rise as boundary conditions, expressions for the values of the variables a_1, β'_1, and β'_2 would be solved for. The process is complicated and arduous, and the resulting expressions are complicated.

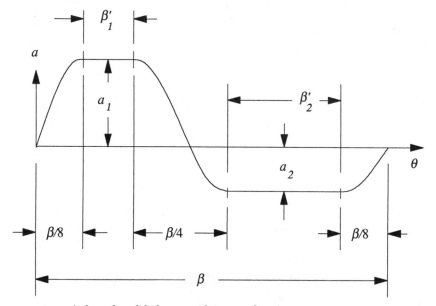

Figure 10.10   A skewed modified trapezoid cam acceleration program.

However, by means of a simpler process, a skewed mod trap program can be derived from the standard mod trap program, resulting in an acceleration curve such as indicated in Fig. 10.11c. Comparing the curve in this figure with that in Fig. 10.9c shows that the "trapezoid" lying between points A and B in Fig. 10.11c has been stretched in height and shrunk in width compared with the corresponding trapezoid in Fig. 10.9c. The reverse has been done to the "trapezoid" lying between points B and C in Fig. 10.11c. This stretching and shrinking have been done in such a manner as to make the area under the "trapezoid" between points A and B equal to that under the "trapezoid" between points B and C, so that the velocity will be the same at start and finish of the rise.

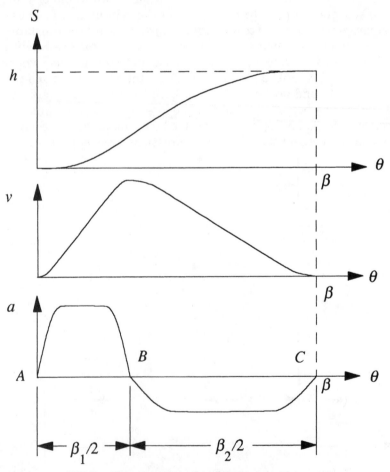

Figure 10.11 SVA diagrams for a skewed modified trapezoid cam program.

Notice that the slope of the curve in Fig. 10.11c changes abruptly at point B, whereas the curve in Fig. 10.10 passes smoothly through the corresponding point. This abrupt change in Fig. 10.11c corresponds to a discontinuity in the jerk curve. Note, however, that this discontinuity will always be smaller than that associated with at least one of the two slope changes at points A and C. The skewed mod trap program corresponding to the curves in Fig. 10.11 has been used very effectively in machines.

It is seen that the "trapezoid" between points A and B is half of a mod trap acceleration program that would have a total rise of h_1 and a total rise angle (or duration) of β_1. Similarly, the "trapezoid" between points B and C is half of a mod trap acceleration program that would have a total rise of h_2 and a total rise angle (or duration) of β_2. Then the total rise of the cam corresponding to the curve in Fig. 10.11c is given by

$$ h = \frac{h_1 + h_2}{2} \tag{10.14} $$

Defining an acceleration ratio $R = a_2/a_1$ and matching velocities for the two half-programs at point B gives (after some manipulation)

$$ h_1 = 2h\left(\frac{R}{R+1}\right) \quad \text{and} \quad \beta_1 = 2\beta\left(\frac{R}{R+1}\right) \tag{10.15a} $$

$$ h_2 = 2h\left(\frac{1}{R+1}\right) \quad \text{and} \quad \beta_2 = 2\beta\left(\frac{1}{R+1}\right) \tag{10.15b} $$

The peak value of acceleration for a mod trap cam motion program is seen from Eqs. (10.13b) and (10.13d) to be given by

$$ a_{\text{peak}} = 4.8881238\,\frac{h}{\beta^2} \tag{10.16a} $$

Then, using Eqs. (10.15), we may write the peak values of acceleration in a skewed mod trap cam as

$$ (a_1)_{\text{peak}} = 4.8881238\,\frac{h_1}{\beta_1^{\,2}} = 4.8881238\,\frac{h}{\beta^2}\left(\frac{1+R}{2R}\right) \tag{10.16b} $$

$$ (a_2)_{\text{peak}} = 4.8881238\,\frac{h_2}{\beta_2} = 4.8881238\,\frac{h}{\beta^2}\left(\frac{1+R}{2}\right) \tag{10.16c} $$

Values for h_1, β_1, h_2, and β_2 can be computed from Eqs. (10.15) and user-prescribed values for h, β, and the desired acceleration ratio $R = a_2/a_1$. Then Eqs. (10.13a), (10.13b), and (10.13c) can be written as follows and used to compute S, v, and a for the first portion of the rise represented by Fig. 10.11:

For $0 \leq \theta \leq \beta_1/8$:

$$S = h_1 \left[0.3889845 \frac{\theta}{\beta_1} - 0.0309544 \sin 4\pi \left(\frac{\theta}{\beta_1} \right) \right]$$

$$v = 0.3889845 \frac{h_1}{\beta_1} \left[1 - \cos 4\pi \left(\frac{\theta}{\beta_1} \right) \right] \qquad (10.17a)$$

$$a = 4.8881238 \frac{h_1}{\beta_1{}^2} \sin 4\pi \left(\frac{\theta}{\beta_1} \right)$$

For $\beta_1/8 \leq \theta \leq 3\beta_1/8$:

$$S = h_1 \left[2.4440619 \left(\frac{\theta}{\beta_1} \right)^2 - 0.2220309 \left(\frac{\theta}{\beta_1} \right) + 0.0072341 \right]$$

$$v = \frac{h_1}{\beta_1} \left[4.8881238 \left(\frac{\theta}{\beta_1} \right) - 0.2220309 \right] \qquad (10.17b)$$

$$a = 4.8881238 \left(\frac{h_1}{\beta_1{}^2} \right)$$

For $3\beta_1/8 \leq \theta \leq \beta_1/2$:

$$S = h_1 \left[1.6110155 \left(\frac{\theta}{\beta_1} \right) - 0.0309544 \sin 4\pi \left(\frac{\theta}{\beta_1} - \frac{1}{4} \right) - 0.3055077 \right]$$

$$v = \frac{h_1}{\beta_1} \left[1.6110155 - 0.3889845 \cos 4\pi \left(\frac{\theta}{\beta_1} - \frac{1}{4} \right) \right] \qquad (10.17c)$$

$$a = 4.8881238 \left(\frac{h_1}{\beta_1{}^2} \right) \sin 4\pi \left(\frac{\theta}{\beta_1} - \frac{1}{4} \right)$$

For the second portion of the rise shown in Fig. 10.11, Eqs. (10.13c), (10.13d), and (10.13e) can be written as follows and used to compute S, v, and a:

For $\beta_1/2 \leq \theta \leq [(\beta_1/2) + (\beta_2/8)]$:

$$S = \left(\frac{h_1 - h_2}{2} \right) + h_2 \left[1.6110155 \left(\frac{\theta}{\beta_2} - \frac{R}{2} + \frac{1}{2} \right) \right.$$

$$\left. - 0.0309544 \sin 4\pi \left(\frac{\theta}{\beta_2} - \frac{R}{2} + \frac{1}{4} \right) - 0.3055077 \right]$$

$$v = \frac{h_2}{\beta_2}\left[1.6110155 - 0.3889845 \cos 4\pi\left(\frac{\theta}{\beta_2} - \frac{R}{2} + \frac{1}{4}\right)\right] \qquad (10.17d)$$

$$a = 4.8881238\left(\frac{h_2}{\beta_2{}^2}\right)\sin 4\pi\left(\frac{\theta}{\beta_2} - \frac{R}{2} + \frac{1}{4}\right)$$

For $[(\beta_1/2) + (\beta_2/8)] \le \theta \le [(\beta_1/2) + (3\beta_2/8)]$:

$$S = \left(\frac{h_1 - h_2}{2}\right) + h_2\left[4.6660928\left(\frac{\theta}{\beta_2} - \frac{R}{2} + \frac{1}{2}\right)\right.$$
$$\left. - 2.4440619\left(\frac{\theta}{\beta_2} - \frac{R}{2} + \frac{1}{2}\right)^2 - 1.2292649\right]$$

$$v = \frac{h_2}{\beta_2}\left[4.6660928 - 4.8881238\left(\frac{\theta}{\beta_2} + \frac{R}{2} - \frac{1}{2}\right)\right] \qquad (10.17e)$$

$$a = -4.8881238\left(\frac{h_2}{\beta_2{}^2}\right)$$

For $[(\beta_1/2) + (3\beta_2/8)] \le \theta \le [(\beta_1/2) + (\beta_2/2)]$:

$$S = \left(\frac{h_1 - h_2}{2}\right) + h_2\left[0.6110155 + 0.3889845\left(\frac{\theta}{\beta_2} - \frac{R}{2} + \frac{1}{2}\right)\right.$$
$$\left. + 0.0309544 \sin 4\pi\left(\frac{\theta}{\beta_2} - \frac{R}{2} - \frac{1}{4}\right)\right]$$

$$v = 0.3889845\frac{h_2}{\beta_2}\left[1 + \cos 4\pi\left(\frac{\theta}{\beta_2} - \frac{R}{2} - \frac{1}{4}\right)\right] \qquad (10.17f)$$

$$a = -4.8881238\left(\frac{h_2}{\beta_2{}^2}\right)\sin 4\pi\left(\frac{\theta}{\beta_2} - \frac{R}{2} - \frac{1}{4}\right)$$

The displacement, velocity, and acceleration diagrams for a skewed mod trap cam program were computed using Eqs. 10.17 with the values $h = 1$, $\beta = 1$, and $R = 0.5$, and the results are presented in Fig. 10.11a, b, and c, respectively. From Eqs. (10.16b) and (10.16c) it can be seen that the peak values of acceleration in this skewed mod trap program with acceleration ratio $R = 0.5$ are, respectively, 1.5 times and 0.75 times the value of peak acceleration in the unskewed mod trap program having the same rise h and duration β. A 25 percent decrease in peak downward acceleration has been achieved at the expense of a 50 percent increase in peak upward acceleration.

The modified sine cam program

In order to decrease the peak value of velocity required during a rise, the *modified sine cam program* (often called *mod sine*) can be used. In this program, the acceleration diagram of the cycloidal program is modified in such a manner as to provide the largest acceleration near the beginning and end of the rise so that the velocity is built up quickly and maintained at its higher values for a longer time.

The acceleration diagram for the mod sine program is shown in Fig. 10.12c. The curve consists of a quarter-cycle of sine wave during the first one-eighth of the rise time, a half-cycle of sine wave during the next three-quarters of the rise time, and another quarter-cycle of sine wave during the last one-eighth of the rise time. The expressions for the velocity and displacement diagrams are derived by integrating the

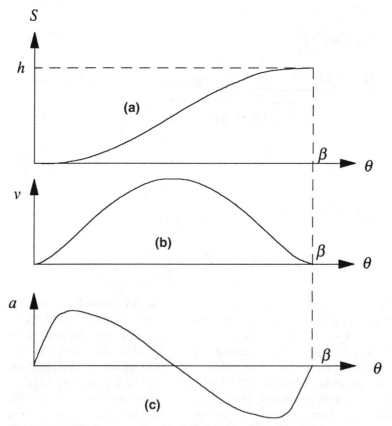

Figure 10.12 SVA diagrams for a modified sine cam program.

acceleration expression and applying the boundary conditions at the transitions and ends of the rise. The results are as follows:

For $0 \le \theta \le \beta/8$:

$$S = h\left[0.4399009\left(\frac{\theta}{\beta}\right) - 0.0350062 \sin 4\pi\left(\frac{\theta}{\beta}\right)\right]$$

$$v = \frac{h}{\beta}\left[0.4399009 - 0.4399009 \cos 4\pi\left(\frac{\theta}{\beta}\right)\right] \qquad (10.18a)$$

$$a = \frac{h}{\beta^2}\left[5.5279571 \sin 4\pi\left(\frac{\theta}{\beta}\right)\right]$$

For $\beta/8 \le \theta \le 7\beta/8$:

$$S = h\left[0.2800496 + 0.4399009\left(\frac{\theta}{\beta}\right) - 0.3150558 \sin \frac{4\pi}{3}\left(\frac{\theta}{\beta} + \frac{1}{4}\right)\right]$$

$$v = \frac{h}{\beta}\left[0.4399009 - 1.3197025 \cos \frac{4\pi}{3}\left(\frac{\theta}{\beta} + \frac{1}{4}\right)\right] \qquad (10.18b)$$

$$a = \frac{h}{\beta^2}\left[5.5279571 \sin \frac{4\pi}{3}\left(\frac{\theta}{\beta} + \frac{1}{4}\right)\right]$$

For $7\beta/8 \le \theta \le \beta$:

$$S = h\left[0.5600992 + 0.4399009\left(\frac{\theta}{\beta}\right) - 0.0350062 \sin 4\pi\left(\frac{\theta}{\beta}\right)\right]$$

$$v = \frac{h}{\beta}\left[0.4399009 - 0.4399009 \cos 4\pi\left(\frac{\theta}{\beta}\right)\right] \qquad (10.18c)$$

$$a = \frac{h}{\beta^2}\left[5.5279571 \sin 4\pi\left(\frac{\theta}{\beta}\right)\right]$$

The displacement, velocity, and acceleration diagrams for such a program are given in Fig. 10.12.

The modified constant-velocity cam program

In the interest of decreasing the maximum value of velocity even further than that associated with the mod sine program, the *modified constant-velocity cam program* (often referred to as *mod CV*) can be used. For this program, the acceleration diagram consists of quarter sine-wave segments with a large center segment in which acceleration is zero, as shown in Fig. 10.13*c*, and in which velocity is constant, as shown in Fig. 10.13*b*. The expressions for displacement, velocity, and acceleration are as follows:

For $0 \leq \theta \leq \beta/16$:

$$S = h\left[0.3188145\left(\frac{\theta}{\beta}\right) - 0.0126852 \sin 8\pi\left(\frac{\theta}{\beta}\right)\right]$$

$$v = \frac{h}{\beta}\left[0.3188145 - 0.3188145 \cos 8\pi\left(\frac{\theta}{\beta}\right)\right] \qquad (10.19a)$$

$$a = \frac{h}{\beta^2}\left[8.0126834 \sin 8\pi\left(\frac{\theta}{\beta}\right)\right]$$

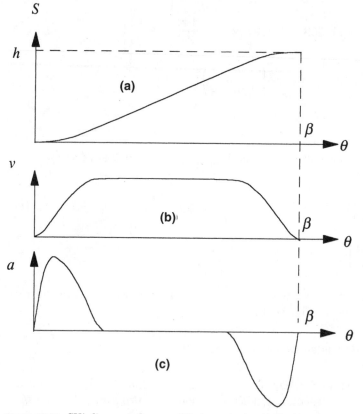

Figure 10.13 SVA diagrams for a modified constant-velocity cam program.

For $\beta/16 \leq \theta \leq \beta/4$:

$$S = h\left[0.1014818 + 0.3188145\left(\frac{\theta}{\beta}\right) - 0.1141671 \sin\left(\frac{8\pi}{3}\frac{\theta}{\beta} + \frac{\pi}{3}\right)\right]$$

$$v = \frac{h}{\beta}0.3188145\left[1 - 3\cos\left(\frac{8\pi}{3}\frac{\theta}{\beta} + \frac{\pi}{3}\right)\right] \tag{10.19b}$$

$$a = \frac{h}{\beta^2}\left[8.0126834 \sin\left(\frac{8\pi}{3}\frac{\theta}{\beta} + \frac{\pi}{3}\right)\right]$$

For $\beta/4 \leq \theta \leq 3\beta/4$:

$$S = h\left[1.2752582\left(\frac{\theta}{\beta}\right) - 0.1376231\right]$$

$$v = 1.2752582\frac{h}{\beta} \tag{10.19c}$$

$$a = 0$$

For $3\beta/4 \leq \theta \leq 15\beta/16$:

$$S = h\left[0.5797036 + 0.3188145\left(\frac{\theta}{\beta}\right) - 0.1141671 \sin\left(\frac{8\pi}{3}\frac{\theta}{\beta} + \pi\right)\right]$$

$$v = \frac{h}{\beta}0.3188145\left[1 - 3\cos\left(\frac{8\pi}{3}\frac{\theta}{\beta} + \pi\right)\right] \tag{10.19d}$$

$$a = \frac{h}{\beta^2}\left[8.0126834 \sin\left(\frac{8\pi}{3}\frac{\theta}{\beta} + \pi\right)\right]$$

For $15\beta/16 \leq \theta \leq \beta$:

$$S = h\left[0.6811855 + 0.3188145\left(\frac{\theta}{\beta}\right) - 0.0126852 \sin\left(8\pi\frac{\theta}{\beta}\right)\right]$$

$$v = \frac{h}{\beta}\left[0.3188145 - 0.3188145 \cos 8\pi\left(\frac{\theta}{\beta}\right)\right] \tag{10.19e}$$

$$a = \frac{h}{\beta^2}\left[8.0126834 \sin 8\pi\left(\frac{\theta}{\beta}\right)\right]$$

The displacement, velocity, and acceleration diagrams for this program are given in Fig. 10.13.

Polynomial cam programs

All the foregoing cam programs are based on fragments of sine and co-sine waves. Instead of using fragments of sine and cosine waves, a cam displacement diagram can consist of a curve that is expressed as a polynomial in the variable θ/β:

$$S = h\left[C_0 + C_1\left(\frac{\theta}{\beta}\right) + C_2\left(\frac{\theta}{\beta}\right)^2 + C_3\left(\frac{\theta}{\beta}\right)^3 + C_4\left(\frac{\theta}{\beta}\right)^4 + \cdots\right] \quad (10.20)$$

This equation can be differentiated with respect to θ to give Eq. (10.21) as the expression for velocity, and that equation again may be differentiated to give Eq. (10.22) as the expression for acceleration.

$$v = \frac{h}{\beta}\left[C_1 + 2C_2\left(\frac{\theta}{\beta}\right) + 3C_3\left(\frac{\theta}{\beta}\right)^2 + 4C_4\left(\frac{\theta}{\beta}\right)^3 + 5C_5\left(\frac{\theta}{\beta}\right)^4 + \cdots\right] \quad (10.21)$$

$$a = \frac{h}{\beta^2}\left[2C_2 + 6C_3\left(\frac{\theta}{\beta}\right) + 12C_4\left(\frac{\theta}{\beta}\right)^2\right.$$
$$\left. + 20C_5\left(\frac{\theta}{\beta}\right)^3 + 30C_6\left(\frac{\theta}{\beta}\right)^4 + \cdots\right] \quad (10.22)$$

As in the cases of the other cam programs, the coefficients are determined by inserting values for the boundary conditions together with the associated values of θ/β into the S, v, and a equations. Doing so for a polynomial program will produce as many linear equations in the coefficients as there are boundary conditions. Therefore, if those equations are to be solved for the values of the coefficients, the polynomial must be chosen to have the same number of coefficients as there are boundary conditions to be satisfied. The simultaneous equations can be solved using any of the many available mathematical software packages.

As an example of a polynomial program and how its coefficients are determined, consider a simple rise of height h during a rise duration β from one dwell to another, as in the previously discussed programs. At the beginning of the rise, the displacement S, velocity v, and acceleration a are all prescribed to be zero. That is, at $\theta/\beta = 0$, $S = v = a = 0$. At the end of the rise, the displacement is h and the velocity and acceleration are zero. That is, at $\theta/\beta = 1$, $S = h$ and $v = a = 0$. These requirements constitute six boundary conditions, so the displacement polynomial must contain six undetermined coefficients. Therefore, we write

$$S = h\left[C_0 + C_1\left(\frac{\theta}{\beta}\right) + C_2\left(\frac{\theta}{\beta}\right)^2 + C_3\left(\frac{\theta}{\beta}\right)^3 + C_4\left(\frac{\theta}{\beta}\right)^4 + C_5\left(\frac{\theta}{\beta}\right)^5\right]$$

$$v = \frac{h}{\beta}\left[C_1 + 2C_2\left(\frac{\theta}{\beta}\right) + 3C_3\left(\frac{\theta}{\beta}\right)^2 + 4C_4\left(\frac{\theta}{\beta}\right)^3 + 5C_5\left(\frac{\theta}{\beta}\right)^4\right] \quad (10.23)$$

$$a = \frac{h}{\beta^2}\left[2C_2 + 6C_3\left(\frac{\theta}{\beta}\right) + 12C_4\left(\frac{\theta}{\beta}\right)^2 + 20C_5\left(\frac{\theta}{\beta}\right)^3\right]$$

For $\theta/\beta = 0$, this equation becomes $S = 0 = hC_0$, so $C_0 = 0$. Also for $\theta/\beta = 0$, the velocity equation becomes $v = 0 = (h/\beta)C_1$, so $C_1 = 0$ also. For $\theta/\beta = 0$, the acceleration equation becomes $a = 0 = (h/\beta^2)2C_2$, so $C_2 = 0$ also.

Then, because $C_0 = C_1 = C_2 = 0$, for the boundary conditions at $\theta/\beta = 1$, the equations become

$$S = h(C_3 + C_4 + C_5) = h$$

$$v = (h/\beta)(3C_3 + 4C_4 + 5C_5) = 0$$

$$a = (h/\beta^2)(6C_3 + 12C_4 + 20C_5) = 0$$

These last three equations, when solved, give values of $C_3 = 10$, $C_4 = -15$, and $C_5 = 6$. Substituting these six coefficient values into Eqs. (10.23) gives

$$S = h\left[10\left(\frac{\theta}{\beta}\right)^3 - 15\left(\frac{\theta}{\beta}\right)^4 + 6\left(\frac{\theta}{\beta}\right)^5\right]$$

$$v = \frac{h}{\beta}\left[30\left(\frac{\theta}{\beta}\right)^2 - 60\left(\frac{\theta}{\beta}\right)^3 + 30\left(\frac{\theta}{\beta}\right)^4\right] \quad (10.24)$$

$$a = \frac{h}{\beta^2}\left[60\left(\frac{\theta}{\beta}\right) - 180\left(\frac{\theta}{\beta}\right)^2 + 120\left(\frac{\theta}{\beta}\right)^3\right]$$

The displacement, velocity, and acceleration diagrams corresponding to Eqs. (10.24) are given in Fig. 10.14.

The determination of values for the coefficients in the foregoing example was made easy by the fact that the boundary conditions were zero at $\theta/\beta = 0$ so that three of the coefficients were easily determined to be zero. In general, polynomial cam programs can be designed for various combinations of nonzero boundary conditions, including conditions not at the ends of the program, provided that a polynomial of sufficiently high degree (i.e., containing a number of coefficients equal to the number of boundary conditions) is used. Finding values for the coefficients in such cases involves solving several simultaneous linear equations, which is easily done using available equation-solving software.

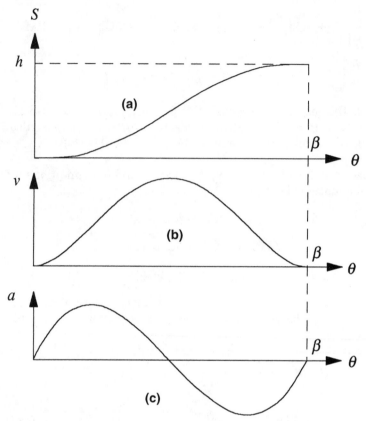

Figure 10.14 SVA diagrams for a 3-4-5 polynomial cam program.

Comparison of the standard cam programs

The displacement diagrams of the cams using the programs discussed in this section to provide transitions between dwells are almost identical to each other, so the comparative advantages of the various programs are not evident in such diagrams. This can be seen by comparing the displacement diagrams in the foregoing figures. Some subjective idea of the comparative smoothness of operation can be obtained by visually comparing the velocity and acceleration diagrams for those programs, however. In addition to being concerned about smoothness, engineers frequently are concerned with the maximum values of velocity and acceleration that will be encountered in the cam follower systems. To assist in comparing these latter features of the various programs, Table 10.1 is presented.

As discussed earlier in this section, the harmonic cam program should be used only in applications in which the motion is slow and/or

TABLE 10.1 Maximum Values of Acceleration and Velocity for Various Standard
Cam Programs

Type of cam program	Maximum acceleration	Maximum velocity
Harmonic	$4.935(h/\beta^2)$	$1.571(h/\beta)$
Cycloidal	$6.283(h/\beta^2)$	$2.0(h/\beta)$
Modified trapezoid	$4.888(h/\beta^2)$	$2.0(h/\beta)$
Skewed mod trap	If $R > 1$: $2.444(h/\beta^2)(1 + R)$ If $R < 1$: $2.444(h/\beta^2)(1 + 1/R)$	$2.0(h/\beta)$
Modified sine	$5.528(h/\beta^2)$	$1.760(h/\beta)$
Modified constant velocity	$8.013(h/\beta^2)$	$1.275(h/\beta)$
3-4-5 Polynomial	$5.774(h/\beta^2)$	$1.875(h/\beta)$

in which the mass of the follower system is very small. The cycloidal
cam program is found to operate smoothly in high-speed applications,
although its peak acceleration values are large. The mod trap program
possesses more transitions in its acceleration diagram than does the
cycloidal program, so it is slightly less smooth in operation, but its
peak acceleration value is about 20 percent lower than that of the cycloidal program. The positive and negative peak values of acceleration
are adjustable in the skewed mod trap program, although the maximum absolute value of acceleration is greater than that of the plain
mod trap program. The modified sine program and the 3-4-5 polynomial program are compromises between the cycloidal program and the
mod trap program.

Notice that the peak acceleration values for the preceding six
programs vary only from a maximum of a little more than $6(h/\beta^2)$
to a little less than $5(h/\beta^2)$, a decrease of only about 22 percent.
Because the factor h/β^2 (note that β is squared) appears in each peak
acceleration value, it can be seen that increasing the rise or fall
time β by as little as 14 percent will decrease the follower peak acceleration by more than will a change in motion program choice.
Engineers should be sure to use rise and fall times that are as long
as possible within the constraints of the machine performance requirements.

The substantially smaller peak velocity attainable with the modified
constant-velocity cam program can be useful when the follower load is
not inertial but is quite large and the cam diameter must be kept
small. This is true because, as discussed in Sec. 10.6, the pressure
angle depends on the velocity and the cam size. Because of the very

large maximum acceleration in this cam program, however, it should not be used in high-speed applications.

Procedure 10.1 and Example 10.1: Design of a Complete Two-Dwell Cam Motion Program   Engineers frequently are required to design cams that will alternately move the follower system between two positions in the general manner indicated by the timing diagram shown in Fig. 10.5. The procedure for doing so is illustrated by the following example, which is sketched in Fig. 10.15a:

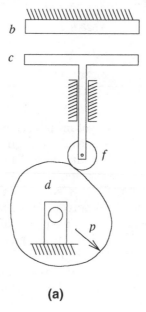

(a)

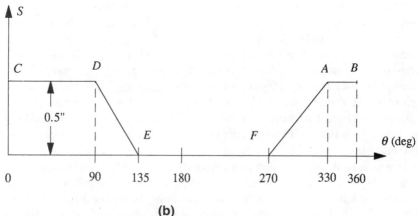

(b)

Figure 10.15   Example design of a cam system.

Example objective: A clamping surface *c* in Fig. 10.15*a* is attached to a cam follower *f* that is in contact with a cam *d* that rotates counterclockwise at a rate of 2 rev/s. It is required that the clamping surface *c* be held in an elevated position in contact with a stationary rubber block *b* during the time that a pointer *p* that is scribed on the side of the cam is at angles greater than 330° and less than 90°, measured from an upward vertical position. During this time, the clamping surface will be holding a sheet of metal stationary. During the time that the pointer *p* is between 135° and 270°, the clamping surface must be retracted 0.5 in below the rubber block to allow the sheet of metal to be removed and to allow protuberances on the sheet to clear the surface *c* and the block *b*.

1. Sketch horizontal line segments on a timing diagram that represent the required positions of the follower system. In this example, the prescribed elevated position of the follower system consisting of follower *f* and clamping surface *c* is represented by the horizontal line segments *AB* and *CD* in the timing diagram in Fig. 10.15*b*. These segments appear separated *in this example* because the duration of the elevated position spans 360° of cam angle. The retracted position of the follower system is represented by line segment *EF*. The vertical displacement between the high dwell (segments *AB* and *CD*) and the low dwell (segment *EF*) is not always shown on a timing diagram, but showing it can be helpful in visualizing what the cam system must do.

2. Sketch slanted line segments on the timing diagram to indicate the rise and fall portions of the motion program. In this example, line segment *FA* represents a rise and segment *DE* represents a fall.

3. From the timing diagram, note the durations of the rise and the fall, and make note of the height of the rise and the depth of the fall (i.e., displacement between high dwell and low dwell). In this example, the rise has a duration of $\beta = 330 - 270 = 60° = 1.0472$ rad and has a height of 0.5 in. The fall has a duration of $\beta = 135 - 90 = 45° = 0.7854$ rad and has a height of 0.5 in.

4. Choose a motion program for the rise and a motion program for the fall. Usually these programs will be chosen from a list of standard programs such as presented in Table 10.1. For the present example, let us choose cycloidal programs for both rise and fall.

5. Use the formulas in Table 10.1 and Eqs. (10.4) and (10.5) to calculate follower system maximum velocities and maximum accelerations. Are these values acceptable in terms of other machine requirements?

For the present example, the maximum acceleration *in the rise* is calculated from Table 10.1 to be

$$a_{max} = 6.283 \left[\frac{0.5}{(1.0472)^2} \right] = 2.865 \text{ in/rad}^2$$

To convert this to inches per second squared, Eq. (10.5) gives

$$A_{max} = a_{max} \omega^2 = 2.865[2.0(2\pi)]^2 = 452.4 \text{ in/s}^2$$

This is seen to be greater than the acceleration of gravity (which is 386.4 in/s²), so a spring must be added to the follower system shown in Fig. 10.15*a* to keep the follower in contact with the cam during the downward acceleration near the top of the rise. By an equivalent computation, Table 10.1 and Eq. (10.4) give the maximum velocity in the rise to be

$$V_{\max} = v_{\max}\omega = 2.0\left(\frac{0.5}{1.0472}\right)\left[2.0(2\pi)\right] = 12.0 \text{ in/s}$$

The corresponding computations *for the fall portion* of the motion program give $A_{\max} = 804.2$ in/s^2 and $V_{\max} = 16.0$ in/s. Although both the rise and fall use a cycloidal program, the acceleration and velocity in the fall are much greater than in the rise because the fall duration is shorter than the rise duration. The spring needed to keep the follower in contact with the cam must be chosen to accommodate the larger acceleration in the fall.

6. If the shop that is to be used to make the cam possesses a stored program for the motion program chosen in step 4, the information generated in steps 1 through 4, plus information about the cam follower geometry, probably will be enough for the shop to make the cam. The follower geometry information would include the follower diameter (if a roller follower is used), whether a sliding follower or a swing follower is to be used, and the cam size desired. These data are discussed in Sec. 10.6.

7. If the shop that is to make the cam does not have a stored program for the motion program chosen in step 4, details of the cam shape will have to be computed using the equations given in this section for the rise and fall programs chosen plus relationships given in Sec. 10.7 to describe the shape or tool path. These computations can be performed using commercially available mathematical software or software written by the reader.

The complete 360° diagrams of follower displacement, velocity, and acceleration for the cam specified in this example are given in Fig. 10.16.

Motion program for a single-dwell cam

In the example discussed in Procedure 10.1 it was essential that the follower system remain stationary (or almost so) in the high-dwell position between $\beta = 330°$ and $\beta = 90°$ in order that the clamping surface perform its clamping function. However, during the remainder of the machine cycle, it was only necessary that the clamping surface be lowered sufficiently far and for a sufficient portion of the cycle to release the clamping action and so that it would not interfere with other moving parts. Consequently, there need not be a low dwell. Although, for the example given, using a low dwell in addition to the high dwell is probably the most practical choice, in many cases a cam with only a single dwell is preferable.

When examining a displacement diagram such as Fig. 10.16a, it would appear that the second dwell could be eliminated by stretching the rise and fall portions of the motion program by moving points E and F closer together until they coincided. Notice, however, that such a stretching also would move points E and F in Fig. 10.16b and c together until they coincided. In Fig. 10.16c this results in the accelera-

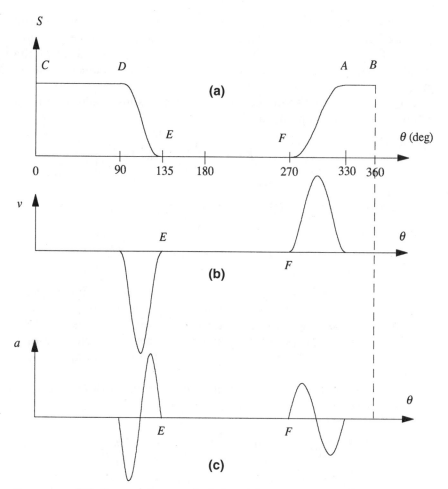

Figure 10.16 SVA diagrams for example design of the cam system in Fig. 10.15.

tion curve having a sharp, downward-directed corner where the points coincide. The occurrence of zero acceleration at this point is not necessary; a smoother acceleration curve is preferable. Unfortunately, all the standard motion programs, when used in this way, produce such a zero acceleration sharp corner. Moving points E and F together in this manner should not be considered to accomplish the elimination of a dwell but rather to reduce the dwell to an *instantaneous dwell*.

A single-dwell motion program, however, can be provided by using a *double-harmonic program*. In such a program, the displacement program for the follower consists of a cosine wave whose period is equal to the total nondwell time plus a second cosine wave whose period is *one-half* or *one-third* the nondwell time and which is added to the first cosine wave. When the period of this second cosine wave is one-half the nondwell time, the program will be called a *1-2 double-harmonic program*. When the period of this second cosine wave is one-third the nondwell time, the program will be called a *1-3 double-harmonic program*.

The equations for the displacement, velocity, and acceleration for a 1-2 double-harmonic program are

$$S = \frac{h}{2}\left[\left(1 - \cos 2\pi\frac{\theta}{\beta}\right) - \frac{1}{4}\left(1 - \cos 4\pi\frac{\theta}{\beta}\right)\right]$$

$$v = \pi\frac{h}{\beta}\left(\sin 2\pi\frac{\theta}{\beta} - \frac{1}{2}\sin 4\pi\frac{\theta}{\beta}\right) \qquad (10.25)$$

$$a = 2\pi^2\frac{h}{\beta^2}\left(\cos 2\pi\frac{\theta}{\beta} - \cos 4\pi\frac{\theta}{\beta}\right)$$

where β is the angular duration of the nondwell portion of the program and h is the maximum follower displacement from its dwell position. The displacement, velocity, and acceleration diagrams for a single-dwell 1-2 double-harmonic program replacement for the example shown in Fig. 10.16 are given in Fig. 10.17. (In this example, h is a negative 1 in.)

The equations for the displacement, velocity, and acceleration for a 1-3 double-harmonic program are

$$S = \frac{9}{16}h\left[\left(1 - \cos 2\pi\frac{\theta}{\beta}\right) - \frac{1}{9}\left(1 - \cos 6\pi\frac{\theta}{\beta}\right)\right]$$

$$v = \frac{9\pi}{8}\frac{h}{\beta}\left(\sin 2\pi\frac{\theta}{\beta} - \frac{1}{3}\sin 6\pi\frac{\theta}{\beta}\right) \qquad (10.26)$$

$$a = \frac{9\pi^2}{4}\frac{h}{\beta^2}\left(\cos 2\pi\frac{\theta}{\beta} - \cos 6\pi\frac{\theta}{\beta}\right)$$

where, again, β is the angular duration of the nondwell portion of the program and h is the maximum follower displacement from its dwell position. The displacement, velocity, and acceleration diagrams for a single-dwell 1-3 double-harmonic program replacement for the example shown in Fig. 10.16 are given in Fig. 10.18. (In this example, h is a negative 1 in.)

Comparing Figs. 10.16 and 10.17 shows that the velocity and acceleration curves for the single-dwell motion in Fig. 10.17 are smoother than those in Fig. 10.16. Also, computation shows that the

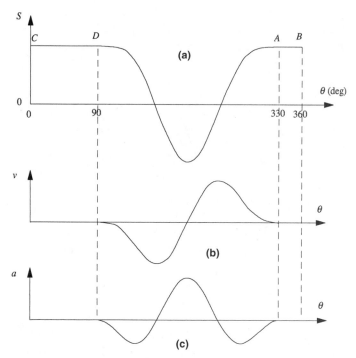

Figure 10.17   SVA diagrams for a single-dwell 1-2 double-harmonic cam program.

peak values of acceleration for the 1-2 double-harmonic program and for the 1-3 double-harmonic program are, respectively, about one-quarter and one-third of those for the system in Fig. 10.16. However, the follower leaves the dwell position more gradually in Fig. 10.17 than it does in Fig. 10.16. As discussed in the example associated with Figs. 10.15 and 10.16, the follower system is used to clamp a sheet of metal against a rubber block. Because the block is elastic, it will be deflected by some of the motion of the follower. Therefore, the clamping action will not be completely released until the deflection has gone to zero and the clamping surface and metal sheet have left the undeflected block. The *duration* of the clamping action will be sensitive to how much the block is deflected and to the rapidity with which the clamping surface moves toward and away from the block. Because the system in Fig. 10.17 moves into contact and out of contact more gradually, its timing will be more sensitive to positioning and elasticity of the rubber block than will that of Fig. 10.16.

The 1-3 double-harmonic program corresponding to Fig. 10.18 provides a compromise between those of Figs. 10.16 and 10.17 because its smoothness of operation lies between those of Figs. 10.16 and

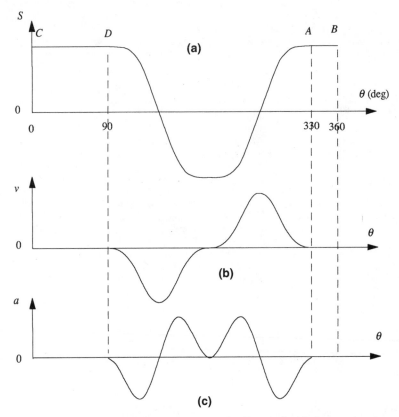

Figure 10.18 SVA diagrams for a single-dwell 1-3 double-harmonic cam program.

10.17 and because it provides a more rapid motion near the ends of the dwell than does that of Fig. 10.17. For a given value of h, its peak acceleration will be about the same as that of the 1-2 double-harmonic program.

If space permits, the maximum displacement of each of these three programs can be increased to increase the rapidity with which the follower approaches and leaves the dwell position. For comparable timing sensitivity, the system in Fig. 10.16 requires the smallest maximum displacement.

Because polynomial functions can be made to fit a wide variety of functions, a polynomial program also can be used for a single-dwell cam. In such a use, the boundary conditions include not only the displacement, velocity, and acceleration at the ends of the program but also the displacement at the midpoint of the motion duration and perhaps also a boundary condition that the velocity be zero at that midpoint.

Motion program for a velocity-tracking cam

There are many machinery and product applications in which it is necessary to provide a cyclic motion, a portion of which consists of a constant-velocity motion. Such applications include those in which an optical system must repeatedly scan along a line at constant velocity, those in which a paint spray gun must be moved repeatedly across an area at constant velocity, and those in which a web of material is moving at constant velocity and some operation (such as cutting, punching, or laminating) must be performed repeatedly on that web without interrupting the web motion. A Hoekens approximate straight-line linkage such as described in Sec. 8.5 can produce a very good approximation to such a constant-velocity output for part of its cycle. However, the use of a cam system allows greater flexibility in the timing of the output.

This type of cam will be referred to as a *velocity-tracking cam* because with it the only portion of the follower motion that is prescribed is that in which follower system output must move at a prescribed velocity for a prescribed time interval or over a prescribed displacement.

A displacement diagram for a velocity-tracking cam is shown in Fig. 10.19. For convenience, the constant-velocity portion AB of this diagram is shown as starting at a cam reference angle of $0°$. The follower displacement during this constant-velocity motion is h_v, and the cam angle during which it occurs is β_v radians as shown. If the constant

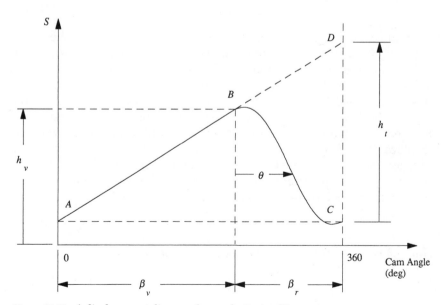

Figure 10.19 A displacement diagram for a velocity-tracking cam.

velocity (in inches per radian) is denoted as v_c, these three variables are related by

$$h_v = \beta_v v_c \qquad (10.27)$$

In any application, two of the variables in this equation will be prescribed, so the third can be computed.

The portion of the displacement diagram from point B to point C is the *return* motion, in which the follower system returns to the displacement point C from which to repeat the constant-velocity displacement. Note that if the constant-velocity motion were to continue for the remainder of the 2π of the cam cycle, the motion would be as indicated by dashed line BD. The total displacement would be

$$h_t = 2\pi v_c \qquad (10.28)$$

If, then, a fall or return motion such as that provided by any of the standard motion programs described previously were *added* to the motion indicated by dashed line BD, the resulting sum-total motion would be as indicated by return curve BC. The standard motion program used must have a maximum displacement of

$$h_r = -h_t = -2\pi v_c \qquad (10.29)$$

and the duration β_r of that program displacement must be

$$\beta_r = 2\pi - \beta_v \qquad (10.30)$$

Because each of the standard motion programs produces the same velocity at the end of its motion as existed at the start of its motion ($v = 0$ in the previous examples), the slope of return curve BC that results from adding the standard program to the motion represented by dashed line BD will be such that the slope of curve BC will be the same at both its ends, as shown. The actual return motion program to be used is thus a sum of a standard program plus a constant-velocity displacement. Thus, if the *standard* program displacement, velocity, and acceleration are denoted by S_s, v_s, and a_s, respectively, and the *return* program displacement, velocity, and acceleration are denoted by S_r, v_r, and a_r, respectively, we may write

$$S_r = S_s + v_c\theta \qquad (10.31)$$

where θ *is measured from the start of the return,*

$$v_r = v_s + v_c \qquad (10.32)$$

$$a_r = a_s \qquad (10.33)$$

The return motion program can be designed by using the following procedure.

Procedure 10.2: Design of a Velocity-Tracking Cam Motion Program

1. Sketch the displacement diagram of the desired cam, and label it in the manner indicated in Fig. 10.19.
2. Determine values for all three variables in Eq. (10.27). Generally, two of these will be prescribed. Recall that β_v is in radians and v_c is in inches per radian.
3. Calculate h_r and β_r using Eqs. (10.29) and (10.30). Be careful of signs. It is generally convenient to use positive values for rising displacements and velocities.
4. Choose one of the standard motion programs. Generally, this will be chosen from the types of programs in Table 10.1.
5. Substitute the value h_r for h and the value β_r for β in the equations for S, v, and a *for the chosen motion program.* These equations are (10.1), (10.2), and (10.3) for the harmonic program, (10.10), (10.11), and (10.12) for the cycloidal program, (10.13) for the modified trapezoidal program, (10.17) for the skewed modified trapezoidal program, (10.18) for the modified sine program, (10.19) for the modified constant-velocity program, and (10.24) for the 3-4-5 polynomial program. The resulting equations are the equations for S_s, v_s, and a_s.
6. Use Eqs. (10.31), (10.32), and (10.33) to obtain the equations for S_r, v_r, and a_r, which are displacement, velocity, and acceleration for the actual return.
7. If desired, use Eqs. (10.4) and (10.5) to find the velocity in inches per second and acceleration in inches per second squared.
8. Size the cam in accordance with Sec. 10.6. If necessary, use relationships in Sec. 10.7 to obtain information for machining the cam.

Figure 10.20 shows displacement, velocity, and acceleration diagrams for a velocity-tracking cam that uses a modified sine motion program for the return portion. That return takes place during 90° of cam rotation. Notice that the acceleration diagram is of the same form as in Fig. 10.12 (although of opposite sign) and the velocity curve is just displaced from that in Fig. 10.12 (although of opposite sign).

Motion programs for other output motions (the half-programs)

The standard motion programs described in the foregoing text are primarily applicable to use for rise or fall between two dwells because the acceleration and velocity are both zero at both the beginning and the end of each of these programs. (However, it was shown in the velocity-tracking cam case that they also can be used between two motion sections in which the velocities are nonzero but equal.)

Notice in each of Figs. 10.8, 10.9, and 10.12 containing the displacement, velocity, and acceleration diagrams for the cycloidal, mod trap, and mod sine programs, respectively, that when $\theta = \beta/2$, the acceleration is zero and the velocity is equal to the maximum value listed for the corresponding program in Table 10.1. Therefore, the first half of

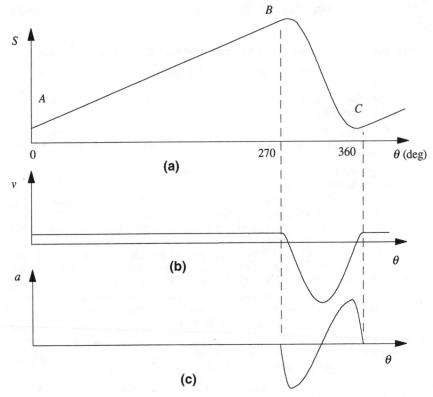

Figure 10.20 SVA diagrams for an example velocity-tracking cam using a modified sine motion program.

each of these programs is suitable for use as a transition from a dwell (velocity equals zero) to a motion section in which the velocity is constant but nonzero. Such a transition is illustrated as the curve from *A* to *B* in Fig. 10.21*a*. Also, the *second half* of each of these programs is suitable for use as a transition from a motion section in which the velocity is constant but nonzero to a dwell. Such a transition is illustrated as the curve from *C* to *D* in Fig. 10.21*b*. A general procedure for designing a transition such as illustrated in Fig. 10.21 is as follows.

Procedure 10.3: The Use of Half-Rise (or Half-Fall) Standard Motion Programs for Transitions between Dwells and Constant-Velocity Motions

1. Sketch a timing diagram for the motion desired. On that diagram, label the transitions between dwells and constant-velocity portions of the motion with symbols Δh_i and $\Delta \beta_i$ in the manner shown in Fig. 10.21, where the subscript i denotes the ith segment of the cam motion.

2. Determine the positions of the ends of the dwells and constant-velocity segments of the motion in terms of cam angle and their height above the lowest pre-

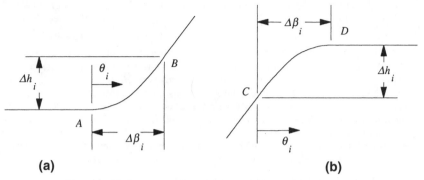

Figure 10.21   The use of half-rise and half-fall standard motion programs.

scribed follower position. That is, the coordinates of the positions of such points as A, B, C, and D in Fig. 10.21 must be specified. The angular coordinates should be specified in radians.

Note that Δh_i and $\Delta \beta_i$ are *not* independent of each other, so only three of the four coordinates associated with points A and B or with C and D can be prescribed. For example, the height *and* cam angle of point B and the height coordinate of point A could be prescribed, and the cam angle of point A would be fixed by the nature of the half-rise program used.

3. From the coordinates determined in step 2, compute either Δh_i or $\Delta \beta_l$, depending on which one of the four previously mentioned coordinates is *not* prescribed.

4. Specify the velocity in the constant-velocity segment adjacent to the ith transition as Δv_i in inches per radian.

5. Choose a half-rise or half-fall program for the ith transition. For a transition from dwell to rising constant velocity, choose a first-half-rise program. For a transition from a rising constant velocity to a dwell, choose a second-half-rise program. For a transition from dwell to falling constant velocity, choose a first-half-fall program. For a transition from a falling constant velocity to a dwell, choose a second-half-fall program.

6. Then note that for the motion program chosen for the transition, the maximum velocity expression given in Table 10.1 is of the form

$$v_{\max} = C_v \frac{h}{\beta}$$

where C_v is called the *velocity coefficient*. Then, because we are dealing with half of a rise or fall program, $\Delta h_i = h_i/2$ and $\Delta \beta_i = \beta_i/2$, and $v_{\max} = \Delta v_i$, so we may write

$$\Delta v_i = C_v \frac{\Delta h_i}{\Delta \beta_i} \tag{10.34}$$

where the value for C_v is inferred from Table 10.1. Then, knowing Δv_i, C_v, and either Δh_i or $\Delta \beta_i$, use Eq. (10.34) to compute either $\Delta \beta_i$ or Δh_i, respectively.

The foregoing steps, when completed for all transitions, define the timing of the cam motion, and if the cam maker is told where *half-programs* are

used, the cam can be machined from these data. If the cam maker does not
have stored standard cam programs, the coordinates of the follower motion
or the shape of the cam must be computed. For this, proceed to step 7.

7. The displacement of the cam follower during the ith transition is given by
Eq. (10.10) for the cycloidal program, (10.13) for the modified trapezoidal program, and by (10.18) for the modified sine program. *However,* these equations
must be modified by the following substitutions: For h substitute $2\Delta h_i$, for β substitute $2\Delta\beta_i$, and if a second-half-program is used, for θ, substitute $\theta + \Delta\beta_i$. In
each transition, θ is measured from the start of the transition. These same substitutions must be made in the equations for velocity and acceleration if those
equations are to be used. Note that the displacement S_i computed in each transition is the displacement *from that at the start of the transition.*

8. If a half-fall is used, the signs of the Δh_i will be negative.

Adventurous readers may want to explore the use of half-rises and half-
falls in transitions between two constant-velocity segments where neither of
the segments has zero velocity. These half-programs can produce such transitions between velocity segments if the programs are augmented by the use
of relationships similar to Eqs. (10.31), (10.32), and (10.33), as was done in
the case of the velocity-tracking cam motion program.

Further discussion of similar techniques of using "building blocks" and
half-programs is given in P. A. Patel, Custom cams from "building blocks,"
Machine Design (May 11, 1978), pp. 86–90; and in the classic M. Kloomok
and R. V. Muffley, Plate cam design—With emphasis on dynamic effects,
Product Engineering (February 1955).

10.5 The Polyharmonic Cam Motion Program

Acceleration frequency spectra as indicators of smoothness

Smoothness of machine operation can, as discussed in Sec. 10.4, be
considered to be characterized by relative absence of shaking or vibration of the machine. Such vibration is produced by varying forces acting on the machine. Usually these forces are inertial in nature, resulting from acceleration of masses in the machine. From the acceleration
diagrams in Sec. 10.4 it can be seen that if a rotary cam is rotated at
constant speed, the acceleration of the follower system will vary periodically. Any unbalanced masses in the follower system will then tend
to produce periodic inertial reaction forces that will cause shaking or
vibration of the machine. The shaking of the overall machine generally
will be small if the mass of the overall machine is large relative to the
mass being accelerated, *unless* the shaking force occurs at or near a
resonant frequency of all or some part of the machine. As will be found
in any book on vibration and/or modal analysis, most machines and
structures have several significant resonant frequencies. This means

that if the machine or some portion of it is subjected to a periodically varying force at or near one of these frequencies, appreciable vibration of the machine or some part of it will result.

Therefore, not only is it important to minimize masses and accelerations in cam follower systems, but it is also important to avoid acceleration variations at frequencies near machine resonant frequencies. Usually, machines and their supporting structures are made sufficiently stiff so that they do not have resonant frequencies near the repetition rate of operations in the machine, and, therefore, the resonances are at frequencies above any camshaft rotation rate. Buildings in which machines are installed often have resonant frequencies of about 10 Hz (i.e., 10 cycles per second), which is also above the cycle rate of most machines. As can be seen, the acceleration variations shown in the acceleration diagrams are not sinusoidal. However, these varying accelerations can be considered to consist of a sum of sinusoidal acceleration variations where each of the sinusoidal variations has a frequency that is an integral multiple of the rotation rate of the cam. That is, the acceleration variation in the acceleration diagram is equivalent to a sinusoidal variation at the rotation rate of the cam, plus another sinusoidal variation at twice the rotation rate of the cam, plus another sinusoidal variation at three times the rotation rate of the cam, plus another sinusoidal variation at four times the rotation rate of the cam, plus another sinusoidal variation at five times the rotation rate of the cam, and so on. These sinusoidal variations are called *harmonics* of the cam rotation rate, and they are represented by terms in a Fourier series.

It is possible for one of these higher harmonic acceleration variations to occur near a resonant frequency of the machine or its support, so one or more of these harmonics could cause significant vibration of all or part of the machine. Therefore, it is important not only to minimize the mass of the follower system and the magnitude of its peak acceleration but also to minimize the magnitudes of higher harmonics of the acceleration variation.

Figure 10.22a shows a displacement diagram of a cam that was used to close and open a set of gripper fingers that gripped a web of material that was intermittently advanced by the gripper. It was required that the fingers close on the web at 330° in the machine cycle and that they open and release the web at 170° in the machine cycle. The dashed horizontal line in Fig. 10.22a indicates the follower displacement at which closing and release occur. The total rise and fall of the cam follower was 0.125 in, in a β of 60°, the machine speed was 220 rpm, and the motion program used for the rise and fall was modified sine. The acceleration diagram is shown in Fig. 10.22b, where the acceleration scale is in inches per second squared.

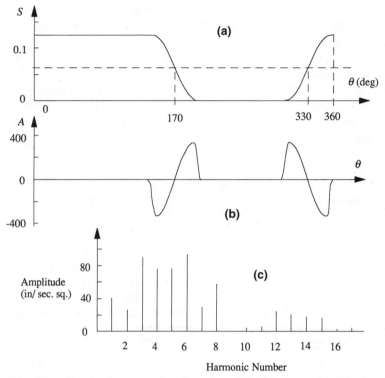

Figure 10.22 S and a diagrams of an example cam motion program and the frequency spectrum of its acceleration curve.

A Fourier analysis of the curve in the acceleration diagram gives the amplitudes of the sine waves which, when summed, are equivalent to that curve. Each of these amplitudes is plotted versus its harmonic number in Fig. 10.22c. That is, the amplitude plotted at harmonic number 1 indicates the amplitude of a sine wave at a frequency equal to the *fundamental frequency* or the rotation rate of the cam. The amplitude at harmonic number 2 corresponds to a sine wave at twice the fundamental frequency, that at harmonic number 3 corresponds to a sine wave at three times the fundamental frequency, and so on. A plot such as this is known as a plot of the *frequency spectrum* of the cam follower acceleration. It will be noticed that harmonics 3 through 8 have magnitudes that are greater than those of the fundamental and second harmonics.

Consider the eighth harmonic: It has an amplitude of approximately 54 in/s² and its frequency is 29.3 Hz. It is quite possible to have a resonant frequency near 30 Hz in some part of a machine or its supporting structure, so these higher harmonics can be of importance.

Using the frequency spectrum of the acceleration diagram as an indication of smoothness is an alternative method of describing smoothness. Unfortunately, like jerk this description also tends to be subjective *unless* it is combined with machine resonance information and/or the results of modal analysis.

Development of the polyharmonic cam program

With the aid of computer software and some trial and error, a cam motion program consisting of the sum of only two sine waves was developed for the application just discussed. Because a program such as this can consist of two or more harmonics (the fundamental plus its second harmonic in this case), it might be called a *polyharmonic motion program*. The resulting displacement and acceleration diagrams are shown in Fig. 10.23a and b. Again, the displacement at which the fingers open and close is indicated by the horizontal dashed line in

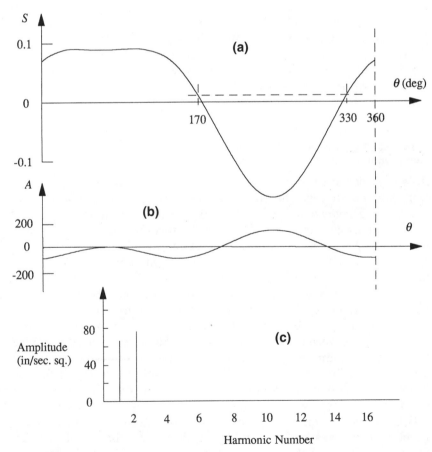

Figure 10.23 *S* and *a* diagrams of a polyharmonic cam program and the frequency spectrum of its acceleration curve.

Fig. 10.23a. It is seen that opening and closing occur at the same cam angles as in Fig. 10.22a. The polyharmonic program displacement provides a good approximation of a dwell during the period when the fingers are gripping the web. The slope of the displacement curve as it crosses the dashed line is quite similar to that of the mod sine program, so the *timing* sensitivity of the gripping and releasing relative to finger adjustment is about the same in both cases. The greater displacement above the dashed line in Fig. 10.23a than in Fig. 10.22a requires that more compliance be provided when the polyharmonic cam is used. The greater displacement below the dashed line in Fig. 10.23a than in Fig. 10.22a requires that more space be provided when the polyharmonic cam is used. Both of these provisions were possible in the application for which this polyharmonic cam was developed.

The frequency spectrum of the acceleration for this polyharmonic cam motion program is presented in Fig. 10.23c. It is seen that this spectrum shows only the amplitudes of the two harmonics from which the motion is constructed. The amplitudes of these sine waves are comparable with those of the third through sixth harmonics in Fig. 10.22c. However, those in Fig. 10.23c occur at frequencies of only 3.7 and 7.3 Hz, which are frequencies probably well below any resonances. This polyharmonic cam system did indeed prove to be a very smoothly running system.

Synthesis procedures for polyharmonic cam motion programs

As stated earlier, the amplitudes and phases of the harmonics that were summed to give the polyharmonic program were found by trial and error. They are *not* values that are obtained by Fourier analysis! Fourier analysis finds the value for each term based on the assumption that there are an infinite number of additional terms. Rather, this synthesis consists of searching for the best fit that can be achieved using a limited number of sine waves.

First, decide what features of the desired cam motion are most important. In the case discussed earlier, the follower displacement and cam angle at the points where the displacement crosses the dashed horizontal line were chosen as prescribed values. Then it was decided to limit the amount of displacement of the follower above these points by attempting to produce an approximate dwell at the top of the follower travel. The remainder of the cam program cycle initially was left free of constraints.

Second, a short computer program was written that summed a small number of sine waves of adjustable amplitude and phase and with frequencies that were integral multiples of some fundamental frequency.

Third, using that computer program, various combinations of sine waves were summed, starting with the fundamental plus the second harmonic, the fundamental plus the third harmonic, the fundamental plus the second and third harmonics, and so on. The values of each of these sums in the region of the requirements chosen in the first step were compared with those requirements. Adjustments were made in the amplitudes and phases until an acceptable fit was achieved.

With the use of the computer program, the search for a satisfactory fit proceeded quite rapidly. With some more effort, an automatic search program could be written for use when many such polyharmonic cams are to be designed.

10.6   Cam Size and Physical Characteristics

The motion programs discussed in Secs. 10.4 and 10.5 dealt only with the motion of the cam follower and not with the nature of the cam that produced that motion. If a sliding cam with a sliding pointed follower such as shown in Fig. 10.1 is used, the shape of the cam will be the same (except for scale factors) as the shape of the displacement diagram. If a plate cam with a pointed follower that slides radially to the cam is used (see Fig. 10.3b), the cam shape will be such that the radius from its center of rotation to its surface will vary in exactly the same manner as does the vertical coordinate of the curve in the displacement diagram. That is, the cam shape will resemble that which would be obtained by "wrapping" the displacement diagram curve around a circle.

This section will deal with plate cams. Pointed followers are *not* used with most plate cams. With types of cam followers other than pointed sliding followers, the nature of the cam shape can differ appreciably from the shape of the displacement diagram. As a result, even though a motion program is chosen for a given application and that program produces a satisfactory displacement diagram, the actual cam shape produced may contain what may be called *kinematic defects* such as excessive pressure angle and undercutting. These defects and methods for avoiding them are the subject of this section. Their effects, causes, and cures are summarized in Table 10.4 in Sec. 10.8.

Cam geometry definitions
(with translating roller follower)

Figure 10.24 depicts a plate cam with a translating roller follower whose radius is R_f. The line l along which the follower is free to translate is offset from the center of rotation O of the cam by an *offset distance* ϵ (sometimes called an *eccentricity*). A closed dashed curve called

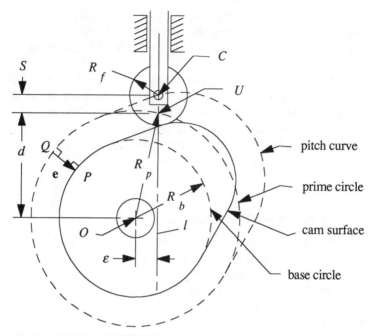

Figure 10.24 Cam geometry definitions (with translating roller follower).

the *pitch curve* is shown surrounding the actual cam surface and passing through the center of the follower roller. This pitch curve is shaped such that if the cam were actually shaped like that curve and if the follower were replaced by a sliding pointed follower whose point was at point c where the roller is pivoted, the motion of the follower as the cam rotated would be as prescribed by the desired displacement diagram.

Each point such as P on the actual surface of the cam is displaced inside the pitch curve by a displacement vector $\mathbf{e}$ from a corresponding point Q on the pitch curve. This vector $\mathbf{e}$ is perpendicular to the tangent to the pitch curve as shown, and its magnitude is R_f. Thus it could be said that the cam surface is parallel to the pitch curve and at a distance R_f inside it. It can be seen that as long as the point of the hypothetical pointed follower remains on the pitch curve as the cam rotates, the roller (whose center C coincides with the tip of the pointed follower) will remain in contact with the actual cam surface, so the roller follower will perform the prescribed motion as it rolls along the actual cam surface. (Notice that the cam surface is "parallel" to the pitch curve and that the radius from point O to each of its points is *not* obtained by subtracting R_f from the radius from O to a corresponding point on the pitch curve.)

The dashed circle that is shown centered at point O and which is internally tangent to the pitch curve at its closest approach to point O is known as the *prime circle* for this pitch curve. Its radius is called the *prime radius,* and it is labeled R_p. It can be seen that if the follower were lowered until its center C was at point U where the prime circle intersects the line l, the follower would be at the lowest position allowed by the cam. Then we may indicate the displacement of the follower from its lowest position by the dimension S as shown.

A dashed circle that is shown centered at point O and which is internally tangent to the actual *cam surface* at its closest approach to point O is known as the *base circle* for this cam. Its radius is called the *base radius,* and it is labeled R_b. The distance R_b is the minimum dimension between the cam surface and the center of rotation of the cam, and it must be great enough to allow room for the cam shaft and for any means for attaching the cam to the shaft. The base radius *is* found by subtracting R_f from the prime radius.

The pitch curve for a translating roller follower

Much of the design of a cam and *roller follower* system is concerned with the design of the pitch curve. The relationships between this curve and the translating cam follower displacement diagram will be derived using variables labeled in Fig. 10.25. In this figure a line OM has been scribed on the cam, which is rotating about point O. This line

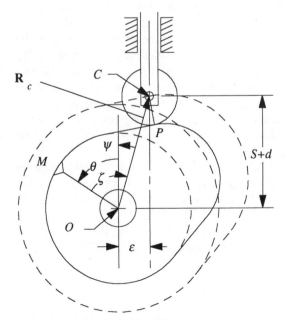

Figure 10.25 Pitch curve relationships for a translating roller follower.

will be vertical (i.e., parallel to the line l along which the follower slides) when the cam is at a rotary position corresponding to $\theta = 0$ on the displacement diagram and when the follower is at the position prescribed for $\theta = 0$ on that diagram.

The cam is shown at a position in which θ is not zero and at which the follower is displaced from its lowest position by a distance S (such as shown in Fig. 10.24), where S is a function of θ as prescribed by the displacement diagram. Point C is on the pitch curve, and its position can be described relative to the cam center O and the line OM scribed on the cam, by polar coordinates ζ and R_c, where R_c is the length of a line segment drawn from O to the point C and ζ is the angle measured from line OM to segment OC as shown.

From dimensions shown in Figs. 10.24 and 10.25, we may write

$$d = \sqrt{R_p^2 - \epsilon^2} \qquad (10.35a)$$

$$R_c = \sqrt{\epsilon^2 + (S + d)^2} \qquad (10.35b)$$

$$\tan \psi = \frac{\epsilon}{S + d} \qquad (10.35c)$$

and $$\zeta = -\psi - \theta \qquad (10.35d)$$

Then, because S is a function of θ as prescribed by the displacement diagram, Eqs. (10.35) can be used to compute the polar coordinates R_c and ζ of points on the pitch curve for all values of θ. Care should be taken to use correct signs of the angle variables. Assume that angles that are measured counterclockwise from one line to another are positive and angles that are measured clockwise from one line to another are negative. For example, in the figure θ and ψ are positive and ζ is negative. Notice that because the cam is rotating counterclockwise, the angle ζ changes in a clockwise direction relative to reference line OM on the cam as θ increases.

Notice that when $\epsilon = 0$ (no offset), $R_c = S + R_p$ and $\zeta = -\theta$, so the relationship of the pitch curve to the displacement diagram (which displays S as the prescribed function of θ) is quite simple when offset is absent.

The pitch curve will be found to be useful in the manufacture of cams for use with roller followers, as discussed in Sec. 10.7.

Cam follower pressure angle for a translating roller follower

In choosing the size for a cam that will be used *with a roller follower*, serious consideration should be given to the maximum value of the cam follower pressure angle that will be encountered. The *pressure angle* is defined as the angle between the force that the cam applies to the follower (neglecting friction) and the direction in which the follower is free to translate, and it is related to the transmission angle in linkages, which was discussed in Sec. 3.6 and Procedure 4.1.

Figure 10.26 shows a plate cam with an offset translating roller follower that is constrained to slide vertically. Because the follower is free to roll along the surface of the cam, the cam can only apply a force $\mathbf{F}$ on the follower in a direction perpendicular to the common tangent t at contact point P between the cam surface and the follower as shown. The pressure angle, then, is the angle ϕ between this direction of $\mathbf{F}$ and the direction of translation of the follower as shown. If this angle becomes large, the component of force on the follower that is perpendicular to the direction in which it can slide (i.e., in the case in Fig. 10.26 it is the horizontal component of $\mathbf{F}$) can become large, possibly causing large friction and binding in the slide.

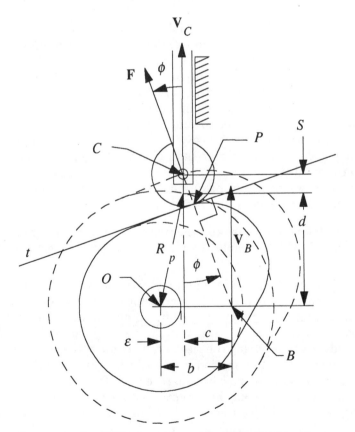

Figure 10.26 Cam follower pressure angle relationships for a translating roller follower.

An expression for the magnitude of this pressure angle can be derived by extending a line parallel to $\mathbf{F}$ from point P downward until it and the horizontal line through point O intersect at point B. Attach point B to the body of the cam. Because the motion of point C *relative to point B* can only be in a direction perpendicular to line CB, and because the velocities of both of these points are parallel to each other (i.e., vertical), the velocity of point C is equal to that of point B. The velocity of point C is given by Eq. (10.4) as $V_C = v\omega$, where v is given for standard motion programs in Sec. 10.4 and ω is the angular velocity of the cam in radians per second. Notice that $V_B = \omega b$. Then, because $V_C = V_B$, it can be concluded that $b = v$.

From the geometry in Fig. 10.26, it is seen that

$$\tan \phi = \frac{c}{S + d} = \frac{b - \epsilon}{S + \sqrt{R_p^2 - \epsilon^2}} = \frac{v - \epsilon}{S + \sqrt{R_p^2 - \epsilon^2}} \qquad (10.36)$$

where, as in Fig. 10.24, S is the displacement of the follower from its lowest position.

By using Eq. (10.36) and a programmable pocket calculator, some reader-generated software, or some commercially available equation-solving software, values for the pressure angle can be computed for values of S and v corresponding to various positions along the displacement and velocity diagrams for the chosen cam motion program and for trial values of R_p and ϵ. Searching in this way, the maximum value of pressure angle can be found and limited to an acceptable value. Generally accepted practice is to avoid cam designs in which the maximum value of ϕ becomes greater than about 30° (i.e., in which $\tan \phi$ is greater than about 0.6).

If computation shows that the maximum pressure angle is unacceptably large, Eq. (10.36) shows that it can be reduced by

1. Increasing R_p, the pitch radius (and therefore the size of the cam). If space permits, this is probably the most effective means for reducing pressure angle.

2. Decreasing v. Because v is proportional to h/β in the standard motion programs, this can be accomplished by increasing β. For this reason, as well as for reasons of smoothness, β should always be made as large as allowed by the prescribed timing. The velocity parameter v also can be reduced by reducing the rise or fall height h. If a certain overall follower system output motion is prescribed, any pressure angle gains obtained by decreasing h by using greater motion amplification in the follower linkage may be deceptive because of changes in the resulting mechanical advantage (see later discussion).

3. Introducing or increasing ϵ. Any offset that improves the pressure angle during rise makes the pressure angle worse during fall, and

vice versa. Therefore, the benefits that can be achieved by using off-set are limited.

4. Selecting a motion program that produces smaller values of v. The improvement obtainable by such selection is quite limited because, as seen in Table 10.1, the maximum velocities produced by the various standard programs differ from each other by a rather limited amount.

If the computation and search process associated with Eq. (10.36) is considered too arduous, the "rule of thumb" embodied in Eq. (10.40) can be used to guide cam geometry selection and provide a first esti-mate of the required prime radius. This rule of thumb is based on the assumption that a standard cam follower motion program is used, and on the conservative assumption that a cam and follower system that has *no offset* and which therefore experiences a least-favorable maxi-mum pressure angle is being used. For such zero offset, Eq. (10.36) becomes

$$\tan \phi = \frac{v}{S + R_p} \tag{10.37}$$

Examination of the velocity and displacement diagrams for the stan-dard motion programs in Figs. 10.7, 10.8, 10.9, 10.12, 10.13, and 10.14 shows that the maximum velocity occurs at $\theta = \beta/2$ and $S = h/2$. Because S appears in the denominator of the right-hand side of Eq. (10.37), the maximum value of $\tan \phi$ for a standard motion pro-gram rise will occur slightly earlier than at $\theta = \beta/2$, but its value will be approximately equal to that which occurs at $\theta = \beta/2$.

Using the condition at $\theta = \beta/2$ as indicative of the condition for ap-proximate maximum ϕ, and assuming that the maximum pressure angle occurs during midrise, Eq. (10.37) becomes

$$\tan \phi = \frac{v_{\max}}{(h/2) + R_p} \tag{10.38}$$

However $v_{\max} = C_v(h/\beta)$, where C_v is one of the values given in Table 10.1. Then, Eq. (10.38) becomes

$$\tan \phi = \frac{C_v(h/\beta)}{(h/2) + R_p}$$

which, when solved for R_p, becomes

$$R_p = h\left(\frac{C_v}{\beta \tan \phi} - 0.5\right) \tag{10.39}$$

If values of ϕ greater than some maximum tolerable level $\phi_{\max}$ are to be prevented, then

$$R_p \geq h \left(\frac{C_v}{\beta \tan \phi_{\max}} - 0.5 \right) \qquad (10.40)$$

where h is the height of the rise, β is the cam angular displacement over which the rise occurs, and C_v is obtained from Table 10.1.

Equation (10.40) then gives the minimum radius from cam center to pitch curve at the start of a rise (or end of a fall) if the pressure angle during that rise (or fall) is not to exceed $\phi_{\max}$. Although this formula pertains to symmetrical motion programs such as the standard programs listed in Table 10.1, it can provide initial estimates for required cam size for a wide variety of cam motion programs. By substituting into Eq. (10.40) $\beta v_{\max}/h$ for C_v, where $v_{\max}$ is follower velocity in inches per radian of cam angle and β is in radians, and by neglecting the 0.5 term in the parentheses, the resulting equation can be used to provide initial estimates of required radius from cam center to the point on the pitch curve where $v = v_{\max}$. An example of the use of Eq. (10.40) for a case using a cycloidal rise program is as follows.

Example 10.1: Sizing a Plate Cam for Acceptable Pressure Angle for a Translating Roller Follower   *Objective:* Find the minimum acceptable size for a plate cam that must have a rise of 2 in in a cam rotation of 60° and which has a radially translating roller follower in which the roller diameter is 2 in. The rise motion program is to be of the modified sine type. The pressure angle must not exceed 30°.

The values to be used in Eq. (10.40) therefore are

$$h = 2.0$$

$$\beta = 60(\pi/180) = 1.0472 \text{ rad}$$

$$\phi_{\max} = 30°$$

$$C_v = 1.76 \qquad \text{(from Table 10.1)}$$

Equation (10.40) gives $R_p \geq 5.82$ in.

This result means that in order to keep the pressure angle from exceeding 30° during the specified rise, the radius measured from the cam center to the pitch curve at the start of that rise must be at least 5.82 in. If this is a double-dwell cam, the rise will start at the low dwell, so this radius will be the pitch circle radius. Because the roller on the follower has a diameter of 2 in (a radius of 1 in), the *base circle* will have a radius of at least $5.82 - 1.0 = 4.82$ in. The high-dwell portion of the cam surface will have a radius that is greater than this by the height of the 2-in rise, or $4.82 + 2.0 = 6.82$ in.

Comments on pressure angle

The foregoing discussion referred to translating roller followers. If readers will consider a translating roller follower in which the bearing in the roller has become locked such as to prevent rotation of the follower, they will note that such a follower system moves in the same manner as though the roller were free to rotate, but the roller would slide along the cam surface instead of rolling on it. This discussion, then, is seen to pertain to translating followers with faces that slide on the cam surface (such as in Fig. 10.3b and c) but in which those faces are portions of circular cylinders. That is, if the point on the follower in Fig. 10.3b were rounded, having a nonzero radius of curvature, and if the flat surface on the bottom of the follower face in Fig. 10.3c were curved, these two cam and follower systems could be treated just as was done above. In such cases, the pitch curve would pertain to the motion of the point that is at the center of curvature of the face of the follower.

If an oscillating follower such as shown in Fig. 10.3d and e is used, the equations become more complicated. However, note that the pressure angle is defined as the angle between the direction of the normal force between the cam and follower and the direction in which the follower is free to move. The swing follower is free to move in the direction tangent to the path along which its roller's center (or center of curvature) travels. Rough estimates of the direction of this freedom and thus of the pressure angle can be obtained by approximating the arc along which the follower swings by the chord of that arc.

Sizing a cam and translating roller follower system to provide a good radius or curvature

As shown in Fig. 10.25, the actual surface of the cam lies inside the pitch curve and parallel to it. The cam surface is *not* just a scaled-down version of the pitch curve, and it will have a shape that is different from that of the pitch curve. Therefore, it is possible that even though the pitch curve is well behaved, the corresponding cam surface curve may possess undesirably sharp curvatures or features that are physically not realizable. Such features can be seen by referring to Fig. 10.27.

Figure 10.27a shows a portion of a pitch curve l and a roller follower whose center is at C on that curve. Corresponding to point C is a point D on the cam surface, where D is on a line perpendicular to the tangent to curve l at C and is at a distance R_f from C. For any point C_i there is a corresponding point D_i on the cam surface, so the cam surface m can be generated by placing point D_i at a perpendicular

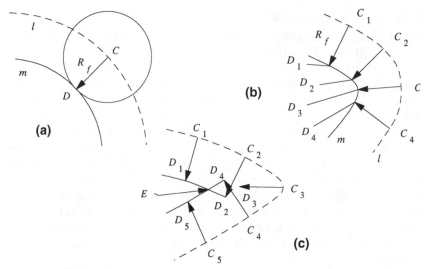

Figure 10.27 Effects of pitch curve radius of curvature on cam shape in a roller follower system.

distance R_f from the pitch curve. The radius of curvature of a curve at any point is defined as the radius of a circular arc that best fits the curve at that point. As drawn, the radius of curvature of the pitch curve in Fig. 10.27a is somewhat larger than the radius of the cam follower roller.

Figure 10.27b shows a portion of a pitch curve l that has a radius of curvature that is only slightly larger than the radius R_f of the follower roller. Constructing lines with lengths R_f and which are perpendiculars to the pitch curve at points C_1, C_2, C_3, and C_4, placing points D_1, D_2, D_3, and D_4 at the ends of those lines, and passing a curve through points D_1, D_2, D_3, and D_4 will generate the corresponding cam surface m. It can be seen that points D_2, D_3, and D_4 are quite close to each other and that the cam shape curve m is very sharply curved near these points; i.e., its radius of curvature is very small.

Figure 10.27c shows a portion of a pitch curve l that has a radius of curvature that is *smaller* than the radius R_f of the follower roller. Attempting to construct the cam surface curve in the same manner as before results in some conflicts. For example, the portion of the surface as defined by points D_1 and D_2 intersects the portion of the surface defined by points D_4 and D_5 at point E. Because these two portions of the surface intersect at point E, there can be no cam material to the right of this intersection point E. Therefore, there will be no cam material for the follower to contact when its center is at points C_2, C_3, and C_4. A cam in which such a needed portion of material is cut away is considered to be *undercut*. The cam surface of an undercut cam will not produce follower displacements corresponding to the given pitch curve and cam follower roller radius.

From Fig. 10.27, then, it may be concluded that the radius of curvature of the pitch curve must always be greater than the radius of the follower roller. The radius of curvature of a pitch curve can be calculated using Eq. (10.45) and commercially available mathematical software or by programming a spreadsheet to perform the calculations.

Figure 10.28 is a reduced-scale tracing of an existing cam surface m and a portion of a corresponding pitch curve l that resulted from a severe cam system design requirement. This cam is one of a set of six different cams on a common shaft in an automatic loom used for weaving corduroy cloth. Several motions of the cam follower system are required during each revolution of the cam shaft, so the cam angle available for rises and and falls is quite limited. The space available for the cam also was quite limited, and yet the magnitudes of the output displacements required of the follower system are substantial. As a result, even though the pitch curve l shown can be seen to be relatively smooth, the cam surface m itself has relatively sharply curved corners, such as at points A through G. These sharp corners occur where the radius of curvature of the pitch curve is not much larger than the radius of the cam follower roller. Because of the severe requirements, it was necessary to use an offset to avoid excessive pressure angles. Notice

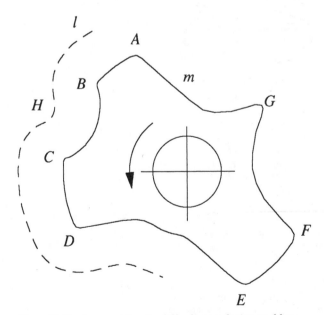

Figure 10.28 An example of a difficult cam design problem.

that where the pitch curve is concave outward as at point H, the radius of curvature of the pitch curve is *smaller* than that of the cam surface, so no cam manufacturing problem can be caused by small concave curvatures in the pitch curve.

The radius of curvature of the pitch curve for a translating roller follower can be calculated using Eq. (10.45), which can be derived as follows: As shown in calculus textbooks, the radius of curvature ρ of a curve is given by

$$\rho = \frac{d\sigma}{d\alpha} \tag{10.41}$$

where σ is the measured displacement along the curve and α is the angle in radians that the tangent to the curve at the point being investigated makes with some fixed reference direction. Expressions for $d\sigma$ and $d\alpha$ can be derived by assuming the pitch curve and the cam represented by curves l and m, respectively, in Fig. 10.29 to be stationary and

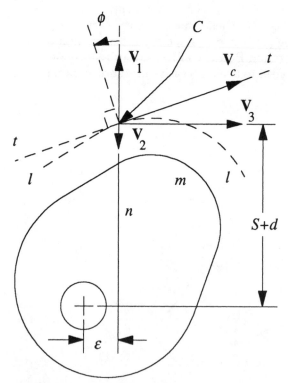

Figure 10.29 Geometry for computing the radius of curvature of the pitch curve for a cam with a translating roller follower.

considering the *base* that carries the line *n* along which the follower and its roller center point *C* normally translate to be rotating clockwise at an angular velocity $d\theta/dt$ about the cam center *O*. The center *C* of the cam roller will then travel along the stationary pitch curve. In Fig. 10.29 the rate of change $d\sigma/dt$ of the displacement σ of *C* along the pitch curve is the magnitude of the velocity vector $\mathbf{V}_c$ of point *C* in the direction of the tangent *t* to the pitch curve. This velocity vector $\mathbf{V}_c$ is the vector sum of the vector slip velocity $\mathbf{V}_1$ of point *C* relative to the rotating base and the velocity vector $(\mathbf{V}_2 + \mathbf{V}_3)$ of a point that is momentarily coincident with *C* but attached to the rotating base. The upward vertical velocity vector $\mathbf{V}_1$ has a magnitude $V = v(d\theta/dt)$ (which is the slip velocity of point *C* along the rotating line *n*). The downward vertical velocity vector $\mathbf{V}_2$ has a magnitude $\epsilon d\theta/dt$, and the horizontal velocity vector $\mathbf{V}_3$ has a magnitude $(S + d)d\theta/dt$. Then, after squaring magnitudes, summing, multiplying by dt, rearranging terms, we may write

$$d\sigma = d\theta \sqrt{\left(\frac{dS}{d\theta} - \epsilon\right)^2 + (S + d)^2} \qquad (10.42)$$

The normal to the tangent to the pitch curve is inclined to the line *n* by the pressure angle ϕ, as shown in Fig. 10.29. As the base carrying line *n* is rotated clockwise by $d\theta$, the normal to the pitch curve tangent will be rotated clockwise by $d\theta - d\phi$, and therefore the tangent itself will be rotated clockwise through a differential angle $d\alpha = d\theta - d\phi$.

From Eq. (10.36) it is seen that

$$\tan \phi = \frac{v - \epsilon}{S + d} \qquad (10.43)$$

Taking the differential of Eq. (10.43), using the geometry of Fig. 10.26, and rearranging terms will give

$$d\phi = \frac{(S + d)dv - (v - \epsilon)dS}{(S + d)^2 + (v - \epsilon)^2} \qquad (10.44)$$

Substituting Eqs. (10.42) and (10.44) into Eq. (10.41), rearranging terms, and noting that $v = dS/d\theta$ and $a = dv/d\theta$ will give the following equation as an expression for the radius of curvature of the pitch curve:

$$\rho_p = \frac{[(S + d)^2 + (v - \epsilon)^2]^{3/2}}{(v - \epsilon)v - (S + d)a + (S + d)^2 + (v - \epsilon)^2} \qquad (10.45)$$

where $d = \sqrt{R_p^2 - \epsilon^2}$.

Although not obvious from Eq. (10.45), major contributors to undesirable smallness of radius of curvature of the pitch curve are a large negative value of a (the follower acceleration) and a small value of R_p (the radius of the pitch circle). To avoid small pitch curve radii of curvature, then, rise and fall durations β_i should be maximized and cam size relative to total follower excursions should be maximized.

Computation of cam surface coordinates and tool path for machining for a translating roller follower

If readers wish to see what the actual cam shape will look like, or if the shop that must make the cam is not equipped to make the cam directly from pitch curve data, the coordinates of the cam surface shape and for the tool path can be computed from Eqs. (10.49) through (10.52). These equations are derived on the basis of the geometry of Fig. 10.30, where the points O, C and P are the same points as shown in Fig. 10.25.

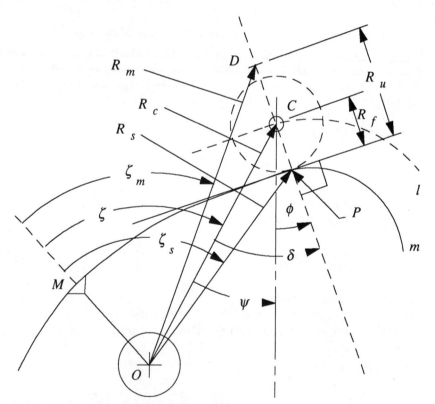

Figure 10.30 Geometry for computation of cam surface coordinates and tool path for machining for a cam with a translating roller follower.

From Figs. 10.25 and 10.26 it can be seen that, as shown in Fig. 10.30, the angle $\delta = \psi + \phi$. Then, having found R_c, ψ, and ζ from Eqs. (10.35) and ϕ from Eq. (10.36), the law of cosines can be written for triangle OCP as

$$R_s^2 = R_c^2 + R_f^2 - 2R_cR_f \cos \delta \qquad (10.46)$$

from which R_s can be computed. Then, again applying the law of cosines to the triangle OCP, we can write

$$R_f^2 = R_c^2 + R_s^2 - 2R_cR_s \cos \alpha \qquad (10.47)$$

from which α can be computed, where α is the angle POC. Then, if the positive value for α is taken,

$$\zeta_s = \zeta - \alpha \qquad (10.48)$$

The polar coordinates relative to line OM scribed on the cam of points P on the cam surface are then given by R_s and ζ_s as computed using Eqs. (10.46), (10.47), and (10.48), and these equations can be applied for any or all of the values of θ around the cam.

Point D in Fig. 10.30 is considered to be the center of a milling cutter whose radius is R_u and which is being used to cut the surface of the cam. The law of cosines can then be applied to triangle OPD to give the following equations:

$$R_c^2 = R_s^2 + R_f^2 - 2R_sR_f \cos \beta \qquad (10.49)$$

$$R_m^2 = R_s^2 + R_u^2 - 2R_sR_u \cos \beta \qquad (10.50)$$

$$R_u^2 = R_m^2 + R_s^2 - 2R_mR_s \cos \gamma \qquad (10.51)$$

$$\zeta_m = \zeta - \alpha + \gamma \qquad (10.52)$$

where γ is the angle POD and β is the angle CPO (not shown in Fig. 10.30).

The polar coordinates relative to line OM scribed on the cam of points on the required path of the cutter are R_m and ζ_m. They can be computed using Eq. (10.49) to find β, then Eq. (10.50) to find R_m, then Eq. (10.51) to find γ, and then Eq. (10.52) to find ζ_m.

The pitch curve for an oscillating roller follower

The parameters involved in the geometry of the pitch curve for an oscillating roller follower are indicated in Fig. 10.31. A reference line OM is scribed on the cam, and that line will coincide with line OA from cam

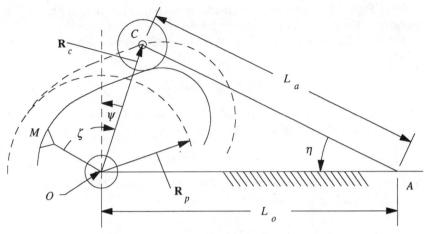

Figure 10.31 Pitch curve relationships for an oscillating roller follower.

center O to follower swing arm pivot A when $\theta = 0$. When an oscillating roller follower is used, the follower motion is described in terms of the angle η of the arm to which the roller is pivoted. This angle consists of a minimum value of η_0 plus a variable increment $\Delta\eta$. It is seen by reference to Fig. 10.31 that by assuming point C to be placed on the pitch *circle,* thereby making $R_c = R_p$, and by using the law of cosines, then η_0 can be computed from

$$R_p^2 = L_o^2 + L_a^2 - 2L_o L_a \cos \eta_0 \qquad (10.53)$$

The value of $\Delta\eta$ is obtained from the follower motion program equations (e.g., for standard programs they are given in Sec. 10.4) by interpreting the motions as angles instead of translations. That is, oscillating follower motions are given by the relationships

$$\Delta\eta = S \qquad (10.54a)$$

$$\frac{d\eta}{d\theta} = v \qquad (10.54b)$$

$$\frac{d^2\eta}{d\theta^2} = a \qquad (10.54c)$$

$$\eta = \eta_0 + \Delta\eta \qquad (10.54d)$$

where, in the motion programs,

$$h = \Delta\eta_{\max} \qquad (10.54e)$$

From Fig. 10.31 we may write

$$\tan \psi = \frac{L_o - L_a \cos \eta}{L_a \sin \eta} \tag{10.55}$$

$$Rc = \frac{L_a \sin \eta}{\cos \psi} \tag{10.56}$$

$$\zeta = \frac{\pi}{2} - \theta - \psi \tag{10.57}$$

The polar coordinates relative to line OM of points on the pitch curve are R_c and ζ, which can then be computed from Eqs. (10.55), (10.56), and (10.57).

Cam follower pressure angle for an oscillating roller follower

The geometry associated with the pressure angle for an oscillating roller follower is shown in Fig. 10.32. Recalling that the pressure angle

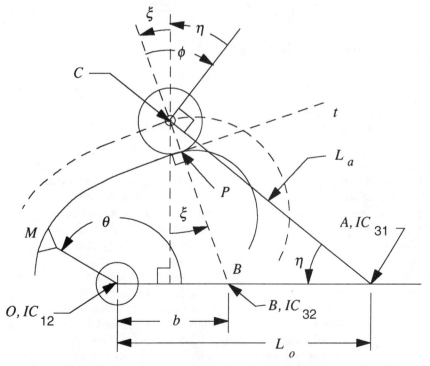

Figure 10.32 Cam follower pressure angle relationships for an oscillating roller follower.

is the angle between the direction of the cam-to-follower force and the direction in which the follower can move, the pressure angle is now the angle ϕ between the normal CB to the cam surface (which is also the normal to the pitch curve) and the normal to line AC as shown. Let the base be called *body 1*, the cam be called *body 2*, and the swing arm AC be called *body 3*. Because point C must move *relative to the cam body* in a direction perpendicular to line CB and point A must move *relative to the cam body* in a direction perpendicular to line AB, point B is the instant center between bodies 2 and 3 and it is so labeled. From the discussion of instant centers and velocity ratios in Chap. 4, and using the symbol b for the distance OB, we may then write

$$b\,d\theta = (L_o - b)\,d\eta$$

from which

$$b = \frac{L_o(d\eta/d\theta)}{1 + (d\eta/d\theta)} \tag{10.58}$$

The angle ξ between the vertical and line CB is seen to be found from

$$\tan \xi = \frac{b - (L_o - L_a \cos \eta)}{L_a \sin \eta} \tag{10.59}$$

The pressure angle ϕ is then given by

$$\phi = -\eta - \xi \tag{10.60}$$

Radius of curvature of a cam with an oscillating roller follower

In many cases the radius of curvature of a pitch curve of a cam used with an oscillating roller follower can be considered to be approximately the same as it would be if the follower were *translating* along a chord of the arc along which the swing follower actually moves. In cases where a more accurate calculation is desired or in which the angular motion of the follower swing arm is large, the radius of curvature of the pitch curve can be calculated using equations contained in this subsection.

Figure 10.33 shows the geometry associated with derivation of expressions for the radius of curvature of the pitch curve. As in the case of the translating roller follower, the radius of curvature of the pitch curve for an oscillating roller follower will be derived by considering the cam fixed while the base is displaced clockwise through a differential angle $d\theta$ about the point O. Such a rotation will displace point A vertically downward by a vector amount $\mathbf{d}_A$ whose magnitude is $L_o d\theta$. The differential rotation of swing arm AC will be $d\theta + d\eta$, so the dif-

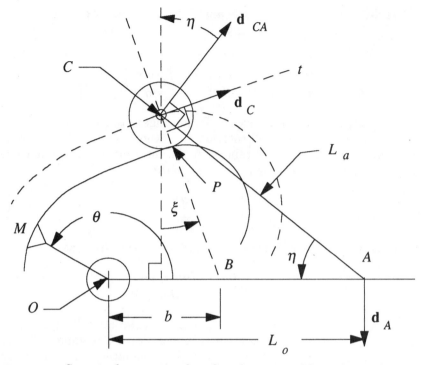

Figure 10.33 Geometry for computing the radius of curvature of the pitch curve for a cam with an oscillating roller follower.

ferential displacement of point C *relative to* an irrotational coordinate system fixed to point A will be the vector $\mathbf{d}_{CA}$, whose magnitude is $L_a(d\theta + d\eta)$ perpendicular to arm AC and upward and to the right as shown. The total differential displacement of point C is vector $\mathbf{d}_C$. It has a magnitude of $d\sigma$, is along the pitch curve, and is the *vector sum* of these two displacements $\mathbf{d}_A$ and $\mathbf{d}_{CA}$. The magnitude of this vector sum, when divided by $d\theta$, gives

$$\frac{d\sigma}{d\theta} = \sqrt{\left[L_o\left(1 + \frac{d\eta}{d\theta}\right) - L_o \cos \eta\right]^2 + (L_o \sin \eta)^2} \qquad (10.61)$$

The differential rotation $d\alpha$ of the tangent will depend on the change in ξ, which, in turn, depends on the change in b, the distance OB. From Eq. (10.58) we may (by differentiating) obtain

$$\frac{db}{d\theta} = \frac{L_o \dfrac{d^2\eta}{d\theta^2}}{\left(1 + \dfrac{d\eta}{d\theta}\right)^2} \qquad (10.62)$$

The differential change in the angle ξ is obtained by differentiating Eq. (10.59) and then dividing by $d\theta$ and appropriately rearranging terms, which gives

$$\frac{d\xi}{d\theta} = \left(\frac{\cos^2 \xi}{L_a \sin \eta}\right)\left\{\frac{db}{d\theta} - \frac{d\eta}{d\theta}\left[\frac{(b - L_o)\cos \eta + L_a}{\sin \eta}\right]\right\} \qquad (10.63)$$

The change $d\alpha$ of the direction of the tangent to the pitch curve is the result of the differential rotation $d\theta$ of the base and the differential change $d\xi$ in the angle ξ relative to the rotating base of the normal to that tangent. Then we may write $d\alpha = d\theta - d\xi$ or

$$\frac{d\alpha}{d\theta} = 1 - \frac{d\xi}{d\theta} \qquad (10.64)$$

Then Eq. (10.41) for radius of curvature ρ can be written as

$$\rho = \frac{d\sigma}{d\alpha} = \frac{(d\sigma/d\theta)}{(d\alpha/d\theta)} \qquad (10.65)$$

Then, using the sequence of Eqs. (10.61) through (10.65), the radius of curvature ρ of the pitch curve can be computed. The values of b, η, and the derivatives of η as used in these equations are given by Eqs. (10.58) and (10.54).

Computation of cam surface coordinates and tool path for machining for an oscillating roller follower

Once the coordinates R_c and ζ of the pitch curve for an oscillating roller follower have been computed using Eqs. (10.55), (10.56), and (10.57), the coordinates of the cam surface and of the tool path can be computed using the same Eqs. (10.46) through (10.52) as used for the translating roller follower *except that* the value of ξ as computed from Eqs. (10.58) and (10.59) must be used to compute $\delta = \psi + \xi$ for use in those equations.

Width requirements for a translating flat-faced cam follower

In the case of a flat-faced follower such as shown in Fig. 10.3c, the point of contact between the cam and the follower moves back and forth across the face of the follower as well as up and down as the cam rotates and the follower moves up and down. Obviously, the contact point cannot move off the edge of the follower face if the prescribed motion is to be produced. This subsection will provide a relationship for finding the follower face-width dimensions required to avoid such a loss of proper contact.

Because of this motion of the cam-to-follower contact point, the shape of the cam surface required to produce the prescribed follower motion does not closely resemble the shape of the prescribed curve of follower displacement versus cam angle. However, unlike in the case of a roller follower, there is no pitch curve that is intermediate between the displacement curve and the cam shape in the case of flat followers.

A cam and right-angled flat-faced follower combination such as shown in Fig. 10.3c is shown in Fig. 10.34. The dashed line l is drawn parallel to the direction of translation of the follower (and therefore perpendicular to the follower face) and through the center of rotation O of the cam. It intersects the follower face at point D. A second line m is drawn parallel to line l and through point C on the follower face where the cam and follower contact. A third line n is parallel to the face of the follower and through the point O, thereby intersecting line m at a point B, which is attached to the cam at a distance b from O. This third line is therefore perpendicular to the first two lines.

The velocity of point C on the follower *relative to point B on the cam* can only consist of sliding perpendicular to line AB. Because, in

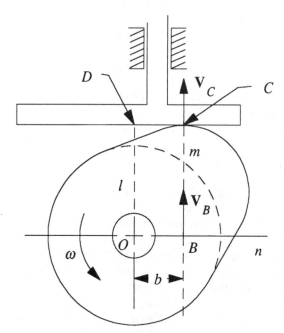

Figure 10.34 Width requirements for a translating flat-faced cam follower.

addition, the velocity of point B is parallel to line CB and the velocity of point C must be vertical, the velocities of these two points are equal in magnitude and parallel. The magnitude of the velocity of point C is given by Eq. (10.4) as $V_C = \omega v$, and the magnitude of the velocity of point B is seen to be $V_B = \omega b$, from which we conclude that $b = v$.

Then, for a cam rotating as shown in the drawing, the v is being experienced during a rise. It may be seen that the face of the follower must extend to the right of point D by a distance equal to v_{max}, where v_{max} is obtained from Table 10.1 or from the equation for whatever motion program is being used. During a fall for the cam rotating as shown in Fig. 10.34, the cam would have to extend to the *left* of point D by a distance equal to the v_{max} that would be experienced during that fall.

Cam radius of curvature for a translating flat-faced cam follower

A flat-faced follower cannot contact a concave cam surface in a single point. Therefore, such concavities are not acceptable in cams for use with flat-faced followers. In addition, just as in the case of roller followers, *undercutting* of cams used with flat-faced followers must be avoided. In the case of a flat-faced follower, such undercutting occurs when the convex radius of curvature of *the cam surface* shrinks beyond zero. Because there is no pitch curve, consideration must be given to the cam surface directly to ensure that its radius of curvature never becomes too small or negative.

Figure 10.35 depicts a portion m of a cam surface and the face f of a flat-faced follower. Initially, the cam and follower are positioned such that the point of contact between cam and follower is at point C, and a line through the center of rotation O of the cam and perpendicular to the face of the follower intersects the follower face at point D. Line CB is drawn parallel to line OD, and line OB is drawn parallel to line CD.

As shown in the discussion in the preceding subsection, the distance from O to B is equal to v. The distance from O to D is equal to the *base* radius R_b (which is the smallest radius from point O to a point on the cam surface) plus the cam follower displacement S.

In deriving an expression for the cam surface radius of curvature, we will consider the cam to be stationary and the base that carries the slide for the follower to be rotated clockwise through a differential angle $-d\theta$, as in the case of the roller followers considered previously.

Rotating the base clockwise through the differential angle $-d\theta$ will displace point D to D' and point C *on the follower* to C'. The point of contact between cam and follower moves from point C to C''. This displacement results in a distance from O to D' of $R_b + S + dS$. Because $b = v$, the distance from D' to C'' becomes $b + db = v + dv$, where dv is negative, as shown in this particular drawing. The x (i.e., horizontal)

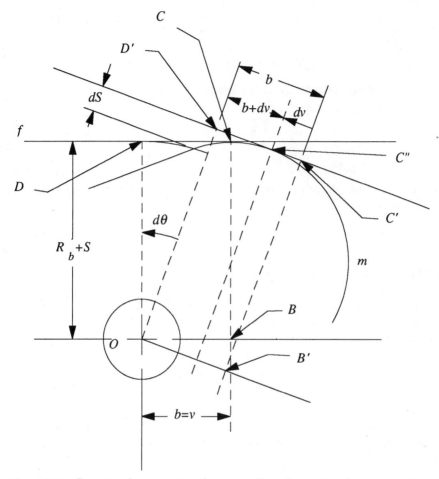

Figure 10.35 Geometry for computing the cam radius of curvature for a cam with a translating flat-faced cam follower.

differential displacement and the y (i.e., vertical) differential displacement of the point of contact are seen to be

$$dx = (R_b + S + dS) \sin d\theta + (v + dv) \cos \theta - v \qquad (10.66)$$

$$dy = (R_b + S + dS) \cos d\theta - (R_b + S) - (v + dv) \sin \theta \qquad (10.67)$$

Because we choose $d\theta$ to be vanishingly small, $\sin d\theta = d\theta$ and $\cos d\theta = 1.0$. Then Eqs. (10.66) and (10.67) become

$$dx = (R_b + S)d\theta + dSd\theta + v + dv - v \qquad (10.68)$$

$$dy = (R_b + S + dS) - (R_b + S) - vd\theta - dvd\theta \qquad (10.69)$$

Neglecting second-order terms $dS d\theta$ and $dv d\theta$ and noting that $v d\theta = dS$ gives

$$dx = (R_b + S)d\theta + dv \qquad \text{and} \qquad dy = 0$$

Then the displacement $d\sigma$ of the point of contact along the cam surface is given by

$$d\sigma = \sqrt{dx^2 + dy^2}$$

or

$$d\sigma = dx = (R_b + S)d\theta + dv \qquad (10.70)$$

Recall that the radius of curvature can be computed from Eq. (10.41) as

$$\rho = \frac{d\sigma}{d\alpha} \qquad (10.71)$$

Also, note that in the present case the change in the angle of the tangent to the cam surface curve is the same as the change in the angular orientation of the follower face, so for a flat-faced follower that has been rotated with the base in this derivation, $d\alpha = d\theta$. Then, combining Eqs. (10.70) and (10.71) gives

$$\rho = \frac{dx}{d\theta} = (R_b + S) + a \qquad (10.72)$$

where $a = (dv/d\theta)$.

Because we are interested in the minimum value of the radius of curvature, Eq. (10.72) can be rearranged to give

$$(R_b)_{\min} = \rho_{\min} - S - a \qquad (10.73)$$

The quantities R_b and S in Eq. (10.72) are always positive, so ρ will have its minimum value when a has its most negative value. Then, for any chosen value for the minimum desired cam surface radius of curvature, Eq. (10.73) can be used with a at its most negative value to compute the required minimum cam size as represented by the size of its base radius R_b. Of course, if a value of zero is chosen for $\rho_{\min}$, a cam made with a base radius equal to the value computed for R_b will possess a sharp (zero radius of curvature) corner.

Computation of cam surface coordinates and tool path for machining for use with a translating flat-faced follower

Figure 10.36 depicts a cam and translating flat-faced follower and also indicates a portion of a circular milling cutter of radius R_u whose cen-

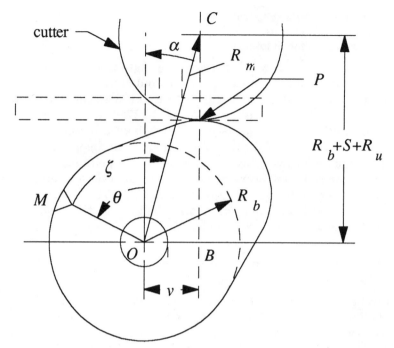

Figure 10.36 Geometry for computation of cam surface coordinates and tool path for machining for use with a translating flat-faced follower.

ter is at point C. From the dimensions shown in this figure it can be seen that the polar coordinates R_m and ζ relative to point O and line OM of point C at the center of the milling cutter are given by the equations

$$R_m = \sqrt{v^2 + (R_b + S + R_u)^2} \tag{10.74}$$

$$\tan \alpha = \frac{v}{R_b + S + R_u} \tag{10.75}$$

$$\zeta = -\alpha - \theta \tag{10.76}$$

Using values of v and S computed from the motion program equations (such as given in Sec. 10.4) and the radius R_u of the milling cutter and a chosen value for R_b for the base radius of the cam, Eqs. (10.74), (10.75), and (10.76) can be used to compute R_m, α, and ζ, respectively.

The foregoing computations give the polar coordinates of points on the path of the center of the milling cutter. The polar coordinates of points on the surface of the cam are computed using the same equations but with $R_u = 0$.

Computation of cam surface coordinates tool path for machining for use with an oscillating flat-faced follower

A plate cam that is used with an oscillating flat-faced follower will have a shape that differs from that which it would have if used with a translating flat-faced follower. The geometry that is associated with an oscillating flat-faced follower is given in Fig. 10.37. The cam center is at point O, the pivot of the flat-faced follower is at a distance of L_O away at point A, the face of the follower is offset from the pivot A by a distance L_d, and the follower face contacts the cam at point P. As in the case of the swing roller follower, the angle of the follower arm and the rates of change of that angle are given by Eqs. (10.54). However, in the case of the flat-faced follower, in accordance with the geometry in Fig. 10.37, the value of η_0 is given by

$$\sin \eta_0 = \frac{R_b - L_d}{L_O} \tag{10.77}$$

A line through contact point P and perpendicular to the follower face intersects line OA at point B. For the same reasons that pertain in the case of the swing roller follower, point B is the instant center of rotation of the cam follower relative to the cam. Then, just as in the case of the roller follower, the magnitude of b is given by

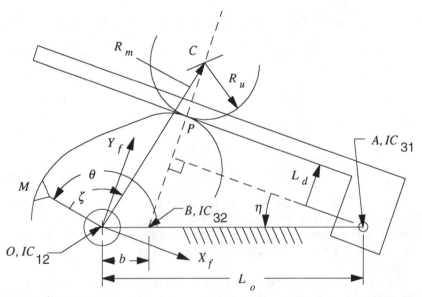

Figure 10.37 Geometry for computation of cam surface coordinates and tool path for machining for use with an oscillating flat-faced follower.

$$b = \frac{L_0(d\eta/d\theta)}{1 + (d\eta/d\theta)} \qquad (10.58)$$

A circular milling cutter with a radius of R_u is indicated tangent to the cam surface at the contact point P so that its center is at point C. A set of axes X_f and Y_f is centered at point O with X_f parallel to the follower face and Y_f perpendicular to that face. The coordinates of point C relative to that set of axes are

$$x_f = b \cos \eta \quad \text{and} \quad y_f = Lo \sin \eta + L_d + R_u \qquad (10.78)$$

A line OM is indicated to be scribed on the cam shown such that when θ in the displacement timing diagram is zero, the line OM is aligned with the line OA. If a positive X_c axis is directed from O toward M and an axis Y_c is rotated 90° counterclockwise from X_c, that axis system will be rotated from the X_fY_f system by the angle $\eta + \theta$. Then the coordinates of point C in this X_cY_c axis system are given by

$$x_c = x_f \cos(\eta + \theta) + y_f \sin(\eta + \theta) \qquad (10.79)$$

$$y_c = -x_f \sin(\eta + \theta) + y_f \cos(\eta + \theta) \qquad (10.80)$$

These are the Cartesian coordinates of points on the path of the center of a milling cutter of radius R_u that would produce the desired cam shape. The polar coordinates R_m and ζ (relative to the X_cY_c axis system) of points on that milling cutter path are given by

$$R_m = \sqrt{x_c^2 + y_c^2} \quad \text{and} \quad \tan \zeta = \frac{y_c}{x_c} \qquad (10.81)$$

The coordinates of points on the surface of the cam are obviously obtained by simply setting $R_u = 0$ and using the same equations as for computing the milling cutter path.

Cam radius of curvature for an oscillating flat-faced cam follower

In many cases the radius of curvature of a cam used with an oscillating flat-faced follower can be considered to be approximately the same as it would be if the follower were *translating*. In cases where a more accurate calculation is desired or in which the angular motion of the follower swing arm is large, the radius of curvature of the cam can be calculated using equations contained in this subsection.

Figure 10.38 presents geometry associated with derivation of expressions for the radius of curvature of a cam with an oscillating flat-faced follower. The cam center is at point O, the pivot of the

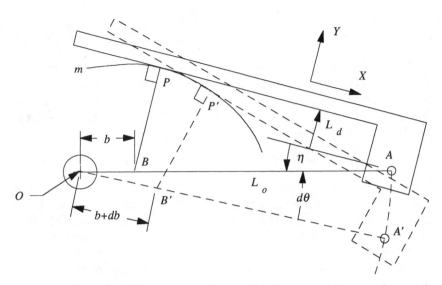

Figure 10.38 Geometry for computation of cam radius of curvature for use with an oscillating flat-faced cam follower.

flat-faced follower is at point A, the face of the follower is offset from the pivot A by a distance L_d, and the follower contacts the cam surface m at point P.

As in the cases of the radius of curvature derivations in previous subsections, the cam will be assumed fixed, and the base (which carries the pivot A) will be angularly displaced clockwise by a differential angle $d\theta$. This displacement moves point A to A', while the follower angle η changes to a value of $\eta + d\eta$, thereby moving the point of contact P to P'.

Then the distance BP is given by

$$BP = L_d + (L_o - b)\sin\eta$$

and the distance $B'P'$ is given by

$$B'P' = L_d + (L_o - b - db)\sin(\eta + d\eta)$$

The distance from P to P' can be found by tracing the path $PBOB'P'$. Assuming an axis system XY as shown in the figure, the X and Y components of each segment of that path can be expressed in terms of the preceding expressions for BP and $B'P'$ and the angles and lengths indicated in Fig. 10.38. Then, by appropriately summing these components, expressions for the X and Y components of the distance from P to P' can be obtained. When these expressions are divided by $d\theta$ and second-order and higher-order terms are eliminated because they vanish as $d\theta$ is made to approach zero, the following expressions are obtained:

$$\frac{dx}{d\theta} = b \sin \eta + \frac{db}{d\theta} \cos \eta + \left(1 + \frac{d\eta}{d\theta}\right)[(L_o - b) \sin \eta + L_d] \quad (10.82)$$

$$\frac{dy}{d\theta} = \left[L_o\frac{d\eta}{d\theta} - b\left(1 + \frac{d\eta}{d\theta}\right)\right] \cos \eta \equiv 0 \quad (10.83)$$

Obviously, evaluation of these equations requires values of the angle η of the follower arm and values of derivatives of that angle and also values of the length b and its derivative. As in the case of the oscillating roller follower, values for the angle η of the follower arm and for the rates of change of that angle are given by Eqs. (10.54). However, in the case of the flat-faced follower, in accordance with the geometry in Fig. 10.37, the value of η_0 is given by Eq. (10.77). The value of b is given by Eq. (10.58), and its derivative is given by Eq. (10.62).

The displacement of point P that corresponds to the derivatives $dx/d\theta$ and $dy/d\theta$ in Eqs. (10.82) and (10.83) is a displacement $d\sigma$ along the cam surface. We may write

$$\frac{d\sigma}{d\theta} = \sqrt{\left(\frac{dx}{d\theta}\right)^2 + \left(\frac{dy}{d\theta}\right)^2} = \frac{dx}{d\theta} \quad (10.84)$$

The differential angular rotation of the tangent to the cam surface is equal to the differential angular displacement of the follower face, which is $d\alpha = d\theta + d\eta$ or

$$\frac{d\alpha}{d\theta} = \left(1 + \frac{d\eta}{d\theta}\right) \quad (10.85)$$

Then Eqs. (10.82), (10.84), and (10.85) can be used together with Eq. (10.65) to compute ρ as

$$\rho = \frac{d\sigma}{d\alpha} = \frac{(d\sigma/d\theta)}{(d\alpha/d\theta)} = \frac{(dx/d\theta)}{1 + (d\eta/d\theta)}$$

$$= \frac{b \sin \eta + (db/d\theta) \cos \eta}{1 + (d\eta/d\theta)} + (L_o - b) \sin \eta + L_d \quad (10.86)$$

Notice that the follower offset distance L_d appears in this equation simply as an additive term. Therefore, if the radius of curvature is found to be unsatisfactory, L_d can be increased, and if the base radius R_b is increased by the same amount, Eqs. (10.77) and (10.86) show that the radius of curvature will be increased by that same amount.

10.7 Follower Forces and Camshaft Torque Fluctuations

The nature of the forces between the cam follower and the cam is important from the standpoint of strength and wear characteristics required of the cam and follower system. The nature of the required cam drive torques is important because of the need to provide some means for driving the camshaft, and because it is usually necessary to keep the cam rotation synchronized with other parts of the machine *despite variations* in the torque needed to drive that cam. The forces produced between the cam and the follower and the torques required to drive the cam will depend on the type of mechanism, materials, and/or bodies that must be acted on by the output motions of the follower system. These forces and torques also will depend on the *motions* required of the things being acted on.

Cam torque and follower force relationships

Figure 10.39 is a flow diagram that indicates that the input to a cam is a torque T and a shaft angular velocity ω in radians per second. The output of the cam is a follower velocity V_f and a force F_f applied to that follower. Note that F_f is the follower force *component* parallel to the direction of follower translation. Total force depends on pressure angle. The follower, then, *acting through a follower mechanism*, produces an output force F_O and an output velocity V_O. Because, as discussed in the subsection "Mechanical advantage in a crank-slider linkage" in Sec. 3.9, the power input to each of the blocks in this figure equals the power out of that block, we may write

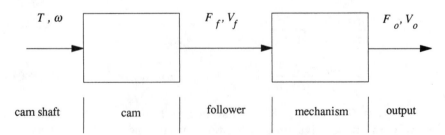

Figure 10.39 Flow diagram showing cam torque and follower force relationships

$$T\omega = F_f V_f = F_O V_O \qquad (10.87)$$

From this equation, and remembering from Eq. (10.4) that $V/\omega = v$, we may write

$$T = \frac{F_f V_f}{\omega} = F_f v \qquad F_f = \frac{F_O V_O}{V_f} \quad \text{and} \quad T = \frac{F_O V_O}{\omega} \qquad (10.88)$$

Note that V_O/V_f and V_O/ω are mechanical advantages as defined in Sec. 3.9.

Using the relationships in Eqs. (10.87) and (10.88), we may make Table 10.2 relating cam follower forces and camshaft torques to various types of system output load forces.

Each type of load force F_o in Table 10.2 is labeled in the left-hand column according to the type of phenomenon that usually causes it. For instance, a *constant load force* such as indicated by W is often the result of lifting or lowering a weight slowly, and this type of load is encountered in low-speed mechanical systems. Such a constant load force also can be produced by dry-sliding friction in the follower output mechanism, although such a constant friction force will go to zero when the velocity goes to zero and will reverse direction when the velocity direction reverses. It will be noticed that the follower force F_f is directly affected by the mechanical advantage in the load mechanism

TABLE 10.2 Cam Torques and Forces for Various Types of Load Forces

Load force F_o	Follower force F_f	Camshaft torque T
Constant W	$W\dfrac{V_o}{V_f}$	$W\dfrac{V_o}{\omega}$
Stiffness kS_o	$kS_o\dfrac{V_o}{V_f}$	$kS_o\dfrac{V_o}{\omega}$
If $\dfrac{S_o}{S_f} = \dfrac{V_o}{V_f}$, then	$kS_f\left(\dfrac{V_o}{V_f}\right)^2$	$kS_f\left(\dfrac{V_o}{V_f}\right)^2 v$
Viscous BV_o	$BV_o\dfrac{V_o}{V_f}$	$B\dfrac{V_o^2}{\omega}$
	or $\quad B\left(\dfrac{V_o}{V_f}\right)^2 v\omega$	or $\quad B\left(\dfrac{V_o}{V_f}\right)^2 v^2\omega$
Inertial mA_o	$mA_o\dfrac{V_o}{V_f}$	$mA_o\dfrac{V_o}{\omega}$
If $\dfrac{A_o}{A_f} = \dfrac{V_o}{V_f}$, then	$ma\omega^2\left(\dfrac{V_o}{V_f}\right)^2\dfrac{V_o}{\omega}$	$mav\omega^2\left(\dfrac{V_o}{V_f}\right)^2\dfrac{V_o}{\omega}$

so that the designer can sometimes control such forces by follower and load mechanism design. However, the camshaft torque T is affected only by the force W itself and the *overall* mechanical advantage between camshaft and load. Because camshaft speed ω is usually determined by overall machine speed, camshaft torques and their variations can be minimized for a given output force W only by minimizing variations in V_o by using the rise and fall times that are as long as possible. Static balancing of the load (see Chap. 12) can be used to minimize both effects.

When the load force $F_o = kS_o$ is caused by a spring or other *stiffness* in the load mechanism, the follower force is again affected by the mechanical advantage in the load mechanism, and the torque is affected by the overall mechanical advantage, just as in the case of the constant load force.

If, as is often the case, the load mechanism consists of gearing or of a multiplying or magnifying linkage such as described in Chap. 8, its mechanical advantage will be constant or relatively constant regardless of its motion. In such a case the displacement ratio S_o/S_f will be equal (or almost so) to V_o/V_f. Then the follower force and camshaft torque caused by a stiffness type of output force can be expressed in terms of the cam motion parameters S_f and v as shown in the table. Of course, if the load spring is connected directly to the follower, $V_o/V_f = 1.0$.

Occasionally, the load will produce a force that is approximately proportional to the velocity of the load, as in the case where the output mechanism is pumping a viscous fluid. As in the case of the stiffness load, the *viscous load* force produces a follower force depending on the load mechanism mechanical advantage and produces a camshaft torque depending on the overall system mechanical advantage. Note, however, the dependencies of these two effects on the *square* of the velocity of the output motion.

In higher-speed machines or in those in which the cam system must move an appreciable mass, *inertial loads* become important. The mechanical advantages are seen to have similar effects to what they had in the other types of loads. In the case of inertial loads, it is seen that the expressions for the general case all include a product of *output acceleration times output velocity*.

In the special but frequently encountered case where the output mechanism mechanical advantage is constant over the range of output motion as mentioned previously, then $A_o/A_f = V_o/V_f$, and the cam follower force and camshaft torque are functions of the appropriate mechanical advantages and the product av. Maximum values of these cam forces and torques will then occur near the tops and bottoms of the rises and falls of the cam motion, where $|av|$ is maximum.

In most cam systems, the load force will be a combination of the types listed in Table 10.2. A most frequent combination is the stiffness and inertial combination. In this case, the spring is used to keep the follower on the cam surface. The spring force therefore tends to lessen the undesirable effect of the inertial force near the tops of the rises and falls and to aggravate the inertial effect near the bottoms of the rises and falls. This aggravated effect can be minimized by using a spring that has a large spring constant and a low preload so that its force is diminished near the bottoms of the rises and falls.

10.8 Summary of the Plate Cam Design Procedure

1. Sketch or draw a timing diagram such as shown in Fig. 10.5 for the desired cam. This diagram must clearly indicate the cam angle θ at which each follower motion segment must start and end. It also should indicate the height of each rise and each fall.

2. Choose a cam type and follower type. A few examples of such types are indicated in Figs. 10.1 through 10.4.

3. For each motion segment in the diagram generated in step 1, define a θ_i and a β_i, where θ_i is cam angle measured from the start of the ith segment and β_i is the angular "duration" of that ith segment. The cam angle θ is then defined as measured from the start of the overall cam timing diagram, and it is expressed in terms of these θ_i's and β_i's.

4. For each motion segment in the diagram generated in step 1, choose a motion program. Each such program can be a standard motion program such as presented in Sec. 10.4, a motion program devised by the reader, a constant velocity, or a dwell.

If the cam is to be fabricated by a shop that specializes in making cams, the information generated in steps 1 through 4 is probably adequate for the shop to use in its manufacture. If the cam base radius is not large compared with the magnitudes of the rises and falls, and if the β_i's of those rises and falls are short, the pressure angle and radius of curvature should be computed at least at suspected critical regions (see subsequent steps).

If the cam fabrication shop requires more than the information provided by steps 1 through 4, the designer must provide either the coordinates of the pitch curve (in cases in which a roller follower is used), the coordinates of the surface of the cam, or the coordinates of the machine tool path. Steps 5 and 6 provide this information.

5. Using the relationships between θ and the θ_i's and β_i's from step 3, and using equations for S, v, and a for the motion programs (such as contained in Sec. 10.4) for the rises and falls, compute values for S,

v, and a in the regions of interest in the cam timing diagram. The region of interest may consist of the entire cam.

6. Compute the coordinates of the pitch curve, and/or the cam surface, and/or the tool path using equations whose numbers are listed in the appropriate cell in Table 10.3. (Familiarize yourself with the subsection containing the equations of interest.) Because some of the equations require input values that are obtained from other equations,

TABLE 10.3 References to Figures and Equations to Be Used for Computing Plate Cam Physical Characteristics

To compute	For translating roller follower	For oscillating roller follower	For translating flat-faced follower	For oscillating flat-faced follower
Pitch curve coordinates	Fig. 10.25 Eq. (10.35)	Fig. 10.31 Eqs. (10.53) to (10.57)		
Cam surface coordinates	Fig. 10.30 Eqs. (10.35), (10.36), (10.46) to (10.48)	Fig. 10.31 Eqs. (10.53) to (10.59), (10.46) to (10.52)	Fig. 10.36 Eqs. (10.54), (10.74) to (10.76) and $R_u = 0$	Fig. 10.37 Eqs. (10.54), (10.77), (10.58), (10.78) to (10.81) and $R_u = 0$
Tool path coordinates	Fig. 10.30 Eqs. (10.35), (10.36), (10.46) to (10.48), (10.49) to (10.52)	Fig. 10.31 Eqs. (10.53) to (10.59), (10.46) to (10.52)	Fig. 10.36 Eqs. (10.54), (10.74) to (10.76)	Fig. 10.37 Eqs. (10.54), (10.77), (10.58), (10.78) to (10.81)
Pressure angle	Fig. 10.26 Eq. (10.36)	Fig. 10.32 Eqs. (10.53), (10.54), (10.58) to (10.60)		
Radius of curvature	Fig. 10.28 Eq. (10.45)	Fig. 10.33 Eqs. (10.53), (10.54), (10.58), (10.59), (10.61) to (10.65)	Fig. 10.35 Eqs. (10.54) and (10.72)	Fig. 10.38 Eqs. (10.54), (10.77), (10.58), (10.62), (10.86)
Follower face width			Fig. 10.34 Eq. (10.4) and $b = v$	

the equations may have to be evaluated in a particular order. This is not necessary if some of the available mathematical software packages are used to perform the computations. There is some duplication of equation numbers because some equations are used for the computation of more than one parameter.

7. Visually inspect a plot of the coordinates obtained in step 6 to see whether the radius of curvature of the cam will be unsatisfactory in any region and whether the pressure angle will be excessive at any point. Such kinematic defects may not be easily seen in such an inspection. If not, it may be advisable to proceed to step 8. If such defects are detected, refer to Table 10.4 for comments.

8. If pressure angle, cam surface radius of curvature, or required width of a flat-faced follower is of concern, compute values of the variable of concern using the equations whose numbers appear in the appropriate cell in Table 10.3. (Familiarize yourself with the subsection containing the equations of interest.) Because some of the equations require input values that are obtained from other equations, the equations may have to be evaluated in a particular order. This is not necessary if some of the available mathematical software packages are used to perform the computations. If values of pressure angle or radius of curvature are not satisfactory, refer to Table 10.4 for comments.

TABLE 10.4 Possible Kinematic Defects in Cam System Designs

Type of defect	Effect on performance	Causes or indicators	Possible cures
Undercutting	Follower displacement errors	Sharp bends in displacement diagram or in pitch curve	Larger prime circle and/or smaller follower radius and/or smoother displacement curve
Excessive pressure angle	Large side force on follower, large normal force on cam	Steep displacement diagram slope, small prime circle	Longer rise or fall times, larger prime circle, larger follower offset
Excessive drive torque fluctuations	Drive timing errors, camshaft windup, electric motor heating	Large follower mass, high speed, large fluctuations in av	Minimize follower system kinetic energy fluctuations
Roughness in the acceleration variation	Vibration, wear	Discontinuous acceleration curve, short rise or fall times	Use smooth acceleration curves, maximize rise and fall times

Table 10.4 summarizes some of the deficiencies or *kinematic defects* that may be encountered when designing a cam and follower system. It also indicates some of the deleterious effects that those defects can have on machine performance, which cam design features can cause or indicate the presence of these defects, and what changes in design tend to eliminate or minimize these defects.

10.9 Some Comments on Barrel Cams and Related Types

Barrel cams of the types indicated in Fig. 10.2*c* and *d* find use in pick-and-place mechanisms and parts-transport mechanisms. Variations of these types are also commercially available as rotary indexing mechanisms. For a given required follower total travel, cam and follower systems of these types sometimes can be made smaller than an equivalent plate cam system. Although the geometry of these systems is different from that of plate cams, the same motion programs can be used, and *analogous* shape problems must be addressed. Insufficient radius of curvature and excessive pressure angle can still be problems.

The barrel cam configuration that uses a roller follower in a groove as shown in Fig. 10.2*c* involves the fewest parts. However, the roller follower must be very accurately fitted to the groove so that it neither "binds" in the groove nor has so much clearance that it experiences impacts as the load forces it from one side of the groove to the other. As the follower load is continually reversed, the roller must continually reverse its direction of rotation as it shifts from one side of the groove to the other. If, as is usually the case, a cylindrical roller is used, some part of it must be slipping against the groove sides at all times, so the roller and groove surfaces are subject to wear and must be kept clean and lubricated. If a crowned roller is used, the slipping is minimized, but the cam-to-follower contact is a point contact rather than a line contact, so contact pressures can be high.

The accuracy requirements for a raised-track barrel cam system such as indicated in Fig. 10.2*d* can be somewhat less demanding than for the grooved type, and the rotation of the rollers can be continuously in one direction. Often one of the two roller followers in the raised-track system will be spring-loaded against the track so that it can accommodate slight inaccuracies in track width. In this configuration also, some part of each roller must be slipping on the cam track at all times, so lubrication and cleanliness are important. In addition, if the cam diameter is small, the width of the track can become unacceptably thin at locations where the follower velocity is high. This is particularly true of the portion of the track that is closest to the axis of rotation of the cam.

Even if conical rollers are used with the preceding types of barrel cams, slippage must occur during some portions of the motion (e.g., during rises and falls). Such slippage can be avoided by using a convex globoidal cam with an oscillating conical roller follower (or followers, if a raised track is used). In this type of system, the follower is constrained to oscillate about an axis that intersects the axis of rotation of the cam, and the apex of the conical surface, of which the roller follower is a frustum, always coincides with that point of intersection.

11

Gears and Gear Trains

11.1 Introduction

This chapter presents the basic terminology and principles involved in the kinematics of gearing, and it provides criteria and procedures for the design of gear trains and involute gears. The treatment starts with the most basic task of gear train design: that of choosing gears from "stock" to provide a coupling of the rotary motion of one shaft to rotary motion of another shaft while providing a given relationship between the speeds and directions of those rotations. This task is discussed in Secs. 11.2, 11.3, and 11.4, and it involves consideration of fundamental parameters such as matching diametral pitches, shaft spacing, and number and sizes of gears in simple, compound, and planetary gear trains.

The other level of design covered is that in which space and speed ratio constraints are more severe, such that the form of the gears must be chosen with care. This level of design is discussed in Secs. 11.5 through 11.8, and it involves consideration of gear tooth addendums, dedendums, pressure angles, interference, undercutting, and contact ratios. Considerations pertinent to timing-belt system design are discussed in Sec. 11.9.

11.2 Principles, Terminology, and Basic Relationships for Spur Gearing

The purpose of gears is to couple two shafts together in such a manner that the rotation of one shaft (the *driven shaft* or *output shaft*) will be a function of the rotation of another shaft (the *driving shaft* or *input shaft*). In most applications in which gears are used, the displacement of the driven shaft is made to be proportional to the displacement of the driving shaft. Because the displacements of these two shafts are proportional to each other, the respective derivatives of the displacements also will be proportional to each other with the same constant of proportionality. This book deals with such *constant-ratio* gearing systems. (Note that *gear shifting* in a gear transmission consists of selecting and engaging different *constant-ratio sets* of gears in that transmission.) However, the reader should be aware that noncircular gears are sometimes used when it is desired that the ratio between the output shaft angular velocity and input shaft angular velocity vary as a function of input shaft angle.

For a given pair of gears, the axis of the input shaft and the axis of the output shaft can be parallel, the axes can intersect at an angle, or the axes can be nonintersecting and nonparallel (skewed). The gearing most often encountered in practice is that in which the shafts are parallel, and this is the type of gearing that is covered in this book. Gearing between nonparallel shafts involves more complicated geometry, and the gears tend to be more expensive. For coverage of gearing with nonparallel shafts, readers are referred to other, more specialized texts such as E. Buckingham, *Analytical Mechanics of Gears* (Dover Publications, New York, 1988).

When the two shafts that are to be coupled rotationally are parallel, the shafts can be represented diagrammatically by their end views, as shown at S_1 and S_2 in Fig. 11.1a. Attached to shaft S_1 and therefore rotating with it is a cyclinder whose diameter is D_1, and attached to shaft S_2 is a cylinder whose diameter is D_2, and the two cylinders are in rolling contact with each other at point P. (Point P is the end view of a *line of contact* between these cylinders.) If there is no slipping between the two cylinders, and if the cylinder attached to shaft S_1 is rotating counterclockwise as indicated, the cylinder attached to shaft S_2 must rotate clockwise as indicated. Also, if the cylinders are not slipping, a point that is on the left-hand cylinder and which is instantaneously at point P must have the same velocity vector $\mathbf{V}_P$ as does a point that is on the right-hand cylinder and which is instantaneously at that same location. The magnitude of the velocity of that point on the left-hand cylinder is $\omega_1 D_1/2$, and the magnitude of the velocity of the coincident point on the right-hand cylinder is $-\omega_2 D_2/2$, where ω_1 and ω_2 are the angular velocities of the respective shafts and where it is to be noted that ω_2 is clockwise and therefore negative. Because these two magnitudes must be equal, we may write

$$\omega_1 \frac{D_1}{2} = -\omega_2 \frac{D_2}{2} \qquad \text{which gives} \qquad \frac{\omega_2}{\omega_1} = -\frac{D_1}{D_2} \qquad (11.1)$$

A drive such as depicted in Fig. 11.1a is known as a *friction drive,* and it is sometimes used for transmitting rotation from one shaft to

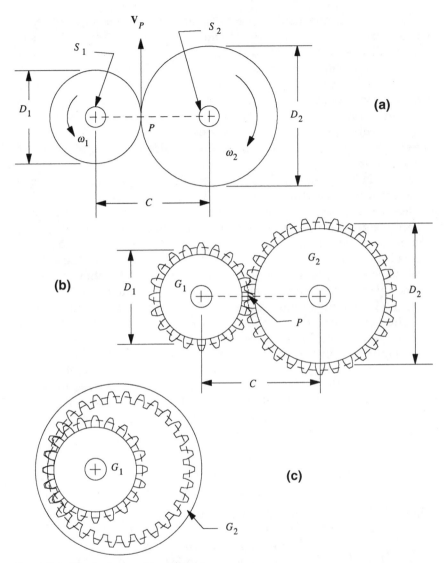

Figure 11.1 The basic operation of gears.

another. Its usefulness is limited by the fact that excessive slippage and wear can occur if appreciable torque or power is to be transmitted from one shaft to the other. Also, because some slippage will inevitably occur, the angular relationship between the shafts cannot be uniquely maintained in such a drive. To overcome these deficiencies, the surfaces of the cylinders can be "roughened" by the addition of gear teeth, as indicated in Fig. 11.1*b*. If these teeth consist of ridges on the surface of the cylinders and the ridges are *parallel* to the axes of the cylinders and shafts, the cylinders are known as *spur gears*. The kinematic requirements that the *shapes* of gear teeth must satisfy are discussed in Sec. 11.5. Often, in order to minimize noise and vibration, the ridges are made to lie along helices on the surfaces of the cylinders, and such gears are known as *helical gears*. Kinematic considerations in the design of gear trains with parallel shafts and helical gears are the same as in the design of gear trains with spur gears, so subsequent discussions will deal with spur gears. When two gears are meshed with each other, the smaller gear is often referred to as a *pinion*.

In Fig. 11.1*b* a dashed circle is drawn on each gear, and the circles are tangent to each other at point *P*. These circles are known as *pitch circles* of the gears, and they are considered to be rotating with their respective gears and to be rolling against each other without slipping, just as were the cylinders in Fig. 11.1*a*. The *pitch diameter* of a gear is defined as the diameter of the pitch circle of that gear. Because these circles are not slipping on each other, the number of inches of circumference of the pitch circle on gear G_1 that pass through point *P* in any time interval *T* must equal the number of inches of circumference of the pitch circle on gear G_2 that pass through that point in the same time interval. Also, because the teeth of the two gears are in *mesh* in the vicinity of point *P*, the number of teeth on gear G_1 that pass by point *P* in that time interval *T* must equal the number of teeth on gear G_2 that pass by that point in the same time interval. This requires that the number of teeth per inch of pitch circle circumference on gear G_1 must equal the number of teeth per inch of pitch circle circumference on gear G_2. Then we may write

$$\frac{N_1}{\pi D_1} = \frac{N_2}{\pi D_2} \tag{11.2}$$

where N_1 = number of teeth on gear G_1
N_2 = number of teeth on gear G_2
D_1 = pitch diameter of gear G_1 in inches
D_2 = pitch diameter of gear G_2 in inches

The *circular pitch* P_c of a gear is defined as the distance along the pitch circle from a point on a tooth on that gear to the corresponding point on the adjacent tooth on that gear. (See arc length *AB* in Fig. 11.2.) Then, as a definition for circular pitch for any gear, we may write

$$P_c = \frac{\pi D}{N} \tag{11.3}$$

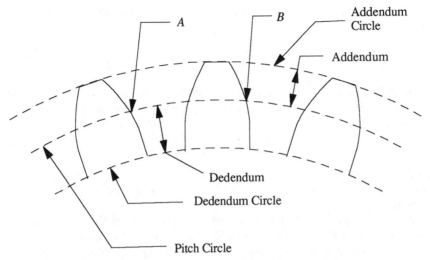

Figure 11.2 Definition of some fundamental gear parameters.

Using this definition for circular pitch, Eq. (11.2) can be written

$$P_{c_1} = P_{c_2}.$$

Thus we see that for any pair of gears to mesh properly with each other, their *circular pitches must be equal.*

  Diametral pitch is a parameter which, although not directly measurable on a physical gear, is usually more convenient for use in gear system design. It is defined as the number of teeth on a gear divided by the pitch diameter of that gear in inches. That is,

$$P_d = \frac{N}{D} = \frac{\pi}{P_c} \tag{11.4}$$

From Eqs. (11.2) and (11.4) we may write $P_{d_1} = P_{d_2}$, so it is seen that for any pair of gears to mesh properly with each other, their *diametral pitches must be equal.*

  The *velocity ratio* of a pair of gears (which is often referred to as the *gear ratio* of the pair) is defined as the ratio of their angular velocities. As described earlier, the two gears in Fig. 11.1b will rotate in such a manner that their pitch circles roll without slipping on each other. Consequently, their velocity ratio is given by Eq. (11.1), just as in the case of the cylinders in Fig. 11.1a. Then, because $P_{d_1} = P_{d_2}$,

$$\frac{D_1}{D_2} = \frac{N_1}{N_2} \tag{11.5}$$

so the velocity ratio of the pair of spur gears in Fig. 11.1b is given by

$$\frac{\omega_2}{\omega_1} = -\frac{D_1}{D_2} = -\frac{N_1}{N_2} \tag{11.6}$$

The distance C between the centers of the two gears in Fig. 11.1b is known as the *center distance,* and it is seen to be given by

$$C = \frac{D_1 + D_2}{2} \tag{11.7}$$

i.e., *the center distance is equal to the sum of the radii of the two pitch circles.*

Figure 11.1b shows that in order for the teeth of the two gears to mesh, the teeth must extend outside the pitch circles. The distance by which a gear's tooth extends outside the pitch circle of that gear is known as the *addendum.* Therefore, the total diameter of a gear is known as the *addendum diameter D_a* and is equal to its pitch diameter plus *twice* the addendum. The relationship of the addendum to a general tooth shape is indicated in Fig. 11.2.

Also, it can be seen in Fig. 11.1b that in order for the teeth of the two gears to mesh, the recesses between the teeth on each gear must extend inside its pitch circle by a sufficient distance to accommodate the tips of the teeth on the other gear. The distance by which the recesses between the teeth on a gear extend inside the pitch circle of that gear is known as the *dedendum.* (See Fig. 11.2.) Therefore, the diameter of a circle that is inscribed tangent to the bottoms of these recesses on a gear is known as the *dedendum diameter* and is equal to the gear's pitch diameter minus *twice* the dedendum. The dedendum of a gear must be greater (by some clearance amount) than the addendum of the gear with which it meshes because the teeth of that meshing gear will extend past the pitch circles of both gears and into the recess of the gear in question by a distance equal to the addendum. The dedendum is shown along with other tooth parameters in Fig. 11.2. Further details of tooth shape are discussed in Sec. 11.5.

Figure 11.1c indicates the *internal meshing* of a small spur gear G_1 with a larger gear G_2 whose teeth are placed on the inside of an annular surface of the gear. A gear such as this larger gear is called an *internal gear, ring gear,* or *annular gear,* and it is seen that the action is analogous to the rolling of the pitch circle of the smaller gear on the *inside* of the pitch circle of the internal gear. Therefore, both gears rotate in the same direction. Equations (11.1) through (11.7) are applicable to this pair of gears also, except that the velocity ratio is *positive* for the case in Fig. 11.1c and a minus sign will be placed in front of the smaller D in Eq. (11.7).

11.3 Simple and Compound Spur Gear Trains

A *gear train* consists of a collection of gears that are meshed with each other. If each of these gears is mounted on a separate shaft, and if each of the shafts is mounted in bearings that are mounted to ground or a common reference platform, the gear train is known as a *simple gear train*. Such a simple gear train is shown in Fig. 11.3. For simplicity, each gear is represented merely by its pitch circle, and its teeth are not shown. Comparison of Figs. 11.1*b* and 11.3 shows that a simple gear train consists of one or more interconnected (meshed) pairs of meshed gears or two-gear trains. Because simple trains of more than two gears and compound gear trains can be treated as combinations of two-gear trains, the latter will be treated first in this section.

Procedure 11.1: Kinematic Design of a Two-Gear Train A velocity ratio is essentially always specified in the design of a gearing system. The velocity ratio is either specified directly or is specified indirectly by some requirement such as a torque relationship. (See "Mechanical Advantage" in Sec. 3.9.) As shown in Sec. 11.2, the velocity ratio (or gear ratio) provided by two gears in mesh such as shown in Fig. 11.1*b* is given by the ratio of their pitch diameters or by the ratio of the numbers of teeth on the two gears, as shown in Eq. (11.6). Then the following steps may be followed in designing a two-gear train:

1. A most intuitive initial approach to the design of a set of two gears in mesh consists of first attempting to choose numbers of teeth for the two gears such that the ratio of those numbers is equal to the desired velocity ratio in accordance with Eq. (11.6). It immediately becomes evident that because the number of teeth on each gear must be an integer, the value of the velocity ratio that can be provided is limited to the ratio of two integers. The smaller of these integers should not be very small, for reasons discussed in Secs. 11.7 and 11.8, and neither of the integers should be so large as to result in an extremely large gear or extremely small teeth. Also, for reasons discussed in Secs. 11.7 and 11.8, the

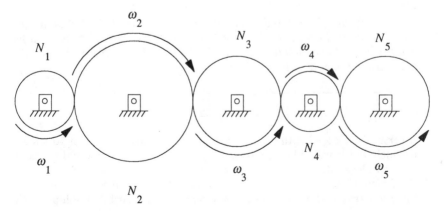

Figure 11.3 A simple spur gear train.

ratio generally should not be smaller than 0.2 nor larger than 5. If these guide-lines cannot be satisfied, perhaps a *gear train* consisting of more than two gears will be required.

2. Once numbers of teeth for each gear have been chosen in step 1, the di-ametral pitch of the gears can be chosen, or the pitch diameter of one of the gears can be chosen, or the center distance can be chosen. Once one of these three pa-rameters has been chosen, the remainder of the parameter values in Eqs. (11.4), (11.5), and (11.7) can be computed by using those equations. Results of these computations should be checked against vendors' catalogs to determine whether standard gears are available for those values. Also, if the results give numbers that are close to the guidelines mentioned in step 1, Secs. 11.7 and 11.8 should be consulted. If the results are not satisfactory, a new value for diametral pitch, pitch diameter, or center distance can be chosen and new results computed. If no satisfactory results are found, perhaps a gear train of more than two gears is re-quired.

Example 11.1: Kinematic Design of a Two-Gear Train To gain some in-sight into the sorts of tradeoffs involved in the choice of gear train parame-ters, consider the synthesis of a two-gear train with parallel shafts 2.0 in apart that is to provide a velocity ratio of

$$\frac{\omega_2}{\omega_1} = -0.3125$$

Although this ratio at first appears to be difficult to match with a ratio of two integers, it is found to be 1/3.2. Therefore, in accordance with Eq. (11.6), we may write

$$\frac{\omega_2}{\omega_1} = -\frac{1}{3.2} = -\frac{10}{32} = -\frac{N_1}{N_2} = -\frac{D_1}{D_2}$$

Equation (11.7) can then be written

$$C = 2.0 = \frac{D_1 + D_2}{2} = \frac{D_1 + 3.2D_1}{2} = 2.1D_1$$

from which the pitch diameters of the two gears are $D_1 = 0.9524$ in and $D_2 = 3.0476$ in. Because of the ratio 10/32 appearing in the expression for velocity ratio above, it is tempting to choose $N_1 = 10$ and $N_2 = 32$. Then the diametral pitch of both gears can be computed as

$$P_d = \frac{N_1}{D_1} = \frac{10}{0.9524} = 10.5$$

It certainly is possible to make a gear with a diametral pitch of 10.5 teeth per inch, but it would be a nonstandard gear that probably would not be in most catalogs. Also, a gear with as few as 10 teeth could experience some of the deficiencies discussed in Secs. 11.7 and 11.8. Even if N_1 were made 20, the diametral pitch of the gears would be a nonstandard 21 teeth per inch.

Therefore, it probably would be advisable to choose a standard value of 20 per inch for the diametral pitch and $N_1 = 20$ teeth. This choice would result in $N_2 = 64$, $D_1 = 1$ in, $D_2 = 3.2$ in, and the center distance $C = 2.1$ in. Unless the originally specified center distance of 2.0 in is critical, this last solution is probably the one to use.

Simple gear trains with more than two gears

Figure 11.3 depicts a simple gear train that consists of more than two gears. Such a train is used when the input shaft and the output shaft are far apart but space limitations forbid the use of two large gears or the use of bevel gears. It is also used when more than two shafts must all be driven at speeds that are in prescribed ratios to each other. For a train such as depicted in this figure, the relationships between velocity ratio and number of teeth for each successive pair of gears can be written in the same manner as was done earlier for the two-gear train. This gives

$$\frac{\omega_2}{\omega_1} = -\frac{N_1}{N_2}, \quad \frac{\omega_3}{\omega_2} = -\frac{N_2}{N_3}, \quad \frac{\omega_4}{\omega_3} = -\frac{N_3}{N_4}, \quad \text{and} \quad \frac{\omega_5}{\omega_4} = -\frac{N_4}{N_5} \quad (11.8)$$

Multiplying all the left-hand sides of these equations together and multiplying all the right-hand sides together respectively produce

$$\frac{\omega_2 \omega_3 \omega_4 \omega_5}{\omega_1 \omega_2 \omega_3 \omega_4} = \frac{(-N_1)(-N_2)(-N_3)(-N_4)}{N_2 N_3 N_4 N_5} \quad (11.9)$$

Canceling like-terms in the numerator and the denominator of each of the left- and right-hand sides gives

$$\frac{\omega_5}{\omega_1} = \frac{N_1}{N_5}(-1)^4 = \frac{N_1}{N_5} \quad (11.10)$$

The numbers of teeth and the angular velocities of the intermediate gears G_2, G_3, and G_4 are seen to have no effect on the overall velocity ratio of the five-gear train. Only the numbers of teeth in the first input gear and the last output gear have any effect. The reason for the lack of effect of these intermediate gears can be seen by tracing the velocity of the circumference of gear G_1 around the peripheries of gears G_2, G_3, and G_4 to gear G_5, as indicated by the curved arrows, and noting that this circumferential velocity is transmitted *without increase or decrease* along that path. These intermediate gears are known as *idler gears* because they are "idle" with respect to any effect on the *magnitude* of the velocity ratio of the overall gear train. However, the angular velocities of the shafts of these idler gears may be of significant use in a particular machine. Also note that the direction of rotation reverses at each external mesh between the gear pairs. Because there are an even number of *external meshes* between the input gear G_1 and the output gear G_5, there are an even number of reversals, and the ratio in Eq. (11.10) is positive. Had there been an odd number of *external meshes,* the ratio would have been negative.

Thus we may conclude that idler gears can be useful for

1. Providing large separation of input and output shafts in a train without requiring large gears

2. Providing velocity ratios for intermediate shafts in the gear train

3. Providing desired polarity (i.e., plus or minus) of the input-to-output velocity ratio

Compound gear trains

For reasons that are discussed in Secs. 11.7 and 11.8, it usually is advisable to avoid using a velocity ratio of greater than 5 or smaller than 0.2 between any pair of meshed gears. This means that a compound gear train should be used when ratios outside this range are required. A *compound gear train* is a gear train that incorporates one or more compound gears, and a *compound gear* consists of two (or more) gears that are rigidly attached to each other so that they must rotate at the same speed. Top and front views of such a compound gear train are shown in Fig. 11.4. The overall velocity ratio of the compound gear train in Fig. 11.4 can be computed by multiplying the velocity ratios of the successive pairs of gears in the chain. Thus

$$\frac{\omega_6}{\omega_1} = \left(\frac{\omega_2}{\omega_1}\right)\left(\frac{\omega_3}{\omega_2}\right)\left(\frac{\omega_4}{\omega_3}\right)\left(\frac{\omega_5}{\omega_4}\right)\left(\frac{\omega_6}{\omega_5}\right) \tag{11.11}$$

Then, because gear 3 is attached to gear 2 and gear 5 is attached to gear 4, the individual velocity ratios can be evaluated as

$$\frac{\omega_2}{\omega_1} = -\frac{N_1}{N_2}, \quad \frac{\omega_3}{\omega_2} = 1.0, \quad \frac{\omega_4}{\omega_3} = -\frac{N_3}{N_4}, \quad \frac{\omega_5}{\omega_4} = 1.0, \quad \text{and} \quad \frac{\omega_6}{\omega_5} = -\frac{N_5}{N_6} \tag{11.12}$$

Combining Eqs. (11.11) and (11.12) gives

$$\frac{\omega_6}{\omega_1} = \frac{-N_1(-N_3)(-N_5)}{N_2 N_4 N_6} = (-1)^3 \frac{N_1 N_3 N_5}{N_2 N_4 N_6} = -\frac{N_1 N_3 N_5}{N_2 N_4 N_6} \tag{11.13}$$

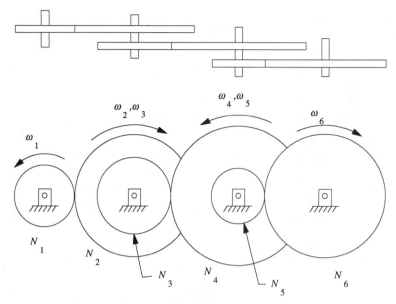

Figure 11.4 A compound spur gear train.

The factors in the numerator of the right-hand side of Eq. (11.13) are the negatives of the numbers of teeth on the *driving* gears at the three external meshes in this train, and the factors in the denominator are the numbers of teeth on the *driven* gears at those meshes. Each of the meshes or meshed gear pairs has a velocity ratio associated with it, and it is often referred to as a *stage* in the gear train. For example, the train in Fig. 11.4 would be called a *three-stage reduction gear train* because, with proportions such as shown, the output angular velocity would be much reduced compared with the input angular velocity. Also, by examining the motions of the gears in the train, it can be seen that the reduction occurs in three increments or stages.

Procedure 11.2: Design of a Compound Gear Train for a Prescribed Overall Velocity Ratio   The design of a compound gear train consists of designing a sequence of two-gear trains or stages such that the product of their velocity ratios equals the desired overall velocity ratio. It involves trying to find ratios of numbers of teeth in the individual stages such that those ratios are relatively similar to each other, such that those ratios are within the range from 0.2 to 5, and such that the numbers of teeth on the individual gears are reasonable. This process tends to be iterative, and it can proceed as follows:

1. Tentatively choose a number n_s for the number of stages that the train will consist of. Unless the desired overall velocity ratio is outside the range from 0.1 to 10, a value of $n_s = 1$ is a good initial estimate.

2. For the desired overall velocity ratio ρ_t, compute a single tentative velocity ratio ρ_s that might be used for all stages by using the formula

$$\rho_s = (\rho_t)^{1/n_s}$$

3. If this ratio is outside the range from 0.2 to 5, increase n_s and return to step 2. Otherwise, continue to step 4.
4. Choose a ratio of integral numbers of teeth such that the value of that ratio is close to the value of ρ_s. The reciprocal of this ratio is a tentative value for the velocity ratio of one of the stages.
5. Attempt to choose values of gear-tooth-number ratios for the other stages in the chain which are close to ρ_s and such that the product of all the corresponding tentative velocity ratios in the chain equals the desired overall velocity ratio ρ_t.
6. Choose a diametral pitch for each gear pair (preferably the same value for all gear pairs). Compute gear pitch diameters using Eq. (11.4).
7. If the resulting parameters are not satisfactory, repeat steps 4 through 7. It may even be necessary to increase n_s and repeat steps 2 through 7.

Gear trains with parallel but noncoplanar shafts and reverted gear trains

The shafts of the gear train in Fig. 11.4 are shown as all lying in the same plane; i.e., their end views all lie along a straight line in the lower view. Space usually can be saved if the end views of these shafts lie at vertices of a polygon, the lengths of whose sides are the center distances of the meshing gear pairs. In some cases it is desired that the output shaft be aligned with the input shaft. Such an arrangement of the four shafts for the gearing in Fig. 10.4 would place the end views of the four shafts at vertices of a triangle with both the input and output shafts lying at the same vertex. A gear train in which the output shaft is coaxial with the input shaft is known as a *reverted gear train*. The kinematic design of gear trains with parallel but noncoplanar shafts is the same as that described for the compound train above.

When a reverted gear train includes only one intermediate shaft between the input shaft and the output shaft, the polygon at whose vertices the end views of the shafts can be located is reduced to a polygon with *only two* vertices; i.e., the end view of the intermediate shaft lies at one end of a straight-line segment and those of the input and output shafts lie coincident at the other end of that line segment. This places the additional constraint on the design that the center distances of the two stages must be equal. An approach to the design of such a reverted gear train is given in R. L. Norton, *Design of Machinery* (McGraw-Hill, New York, 1992). A variation of this approach is used in the discussion of Procedure 11.4 in the next section.

11.4   Planetary Gear Trains

In the gear trains in Figs. 11.3 and 11.4, the bearings for the shafts were all grounded so that although the shafts could rotate, they could not translate. A very useful class of gear trains known as *planetary gear trains* or *epicyclic gear trains* exists in which the shafts of some of the gears, called *planet gears,* are mounted in bearings that are attached to a body called the *arm* or *planet carrier* that rotates about the axis of one of the gears that is known as a *sun gear.* Such trains are used as speed reducers and as differentials. An additional arm attached to and rotating with one of the shafts whose bearings are mounted on the planet carrier can quite easily be made to trace a rather complicated path. These trains therefore also can be used as pick-and-place mechanisms and other manipulating and positioning mechanisms. Two such applications are discussed in connection with Fig. 11.6.

Figure 11.5*a* shows a planetary gear train. The shaft of one of the gears *F*, the *sun gear,* is pivoted to ground by grounded bearings, and the shafts of other gears, the *planet gears,* are in bearings that are attached to a body *A*, called the *planet carrier* or *arm,* and this arm rotates relative to ground and relative to the sun gear. It will be noted that in order for the sun gear *F* to be in continuous mesh with the next gear in the train, the planet carrier or arm *A* must be pivoted to ground at the center of the sun gear. The dashed circles indicate a gear train that can consist of any number of gears connecting sun gear *F* to the last planet gear *L*. If the arm *A* is held stationary, the train from *F* to *L* is seen to be of the same form as the trains in Figs. 11.3 and 11.4. This train can be either simple or compound, and it can include external and/or internal meshes. As in the cases of the trains in Figs. 11.3 and 11.4, the centers of the gears as seen in the view in Fig. 11.5*a* need not lie in a straight line; i.e., they can lie at the vertices of a polygon. This train can, and often is, reverted so that the center of the last planet gear *L* is coincident with the center of gear *F*, as shown in Fig. 11.5*b*, so that the shaft of gear *L* will not translate and can easily be connected to other parts of a machine. In such a configuration, the gear *L* ceases to be a planet gear and becomes instead a second sun gear. Note that because gear *L* is pivoted to ground, it need not be physically pivoted to the arm. In this figure gears *F* and *L* are *not attached* to each other!

Inputs, outputs, and degrees of freedom of planetary gear trains

Planetary gear trains can consist of many different combinations of arms and gears, and as a consequence they can be made to accept and

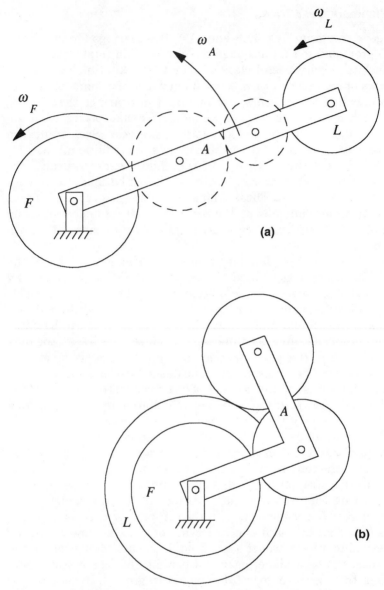

Figure 11.5 Two planetary spur gear trains.

combine the effects of different numbers of input rotations. The number of independent input rotations or drives that can be accommodated
by (and are required by) a given planetary gear train is equal to the
number of *degrees of freedom* of that train. Degrees of freedom of
mechanisms were discussed in Sec. 1.12 of this book, and this subsection will make use of that background.

The arm and the gears in Fig. 11.5 are rigid bodies, and these rigid bodies are connected to each other by bearings and by meshes between the gears. Just as was done in the discussion in Sec. 1.12, these rigid bodies can be referred to as *links* in a mechanism, and the bearings and meshes can be considered as *joints*. Then a slightly modified form of Gruebler's formula [Eq. (1.2)] can be used to determine how many degrees of freedom a given gear train possesses, and that modified form is given as Eq. (11.14).

Each of the gears and arms in a planetary train (including the ground) is considered to be a link in determining the quantity L for use in Gruebler's formula. Each of the bearings is a single-degree-of-freedom revolute joint. Each mesh, if it were considered to be rolling contact without slipping between pitch circles, also would be a single-degree-of-freedom joint. Unfortunately, for each mesh in the train that is treated as a single-degree-of-freedom joint when Gruebler's formula is used, the resulting *computed* number of degrees of freedom will be one fewer than the *practical* (actual) degrees of freedom. This paradox is analogous to the paradoxes described in Sec. 1.13, but it disappears when it is noted that the rolling-slipping contact *between meshed gear teeth* actually is a *two-degree-of-freedom joint*. (Some texts substitute the kinematically equivalent link and two revolute joints for the tooth contact at each mesh.) Then Eq. (1.2) becomes

$$F = 3(L - 1) - 2B - M \qquad (11.14)$$

where F = number of degrees of freedom in the planetary gear train
  L = number of rigid bodies in the train (gears, arms, and ground)
  B = number of bearings (revolute joints) in the train
  M = number of meshes between the gears in the train

Use of this formula will be demonstrated by computing the number of degrees of freedom in the planetary gear train shown in Fig. 11.5b. In this train the rigid bodies consist of four gears, one arm, and ground for a total of six rigid bodies, so $L = 6$. Gear F, gear L, and the arm A are each connected to ground by a separate bearing, and the other two gears are each connected to the arm by a bearing, so there are five bearings, so $B = 5$. There are three meshes (points of tangency between the pitch circles), so $M = 3$. Thus

$$F - 3(6 - 1) - 2(5) - 3 = 2$$

The resulting two degrees of freedom means that (considering ground to be fixed) the displacement or angular velocity of any two of the gears or of any gear and the arm can be specified and the position or motion of the entire train will be determined. This also means that two drives will be required or that if one drive is provided and one of the outputs is known, the other output will be determined.

Calculation of velocity ratios for a planetary gear train

An equation relating the angular velocities of the gears F and L and the arm A in planetary gear trains of the general type shown in Fig. 11.5a and b is quite easily derived as follows.

Procedure 11.3: Derivation of Velocity Ratios for a Planetary Gear Train

1. Designate the sun gear (which is pivoted to ground) as the "first" gear in the train, and label it F. If two gears are pivoted to ground as in Fig. 11.5b, either one can be chosen as the first gear in the train.

2. Designate the gear at the opposite end of the train from gear F by the letter L. The last gear in the train is found by tracing from mesh to mesh along the train from gear F until the last gear is found. Examples of gear L are shown in Fig. 11.5a and b.

3. Considering the arm A to be nonrotating, calculate the velocity ratio

$$\rho_T = \frac{\omega_{LA}}{\omega_{FA}} \tag{11.15}$$

for the gear train in a manner analogous to Eq. (11.10) (for a simple gear train) or Eq. (11.13) (for a compound gear train). Note that ω_{LA} and ω_{FA} are angular velocities of gears L and F, respectively, *relative to the arm* (which is temporarily considered stationary).

4. Then observe that if the arm is rotating at an angular velocity of ω_A, the total angular velocity ω_L of gear L and the total angular velocity ω_F of gear F are given by

$$\omega_L = \omega_{LA} + \omega_A \quad \text{and} \quad \omega_F = \omega_{FA} + \omega_A \tag{11.16}$$

5. Combining Eqs. (11.15) and (11.16) gives the following relationship between the angular velocities of the two gears F and L and the angular velocity of the arm A:

$$\omega_L = \rho_T \omega_F + (1 - \rho_T)\omega_A \tag{11.17}$$

Equation (11.17) can be used for the synthesis of a planetary gear train for generating a prescribed path, for providing a speed reduction, or for functioning as a differential.

Synthesis of a planetary gear train to trace a prescribed path

The paths that can be traced by planetary gear trains are limited to epicycloids, hypocycloids, pericycloids, and combinations of these curves. The simpler forms of these motions can be quite useful for pick-and-place actions in automatic assembly machinery and for the generation of straight-line motion. More complex forms are described in J. J. Holtzapffel, *The Principles and Practice of Ornamental or Complex Turning* (Dover Publications, New York, 1973), where the use of these gear trains for production of highly complex ornamental patterns is described.

Figure 11.6a depicts the mechanism that is also shown in Fig. 13.11 and which is used to produce a sinusoidal reciprocating motion of point

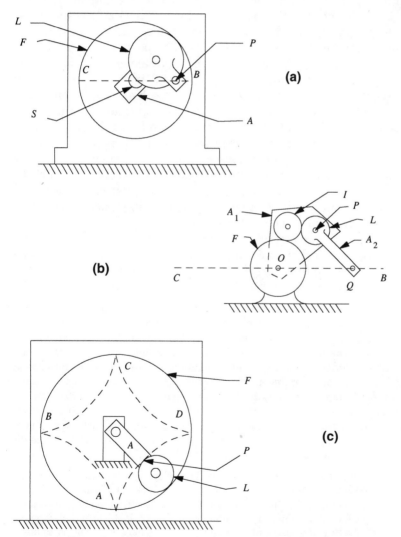

Figure 11.6 Examples of planetary gear trains that trace prescribed paths.

P (attached to gear L) along the dashed straight-line segment BC. This uses an internal gear for the first gear F and a smaller gear L that is pivoted to arm A and which rolls internally in gear F as arm A rotates on shaft S. The diameter of L is half that of F, so according to step 3 of Procedure 11.3, the velocity ratio ρ_T of the gear train with the arm fixed is $\rho_T = +2.0$. The mechanism as shown employs an internal gear, and such gears tend to be expensive. This same motion also can be produced by a planetary train consisting of only external spur gears. Such a planetary train must have a value of $\rho_T = +2.0$, just as did that shown in Fig. 11.6a. Figure 11.6b shows such a mechanism, in which the distance OP equals distance PQ, and the gear F is fixed to ground (i.e., does not rotate) and has twice as many teeth as does gear L. The idler gear I can be of any convenient size. The point Q on arm A_2 that is attached to gear L will travel back and forth along line segment BC as arm A_1 is rotated about point O. The total travel BC is given by $BC = 4(OP)$. Two small spur gears have been substituted for one large internal gear, and the resulting, more compact mechanism produces a larger output motion. This mechanism also can be used in a pick-and-place mechanism operating between stations at points B and C.

Figure 11.6c shows a planetary gear train in which point P on gear L follows a four-cusped path (shown dashed). A mechanism such as this can be used in a pick-and-place mechanism that could pick up a part at point A and place it at point B, or it could present it for some successive operations at points B and C and release it at point D. Such an action has proven useful in assembly machinery. Although Fig. 11.6c shows this mechanism using an internal gear F, it also could be implemented with only external gears. For this mechanism, $\rho_T = +4.0$.

Procedure 11.4: Synthesis of a Speed Reducer   The output shaft of an electric motor is often connected to a speed reducer, usually for the purpose of providing a greater torque to some load than is available directly from the motor. From the discussion of mechanical advantage in Chap. 3 it is seen that the magnification of the torque is equal to the reduction in shaft speed provided by the speed reducer. For example, if the ratio of a speed reducer's output shaft angular velocity to its input angular velocity is $1:20$, then the increase in torque that that reducer provides is $20:1$. The procedure for synthesizing a planetary gear train speed reducer will be illustrated by synthesizing a $20:1$ speed reducer. The output shaft is to rotate in the same direction as the input shaft and is to be coaxial with it.

1. Perform steps 1 and 2 in Procedure 11.3. Note which of the gears in the prospective planetary train pertain to the terms in Eqs. (11.15) and (11.17).

2. One of the three angular velocities in Eq. (11.17) will be made zero because the body to which it pertains will be clamped to ground. If the arm is clamped to ground, the train will be an ordinary reverted train. Therefore, in order to make the train a planetary train, either gear F or gear L will be clamped to ground, and ω_F or ω_L, respectively, will be zero. Because the choice of which gear is to be

F and which is to be L is arbitrary, it makes no difference as long as the choice is not changed during synthesis. The result of the grounding of F or L is that the prescribed speed reducer speed ratio must be the ratio between the angular velocity of the arm and the angular velocity of either L or F. For the example given here, choose L to be grounded so that $\omega_L = 0$.

3. Write Eq. (11.17) as it applies to the choices made above, and solve for the required value of ρ_T (i.e., the velocity ratio with the arm fixed). For the present example, Eq. (11.17) becomes

$$0 = \rho_T \omega_F + (1 - \rho_T)\omega_A$$

If the input is chosen as ω_F and the output is chosen as ω_A, then this gives for the velocity ratio of the reducer

$$\frac{\omega_A}{\omega_F} = \frac{\rho_T}{(\rho_T - 1)} = \frac{1}{20}$$

which, in turn, gives

$$\rho_T = -1/19$$

4. Using techniques presented in Procedures 11.1 and 11.2, devise a gear train that will have the value of ρ_T computed in step 3.

The negative sign of the required value for ρ_T means that the gear train must have an odd number of external meshes. If one gear is to be meshed between gear F and gear L, then there will be two meshes, and one of them must be an internal mesh. Figure 11.7a shows an end view and a schematic side view of a possible arrangement. Because the planet I is an idler, a velocity ratio of $\rho_T = -1/19$ requires that the ratio N_L/N_F must be 19, which would require N_L to be very large or N_F to be very small. It appears desirable to use a compound gear P as shown in Fig. 11.7b instead of the simple gear I shown in Fig. 11.7a.

Using the configuration in Fig. 11.7b, and considering the arm to be stationary, it is seen that the required conditions are

$$\rho_T = \left(\frac{-N_F}{N_{P_1}}\right)\left(\frac{N_{P_2}}{N_L}\right) = \frac{-1}{19}$$

These conditions could be satisfied if

$$\left(\frac{N_F}{N_{P_1}}\right) = \left(\frac{N_{P_2}}{N_L}\right) = \frac{1}{\sqrt{19}} = \frac{1}{4.36}$$

However, this ratio is an irrational number. Therefore, choose ratios of integral tooth numbers that will give approximately this ratio *and* the required value of ρ_T. For example, try $N_F = 4m_1$, $N_{P_1} = 19m_1$, $N_{P_2} = m_2$, and $N_L = 4m_2$, which gives

$$\left(\frac{N_F}{N_{P_1}}\right) = \frac{4m_1}{19m_1} \quad \text{and} \quad \left(\frac{N_{P_2}}{N_L}\right) = \frac{m_2}{4m_2}$$

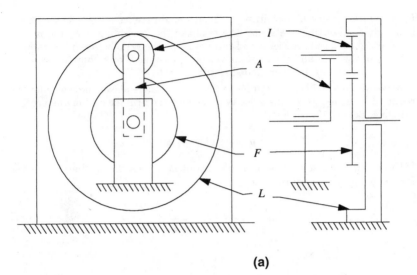

(a)

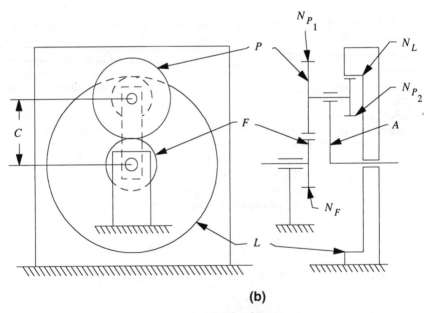

(b)

Figure 11.7 Planetary gear trains used as speed reducers.

where m_1 and m_2 are factors yet to be determined. The center distance between gears F and P_1 must be the same as the center distance between gears L and P_2. By examining Fig. 11.7b it can be seen that this distance C is given by

$$C = \frac{(D_F + D_{P_1})}{2} = \frac{(D_L - D_{P_2})}{2}$$

Then, using the definition of diametral pitch given in Eq. (11.4),

$$C = \frac{(N_F + N_{P_1})}{2P_d} = \frac{(N_L - N_{P_2})}{2P_d}$$

Substituting the foregoing trial tooth-number expressions into this expression for C gives

$$2CP_d = (4 + 19)m_1 = (4 - 1)m_2 \quad \text{or} \quad 2CP_d = 23m_1 = 3m_2$$

Thus $m_1/m_2 = 3/23$. As an initial attempt, try $m_1 = 3$ and $m_2 = 23$. Then $N_F = 12$, $N_{P_1} = 57$, $N_{P_2} = 23$, and $N_L = 92$. These numbers seem reasonable for an initial attempt, although 12 teeth on the gear F is marginally small. Go to step 5.

5. Choose a diametral pitch value for these gears, and compute gear sizes. Choose a diametral pitch that will give the desired gear train size and load capacity. Check catalogs for availability and/or check Secs. 11.7 and 11.8 for limitations on feasibility.

If a diametral pitch of 32 teeth per inch is chosen in the present example, the pitch diameters of the gears will be $D_{PF} = 12/32 = 0.375$ in, $D_{PP_1} = 57/32 = 1.782$ in, $D_{PP_2} = 23/32 = 0.719$ in, and $D_{PL} = 92/32 = 2.875$ in. Considering center distances of $69/32 = 2.156$ in, this gear reducer will have a diameter of $(2)(2.156) + 1.782 + (2)(\text{addendum}) = 6.158$ in. This could be halved by choosing a diametral pitch of 64 teeth per inch (i.e., 64-pitch gears).

If the numbers of teeth found in step 4 are considered unsatisfactory, they can each be multiplied by a common factor without changing any of the ratios. In the present example, all the numbers computed happened to be products of prime numbers, so they can only be multiplied by integers because the results must be integral numbers of teeth.

Differential gear trains

Differential gear trains are useful for allowing the phasing of rotating machinery to be adjusted continuously while the machine is running by adding or subtracting an additional rotation to the main rotation. Their most common use, however, is in automobiles, where they allow two wheels of the car to be driven from the engine even though the two wheels must rotate at different angular velocities as in driving around a curve. These differential gear trains are planetary gear trains, and their kinematics are explained in the following paragraphs.

Equation (11.17) states that in a planetary gear train the angular velocities of the first gear F, of the last gear L, and of the arm A are all related to each other by addition and subtraction and by the multiplying *constants* ρ_T and $(1 - \rho_T)$. The special cases in which ρ_T is equal to -1 or equal to $+2$ or equal to $+0.5$ are of particular interest because it can be seen that

For $\rho_T = -1$,
$$\omega_A = \frac{\omega_F + \omega_L}{2} \tag{11.18}$$

For $\rho_T = +2.0$,
$$\omega_F = \frac{\omega_A + \omega_L}{2} \tag{11.19a}$$

For $\rho_T = +0.5$,
$$\omega_L = \frac{\omega_A + \omega_F}{2} \tag{11.19b}$$

Each of these equations states that the angular velocity on the left-hand side is equal to half the sum of the other two angular velocities, or in other words, it is equal to the average of those other two angular velocities. Automotive differentials that couple the drive shaft from the engine transmission to the axles driving the wheels take advantage of this phenomenon so that the two driven wheels can rotate at *different* angular velocities as the automobile travels along a curved path and the drive shaft rotates at the *average* of those two angular velocities. Automotive differentials usually employ bevel gears in their trains, as shown schematically in Fig. 11.8. In this figure a gear A is driven from the transmission, and gear A carries the bearings for two intermediate bevel gears I_1 and I_2. Gear I_1 is a planet in the gear train FI_1L, and

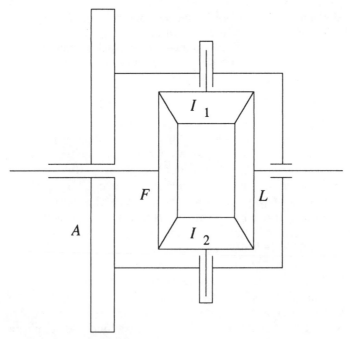

Figure 11.8 Schematic diagram of a differential gear train.

gear I_2 is a planet in the gear train FI_2L. The kinematics of the differential can be visualized by considering only one of these two trains, because the second one is used only to balance bearing forces. By tracing the gear train FI_1L, it can be seen that if arm A is held stationary and an input angular velocity ω_F is applied to gear F and its wheel axle, gear L and its axle at the other end of the train will have an angular velocity of $\omega_L = -\omega_F$, so that indeed $\rho_T = -1$.

Although most current automotive differentials use bevel gears, more compact spur gear planetary differentials are being made and used. Recall that for such use it is only necessary to make ρ_T equal to -1 or equal to $+2$ or equal to $+0.5$, and then the action is dictated by Eq. (11.18), (11.19a), or (11.19b), respectively.

Spacing of planets in a planetary gear train

Figure 11.7a and b showed only one planet in the train. (It was a compound gear in Fig. 11.7b). If the speed of operation or the torque transmitted is expected to be large, more than one planet is used. Such a train might look as shown in Fig. 11.9, where three planets are shown equally spaced around the three-branched arm. These planets *cannot* be spaced arbitrarily around the arm, and their possible locations will depend on the numbers of teeth in the gears in the train. A method for

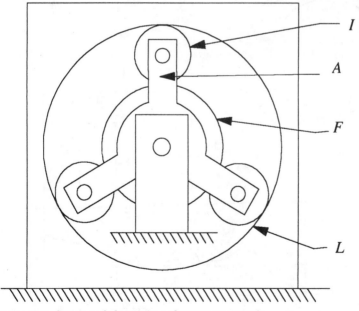

Figure 11.9 Spacing of planets in a planetary gear train.

determining the possible spacing was developed by Professor G. B. DuBois of Cornell University and is described in H. H. Mabie and C. F. Reinholtz, *Mechanisms and Dynamics of Machinery,* 4th ed. (Wiley, New York, 1987).

Planetary gear train velocity analysis

When the numbers of teeth on the gears in a planetary gear train are given, the relationships between the angular velocities of the gears can be computed using Eq. (11.17), as described in Procedure 11.3. An alternate method using velocity-distribution triangles such as described in Sec. 2.5 provides vivid visualization of the operation of the train. This method is essentially identical to the method described as "the instant center method (or tangential velocity method)" in A. G. Erdman and G. N. Sandor, *Mechanism Design, Analysis and Synthesis,* Vol. 1, 2d ed. (Prentice-Hall, Englewood Cliffs, NJ, 1991).

The use of velocity-distribution triangles in the angular velocity analysis of a planetary gear train will be illustrated by analyzing the differential shown schematically in Fig. 11.10. This differential was synthesized for the condition $\rho_T = +2$, as stated in

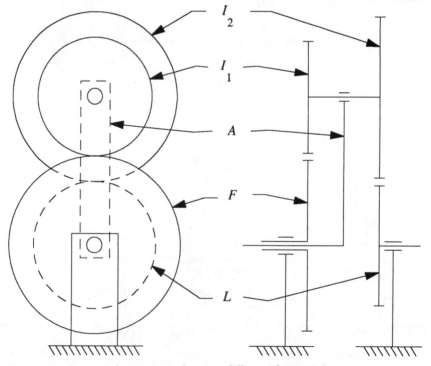

Figure 11.10 An example spur gear planetary differential gear train.

Eq. (11.19a) using the methods of Procedure 11.4. The resulting numbers of teeth are

$$N_F = 21$$

$$N_{I_1} = 14$$

$$N_{I_2} = 20$$

$$N_L = 15$$

For use in a drive of a wheeled vehicle, the *input* to the differential would be applied to the hollow shaft connected to gear F. (The hollow shaft is not shown in the left-hand view.)

The velocity ratio $\rho_T = +2$ is easily determined by tracing the velocity ratio from gear F through compound gear $I_1 I_2$ to gear L with arm A locked to ground. It is less obvious, however, that the ratio $\omega_L/\omega_A = -1$ is provided, as is required for differential action, if sun gear F is locked to ground. (For this condition, the fact that $\omega_F = 0$ must be equal to the average of ω_A and ω_L dictates that $\omega_L = -\omega_A$.)

To determine ω_L/ω_A with sun gear F locked to ground, refer to Fig. 11.11 and proceed as follows:

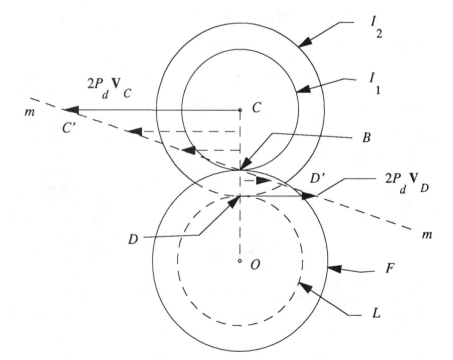

Figure 11.11   Velocity analysis of the gear train in Fig. 11.10.

1. Assume an input angular velocity of ω_A for arm A (shown as line OC), which is the input member of the train being analyzed. (A value of $\omega_A = 1$ rad/s is convenient.) The velocity of point C on the arm is then given by

$$V_C = (R_F + R_{I_1})\omega_A = \left(\frac{N_F}{2P_d} + \frac{N_{I_1}}{2P_d} \right)\omega_A$$

where R_F and R_L are the radii of the pitch circles of gears F and L, respectively, N_F and N_L are the numbers of teeth on gears F and L, respectively, and P_d is the diametral pitch of all the gears in the train. Notice that because of the definition of diametral pitch, the radius of the pitch circle of each of the gears can be expressed as the number of teeth on that gear divided by 2 times the diametral pitch. Readers will find that because the velocities of points in the train are all expressed in terms of angular velocities times pitch circle radii, when we express these radii in terms of numbers of teeth, every term in the resulting expressions contains the factor $2P_d$ in the denominator. We will therefore write equations in terms of $2P_d$ times the velocities of points. Then the preceding equation becomes

$$2P_dV_C = (N_F + N_{I_1})\omega_A$$

Draw the vector $2P_dV_C$ perpendicular to line OC on the drawing of the gear train and with its tail at point C and at some convenient scale.

2. The point C is not only attached to the arm, it is also attached to the rigid body that is the compound gear I_1I_2. Because gear F is considered fixed, the velocity of point B on gear I_1I_2 where that gear meshes with gear F is zero. Therefore, in the manner of Sec. 2.5, draw the velocity-distribution triangle consisting of sides BC, the vector $2P_dV_C$, and the dashed line mm through point B and the tip of $2P_dV_C$.

3. Extend the lines mm and BC through point B. The tails of all the velocity vectors for points that are attached to compound gear I_1I_2 and which are instantaneously on the extended line BC will lie on that line, these vectors will be perpendicular to that line (as is $2P_dV_C$), and the tips of these vectors will lie on line mm. This includes velocities of points below point B as shown.

4. Draw the velocity vector $2P_dV_D$ perpendicular to line BC extended with its tail at point D and its tip on line mm.

5. The magnitude of the vector $2P_dV_D$ can be determined by measuring its length on the drawing and using the drawing scale factor. Because the triangles BCC' and BDD' are similar, the magnitude of

$2P_dV_D$ also can be computed from the magnitude of $2P_dV_C$ as found in step 1 and the ratio of the lengths BD and BC as

$$2P_dV_D = 2P_dV_C\left(\frac{BD}{BC}\right) = (N_F + N_{I_1})\omega_A\left(\frac{N_{I_2} - N_{I_1}}{N_{I_1}}\right)$$

This vector magnitude represents the magnitude of the velocity of the periphery of gear I_1I_2 at location D.

6. Because the compound gear I_1I_2 and gear L are in mesh at point D, the velocity of a point at D on gear L is equal to the preceding computed velocity $2P_dV_D$ of the periphery of I_1I_2 at that point. Note that in this example this peripheral velocity is rightward, so the angular velocity of gear L is clockwise (i.e., negative). This peripheral velocity is also given by the angular velocity of gear L multiplied by the pitch radius of gear L. Thus we can write

$$2P_dV_D = \omega_L N_L$$

which gives

$$\omega_L = \frac{2P_dV_D}{N_L}$$

Using this equation, compute ω_L from the value of $2P_dV_D$ computed in step 5. Then the desired ratio ω_L/ω_A is equal to ω_L because ω_A was originally assumed to be 1 rad/s in step 1. Alternatively, this ratio can be computed by combining this last equation with the equation in step 5. This combination gives

$$\frac{\omega_L}{\omega_A} = -\left(\frac{N_F - N_{I_1}}{N_{I_1}}\right)\left(\frac{N_{I_2} - N_{I_1}}{N_{I_L}}\right)$$

Substituting the tooth numbers given earlier for this train into this equation gives $\omega_L/\omega_A = -1$, which is consistent with operation as a differential.

11.5 The Involute Gear Tooth Profile and the Fundamental Law of Gear Tooth Action

For most machine and mechanism design tasks, relationships discussed in the preceding sections provide adequate guidance for the choice of gears and gear train configuration. There are, however, occasions on which space limitations, load conditions, center distance variability, gearing ratio requirements, or other prescribed conditions require that engineers have an understanding of the principles discussed in this section and the three subsequent sections.

Because the meshing of two gears involves the *intermittent* contact between pairs of discrete teeth on those two gears, the manner in which the teeth make, maintain, and cease contact with each other must be designed with care if this intermittent action is not to result in undesirable fluctuations in the motion of the gears. It is pointed out in E. Buckingham, *Analytical Mechanics of Gears* (Dover Publications, New York, 1988), that the required "conjugate action" can be provided by many different tooth shapes. In practice, however, two types of shapes occur almost exclusively: the involute profile and the cycloidal profile. The cycloidal profile is used ordinarily in only very special applications. The involute profile is used in the vast majority of applications that engineers will encounter, and it has the unique and important feature that it allows a pair of gears that are based on it to provide smooth operation even if the center distance between those gears is varied. Involute gear kinematics will be the subject of this and subsequent sections.

Principles of involute gear tooth action

If a length of string that is wrapped around the periphery of a circular disk is then gradually unwrapped, a point at the end of that string will follow a curved path that is known as an *involute of the circle* that is represented by the disk. Figure 11.12 shows a string *s* being unwrapped from around a circle *c*. Successive positions of the string as it is unwrapped are shown as dashed lines, and the path followed by the end of that string is shown as the curve *i*. That curve is an involute of the circle *c*. All other involutes generated on that circle will be congruent with the curve *i*; i.e., they will have shapes identical to that of *i*. Indeed, there is nothing special about the end of the string, and if points that are fixed to the string and are equally spaced along it are considered, then, as the string is unwound from the circle, these points will trace identical involutes that are equally spaced along the string and equally spaced around the circle.

For readers who wish to compute and plot an involute, dimensions, parameters, and coordinate axes are indicated in Fig. 11.12, from which, by referring to right triangle OTQ, the following equations may be derived:

$$\theta = \sqrt{\left(\frac{R_i}{R_b}\right)^2 - 1} \tag{11.20}$$

$$\gamma = \theta - \tan^{-1}\theta \tag{11.21}$$

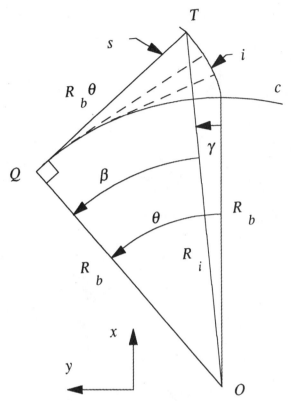

Figure 11.12 Geometric relationships in the involute of a circle.

where θ and γ are in radians and R_i is the distance OT. These equations relate the polar coordinates R_i and γ of the end point T of the string to each other, so that by choosing a value of R_i/R_b, a value of γ may be computed. If Cartesian coordinates are desired relative to axes X and Y shown, they can be computed from $x_i = R_i \cos \gamma$ and $y_i = R_i \sin \gamma$.

The manner in which the involute serves so well as a shape for gear teeth can be seen by reference to Fig. 11.13. This figure shows portions of two circles b_1 and b_2 and a taut string that is unwinding from circle b_1 at point Q_1 and simultaneously being wound onto circle b_2 at point Q_2 as the circles rotate about O_1 and O_2, respectively. If a pen point were attached to the string at a point such as point T, then as the string unwound from circle b_1, the pen point would draw an involute i_1 on a sheet of paper attached to circle b_1 as shown. Simultaneously, that pen point would draw an involute i_2 on a sheet of paper attached to circle b_2 as the string wound onto circle b_2. Those two involutes would

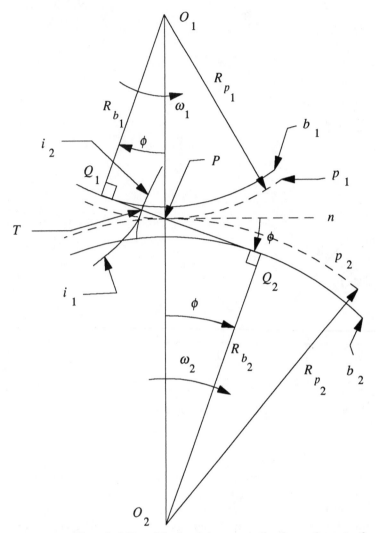

Figure 11.13 The suitability of the involute curve to the shape of gear teeth.

always be tangent to each other at the location T of the pen point as that point moved with the moving string.

The string is tangent to the two circles at points Q_1 and Q_2. The velocity of point Q_1 is equal to $R_{b_1}\omega_1$, and the velocity of point Q_2 is equal to $R_{b_2}\omega_2$. Because the inelastic string is stretched between these two points, the two velocities must be equal. Then

$$R_{b_1}\omega_1 = R_{b_2}\omega_2 \quad \text{or} \quad \frac{\omega_1}{\omega_2} = \frac{R_{b_2}}{R_{b_1}} \qquad (11.22)$$

The involutes i_1 and i_2 remain tangent to each other as the circles b_1 and b_2 rotate with a constant ratio R_{b_2}/R_{b_1} of angular velocities. Therefore, teeth shaped and located like these involutes but which are attached to gears that are in turn attached to the circles (and which therefore rotate with the circles) would remain coincident with these involutes and also would remain tangent to each other as the circles and gears rotate with that constant ratio of angular velocities. By appropriately equally spacing pen points along the string, the pens could be made to trace identical shapes for all the other teeth required on the gears. By reversing the drawing in Fig. 11.13 right for left, the pens could be made to draw the other side of each of the teeth on both gears.

Notice that the point of tangency of such involute gear teeth will always lie somewhere on the line segment Q_1Q_2. This segment Q_1Q_2 will always be perpendicular to the line which is tangent to the surfaces of both teeth at their point of tangency T. The force that one tooth transmits to the other tooth is applied at point T, and except for the effects of friction, that force must be perpendicular to the tooth surfaces at T. That force therefore must be directed along line segment Q_1Q_2, and this segment is referred to as the *axis of transmission* or the *pressure line*.

The axis of transmission intersects the *line of centers* O_1O_2 at a *fixed point P*. This point is called the *pitch point*. A circle p_1 is drawn through pitch point P and with its center at point O_1. Another circle p_2 is drawn through point P and with its center at point O_2. The circle p_1 is the *pitch circle* of gear 1, and it has a radius of R_{p_1}. The circle p_2 is the pitch circle of gear 2, and it has a radius of R_{p_2}. The angle at the vertex P of triangle O_1PQ_1 is equal to the angle at the vertex P of triangle O_2PQ_2, so these two right triangles are similar. Therefore,

$$\frac{R_{p_2}}{R_{p_1}} = \frac{R_{b_2}}{R_{b_1}}. \tag{11.23}$$

From Eqs. (11.22) and (11.23) it can be seen that these two pitch circles do indeed roll on each other without slipping, just as we have considered them to do in previous sections.

A line Pn is drawn through the pitch point P and perpendicular to the line of centers O_1O_2. This line Pn is tangent to the pitch circles at their point of tangency P. The angle ϕ between this line Pn and the axis of transmission Q_1Q_2 is the angle by which the direction Q_1Q_2 of the force (neglecting friction) that one tooth transmits to a tooth on the mating gear deviates from a direction Pn tangent to the pitch circles. This angle ϕ is seen to be equal to the angles PO_1Q_1 and PO_2Q_2, and it is called the *pressure angle*. Because this angle is never zero, the transmitted force always tends to push centers of the gears apart.

As seen in triangles PO_1Q_1 and PO_2Q_2, the distance $O_1P = R_{p_1}$ and the distance $O_2P = R_{p_2}$. Then the pressure angle ϕ is related to the radii of the base circles and the radii of the pitch circles by

$$\frac{R_{b1}}{R_{p1}} = \frac{R_{b2}}{R_{p2}} = \cos \phi \qquad (11.24)$$

The fundamental law of gear tooth action

Readers who wish to *design gear tooth shapes* other than the involute must be aware that all tooth shapes must obey the *fundamental law of gear tooth action.* This law states:

> In order for the ratio between the angular velocities of two meshed gears to be constant as the gears rotate, the line perpendicular to the gear tooth surfaces of the two gears at their point of contact must always intersect the line connecting the centers of those gears at a fixed point (the pitch point).

Use of the involute gear tooth profile automatically ensures that the axis of transmission (which is perpendicular to the gear tooth surfaces at their point of tangency) will at all times during the gears' rotation intersect the line of centers at a *fixed point P*, thereby ensuring a constant ratio of the angular velocity of the gears. To show that *any* tooth profile shape must obey this law if the associated gears are to rotate with a constant velocity ratio, refer to Fig. 11.14. In this figure, fragments of two gears G_1 and G_2 are shown with their surfaces contacting each other at their respective points T_1 and T_2. These gears are rotating about centers O_1 and O_2, respectively. Through the coincident points T_1 and T_2, draw the line s perpendicular to the surfaces at those points. From the center point O_1, draw a perpendicular to the line s intersecting s at point Q_1. From the center point O_2, draw a perpendicular to the line s intersecting s at point Q_2. Points T_1 and T_2 must have velocities that have components along line s that are equal. Because gear G_1 is a rigid body, all points on it that are along line segment T_1Q_1 must have velocities whose components along that line are equal. Similarly, because gear G_2 is a rigid body, all points on it that are along line segment T_2Q_2 must have velocities whose components along that line are equal. Then it is concluded that the total velocity of point Q_1 on gear G_1 must be equal to the total velocity of point Q_2 on gear G_2. Then the ratio of angular velocities of the two gears depends on the ratio between radii O_1Q_1 and O_2Q_2. Using the same similar-triangles argument as used in the discussion of the operation of involute teeth, it can be shown that this ratio is the same as the ratio between O_1P and O_2P. The only way to keep this latter ratio constant, if the distance O_1O_2 is constant, is to keep the location of point P fixed. Thus, al-

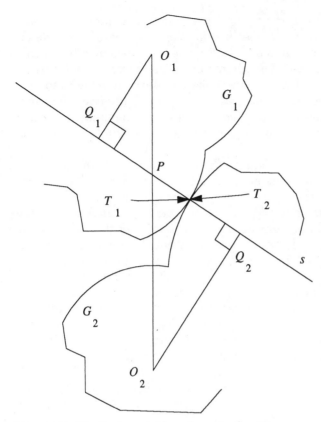

Figure 11.14 The fundamental law of gear-tooth action.

though the *inclination* of line s may vary (and thus the pressure angle may vary) as the gears rotate, the intersection of line s with the line of centers must remain fixed if the velocity ratio of the gears is to remain constant.

11.6 Effects of Varying the Center Distance

As was shown in the preceding section, the shape of the involute teeth on a gear is determined by the size of the base circle because the involutes can be considered to be generated by unwinding a string from around that base circle. The pitch circles were then derived from the location of the pitch point, which, in turn, is determined by the intersection of the line of centers of *two meshed gears* and the normal to the surfaces of the meshed teeth of *those two gears*. Thus the pitch diameter of a gear appears to be *not* determined by the geometry of an *individual* gear. However, gears are usually specified in terms of and listed in catalogs according to diametral pitch and number of teeth. As Eq. (11.4) shows, values specified for these two parameters determine a value for pitch diameter, and indeed catalogs also list such a value for pitch diameter. This diameter value is known as the *standard pitch diameter* for that gear. These gears are manufactured with widths of their teeth (as measured along the pitch circle) such that when two such gears are installed with a center distance consistent with their standard pitch diameters, the teeth of the two gears will mesh "well."

Refer to Fig. 11.13 and visualize the center points O_1 and O_2 moved slightly further apart while keeping the base circle radii R_{b_1} and R_{b_2} unchanged. The base circles would be moved slightly further apart, and the inelastic string would cause the circles to rotate slightly relative to each other. However, the point T attached to the string would still remain on the same two involutes, although the involutes would have rotated slightly. The point T would just be moved along either or both of these involute curves. The involutes would still be tangent to each other, but line s would be more steeply inclined; i.e., the pressure angle would be increased. Triangles O_1PQ_1 and O_2PQ_2, although changed, would still be similar to each other, so the velocity ratio would be the same as before increasing the center distance. Increasing the center distance would have increased the pitch diameters of the two gears proportionately, and the velocity ratio would be unchanged. This is a feature that is unique to involute gearing. It allows some tolerance in the design and fabrication of the location of bearings for the gears and their shafts.

When the center distance for two gears is increased, the points of contact between the teeth of the two gears occur further from the centers of the two gears and nearer to the tips of the teeth. The teeth are narrower near their tips, and, therefore, the teeth of each gear do not completely fill the spaces between the teeth of the other gear. The leftover space produces what is known as *backlash*. That is, if one gear is held stationary, the other gear can be rotated back and forth through a small angle as its teeth "rattle" back and forth in the space between the teeth in the stationary gear. Excessive backlash can cause noise and vibration and uneven running of the driven gear.

Conversely, if the center distance is decreased to a value below the sum of the standard pitch radii of the two gears, there is not enough space between the teeth of one gear to accommodate the teeth of the other gear, and *binding* can occur. Excessive binding can cause noise, vibration, and rapid wear.

In applications where it is possible to make the position of the gear shafts adjustable, such adjustment can be used to adjust the amount of backlash at the gear mesh.

11.7   Tooth Interference and Undercutting and Nonstandard Gears

As pointed out in Fig. 11.13, the path of the point of contact between the involute surfaces on the teeth of the two gears travels along the path between points Q_1 and Q_2. Note that the two involutes *exist* for the two base circles *only* between points Q_1 and Q_2 because outside these points the "string" is wrapped on one or the other of the base circles. The geometry shown in Fig. 11.13 is repeated in Fig. 11.15, and in this latter figure the addendum circles are shown. The addendum circles are the paths along which the tips of the teeth on the respective gears travel.

In Fig. 11.15 parts of two teeth are shown in contact with each other at point T. If the upper gear G_1 were rotated clockwise and the lower

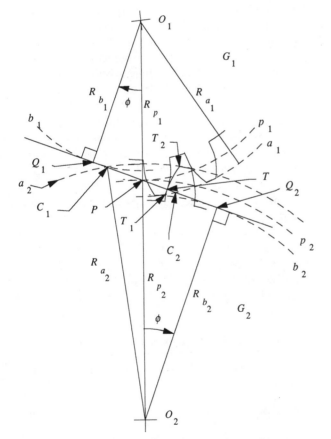

Figure 11.15 Geometry associated with tooth interference, undercutting, and nonstandard gears.

gear G_2 were rotated counterclockwise while keeping these teeth in contact, the point of contact would move leftward toward point Q_1 until, at some position of the gears, the tip T_2 of the tooth on gear G_2 would reach point C_1 where the axis of transmission Q_1Q_2 intersects the addendum circle a_2 of gear G_2. At that time, the point of contact between the teeth would be at the tip T_2 of the tooth on gear G_2. Further rotation of the gears would cause the tooth on gear G_2 to cease contact with the tooth on gear G_1. Similar reasoning shows that the tip T_1 of the tooth on gear G_1 would cease contact with the tooth on gear G_2 if gear G_1 were rotated counterclockwise until tip T_1 moved rightward beyond point C_2 where the addendum circle a_1 intersects line Q_1Q_2. Contact between the involute-shaped surfaces on the two gears is thus limited to line segment C_1C_2, and the length of this segment is called the *contact length*. This length is discussed in Sec. 11.8.

If the addendum circle a_2 of gear G_2 were sufficiently large that it intersected the axis of transmission Q_1Q_2 at a point to the left of point Q_1, tip T_2 would attempt to contact gear G_1 in an area in which the involute for base circle b_1 of gear G_1 is not defined. The result would be that tip T_2 of the tooth would attempt to occupy space that was *not cut away to form the involute* on gear G_1. That is, there would be mechanical *interference,* and the gears would "jam." Because the contact between teeth must be limited to the region between Q_1 and Q_2 if interference is to be avoided, these points are called *interference points.*

Notice that if the addendum distances on the two gears are equal (as is usually the case), the teeth on the larger gear (gear G_2 in this figure) are more likely to interfere with the smaller gear than are the teeth on the smaller gear to interfere with the larger gear. Interference with the smaller gear is thus the limiting condition so that if it is avoided, all interference is avoided.

The conditions that must be met if interference with the smaller gear is to be avoided are derived as follows: Call the distance C_1P by the symbol L_a, where P is the pitch point as previously defined. This length C_1P is given by the difference between the lengths C_1Q_2 and PQ_2. Then, using the relationships in right triangles $O_2Q_2C_1$ and O_2Q_2P, we may write

$$L_a = \sqrt{R_{a_2}^2 - R_{b_2}^2} - R_{p_2} \sin \phi \qquad (11.25)$$

In order to avoid interference, this length must be equal to or smaller than the length PQ_1, where

$$PQ_1 = R_{p_1} \sin \phi \qquad (11.26)$$

Then we may write as a condition for noninterference

$$\sqrt{R_{a_2}^2 - R_{b_2}^2} - R_{p_2} \sin \phi \leq R_{p_1} \sin \phi \qquad (11.27)$$

This equation can be somewhat simplified by using the following relationships:

$$R_{b_2} = R_{p_2} \cos \phi \tag{11.28}$$

The radius R_{a_2} of the addendum circle of gear G_2 is equal to the pitch radius R_{p_2} of that gear plus the addendum distance a. The addendum distance a as specified by the AGMA and as quoted in most gear catalogs can be expressed as

$$a = \frac{k_a}{P_d}$$

where k_a will be called the *addendum constant*. Then we may write

$$R_{a_2} = R_{p_2} + \frac{k_{a_2}}{P_d} \tag{11.29}$$

Then, by using the definition of diametral pitch as given in Eq. (11.4), by substituting Eqs. (11.28) and (11.29) into Eq. (11.27), and by multiplying each term in the resulting expression by $2P_d$, we obtain for the condition for noninterference:

$$\sqrt{(N_2 + 2k_{a_2})^2 - N_2^2 \cos^2 \phi} - N_2 \sin \phi \leq N_1 \sin \phi \tag{11.30}$$

where N_1 and N_2 are the numbers of teeth on the respective gears, k_{a_2} is the constant determining the addendum distance on gear G_2, and ϕ is the pressure angle.

For AGMA standard gears, $k_a = 1.0$. Then, for the most popular pressure angle, 20°, we may choose numbers N_2 of teeth for the large gear and use Eq. (11.30) to compute the minimum required number N_1 of teeth on the smaller gear if interference is to be avoided. A few example results of such a calculation are given in Table 11.1, where the numbers of teeth have been rounded off to the next larger integer.

TABLE 11.1 Numbers of Gear Teeth Required for Avoiding Interference (Pressure Angle = 20°)

N_2	$\left(\dfrac{N_1}{N_2}\right)_{\min}$	$(N_1)_{\min}$ $(k_{a_2} = 1.0)$	$(N_1)_{\min}$ $(k_{a_2} = 1.25)$
20	0.7	14	—
50	0.32	16	—
100	0.16	16	—
16	—	—	16

This table indicates that if interference is to be avoided, no fewer than 14 teeth can be used on the smaller gear, even if the larger gear possess only 20 teeth. As the number of teeth on the larger gear is increased, the number of teeth required on the smaller gear increases, and if the gear is to mesh with a rack, it is found that 18 teeth are required on the gear. Notice also that a gear *reduction* of more than 6:1 (i.e., $N_1/N_2 < 0.16$) requires that the large gear contain more than 100 teeth if interference is to be avoided.

Table 11.2 presents some numbers of teeth required if interference is to be avoided when a pressure angle of 25° is used. By comparing the first rows in these two tables, it can be seen that increasing pressure angle decreases the minimum number of teeth required on the smaller gear if interference is to be avoided.

Examination of Eq. (11.30) shows that decreasing the addendum of the larger gear G_2 as would result from a smaller value of k_{a_2} also would allow a smaller value of N_1 to be used without interference. Such a decreased addendum gear is known as a *stub gear,* and such gears commonly are made with an addendum constant of 0.8. By using Eq. (11.30), tables of required numbers of teeth can then be made for stub-toothed gears.

Undercutting of gear teeth

Most metal gears are produced by *hobbing* or by *shaping.* If hobbing is used, the cutter used in machining would tend to interfere with a small gear made in the same way that a rack would interfere with that gear. As a consequence, hobbing will cut away any tooth material with which a rack would interfere. The result is that any gear produced by hobbing and having fewer than 18 teeth would have material removed from the roots of its teeth, thereby weakening those teeth. Some material also might be removed from the useful involute surface. This cutting away of tooth-supporting material is known as *undercutting.* Although some undercutting often can be tolerated, in practice it should be avoided when possible.

If a gear is manufactured by *shaping* using a cutter that is shaped like a gear, it is possible to produce gears with fewer teeth with-

TABLE 11.2 Numbers of Gear Teeth Required for Avoiding Interference (Pressure Angle = 25°)

N_2	$\left(\dfrac{N_1}{N_2}\right)_{\min}$	$(N_1)_{\min}$ $(k_{a_2} = 1.0)$	$(N_1)_{\min}$ $(k_{a_2} = 1.25)$
20	0.5	10	—
12	—	—	11

out undercutting than is possible with hobbing. In the last row in Table 11.1 it is seen that a gear with $N_2 = 16$ teeth could mesh with a gear having $N_1 = 16$ teeth without interference. Therefore, a shaping cutter with 16 teeth could machine a 16-tooth gear without undercutting its teeth. The addendum constant k_{a_2} as used in this last row was increased to 1.25 for this computation because the cutter would have to have more than the standard addendum in order to cut spaces between the teeth that were deeper than the addendum distance of the gear with which it is to mesh.

Comparing the last rows in Tables 11.1 and 11.2 shows that increased pressure angle allows smaller gears *to be made* without undercutting.

11.8   Contact Ratio, Approach Ratio, and Recess Ratio

As a pair of gears that are in mesh rotate, there must at all times be contact between one or more teeth on one gear and corresponding teeth on the other gear. The *average* number of tooth pairs that are simultaneously in contact as the gears rotate is known as the *contact ratio* for that pair of meshed gears. The higher this ratio, the more teeth pairs will be sharing the load at any time and the smoother and quieter will be the operation of the gears. In any event, this ratio must always be 1.0 or greater and preferably on the order of 1.2 or 1.4 or greater. The contact ratio can be calculated from Eq. (11.37), which is derived in the following paragraphs.

Calculation of the contact ratio

To derive an expression for this ratio, refer to Fig. 11.15 and note that as the smaller gear G_1 rotates counterclockwise and the larger gear G_2 rotates clockwise, any given pair of teeth will be in contact with each other as their contact point moves along the axis of transmission from point C_1 to point C_2. Then the distance from C_1 to C_2 is known as the *contact length*. In the preceding section the distance from C_1 to P was denoted L_a. This is the distance the contact point moves as it *approaches* the pitch point P. It is called the *approach length,* and its length was found to be expressed by Eq. (11.25). The remaining portion of the distance from C_1 to C_2 is the distance from P to C_2. This is the distance that the contact point travels as it *recedes* from the pitch point P. It is known as the *recess length,* and we will denote this length by L_r. Then, in a manner similar to that used in deriving Eq. (11.25), we may write

$$L_r = \sqrt{R_{a_1}^{\,2} - R_{b_1}^{\,2}} - R_{p_1} \sin \phi \qquad (11.31)$$

The total contact length is then given by

$$L_c = L_a + L_r = \sqrt{R_{a_2}^2 - R_{b_2}^2} + \sqrt{R_{a_1}^2 - R_{b_1}^2} - (R_{p_2} + R_{p_1}) \sin \phi$$

$$(11.32)$$

Then, by using the definition of diametral pitch as given in Eq. (11.4) and by substituting Eqs. (11.28) and (11.29) and their equivalent relationships for gear G_1 into Eq. (11.32), we get

$$L_c = \left(\frac{1}{2P_d} \right) \Bigg[\sqrt{(N_2 + 2k_{a_2})^2 - N_2^2 \cos^2 \phi}$$
$$+ \sqrt{(N_1 + 2k_{a_1})^2 - N_1^2 \cos^2 \phi} - (N_2 + N_1) \sin \phi \Bigg] \quad (11.33)$$

The number of tooth pairs that can be in contact in any given time along this length L_c is found by dividing this length L_c by the length of the space between successive points of contact along this line. The result of this division is called the *contact ratio* ρ_c, and it can be expressed as

$$\rho_c = \frac{L_c}{P_{b_1}} = \frac{L_c}{P_{b_2}} = \frac{L_c}{P_b} \qquad (11.34)$$

where P_b is known as the *base pitch* because it is the spacing from one involute to the next as measured along the base circles (and therefore also along the string from Q_1 to Q_2 that generates the involutes). This space is the same for both gears in a mesh, and it is equal to the circumference of the base circle of a gear divided by the number of teeth on that gear. Thus we may write

$$P_b = 2\pi R_b / N \qquad \text{for any gear} \qquad (11.35)$$

In Fig. 11.15, right triangle O_1Q_1P shows that $R_{b_1} = R_{p_1} \cos \phi$. Then, using this fact and Eqs. (11.4) and (11.35), we may write

$$P_b = \frac{\pi \cos \phi}{P_d} \qquad (11.36)$$

As dictated by Eq. (11.34), we divide Eq. (11.33) by Eq. (11.36) to obtain

$$\rho_c = \left(\frac{1}{2\pi \cos \phi} \right) \Bigg[\sqrt{(N_2 + 2k_{a_2})^2 - N_2^2 \cos^2 \phi}$$
$$+ \sqrt{(N_1 + 2k_{a_1})^2 - N_1^2 \cos^2 \phi} - (N_2 + N_1) \sin \phi \Bigg] \quad (11.37)$$

As can be seen in this equation, the contact ratio depends heavily on the numbers of teeth on the gears, so problems in obtaining adequate contact ratio can arise when gears with small numbers of teeth are used. In Fig. 11.15 it can be seen that if the pressure angle ϕ were decreased and the addendum circles were left unchanged, the contact length C_1C_2 would be increased, which would allow a larger contact ratio. However, the change in ϕ also would decrease the length between the interference points Q_1 and Q_2, so that the chances of interference would be increased. The effects of changing the addendum circles by changing the addendum constants k_{a_1} and k_{a_2} are discussed in subsequent paragraphs on approach ratio and recess ratio.

Approach ratio and recess ratio

Recall that in deriving the equation for contact ratio, use was made of the approach length L_a that is given by Eq. (11.25) and the recess length L_r that is given by Eq. (11.31). Each of these equations can be treated in the same manner as were Eqs. (11.32) and (11.33). Such treatment gives

$$\rho_a = \left(\frac{1}{2\pi \cos\phi} \right)\left[\sqrt{(N_2 + 2k_{a_2})^2 - N_2{}^2 \cos^2\phi} - N_2 \sin\phi \right] \qquad (11.38)$$

where ρ_a is known as the *approach ratio* and

$$\rho_r = \left(\frac{1}{2\pi \cos\phi} \right)\left[\sqrt{(N_1 + 2k_{a_1})^2 - N_1{}^2 \cos^2\phi} - N_1 \sin\phi \right] \qquad (11.39)$$

where ρ_r is known as the *recess ratio*. As would be expected by analogy with the concept of contact ratio, these ratios give the average numbers of teeth that will be in mesh along the approach length and along the recess length, respectively, as the gears rotate.

Ideally, the teeth that are in contact in a gear mesh would slide on each other frictionlessly, and therefore the force transmitted between each pair of teeth would lie normal to their surfaces at the point of contact. That is, the transmitted forces would all lie along the axis of transmission. Unfortunately, this is not the case, and the resulting forces at the points of contact are not directed exactly along the axis of transmission. Because the teeth are coming into mesh in the approach phase of the mesh and are withdrawing from each other during the recess phase, the effects of friction will incline the resulting transmitted force at an angle that is *steeper* (in a view such as Fig. 11.15) than the angle of the axis of transmission during contact travel along the approach length, and at an angle that is *less steep* than the angle of the axis of transmission during contact travel along the recess length.

Visualizing gear G_1 as the driving gear while viewing Fig. 11.15 shows that gear teeth in contact during the approach from C_1 to P would apply transmitted forces at a steep downward angle, and during the recess from P to C_2, contacting teeth would apply forces that are more horizontal. The more horizontal forces will be more effective in transmitting torque between the gear shafts. It therefore generally would be preferable that the recess portion of contact be maximized.

By examining Eqs. (11.38) and (11.39), it can be seen that if gear G_1 is the smaller gear, then $N_2 > N_1$, and if the addendums are equal ($k_{a_1} = k_{a_2}$), then $\rho_a > \rho_r$ and there will be more teeth in contact during the approach phase (where their torque transmission is less efficient) than during the recess phase. It is possible, of course, to choose different addendum constants k_{a_1} and k_{a_2} and thereby produce a desired balance between the approach ratio and the recess ratio. Indeed, by appropriately choosing values for k_{a_1} and k_{a_2} it is possible to produce approach and recess lengths as determined by Eqs. (11.25) and (11.31) that will place the points C_1 and C_2 at which tooth contact ends at points Q_1 and Q_2, respectively, thereby providing the maximum amount of tooth contact that can be provided without encountering interference or undercutting. For this condition, $\rho_a P_b = L_a = R_{p_1} \sin \phi$, so the limiting value of ρ_a is given by $(\rho_a)_{\lim} = (R_{p_1} \sin \phi)/P_b$. Then, using Eqs. (11.36) and (11.4), we can write

$$(\rho_a)_{\lim} = \frac{N_1 \tan \phi}{2\pi} \tag{11.40}$$

In a similar manner, an expression for the limiting value for the recess ratio ρ_r can be derived as

$$(\rho_r)_{\lim} = \frac{N_2 \tan \phi}{2\pi} \tag{11.41}$$

For a given pair of gears, Eqs. (11.40) and (11.41) can be used together with Eqs. (11.38) and (11.39) to find the unequal addendums that would provide maximum tooth contact without interference or undercutting.

If the efficiency or wear characteristics of gears in a particular application are very important, and/or if a large number of gear sets are to be produced for that application, it may be worthwhile to use an *unequal addendum* gear set designed according to the preceding equations. Obviously, dedendums must be adjusted to accommodate these addendums.

11.9   Timing-Belt and Pulley Systems

Many products and machines now use timing belts and pulleys to perform functions that traditionally were performed by gears, sprockets, and chains. Common examples of such uses occur in office machines (such as electric typewriters, copy machines, etc.) and in automobiles. Timing belts in automobile engines routinely serve for many millions of engine revolutions (e.g., 60,000 miles or more) with no maintenance or adjustment and under harsh environmental conditions. When properly applied, timing-belt systems can provide economical, space-saving operation with low output of noise and vibration. This section will deal with the kinematic considerations in the use of timing-belt systems. Details of power-transmission capacities, of belt and pulley materials, and of dimensions are covered in vendors' catalogs and handbooks.

Timing belts differ from common flat belts and vee belts by having teeth molded into their surfaces, so that when they mesh with teeth on the pulleys with which they are used, no slippage is possible and synchronism is maintained between all the pulleys that engage a given belt. The sizes and shapes of the teeth used vary, and Fig. 11.16 indicates some example profile views of timing belts. The pulleys used with these belts possess teeth that match the profiles of the belts so that the pulley teeth mesh well with the belt teeth. Notice that in

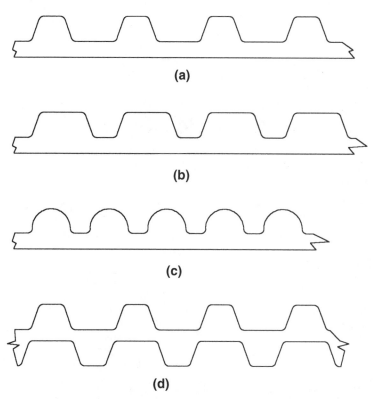

(a)

(b)

(c)

(d)

Figure 11.16   Examples of profile shapes of timing belts.

Fig. 11.16a, b, and c the teeth lie on only one face of the belt. The most common use of such belts consists of coupling two or more pulleys by meshing the pulleys with a loop of belt that surrounds the pulleys in a manner such as shown schematically in Fig. 11.17a and b (where,

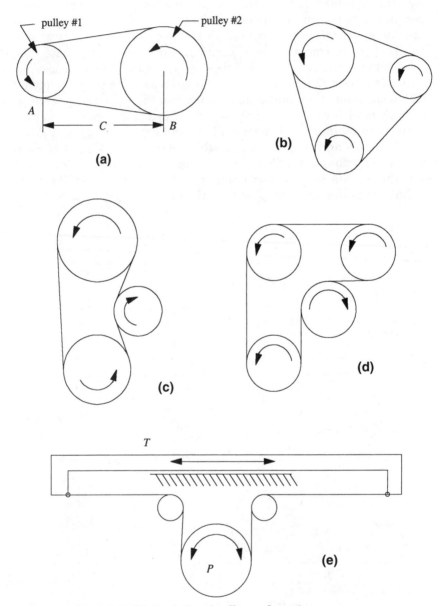

Figure 11.17 Examples of timing-belt and pulley configurations.

for simplicity, the teeth are not shown). In such a configuration, the belt could be visualized as a flexible internal gear, and as discussed in preceding sections, all the meshes would be internal meshes. Consequently, all pulleys would rotate in the same direction. As shown in Fig. 11.16d, belts are also available with teeth on both surfaces so that belt and pulley configurations such as shown in Fig. 11.17c and d can be used. With these arrangements, the pulley that is meshed with the external surface of the belt will be in an external mesh with the belt, and therefore will rotate in a direction that is opposite to that of the other pulleys.

The belts shown in Fig. 11.17c and d experience a reverse bending as they pass over the external pulleys. Such a reversal of bending is known as *contraflexure*. Configurations involving contraflexure may result from either the need to provide a negative angular velocity ratio or the need to use an "idler pulley" to adjust the belt tension. An external idler pulley used with a belt that has teeth only on its inside should be a smooth (no teeth) pulley.

Timing-belt systems are also frequently used as conveyor belts, in which case fittings of various shapes are attached to the belt or are molded into the belt so that the parts being conveyed will be positioned accurately as they are being conveyed. Timing belts are available in the form of loops of prescribed lengths and in open lengths. Open lengths can be applied to linear drives, an example of which is indicated in Fig. 11.17e, where a flat table T is caused to slide back and forth by rotating pulley P back and forth.

Terminology and basic relationships

The spacing of the teeth along a belt must equal the spacing of the teeth around the *pitch circumference* of any pulley with which it meshes. This spacing is measured from a point on a tooth to the corresponding point on the next tooth and is called the *pitch P* of the belt and pulley. Standard belts and belting are available in various pitches. Some common pitches are 0.2 in, 3/8 in, 1/2 in, 7/8 in, 5 mm, 10 mm, and 20 mm.

The length of a belt is given by

$$L_B = N_B P \tag{11.42}$$

where L_B = total length of the belt
         N_B = number of teeth on the belt
         P = pitch

The pitch circumference of a pulley i that meshes with a belt can be computed in a similar manner, and because the pitch diameter of the pulley is $1/\pi$ times this circumference, the *pitch diameter* is given by

$$D_i = \frac{N_i P}{\pi} \qquad (11.43)$$

where D_i = pitch diameter of the ith pulley
        N_i = number of teeth on the ith pulley

The *pitch diameter* of a pulley is analogous to the pitch diameter of a gear, in that it is not measurable on the pulley itself. Rather, it is a quantity that is useful in calculating velocity ratios and distances between the centers of pulleys in a system.

Timing-belt applications

A general procedure for the kinematic design of a timing-belt system is given as Procedure 11.5 later in this section. Some example configurations will now be described, and some relationships will be derived.

The simplest timing-belt system configuration is that in which two parallel shafts are to be coupled in such a manner that the ratio between their angular velocities has a prescribed value. Such a configuration is shown in Fig. 11.17a, and the shafts being coupled are indicated to be a distance C apart. Pulley number 1 possesses N_1 teeth and pulley number 2 possesses N_2 teeth. The velocity of the belt teeth that leave mesh with pulley 1 at point A must equal the velocity of teeth that enter mesh at point B with pulley 2 because the length of belt between points A and B must not change if the belt is not being stretched or being allowed to go slack. Therefore, $\omega_1 D_1/2 = \omega_2 D_2/2$ or, using Eq. (11.43),

$$\frac{\omega_2}{\omega_1} = \frac{D_1}{D_2} = \frac{N_1}{N_2} \qquad (11.44)$$

Equation (11.44) shows that the angular *velocity ratio* is dependent on only the ratio of the numbers of teeth on the pulleys. The number of teeth on the *belt* and the length of the belt are dependent on the *center distance* C between the centers of the pulleys as well as on the sizes of the pulleys (and thus on the numbers of teeth on the pulleys). The portions of the belt that are meshed with pulleys and which are therefore "wrapped" around those pulleys will be referred to as *wrap lengths,* and the portions of the belt that are not in contact with pulleys will be referred to as *unsupported lengths.* Pulley placement, center distances, and belt length are discussed in the next subsection.

Notice that the center distance can be made quite large without using large pulleys, although care must be taken to avoid long, unsupported lengths of belt, which can sag, vibrate, or flap.

Comparing Eq. (11.44) with Eq. (11.6) shows that two meshed gears could be replaced by two timing pulleys and a timing belt to give a prescribed velocity ratio. However, the difference in signs in these two equations indicates that the angular velocity direction reversal produced by the gears in external mesh would not occur with the timing-belt system. If this reversal were required, a double-sided belt such as shown in Fig. 11.16d and an additional pulley would be required, as indicated in Fig. 11.17c and d.

It is generally recommended that each pulley that experiences appreciable torque must have at least six teeth meshed with the belt. This means that unless a pulley possesses a large number of teeth, the belt must be wrapped around it for a large angle, called the *wrap angle*. As seen in Fig. 11.17c, it can be difficult to accomplish this when trying to provide angular velocity reversal with a small number of pulleys. An additional pulley, as indicated in Fig. 11.17d, may be required.

The use of timing-belt systems also can be extended to providing analogies of planetary gear trains. For example, the motion indicated in Fig. 11.6c can be provided by a *timing-belt planetary system* such as shown schematically in Fig. 11.18, where pulley 1 is attached to

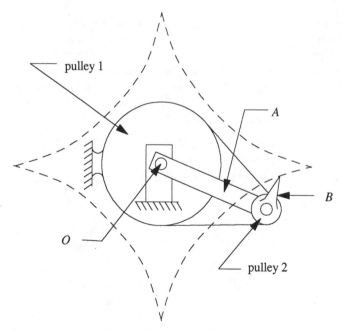

Figure 11.18 Use of a planetary timing-belt system for tracing a prescribed path.

ground and does not rotate. As arm A rotates, the pointed tip of arm B that is attached to pulley 2 follows the dashed four-cornered hypocycloidal path shown. The ratio of the number of teeth on pulley 1 to the number of teeth on pulley 2 is $4:1$. The ratio of the length of arm A to the length of arm B is $3:1$. A system such as this, with a vacuum chuck attached to the tip of arm B, has been used very effectively in a high-speed pick-and-place mechanism in a film assembly machine. In this application, arms A and B and pulley 2 and the vacuum chuck were duplicated at 90° intervals around the pivot O so that the total assembly picked and placed a part for each 90° of rotation of the shaft at point O.

Of course, the action depicted in Fig. 11.6a and b also could be provided by a timing-belt system similar to that in Fig. 11.18 but with a pulley tooth ratio of $2:1$ and with the length of arm A being equal to the length of arm B. Compound pulleys (i.e., two pulleys keyed to the same shaft) also can be used as planets, and a variety of configurations, including differentials, can be constructed.

The synthesis and analysis of planetary timing-belt systems can be performed in the same manner as was used for planetary gear trains using Procedure 11.3 and Eq. (11.17).

Computation of center distances and belt lengths

The ratios between the angular velocities of the pulleys in a timing-belt system depend only on the ratios of the numbers of teeth on the pulleys in the manner discussed above. The required length of the belt and thus the number of teeth on that belt will depend on the number of teeth on the pulleys and on how far apart and in what directions the pulleys are placed. If the system contains just two $equal$ pulleys, as shown in Fig. 11.19a, the required length of the belt is given by

$$L_B = 2C + \pi D_P \qquad (11.45)$$

Substituting Eqs. (11.42) and (11.43) into Eq. (11.45) and dividing by the pitch P gives

$$N_B = \frac{2C}{P} + N_P \qquad (11.46)$$

where L_B = length of the required belt
C = distance between the centers of the two pulleys
D_P = pitch diameter of the two identical pulleys
N_B = number of teeth on the required belt
P = pitch of the pulleys and of the belt
N_P = number of teeth on each of the two identical pulleys

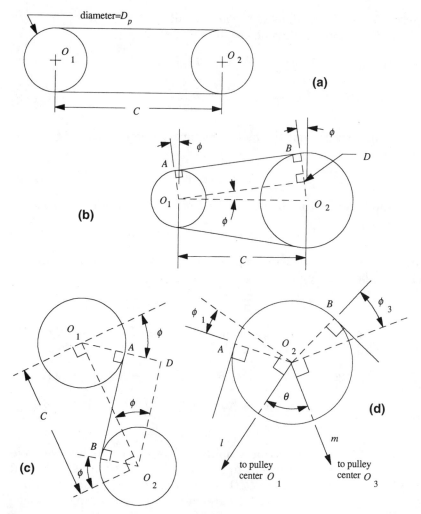

Figure 11.19 Timing-belt and pulley geometric relationships used in computing belt length.

The slightly more complicated configuration in which the pulley pitch radii are not equal is indicated in Fig. 11.19b. In this figure the lines O_1A and O_2B have been drawn from the centers of pulleys 1 and 2 to points A and B at which the belt is tangent to those respective pulleys. The line O_1D is drawn from O_1 perpendicular to line O_2B, ending at point D on line O_2B, thereby forming rectangle O_1DBA. It is seen that the angle O_2O_1D can be computed from

$$\sin \phi = \frac{R_{P_2} - R_{P_1}}{C} \tag{11.47}$$

where R_{P_2} = pitch radius of the larger pulley
R_{P_1} = pitch radius of the smaller pulley

Substituting Eq. (11.43) into Eq. (11.47) gives

$$\sin \phi = \frac{(N_2 - N_1)P}{2\pi C} \qquad (11.48)$$

It also can be seen that the belt's unsupported length $L_u = AB = O_1D$ is given by

$$L_u = C \cos \phi \qquad (11.49)$$

The wrap length W_2 of the belt around pulley 2 is seen to be given by

$$W_2 = \pi R_{P_2} + 2R_{P_2}\phi = R_{P_2}(\pi + 2\phi) \qquad (11.50)$$

where ϕ is in radians. The wrap length W_1 of the belt around pulley 1 is seen to be given by

$$W_1 = \pi R_{P_1} - 2R_{P_1}\phi = R_{P_1}(\pi - 2\phi) \qquad (11.51)$$

The total required length L_B of the belt is then $L_B = 2AB + W_1 + W_2$ or, using Eqs. (11.49), (11.50), and (11.51),

$$L_B = 2C \cos \phi + \pi(R_{P_2} + R_{P_1}) + 2\phi(R_{P_2} - R_{P_1}) \qquad (11.52a)$$

or $$N_B = \frac{2C \cos \phi}{P} + \frac{N_{P_2} + N_{P_1}}{2} + \frac{(N_{P_2} - N_{P_1})\phi}{\pi} \qquad (11.52b)$$

Figure 11.19c depicts part of a timing-belt system that involves contraflexure. The belt is tangent to pulley 1 at point A and it is tangent to pulley 2 at point B. Line O_2D is constructed perpendicular to line O_1A, ending at point D on O_1A extended, thereby creating rectangle DO_2BA. It is now seen that

$$\sin \phi = \frac{(R_{P_2} + R_{P_1})}{C} = \frac{(N_2 + N_1)P}{2\pi C} \qquad (11.53)$$

and the unsupported belt length $L_u = AB = DO_2$ is given by

$$L_u = C \cos \phi$$

Comparing this last expression with Eq. (11.49) shows that the unsupported length of belt between *any* two pulleys can be written as

$$L_{uj} = C_j \cos \phi_j \qquad \text{or} \qquad N_{uj} = \frac{(C_j \cos \phi_j)}{P} \qquad (11.54)$$

where L_{uj} = the jth unsupported length of belt in the system (i.e.,
             between the jth pair of pulleys)
       N_{uj} = number of teeth in the jth unsupported length of belt
             in the system (i.e., between the jth pair of pulleys)
       C_j = distance between the centers of the jth pair of pulleys
       ϕ_j = angle computed from Eq. (11.47) or (11.48) if there is *no*
             contraflexure between the jth pair of pulleys or the
             angle computed from Eq. (11.53) if there *is* contraflex-
             ure between the jth pair of pulleys
       P = pitch of the teeth

Figure 11.19d depicts a pulley with a length of belt wrapped around
it from tangent point A to tangent point B. Line l is the line connecting the center O_2 of pulley 2 to the center O_1 of pulley 1 (which is not
shown), and line m is the line connecting the center O_2 of pulley 2 to
the center O_3 of pulley 3 (which is not shown). The angle θ is the angle
in radians smaller than π between lines l and m. The angles ϕ_1 and ϕ_3
are angles in radians that are calculated from Eqs. (11.47), (11.48),
or (11.53) depending on the relationships between the pulleys 1, 2,
and 3. (These relationships will be elaborated on in the following discussion.)

The length of belt that is wrapped around this pulley is seen to be
the length of the arc from A to B, and this length *in this particular figure* is given by

$$L_W = (\pi - \theta - \phi_3 + \phi_1)R_2 \tag{11.55}$$

Notice that the value ϕ_3 is preceded by a negative sign, whereas ϕ_1 is
preceded by a positive sign. By comparing the effects that ϕ has on the
amount of wrap on pulleys 1 and 2 in Fig. 11.19b as compared with the
wrap on the pulleys in Fig. 11.19a, it is seen that ϕ in Fig. 11.19b contributes to (i.e., increases) the wrap on the larger pulley but that ϕ
detracts from the wrap on the smaller pulley. Then we can rewrite
Eq. (11.55) in general form for the ith pulley in a system involving several pulleys as

$$L_{W_i} = [\pi - \theta_i + s_{(i-1)}\phi_{(i-1)} + s_{(i+1)}\phi_{(i+1)}]R_i \tag{11.56}$$

where L_{W_i} = length of belt wrap around the ith pulley
       θ_i = angle in radians (and smaller than π) between the
             lines connecting the center of the ith pulley with the
             centers of the $(i-1)$th and $(i+1)$th pulleys in the
             system

$\phi_{(i-1)}$ = angle in radians computed from the equivalent of Eq. (11.47) or (11.48) if there is no contraflexure between the ith pulley and the $(i-1)$th pulley or computed from the equivalent of Eq. (11.53) if there *is* contraflexure between the ith pulley and the $(i-1)$th pulley

$\phi_{(i+1)}$ = angle in radians computed from the equivalent of Eq. (11.47) or (11.48) if there is no contraflexure between the ith pulley and the $(i+1)$th pulley or computed from the equivalent of Eq. (11.53) if there *is* contraflexure between the ith pulley and the $(i+1)$th pulley

$s_{(i-1)}$ = +1 if there *is* contraflexure between the ith pulley and the $(i-1)$th pulley or if the $(i-1)$th pulley is smaller than the ith pulley; otherwise, $s_{(i-1)} = -1$

$s_{(i+1)}$ = +1 if there *is* contraflexure between the ith pulley and the $(i+1)$th pulley or if the $(i+1)$th pulley is smaller than the ith pulley; otherwise, $s_{(i+1)} = -1$

Substituting Eqs. (11.42) and (11.43) into Eq. (11.56) and dividing by P, we obtain

$$N_{W_i} = \frac{(\pi - \theta_i + s_{i-1} + s_{i+1}\phi_{i+1})N_i}{2\pi} \qquad (11.57)$$

where N_{W_i} is the number of teeth in the belt that are in mesh with (i.e., wrapped around) the ith pulley and N_i is the number of teeth on the ith pulley.

Procedure 11.5: Kinematic Design of a Timing-Belt System   The design of a system consisting of a timing belt and pulleys will start with a knowledge of prescribed values for the angular velocities of the pulleys (or of the relationships between the torques that must be transmitted from and to those pulleys) and a knowledge of the locations of the shafts on which the pulleys must be mounted. The first step in the kinematic design consists of choosing the pitch of the teeth for the belt and pulleys. The pitch chosen will depend on the torque and power that are to be transmitted from one pulley to another. Higher torques and powers require larger pitches. Data for use in making the proper choice of pitch can be obtained from vendors' catalogs and handbooks. Once the pitch has been chosen, the pulleys can be sized.

1. Choose a number of teeth for the smallest pulley. This will be the pulley that has the largest prescribed angular velocity or which experiences the smallest prescribed torque. The minimum value that this may have depends on the pitch and the belt flexibility, and a recommended value is usually given by the belt manufacturer. This value usually is in the range from 10 to 20 teeth.

2. Choose a number of teeth for each of the other pulleys in the system. The ratio of the number of teeth on a given pulley to the number of teeth on the

smallest pulley is equal to the ratio of the torques that will be experienced by the pulleys. It is also equal to the *reciprocal* of the ratio of the angular velocities that the two pulleys will experience. Some of these calculations will result in noninteger values for the number of teeth. Because the number of teeth on any pulley must be an integer, step 1 may have to be repeated by choosing a different number of teeth for the smallest pulley. If iterating steps 1 and 2 does not result in an acceptable set of teeth numbers, it may be necessary to review the design requirements.

3. *If* there are only two pulleys in the system, compute the belt length L_B and/or number of teeth on the belt N_B using Eq. (11.47) or (11.48) and Eq. (11.52a) or (11.52b). If the total number of teeth computed is not an integer, the center distance will have to be changed so that the total number of teeth is an integer. If the center distance cannot be changed, an idler pulley must be added. Many belt vendors recommend that the location of the center of at least one pulley in a system be adjustable to facilitate installation and replacement of the belt and to allow adjustment of belt tension. The belt tension preload need not be high but must be sufficient to keep the belt teeth meshed with the pulley teeth and sufficient to prevent the tension in any unsupported length of the belt from dropping to zero. Check catalogs for availability of a suitable belt. Iterate steps 1, 2, and 3 if necessary to satisfactorily complete the synthesis.

4. If there are more than two pulleys in the system, number the pulleys in the system. These numbers will serve as values for the subscripts i for quantities in Eqs. (11.47), (11.48), (11.53), (11.56), and (11.57).

5. Number the unsupported lengths of belt in the system. These numbers will serve as values for the subscripts j for quantities in Eq. (11.54).

6. For each unsupported length of belt, compute the value of ϕ from Eq. (11.47) or (11.48) if there is *no* contraflexure in that length or from Eq. (11.53) if there *is* contraflexure in that length.

7. Compute the length or number of teeth for each unsupported length of belt using Eq. (11.54) (using appropriate j subscript values from step 5).

8. For each pulley, compute the length or number of teeth in the wrap around that pulley using Eq. (11.56) or (11.57) (using appropriate i subscript values from step 4).

9. The total length of the belt or the total number of teeth on the belt is the sum of all lengths or tooth numbers, respectively, as computed in steps 6 and 7. If the total number of teeth computed is not an integer, one or more of the center distances will have to be changed so that the total number of teeth is an integer. If none of the center distances can be changed, an idler pulley must be added. Many belt vendors recommend that the location of the center of at least one pulley in a system be adjustable to facilitate installation and replacement of the belt and to allow adjustment of belt tension. The belt tension preload need not be high but must be sufficient to keep the belt teeth meshed with the pulley teeth and sufficient to prevent the tension in any unsupported length of the belt from dropping to zero. Check catalogs for availability of a suitable belt. Iterate steps 1, 2, and 6 through 9 if necessary to satisfactorily complete the synthesis.

12

Forces, Torques, and Static Balancing

12.1 Introduction

This chapter presents procedures for finding the static forces in mechanisms and procedures for static balancing of mechanisms. *Static balancing* consists of providing a distribution of mass in the mechanism such that if the mechanism is placed at rest in any position, the effects of gravity will not disturb it from that position, even in the absence of other external forces or torques (other than those support forces supplied by "ground"). That is, a statically balanced mechanism will be in *static equilibrium* in all positions that are permitted by the connections between its masses. Static balancing is used on mechanisms such as articulated desk lamps, lids and covers, and suspended tools to allow them to remain in set positions with the aid of only a little friction in the joints. It is also used on cranes and drawbridges and other heavy mechanisms to minimize or eliminate the effort otherwise required to lift and lower the heavy mechanism masses involved. This chapter also includes procedures for using springs as alternatives to masses for static balancing.

A statically balanced mechanism is *not* necessarily dynamically balanced. That is, it may still produce large vibratory torques when it is caused to move. Dynamic forces and dynamic balancing are covered in Chap. 13.

12.2 A Brief Review of Some Statics

Forces and couples

Consider the rigid body L that is pivoted to ground at point O in Fig. 12.1a. A force $\mathbf{F}$ is applied to that body at point P. Obviously, this

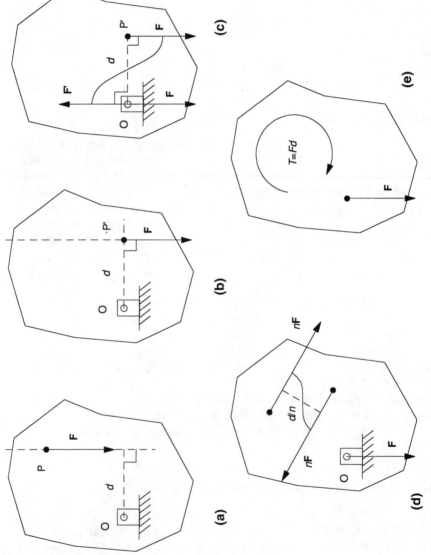

Figure 12.1 Force vectors and couples.

force will tend to rotate the body clockwise about the pivot at O. A line through point P and parallel to force $\mathbf{F}$ is known as the *line of action* of that force. Such a line of action is shown for force $\mathbf{F}$ as the (vertical) dashed line in Fig. 12.1a.

It is shown in the study of statics that a force can be applied at any point along its line of action without changing its effect on the body to which it is applied. In Fig. 12.1b the same force has been applied at point P' that is on the line of action of the original force and is at the foot of a perpendicular from point O to the line of action. It is easily seen in the figure that the torque applied about the pivot at point O and therefore tending to rotate the body is $T = Fd$.

In Fig. 12.1c a force $\mathbf{F}$ equal to and parallel to the original $\mathbf{F}$ has been added at point O and a force $\mathbf{F}'$ that is equal to and in the opposite direction to the original $\mathbf{F}$ also has been applied at point O. These two new forces obviously cancel, so the effect on the body has not changed. The force $\mathbf{F}$ at point O constitutes a load on the pivot bearing at O. The pair of forces $\mathbf{F}'$ at O and $\mathbf{F}$ at P constitutes what is known as a *couple*. A couple applies no net force; it only applies a torque tending to rotate the body. The couple therefore could consist of any pair of equal and opposite forces of magnitude nF whose lines of action were separated by a suitable distance d/n such that the product Fd is preserved. Because the product Fd and the parallelism of the forces (and the direction of the torque) are the only characteristics of importance, the couple could be oriented and located arbitrarily, as shown in Fig. 12.1d, and it could be denoted simply, as shown in Fig. 12.1e.

Note that in Fig. 12.1e the point O is not indicated to be a pivot. It is merely considered to be an arbitrary reference point. Thus we may conclude that

> The effect of any force applied at a given point is equivalent to an equal and parallel force applied at any other point plus an appropriate couple. The magnitude of that couple is equal to the magnitude of the original force multiplied by the perpendicular distance from the second point to the line of action of the original force. The sense of the couple is such as to produce a torque in the same direction as that which would be produced around that second point by the original force.

Static equilibrium

Newton's laws tell us that if a body is not acted on by any net force or torque (couple), its velocity will remain constant, and if it is at rest, it will remain at rest. Such a condition is known as a condition of *static equilibrium*.

Consider the body L that is acted on by two equal and opposite forces $\mathbf{F}$ at points P and Q in Fig. 12.2a. As drawn, the lines of action of these forces are coincident. Therefore, the force at point Q could be moved along the line of action to point P, where it is shown as a dashed vector, without changing its effect. It is seen that the solid vector at P and the dashed vector at P cancel each other, so there is no net effect on the body as the result of applying the original forces. The body is in static equilibrium.

If one of the forces in Fig. 12.2a were displaced away from the line of action of the other force, it would be equivalent to a force on the line of action of the other force plus a couple. The net effect of this displaced force plus the undisplaced force would then be no residual force,

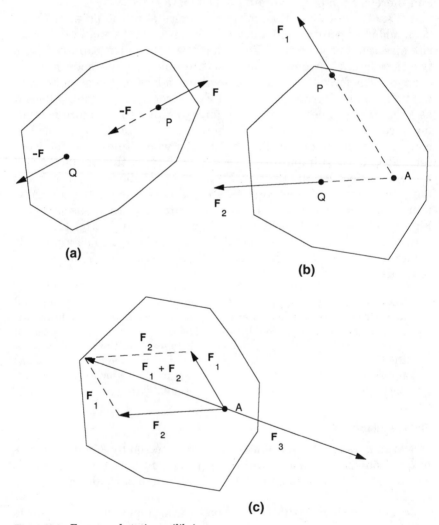

Figure 12.2   Forces and static equilibrium.

but there would be a residual couple. The body would *not* be in static equilibrium.

In Fig. 12.2*b* a body *L* is acted on by a force $\mathbf{F}_1$ applied at point *P* and a force $\mathbf{F}_2$ applied at point *Q*. The lines of action of these forces intersect at point *A*. If each of these forces is moved along its line of action until it is applied at point *A*, it is seen that the net effect of these two forces is the vector sum $\mathbf{F}_1 + \mathbf{F}_2$ of these two force vectors, as shown in Fig. 12.2*c*. This vector sum has a line of action through point *A* and is not in general (as shown) equal to zero, so the body would not in general be in static equilibrium.

From the foregoing discussion, then, we may conclude that

> A body that is acted on by two forces will be in static equilibrium if and only if the two forces are equal in magnitude and opposite in direction and their lines of action are coincident.

The body in Fig. 12.2*c* could be placed in static equilibrium by adding a force $\mathbf{F}_3$ equal and opposite to $\mathbf{F}_1 + \mathbf{F}_2$ and with a line of action through point *A* as shown. Thus the force vectors $\mathbf{F}_1 + \mathbf{F}_2$ and $\mathbf{F}_3$ constitute two equal and opposite forces with coincident lines of action, which, as shown above, produce static equilibrium. The vector triangles in Fig. 12.2*c* show that $\mathbf{F}_1 + \mathbf{F}_2 + \mathbf{F}_3 = 0$. That is, drawing the three vectors $\mathbf{F}_1$, $\mathbf{F}_2$, and $\mathbf{F}_3$ tail to head in succession produces a closed triangle. If the line of action of $\mathbf{F}_3$ were not to pass through point *A*, a residual couple would result. Thus it is seen that

> A body that is acted on by three forces will be in static equilibrium if and only if the lines of action of the three forces intersect in a single point and the vector sum of the three forces is zero.

12.3 Static Balancing of Rotors

The motion of a rigid body that is pivoted to ground consists of only pure rotation about that pivot, and such a body will be referred to as a *rotor*. The rotor indicated in Fig. 12.3*a* consists of two masses m_1 and m_2 that are connected together to form a single rigid body that is pivoted to ground at point *O*. Mass m_1 has a mass of m_1 and its center of mass or center of gravity[1] is located relative to the pivot as indicated by the vector $\mathbf{R}_1$. Mass m_2 has a mass of m_2 and its center of mass is located relative to the pivot as indicated by the vector $\mathbf{R}_2$. These two masses will be acted on by vertical gravity forces m_1g and m_2g, respectively, which can be considered to be acting at their centers of mass as indicated.

---

[1]For the systems discussed in this book, the center of mass of a body can be considered to be identical to its center of gravity, which will be abbreviated as c.g.

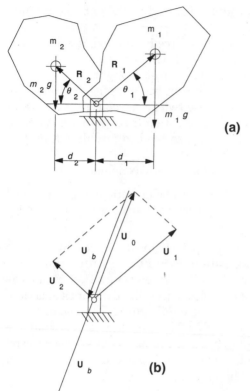

(a)

(b)

Figure 12.3 Static imbalance of a rotor and static balancing of that rotor.

It can be seen that gravity acting on mass m_1 will produce a clockwise torque of $T_1 = m_1 g R_1 \cos \theta_1$, and on mass m_2 it will produce a counterclockwise torque of $T_2 = -m_2 g R_2 \cos \theta_2$. Both these torques are functions of the angles θ_1 and θ_2, which, of course, vary as the rotor is rotated. At some position of the rotor, the values of these angles will be such that the torques cancel. However, in general, they will not cancel, and gravity will tend to rotate the rotor from its position. That is, there will be a static imbalance.

In dealing with static imbalance and attempts to eliminate it (i.e., attempts to provide static balance), it is useful to define an imbalance vector **U** (for unbalance). This vector has the same direction as the vector describing the position of the individual center of mass relative to the point about which it is desired to provide static balance, and has a magnitude equal to the individual mass times the length of that position vector. Then, for mass m_1 in Fig. 12.3a, $\mathbf{U}_1 = m_1 \mathbf{R}_1$, and for mass m_2, $\mathbf{U}_2 = m_2 \mathbf{R}_2$. These vectors and their vector sum $\mathbf{U}_o$ are drawn in

Fig. 12.3b. It will be noted in this figure that the horizontal component of $\mathbf{U}_o$ is equal to the vector sum of the horizontal components of $\mathbf{U}_1$ and $\mathbf{U}_2$. The magnitudes of these horizontal components of $\mathbf{U}_1$ and $\mathbf{U}_2$, if each were multiplied by g, would have magnitudes of T_1 and T_2, respectively, as can be seen from the expressions in the preceding paragraph. Then the horizontal component of $\mathbf{U}_o$, when multiplied by g, is the total imbalance torque, and $\mathbf{U}_o$ is the total imbalance vector.

Note also that if the total imbalance vector $\mathbf{U}_o$ is divided by the total mass $m_1 + m_2$, the result is the position vector $\mathbf{R}_o$, which is a vector from the pivot point O to the center of gravity of the total rotor. This can be seen because $\mathbf{U}_o = (m_1 + m_2)\mathbf{R}_o$.

Obviously, the total imbalance torque of any rotor will differ from zero for all but two angular positions of the rotor unless the total imbalance vector is zero (or unless the pivot or shaft axis is vertical). Thus, for static balance, the total imbalance vector of the untreated rotor must be compensated for by a balancing vector $\mathbf{U}_b$ such that $\mathbf{U}_o + \mathbf{U}_b = 0$, where $\mathbf{U}_o$ is the imbalance vector for the untreated rotor. Then, if there are n masses, where each mass m_i is at a position vector of $\mathbf{R}_i$ from the pivot, the total imbalance vector $\mathbf{U}_o$ is given by

$$\mathbf{U}_o = \sum_i m_i \mathbf{R}_i$$

Thus, for complete static balance,

$$\mathbf{U}_b + \mathbf{U}_o = \mathbf{U}_b + \sum_i m_i \mathbf{R}_i = 0 \qquad (12.1)$$

The total imbalance vector of the balanced rotor is

$$\mathbf{U}_t = \mathbf{U}_b + \mathbf{U}_o = 0$$

But $\mathbf{U}_t = m_t \mathbf{R}_t$, where m_t is the total mass of the balanced rotor and $\mathbf{R}_t$ is the vector from the pivot to the center of gravity of the balanced rotor. Then, because $\mathbf{U}_t$ is zero but m_t is not zero, $\mathbf{R}_t$ is zero and the total center of gravity lies at the pivot. Therefore, no matter how the balanced rotor rotates, its center of gravity will not move.

Procedure 12.1: Graphical Static Balancing of a Rotor The procedure consists of finding a balancing vector $\mathbf{U}_b = m_b \mathbf{R}_b$ such that when it is added to the sum of all the imbalance vectors of the rotor produces a zero vector sum. Some computer-aided design (CAD) systems have search or optimization routines that will do this automatically. If such a system is not available, the computations implied by Eq. (12.1) can be performed graphically (manually or on CAD) or the calculations may be performed manually in terms of the x and y components of the vectors. The following procedure is based on a graphical approach.

1. Find the weight of each of the separate parts of the rotor. Many CAD systems can calculate this automatically. (Using the weights instead of the masses will not affect the results because each of the weights is simply mass times the factor g, and since this factor appears in all terms, its effects cancel out.)

2. Find the center of mass (center of gravity) of each of the separate parts of the rotor. Many CAD systems can calculate this automatically. Centers of volume of many simple shapes can be found in handbooks. For a uniform-density part, the center of the volume is the center of gravity.

3. On paper or on a CAD system, for each separate part, draw its imbalance vector. This vector will be a vector from an origin that represents the rotor pivot, it will be in the direction from that pivot to the center of gravity of the part, and it will have a magnitude equal (at some convenient scale) to the distance from the pivot to the part's center of gravity times the part's mass. Figure 12.4a shows an example rotor with the centers of gravity of three separate parts indicated. The mass m_1 is large, so for the scale used the imbalance vector U_1 is long and extends beyond the center of gravity of the mass m_1. The mass m_3 is small, so at the chosen scale its imbalance vector U_3 is short and does not even reach the center of gravity. On a CAD system this can take the form of drawing a line from the pivot to the center of gravity of the part and then scaling its length until it, at some convenient scale, represents the imbalance vector for that part.

4. Choose one of the imbalance vectors (preferably the longest) and successively move each of the other vectors until they are all connected tail to tip, starting at the tip of the chosen vector. Figure 12.4b shows such a construction corresponding to the vectors in Fig. 12.4a. For clarity, readers may wish to draw the diagram corresponding to Fig. 12.4b on a transparency over the original drawing or on a separate CAD layer.

5. From the tip of the last vector connected to step 4, draw a vector to the pivot, thereby closing the vector polygon, as shown by vector U_b in Fig. 12.4. This is the balancing vector U_b that satisfies Eq. (12.1).

6. Move the vector drawn in step 5 until its tail is at the pivot. Measure its length. This length, when divided by the vector scale factor used, is equal to the balance *weight* multiplied by the distance of that weight from the pivot. The direction of the location of that balance weight is, of course, in the direction of the vector. Note that the important quantity is the product of weight times distance. Therefore, a small weight at a large distance has the same effect as a large weight at a small distance. In any event, the balance weight m_b will be located along the balance vector U_b, as shown in Fig. 12.4c, or along an extension of that vector.

Comments and suggestions

The foregoing assumed that a balancing mass would be added to the rotor. In some cases, balancing can be better achieved by removing mass such as by drilling a hole or cutting away material. Such a modification of the rotor can be considered to be the adding of a "negative" mass. When the position vector of such a negative mass is multiplied by a negative mass value, it results in a balancing vector U_b that is in

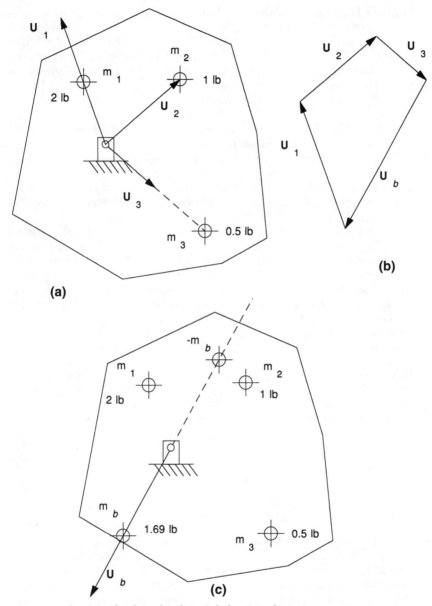

Figure 12.4 An example of graphical static balancing of a rotor.

the opposite direction from the position vector $\mathbf{R}_b$ from the pivot to the removed mass. Thus, as shown in Fig. 12.4c, balancing could have been achieved by removing mass (indicated as $-m_b$) with a center of gravity along the dashed line, and the balancing vector $\mathbf{U}_b$ would have been unchanged from the value shown.

Example 12.1: Static Balancing a Rotor   Figure 12.4a schematically depicts a rotor that is pivoted to ground and which is to be statically balanced. It consists of three masses (m_1, m_2, and m_3), whose centers of gravity are located as shown. Mass m_1 weighs 2 lb and its center of gravity is 1 in from the pivot, mass m_2 weighs 1 lb and its center of gravity is 1.5 in from the pivot, and mass m_3 weighs 0.5 lb and its center of gravity is 2 in from the pivot. The imbalance vectors for these masses are therefore $\mathbf{U}_1 = (1 \text{ in})(2 \text{ lb}) = 2$ in · lb, $\mathbf{U}_2 = (1.5 \text{ in})(1 \text{ lb}) = 1.5$ in · lb, and $\mathbf{U}_3 = (2 \text{ in})(0.5 \text{ lb}) = 1$ in · lb, respectively. These vectors are drawn in Fig. 12.4a to a scale of 1 in · lb per inch.

Then, in accordance with Eq. (12.1), these imbalance vectors are added as shown in Fig. 12.4b, where they are drawn tail to head starting with $\mathbf{U}_1$. Because Eq. (12.1) requires the sum of these vectors plus $\mathbf{U}_b$ to equal zero, the vector that closes the vector polygon is, as shown, the balancing vector $\mathbf{U}_b$.

The vector $\mathbf{U}_b$ is drawn in Fig. 12.4c, which is a redrawing of the original rotor. This vector indicates the direction from the pivot to the location of the required balance mass m_b. The measured length of $\mathbf{U}_b$ as drawn in Fig. 12.4b and Fig. 12.4c is 2.54 in, which at the chosen drawing scale of 1 in · lb per inch gives $U_b = 2.54$ in · lb. A distance of 1.5 in from the pivot was chosen as a convenient location for the balance mass, and that location is shown in Fig. 12.4c. At that distance from the pivot, the balance mass must weigh 1.69 lb because $U_b = r_b w_b = (1.5 \text{ in})(1.69 \text{ lb}) = 2.54$ in · lb, which is the magnitude of the required balance vector.

Procedure 12.2: Balancing of a Rotor by Calculation   The static balancing of a rotor can be performed very easily by calculation as well as by the foregoing graphical method. To do this, solve Eq. (12.1) for $\mathbf{U}_b$ and resolve the vectors into x and y components to give

$$U_{bx} = U_b \cos \theta_b = -\sum_i U_{ix} = -\sum_i U_i \cos \theta_i$$

$$U_{by} = U_b \sin \theta_b = -\sum_i U_{iy} = -\sum_i U_i \sin \theta_i$$

(12.1a)

To see how these equations are used in calculation, consider the rotor in Example 12.1. The magnitudes of the imbalance vectors for the three masses were computed as $U_1 = 2.0$ in · lb, $U_2 = 1.5$ in · lb, and $U_1 = 1.0$ in · lb. If an x axis is assumed to extend horizontally to the right in Fig. 12.4a and a y axis is assumed to extend vertically upward, the angles of the position vectors and thus of the imbalance vectors for the three masses can be measured as $\theta_1 = 110°$, $\theta_2 = 40°$, and $\theta_3 = -40°$.

Then we may calculate

$$U_{1x} = U_1 \cos 110° = (2.0)(-0.342) = -0.68 \text{ in · lb}$$

$$U_{1y} = U_1 \sin 110° = (2.0)(0.940) = 1.879 \text{ in · lb}$$

In the same manner, it is calculated that

$$U_{2x} = 1.149 \text{ in} \cdot \text{lb}, \quad U_{2y} = 0.964 \text{ in} \cdot \text{lb}, \quad U_{3x} = 0.766 \text{ in} \cdot \text{lb},$$

$$\text{and} \quad U_{3y} = -0.623 \text{ in} \cdot \text{lb}$$

Equation (12.1a) then gives

$$U_{bx} = -(-0.68) - 1.149 - 0.766 = -1.235 \text{ in} \cdot \text{lb}$$

$$U_{by} = -1.879 - 0.964 - (-0.623) = -2.220 \text{ in} \cdot \text{lb}$$

These components of the balancing vector can then be combined to give the magnitude and angle of the balancing vector as

$$U_b = \sqrt{U_{bx}^2 + U_{by}^2} = 2.54 \text{ in} \cdot \text{lb} \quad \text{and} \quad \theta_b = \tan^{-1}\left(\frac{U_{by}}{U_{bx}}\right) = -119.1°$$

The results of these calculations are seen to agree with the results of the graphical construction in Procedure 12.1 as applied in Example 12.1.

12.4 Static Forces and Moments in a Mechanism

When a mechanism is in actual use, it usually will be subjected to externally applied forces and torques from gravity and/or inertial effects and/or from other bodies or mechanisms with which it must interact when performing its functions. These forces and torques will be applied to individual parts (links) of the mechanism, and equilibrium equations can be written for each of the links individually to relate these externally applied forces and torques to the loads within the mechanism. These equations may be solved simultaneously for the internal effects, as is described later in this section. However, the graphical technique is useful in allowing a visualization of the interactions between the various forces and torques and the geometry of the mechanism.

The graphical procedure described next uses the statics principles discussed in Sec. 12.2 and the principle of superposition. In this procedure, use of this principle consists of considering only the external forces and torques that are applied to a single link, while all other links are assumed to be massless and unaffected by external forces and torques. The interactions of this chosen link with other links in the mechanism can then be calculated and the resulting forces and torques calculated. These calculations provide partial answers. Then another link is chosen, and the same process is repeated for that link. When all the links have been treated in this manner, the partial answers are superimposed. That is, at each pertinent location in the linkage, the partial answers for that location are added (vectorially) to give the total answer for that location.

Procedure 12.3: Graphical Force and Torque Analysis of a Four-Bar Pin-Jointed Linkage This procedure will be described by using an analysis of the pin-jointed four-bar linkage shown in Fig. 12.5. However, the same steps and general principles can be used for other linkages.

Figure 12.5 is a schematic drawing of a mechanism for opening and closing a heavy industrial oven door. The door is shown in the closed position by dashed lines and in a half-opened position by solid lines. The door constitutes a link L_2, and it weighs 100 lb and its center of gravity is at the point indicated. It is connected to a link L_1 by a pivot at point A, and the other end of link L_1 is pivoted to the oven (ground) at point O_1. Link L_1 is 13.2 in long, it weighs 20 lb, and its center of gravity is midway along its length. The door is also connected to a link L_3 by a pivot at point B, and the other end of link L_3 is pivoted to the oven (ground) at point O_2. Link L_3 is 27 in long, it weighs 50 lb, and its center of gravity is midway along its length.

Objective: Find the bearing load at each of the pivot points, and find the torque required to be applied at the pivot point O_1 in order to hold the door at the half-open position shown.

1. Draw the subject linkage to a suitable scale. Include the locations of the points of application of the external forces, and indicate the directions of those forces. (In this example, these forces are gravity forces $\mathbf{W}_1$, $\mathbf{W}_2$, and $\mathbf{W}_3$, and they are applied vertically at the centers of gravity of the parts, as shown in Fig. 12.5a.)

Because several subsequent vector diagrams also will be drawn using the directions of the lines in this linkage drawing, it may be desirable to make *those* diagrams on transparencies that may be placed over this basic drawing. If a CAD system is used, the diagrams can be made on separate layers.

2. Choose one of the links and draw it (on a transparency or separate CAD layer) along with each of the external forces applied to this link, as a vector to a convenient scale and with its tail at the point of application. (In this example, the single force is gravitational, so it is vertically downward and is applied at the center of gravity.) Steps 2 through 7 assume that no external forces or torques are applied to any link except the chosen link. (The separate drawing for the chosen link L_2 of the example is shown in Fig. 12.5b as the line AB.)

2a. If more than one external force is applied to this link, those forces can be vectorially added in pairs, and the sum of each pair can be assumed to be applied at the intersection of their lines of action. This pairing can be repeated until only one net resulting force is found. If parallelism of forces is encountered in the process, the two parallel forces should be reduced to one force and line of action plus a couple using the principle described in Sec. 12.2.

2b. If a couple (torque) is applied externally to this link in addition to a net external force, this combination should be replaced by a force that is equal to and parallel to the net force and whose line of action is displaced from that of the net force by a distance $d = T/F$, where T is the magnitude of the couple and F is the magnitude of the net externally applied force. The direction of displacement d can be determined by *imagining* the net force vector to be rotated 90° in the direction *opposite* to the direction of the rotation indicated by the couple (where the couple is denoted in the form indicated in Fig. 12.1e). This imagined rotated force vector then points in the direction in which the line of action must be *displaced.*

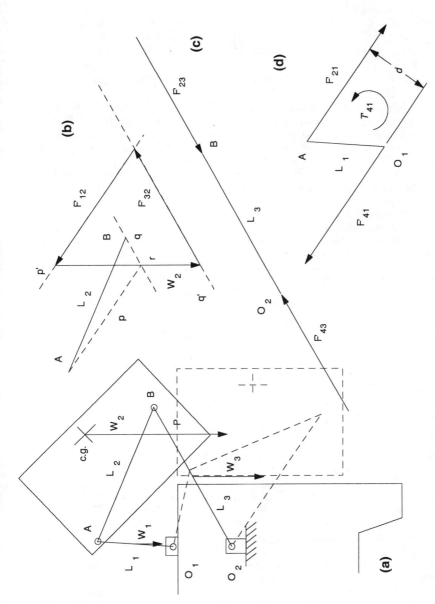

Figure 12.5 Schematic diagram of a hinged oven door linkage, and vector diagrams used for obtaining the first partial answers for graphical analysis of the forces involved.

It can be seen from steps 2, 2a, and 2b that any combination of externally applied forces and couples can be reduced to a single net applied force along some line of action or to a single net applied couple.

3. On the drawing produced in step 2, draw the line of action of the net externally applied force. This line, of course, passes through the point of application of the force. For this example it is the vector $\mathbf{W}_2$ in Fig. 12.5b, and its point of application is the center of gravity of the link, which in Fig. 12.5b is at point p'.

3a. If the link is subjected to only a single net externally applied couple, go to step 5b.

4. Consider the chosen link to be a free body. It will be acted on by the external force and by the forces at its two pivots. As pointed out in Sec. 12.2, this body will be in static equilibrium only if the three forces have lines of action that intersect in a single point and if the vector sum of the forces is zero. The pivot forces will be applied at the pivot points, and the purpose of this step is to find directions for these lines of action such that they will intersect the line of action of the applied force in the same point. In this example, the pivot points are points A and B, and if no external forces or moments are applied to links L_1 and L_3 and the pivots are frictionless, the forces at these pivots must be parallel to the links, respectively. It is seen that these three lines of action will not intersect in a single point. An external force or torque (couple) must be applied to L_1 or L_3 to cause such coincidence. The problem statement assumed that an external torque (couple) would be applied to link L_1 at point O_1 to hold the door in the position shown. No such torque or force is to be applied to link L_3, so the pivot force at point B will indeed be parallel to link L_3, as shown by the dashed line q through point B in Fig. 12.5b. This line of action intersects the line of action of the external force $\mathbf{W}_2$ at point r. Therefore, a line p through points r and A is the line of action of the pivot force that must be supplied at point A to hold link L_2 in static equilibrium.

5a. Determine the partial answer for the pivot forces on the chosen link based on the directions of their lines of action as found in step 4. Draw a vector parallel to the line of action of the net applied external force and with a magnitude representing that force to a suitable scale. In this example, this vector is shown as $\mathbf{W}_2$ in Fig. 12.5b. At the *tip* of this vector draw a line (such as q' that is parallel to line q) parallel to the line of action of one of the pivot forces from step 4. At the *tail* of the external force vector draw a line parallel to the line of action of the other pivot force from step 4. (This is line p' parallel to line p for this example.) The partial pivot forces are then drawn on these lines to close the vector triangle as shown in Fig. 12.5b for this example. These pivot forces are labeled with a single prime to indicate that they are the first partial answers for the pivot forces. The first subscript number refers to the link *applying* the force, and the second subscript refers to the link *to which* the force is being applied by the pivot.

5b. If the chosen link is subjected to only an externally applied torque (couple), the partial answers for the pivot forces will be equal forces in opposite directions. The direction of the line of action of one of these forces must be known, just as discussed in step 4. Then the magnitude of each force is equal to $F = T/d$, where F is the magnitude of each of the pivot forces, T is the magnitude of the net externally applied couple, and d is the perpendicular distance between the lines of action of the two pivot forces.

6. Determine the pivot forces on one of the other movable links by considering it as a free body. In this example link L_3 was assumed to have no externally

applied forces or torques in this iteration, so it is subjected to only the two equal opposing pivot forces shown in Fig. 12.5c. Label the forces with a prime to indicate that they are the first partial answers.

7. Determine the pivot forces on the remaining movable link by considering it as a free body. In this example link L_1 was assumed to have no externally applied forces in this iteration, but it was assumed to have a torque applied at pivot O_1 as discussed in step 4. The pivot forces are then equal and opposite, as shown in Fig. 12.5d, and the torque that must be applied at point O_1 is $T_{41} = F'_{21}d$ and is in the direction shown. Label the forces and torques with a prime to indicate that they are the first partial answers.

8. Choose a second movable link and repeat steps 2 through 7, assuming that no external forces or torques are applied to any but the chosen link. This will produce a second set of partial answers that will be labeled with double primes. In this example link L_3 was chosen for this iteration, and the results are shown in Fig. 12.6a through c.

9. Choose the third and remaining movable link and repeat steps 2 through 7, assuming that no external forces or torques are applied to any but the chosen link. This will produce a third set of partial answers that will be labeled with triple primes. In this example the third link chosen was link L_1, and because no external forces were assumed to be applied to links L_2 and L_3 for this iteration, no pivot forces were generated on those links. This step for the example is illustrated in Fig. 12.6d.

10. At each pivot and/or point of interest where partial answers have been found, vectorially add the partial answers. This will provide the total forces and moments at those points. In this example, step 10 is illustrated in Fig. 12.7.

Example 12.2: Forces and Moments in a Four-Bar Linkage   A quantitative force and moment analysis of the oven door mechanism of Fig. 12.5a is used in this example to illustrate the use of Procedure 12.3. The dimensions and weights of the mechanism are given at the beginning of Procedure 12.3, and the door is assumed to be in the position shown by the solid lines in Fig. 12.5a. The scales that will be used in the graphical constructions are 10 in per inch and 40 lb per inch. As a consequence, the door weight vector $\mathbf{W}_2$ is drawn as 2.5 in long in Fig. 12.5a and b to represent the 100-lb weight of the door, and the length of the door link L_2 is drawn 2.457 in long to represent its true length of 24.57 in.

The link L_2 is chosen as the first link to be analyzed for partial answers for the forces and torques. According to Procedure 12.3, no external forces or torques are assumed to be applied to any of the other links (except for the torque that must be applied to link L_1 at point O_1 to hold the door in the position shown). The applied force $\mathbf{W}_2$ is considered to be attached to link L_2 (line AB), and together they are copied from Fig. 12.5a to Fig. 12.5b as shown. The dashed line q is constructed through point B in Fig. 12.5b and parallel to link L_3 that passes through point B in Fig. 12.5a. This construction represents the fact that no torque (couple) is applied to link L_3, so the forces at its pivots must be aligned with the link itself.

Line q intersects the line of action of $\mathbf{W}_2$ at point r, so line Ar (labeled p in Fig. 12.5b) represents the direction of the line of action of the pivot force at point A. Line p' is drawn through one end of vector $\mathbf{W}_2$ parallel to line p, and line q' is drawn through the other end of $\mathbf{W}_2$ parallel to line p'. The force $\mathbf{F}'_{32}$, which as stated above is parallel to link L_3 and therefore is parallel to line q, is drawn as the vector $\mathbf{F}'_{32}$ from the tip of $\mathbf{W}_2$ to the intersection of lines p' and q' in Fig. 12.5b. Its length is measured to be 2.259 in, so $F'_{32} = (40 \text{ lb/in})(2.259 \text{ in}) = 90.38 \text{ lb}$.

For link L_2 to be in static equilibrium, the vector triangle must be closed by the vector $\mathbf{F}'_{12}$, which represents the force that link L_1 applies to link L_2. Thus $\mathbf{F}'_{12}$ is drawn as shown in Fig. 12.5b. As drawn, its length is 2.377 in, so $F_{12} = 95.06 \text{ lb}$.

Because link L_3 acts on link L_2 by applying force $\mathbf{F}'_{32}$ upward and to the right at pivot point B, link L_2 reacts by applying an equal and opposite force $\mathbf{F}'_{23}$ downward and to the left to link L_3 at point B. Then the free-body diagram for link L_3 appears as shown in Fig. 12.5c. This link is seen to be in pure compression.

Link L_1 acts on link L_2 by applying force $\mathbf{F}'_{12}$ upward and to the left at pivot point A, so link L_2 reacts by applying an equal and opposite force $\mathbf{F}'_{21}$ downward and to the right to link L_1 at point A. Then the free-body diagram for link L_1 is drawn 1.32 in long and oriented as shown in Fig. 12.5d. In order for link L_1 to be in static equilibrium, $\mathbf{F}'_{21}$ must be balanced by an equal and opposite force $\mathbf{F}'_{41}$ that is applied to link L_1 at point O_1 by the ground. The couple that is seen to be produced by these opposing, *noncollinear* forces must be opposed by a counterclockwise torque T'_{41}, as indicated in the figure. The magnitude of this torque is $T'_{41} = F'_{41}d = F'_{21}d$. The distance d is measured from the diagram to be 1.138 in, which, according to the scale used, represents an actual d of 11.38 in. Therefore, $T'_{41} = F'_{21}d = (95.06 \text{ lb})(11.38 \text{ in}) = 1082 \text{ in} \cdot \text{lb}$ of torque (counterclockwise).

This completes the computations for the first chosen link L_2. Next, link L_3 is chosen, and all other links are assumed to have no applied external loads (except for the required door-support torque on link L_1 at point O_1).

The free-body diagram for link L_3 is given in Fig. 12.6a. Link L_3 is drawn 2.70 in long as line O_2B, as shown, and its weight (50 lb) vector $\mathbf{W}_3$ is drawn vertically downward from its center and with a length of 1.25 in. The line of action of $\mathbf{W}_3$ is shown as the vertical dashed line. Because link L_2 is assumed to have no externally applied loads, it must be in pure tension or compression, so a dashed line is drawn through point B and parallel to link L_2 to intersect the line of action of $\mathbf{W}_2$. A line from point O_2 to this intersection determines the direction of the force $\mathbf{F}''_{43}$ that the ground must apply to link L_3 at point O_2. A line parallel to this line is also drawn through the tip of vector $\mathbf{W}_3$. A line q' parallel to the dashed line through point B is drawn through the tail of $\mathbf{W}_3$. The vectors $\mathbf{F}''_{43}$ and $\mathbf{F}''_{23}$ are then drawn on these latter dashed lines to complete the closed vector triangle shown.

The vector $\mathbf{F}''_{43}$ is measured to be 1.147 in long, so $F_{43} = (1.147 \text{ in})(40 \text{ lb/in}) = 45.9 \text{ lb}$. The vector $\mathbf{F}''_{23}$ is measured to be 0.669 in long, so $F_{23} = (0.669 \text{ in})(40 \text{ lb/in}) = 26.8 \text{ lb}$.

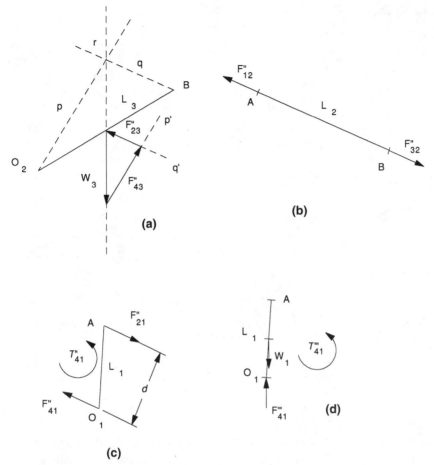

Figure 12.6 Vector diagrams used for obtaining the second and third partial answers in the graphical analysis of the forces involved in the oven door example.

The free-body diagram for link L_2 is shown in Fig. 12.6*b*, where it is seen that the link is in pure tension because $\mathbf{F}_{32}'' = -\mathbf{F}_{12}''$. Then, knowing $\mathbf{F}_{12}''$ from Fig. 12.6*b*, the free-body diagram for link L_1 can be drawn as Fig. 12.6*c*. Figure 12.6*c* is treated the same as was Fig. 12.5*d*, so $T_{41}'' =$ (26.8 lb)(13.07 in) = 349.7 in · lb counterclockwise.

The third and last link that is chosen for analysis to give a partial answer is link L_1. The free-body diagram for this link is shown in Fig. 12.6*d*. Because no external loads are assumed to be applied to any link other than L_1, the only forces acting on L_1 are its weight $\mathbf{W}_1$ of 20 lb and the equal and opposite support force $\mathbf{F}'''$ at O_1. These two noncolinear forces produce a couple that must be opposed by a torque of magnitude $T''' =$ (20 lb)(13.2/2 in)(cos 86.2°) = 8.8 in · lb.

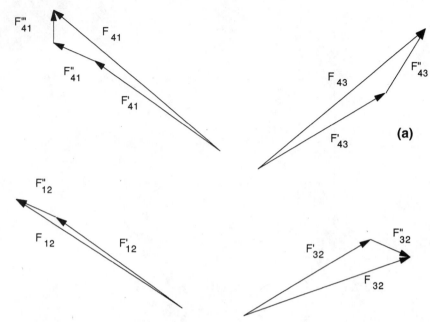

Figure 12.7 Vector diagrams used for obtaining the total answers in the graphical analysis of the forces involved in the oven door example.

The total torque that must be applied at pivot O_1 to hold the door in the half-open position shown in Fig. 12.5a is given by

$$T_{41} = T'_{41} + T''_{41} + T'''_{41} = 1082 + 349.7 + 8.8 = 1440 \text{ in} \cdot \text{lb}$$

The forces at the pivots are $\mathbf{F}_{41} = \mathbf{F}'_{41} + \mathbf{F}''_{41} + \mathbf{F}'''_{41}$ at O_1, $\mathbf{F}_{43} = \mathbf{F}'_{43} + \mathbf{F}''_{43}$ at O_2, $\mathbf{F}_{12} = \mathbf{F}'_{12} + \mathbf{F}''_{12}$ at A, and $\mathbf{F}_{32} = \mathbf{F}'_{32} + \mathbf{F}''_{32}$ at B. These vector additions are shown in Fig. 12.7.

Simultaneous solution for the entire mechanism

The superposition process used in Procedure 12.3 required that all the joint forces and torques be evaluated once *for each link* to which an external force and/or torque is applied. In Example 12.2 this involved evaluating these joint forces and torques three times. It is possible to solve analytically for all these forces and torques simultaneously. This analytical procedure can be useful for checking the accuracy of results of other procedures. However, because for use on even a four-bar linkage it involves the solution of nine simultaneous equations, it requires the use of a computer or calculator with simultaneous equation-solving ability. If convenient means are available for solving these equations repeatedly for successive sets of input values, it can be much more practical than graphical procedures for evaluating forces and torques if such evaluation is required at many positions of a linkage.

To see how the equations for use in the analytical force and torque analysis are derived, consider the individual link shown in Fig. 12.8. This figure is a free-body diagram of the ith link in a linkage, as indicated by the subscripts i. It is connected to the jth link by a pin joint at point A, and a joint force $\mathbf{F}_{ji}$ is applied to this ith link by the jth link at that joint. It is connected to the kth link by a pin joint at point B, and a joint force $\mathbf{F}_{ki}$ is applied to this ith link by the kth link at that joint. An external force $\mathbf{F}_i$ is applied to this link i at the point C. The distance from point C to point A is the vector $\mathbf{r}_{ij}$, and the distance from point C to point B is the vector $\mathbf{r}_{ik}$. An external couple of magnitude T_i is applied to link i as shown.

The force balance equation for this link is

$$\mathbf{F}_{ji} + \mathbf{F}_{ki} + \mathbf{F}_i = 0 \quad \text{or} \quad \mathbf{F}_{ji} + \mathbf{F}_{ki} = -\mathbf{F}_i$$

The vectors in this equation can be resolved into their x and y components to give the two scalar equations:

$$F_{xji} + F_{xki} = -F_{xi} \quad \text{and} \quad F_{yji} + F_{yki} = -F_{yi} \tag{12.2}$$

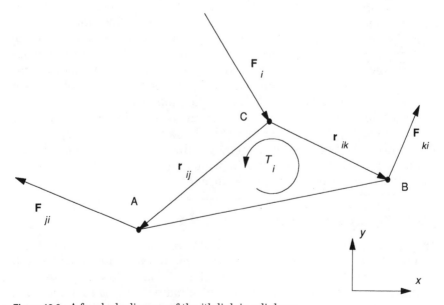

Figure 12.8 A free-body diagram of the ith link in a linkage.

The moment balance equation in terms of the x and y components of the joint forces and the x and y components of the $\mathbf{r}$ vectors for this link is

$$T_i + r_{xik}F_{yki} + r_{xij}F_{yji} - r_{yik}F_{xki} - r_{yij}F_{xji} = 0$$

$$\text{or} \quad r_{xik}F_{yki} + r_{xij}F_{yji} - r_{yik}F_{xki} - r_{yij}F_{xji} = -T_i \tag{12.3}$$

If the externally applied force and couple are known, Eqs. (12.2) and (12.3) constitute three equations in four unknowns. These unknowns are the x and y components of each of the two joint forces.

If link i were the coupler in a four-bar linkage, a similar set of three equations can be written for each of the other two movable links in the linkage. This will provide a total of nine equations in twelve unknowns. However, note that in these six equations for the links that are connected to the coupler, the forces at the joints that connect them to the coupler are simply the negatives of the joint forces that appear in the equations for the coupler. For example, $F_{xij} = -F_{xji}$. Therefore, the equations for each of these noncoupler links add only two unknowns, so the nine equations contain only eight unknowns. Obviously, if the equations are to be solved, one of the external forces or torques must be considered to be an unknown. In Example 12.1 the external moment at point O_1 was considered to be unknown, and it was solved for along with the joint forces.

The procedure for analytical simultaneous solution of the force and moment problem, then, consists of writing Eqs. (12.2) and (12.3) for equilibrium of each individual movable link as a free body. In doing so, be sure to write the joint force terms so that they act equally and in opposite directions on the two links that are joined at each joint. Substitute values for the known external forces and/or couples. Then use appropriate software to solve for the unknown values.

Some software requires all terms in the unknowns to be on one side of the equations and all terms in the known quantities to appear on the other side. Some software requires the variables to appear in the same order in each equation in which they appear, leaving blank spaces where a variable does not appear.

12.5 Static Balancing of Four-Bar Linkages

Figure 12.9a shows a four-bar linkage in which the location of the center of gravity of each link is indicated. Temporarily, consider link L_1 to be disconnected from the coupler so that it becomes an individual rotor, just as discussed in connection with Figs. 12.3 and 12.5. An infinitesimal counterclockwise rotation of L_1 would slightly lower

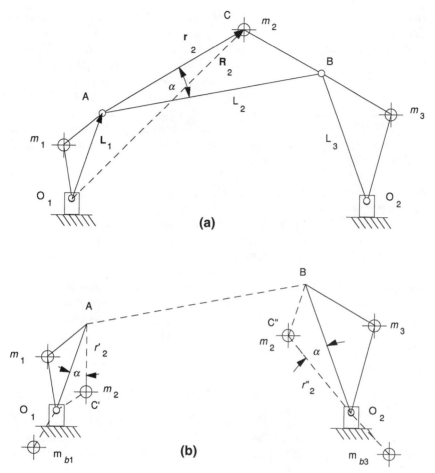

Figure 12.9 Schematic diagrams of a four-bar linkage with the locations of the centers of gravity of its links indicated.

its center of gravity. If no external forces or torques other than those applied by gravity and the joint force at O_1 are present, gravity will act such as to continue to lower the center of gravity. That is, the link will rotate until the center of gravity is directly below O_1. After the ensuing oscillations die out, the link will come to rest with the center of gravity directly below O_1, from which position an infinitesimal rotation would not raise or lower its center of gravity. This link would not be statically balanced because it could not be placed in an *arbitrarily chosen* position and released without having gravity move it from that position. As described in the preceding section, a rotor (such as this individual link) can be statically balanced by causing its total center of gravity, including a balance mass, to be placed at the pivot point so that no rotation will raise or lower its center of gravity.

This observation can be generalized to linkages in general. That is, if the parts of a linkage can be moved infinitesimally (without disconnecting them) in such a manner that the *total* linkage center of gravity is lowered, and if that linkage is subjected to only the forces of gravity and the support forces at its grounded pivots (or slides), then gravity *will* move that linkage in such a manner as to lower its total center of gravity. Such a linkage will not be in static equilibrium; it will not be statically balanced.

The objective in balancing a linkage, then, is to distribute the masses such that no matter how the linkage moves through its possible positions, the total center of gravity of that linkage will neither rise nor fall. Actually, it is easier to prevent that center of gravity from moving in *any* direction. Doing so amounts to causing the total imbalance vector to remain constant, regardless of the motion of the linkage. As will be shown in the derivation following Example 12.3, this can be done in a four-bar linkage by considering the coupler mass to be equivalent to four masses. Two of these masses are stationary, so their effects need not be balanced. One of the equivalent masses moves with link L_1, and the other moves with link L_3. Then links L_1 and L_3 can be balanced as rotors, considering them to include the respective equivalent masses.

Procedure 12.4: Static Balancing of Four-Bar Linkages   Figure 12.9a shows a four-bar linkage consisting of moving links L_1, L_2, and L_3 having masses m_1, m_2, and m_3, respectively, and the ground link L_4. The links L_1 and L_3 can only rotate, so they could be balanced as rotors. But the motion of the coupler is complicated, so it must be represented for balancing purposes by masses attached to the other two moving links.

1. Draw the mechanism to scale in one of its convenient positions, and show the locations of the centers of gravity of the three moving links, as shown, for example, in Fig. 12.9a.

2. With point C at the location of the center of gravity of the coupler, and point A at the pivot point connecting the coupler and link L_1, and point B at the pivot point connecting the coupler and link L_3, draw the triangle ABC. A vector from A to C will be labeled $\mathbf{r}_2$, and its length will be r_2.

The objective of steps 3a and 3b is to construct a triangle that is similar to triangle ABC but rotated and scaled so that its side that corresponds to AB coincides with link L_1.

3a. If a CAD system is used, duplicate triangle ABC in place, and rotate the duplicate about point A until its base AB is parallel to link L_1. Then magnify or diminish (rescale) the rotated triangle by a factor L_1/L_2, keeping the point at A stationary. For the linkage in Fig. 12.9a, this will place a point C' (a vertex of the rotated, rescaled triangle) in the position shown in Fig. 12.9b. The triangle AO_1C' should then be geometrically similar to triangle ABC. Alternatively, step 3b can be used on a CAD system also.

3b. If a manual graphical construction is being used, measure the angle BAC and call it α, as shown in Fig. 12.9a. Draw a line of length r_2L_1/L_2 through the pivot A of links L_1 and L_2 such that this line is rotated from L_1 by an angle α in the same direction as AC is rotated from AB. For the example in Fig. 12.9a this line is shown dashed as AC' in Fig. 12.9b. The triangle AO_1C' should then be geometrically similar to triangle ABC.

The objective of steps 4a and 4b is to construct a triangle that is similar to triangle ABC but rotated and scaled so that its side that corresponds to AB coincides with link L_3.

4a. If a CAD system is used, duplicate triangle ABC in place, and rotate the duplicate about point B until its base AB is parallel to link L_3. Then magnify or diminish (rescale) the rotated triangle by a factor L_3/L_2, keeping point B stationary. For the linkage in Fig. 12.9a, this will place a point C'' (a vertex of the rotated, rescaled triangle) in the position shown in Fig. 12.9b. The triangle O_2BC'' should then be geometrically similar to triangle ABC. Alternatively, step 4b can be used on a CAD system also.

4b. Draw a line of length r_2L_3/L_2 through the grounded pivot O_2 of link L_3 such that this line is rotated from L_3 by an angle α in the same direction as AC is rotated from AB. For the example in Fig. 12.9a this line is shown dashed as O_2C'' in Fig. 12.9b. The triangle O_2BC'' should then be geometrically similar to triangle ABC.

5a. Temporarily place at point C' a point mass equal to the mass of the coupler. Consider the mass at C' plus the mass of link L_1 to constitute a rotor that is pivoted at point O_1. Statically balance that rotor using Procedure 12.1 or 12.2. Then remove the temporary mass at point C'.

5b. Temporarily place at point C'' a point mass equal to the mass of the coupler. Consider the mass at C'' plus the mass of link L_3 to constitute a rotor that is pivoted at point O_2. Statically balance that rotor using Procedure 12.1 or 12.2. Then remove the temporary mass at point C''.

Now the four-bar linkage is statically balanced.

Example 12.3: Static Balancing a Four-Bar Linkage Figure 12.10 shows a mechanism for controlling the motion of a large industrial oven door. This is the same mechanism that was shown in Fig. 12.5a. In Fig. 12.10 the door is shown in the closed position. In Example 12.2 it was assumed that the mechanism was driven by a motor at pivot O_1. In the interest of safety, it would be preferred that, in the event of a failure of the drive system, the door would not move. That is, if the mechanism were acted on only by gravity, the door should neither rise nor fall; it should remain at rest in any position in which it is placed at rest. It should be statically balanced.

The scales that will be used in the graphical constructions for statically balancing this system will be 10 in per drawing inch and 500 in · lb per drawing inch. In accordance with steps 1 and 2 of Procedure 12.3, centers of gravity of the links have been located and the triangle ABC has been drawn in Fig. 12.10a. Then the diagram in Fig. 12.10b is produced in steps 3a and 4a by rotating and rescaling triangle ABC to produce triangles AO_1C' and O_2BC''', where these latter triangles are similar to ABC. (This diagram also can be produced by steps 3b and 4b.) This diagram in Fig. 12.10b is perhaps most easily drawn on a transparency over the drawing of steps 1 and 2 or on a separate CAD layer.

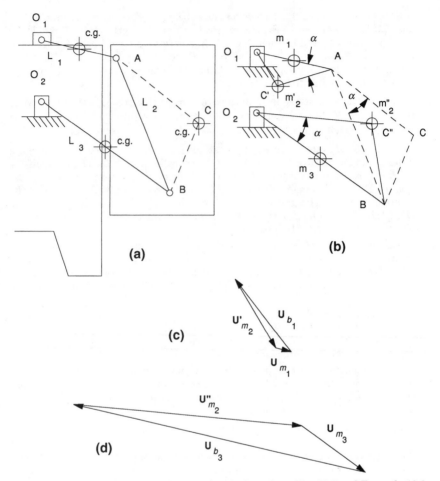

Figure 12.10 Static balancing of the oven door linkage from Fig. 12.5 and Example 12.2.

Consider the triangle AO_1C' in Fig. 12.10b with the attached masses m_1 and m_2' to be a rotor. The mass m_2' is at a distance of 6.78 in from O_1 as scaled off Fig. 12.10b. Then, because $m_2' = m_2$ weighs 100 lb the imbalance vector for the mass at C' is $\mathbf{U}_{m_2}'$ of magnitude $U_{m_2}' = (6.78 \text{ in})(100 \text{ lb}) = 678 \text{ in} \cdot \text{lb}$. This vector is drawn with a length of 1.358 in in Fig. 12.10c parallel to line O_1C'. The mass m_1 is at a distance of $l_1/2 = 6.6$ in from O_1 as scaled off Fig. 12.10b. Then, because m_1 weighs 20 lb, the imbalance vector for the mass of link L_1 is $\mathbf{U}_{m_1}$ of magnitude $U_{m_1} = (6.6 \text{ in})(20 \text{ lb}) = 132 \text{ in} \cdot \text{lb}$. This vector is drawn with a length of 0.26 in in Fig. 12.10c parallel to line O_1A. These two vectors are shown added in this figure, and the balance vector needed to close the triangle is $\mathbf{U}_{b_1}$, as shown.

The length of the vector $\mathbf{U}_{b_1}$ is measured in Fig. 12.10c to be 1.559 in long. This therefore represents an imbalance magnitude of $U_{b_1} = (1.559 \text{ in})(500 \text{ in} \cdot \text{lb/in}) = 779 \text{ in} \cdot \text{lb}$. The balancing mass m_{b_1} will be located relative to the pivot O_1 in a direction indicated by the vector $\mathbf{U}_{b_1}$, and its distance from O_1 times its weight will equal 779 in · lb. In Fig. 12.11 such a balance mass m_{b_1} weighing 65 lb is shown located at a distance of 12 in from O_1 in a direction parallel to the direction of vector $\mathbf{U}_{b_1}$ in Fig. 12.10c. This places the mass m_{b_1} 38.0° off the extension of the line through points O_1 and A as shown in Fig. 12.11.

Next, consider the triangle BO_2C'' with the attached masses m_3 and m_2'' to be a rotor. The mass m_2'' is at a distance of 19.764 in from O_2 as scaled off Fig. 12.10b. Then, because $m_2'' = m_2$ weighs 100 lb, the imbalance vector for the mass at C'' is $\mathbf{U}_{m_2}''$ of magnitude $U_{m_2}'' = (19.764 \text{ in})(100 \text{ lb}) = 1976.4$ in · lb. This vector is drawn with a length of 3.953 in in Fig. 12.10d parallel to line O_2C''. The mass m_3 is at a distance of $L_3/2 = 13.5$ in from O_2 as scaled off Fig. 12.10b. Then, because m_3 weighs 50 lb, the imbalance vector for the mass of link L_3 is $\mathbf{U}_{m_3}$ of magnitude $U_{m_3} = (13.5 \text{ in})(50 \text{ lb}) = 675$ in · lb. This vector is drawn with a length of 1.35 in in Fig. 12.10d parallel to line O_2B. These two vectors are shown added in this figure, and the balance vector needed to close the triangle is $\mathbf{U}_{b_3}$ as shown.

The length of the vector $\mathbf{U}_{b_3}$ is measured in Fig. 12.10d to be 5.163 in long. This therefore represents an imbalance magnitude of $U_{b_3} = (5.163 \text{ in})(500 \text{ in · lb/in}) = 2581$ in · lb. The balancing mass m_{b_3} will be located relative to the pivot O_2 in a direction indicated by the vector $\mathbf{U}_{b_3}$, and its distance from O_2 times its weight will equal 2581 in · lb. In Fig. 12.11

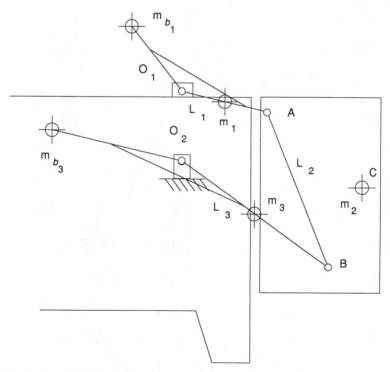

Figure 12.11 The statically balanced oven door.

such a balance mass m_{b_3} weighing 129 lb is shown located at a distance of 20 in from O_2 in a direction parallel to the direction of vector $\mathbf{U}_{b_3}$ in Fig. 12.10d. This places the mass m_{b_3} 22.4° off the extension of the line through points O_2 and B as shown in Fig. 12.11.

Derivation of Procedure 12.4

As stated at the beginning of Procedure 12.4, in a four-bar linkage the two links that are pivoted to ground can only rotate, so they can be balanced as rotors. But the motion of the coupler consists of both rotation and translation, so balancing of the effects of the motion of its mass is more complicated. The effects of the motion of the coupler mass can be studied by reference to Fig. 12.12a, which shows a vector representation of the links in a four-bar linkage.

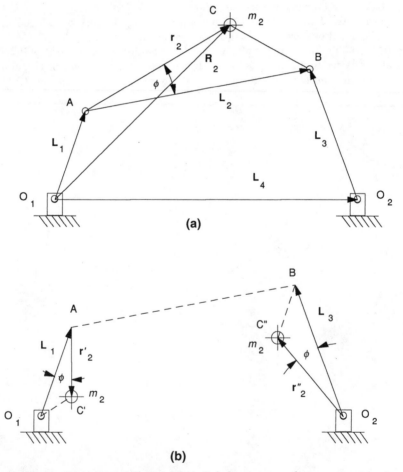

Figure 12.12 Derivation of the linkage static balancing procedure.

The coupler link is represented by the vector $\mathbf{L}_2$, and the center of gravity of the mass m_2 of that coupler is located at point C. We wish to study the variation of the imbalance vector of this mass as the linkage moves in order to find how we may cancel out the effects of that variation. The imbalance vector for mass m_2 is m_2 times the position vector $\mathbf{R}_2$ of the center of gravity of the mass of the coupler. That is,

$$\mathbf{U}_2 = m_2 \mathbf{R}_2$$

It can be seen in Fig. 12.12a that $\mathbf{R}_2 = \mathbf{L}_1 + \mathbf{r}_2$, so

$$\mathbf{U}_2 = m_2(\mathbf{L}_1 + \mathbf{r}_2) \tag{12.4}$$

The vector $\mathbf{r}_2$ bears a fixed relationship to the coupler vector $\mathbf{L}_2$. It is always at a fixed angle ϕ from $\mathbf{L}_2$, and the ratio of its length to the length of $\mathbf{L}_2$ is always r_2/L_2 regardless of the motion of the linkage. This relationship can be stated in terms of the complex variable form discussed in Chap. 2 as

$$\mathbf{r}_2 = \frac{r_2}{L_2} \mathbf{L}_2 e^{j\phi} \tag{12.5}$$

Then, substituting Eq. (12.5) into Eq. (12.4) gives

$$\mathbf{U}_2 = m_2 \left(\mathbf{L}_1 + \frac{r_2}{L_2} \mathbf{L}_2 e^{j\phi} \right) \tag{12.6}$$

The vector loop closure equation for this linkage is

$$\mathbf{L}_1 + \mathbf{L}_2 = \mathbf{L}_3 + \mathbf{L}_4$$

or when solved for $\mathbf{L}_2$, it is

$$\mathbf{L}_2 = -\mathbf{L}_1 + \mathbf{L}_3 + \mathbf{L}_4 \tag{12.7}$$

Substituting Eq. (12.7) into Eq. (12.6) gives

$$\mathbf{U}_2 = m_2 \left[\mathbf{L}_1 + \left(\frac{r_2}{L_2} \right) e^{j\phi} (-\mathbf{L}_1 + \mathbf{L}_3 + \mathbf{L}_4) \right] \tag{12.8}$$

Grouping the terms of Eq. (12.8) and both adding and subtracting a term $m_2 \mathbf{L}_4$ will give

$$\mathbf{U}_2 = m_2 \left[\mathbf{L}_1 - \mathbf{L}_1 (\frac{r_2}{L_2}) e^{j\phi} \right] + m_2 \mathbf{L}_4 \left(\frac{r_2}{L_2} \right) e^{j\phi}$$

$$+ m_2 \left[\mathbf{L}_4 + \mathbf{L}_3 \left(\frac{r_2}{L_2} \right) e^{j\phi} \right] + m_2(-\mathbf{L}_4) \tag{12.9}$$

Equation (12.9) contains four terms in m_2, each of which can be considered to represent the imbalance vector of a separate mass of magnitude m_2. The second and fourth terms represent vectors that are seen to involve no variables; they are tied to the stationary vector $\mathbf{L}_4$. Because these terms represent the imbalance of masses that are not moving, there is no need to provide static balancing for them. They will not contribute to the *variation* in the total imbalance vector of the linkage regardless of how the linkage moves.

The second term within the brackets in the first term in Eq. (12.9) represents a vector $\mathbf{r}_2'$ that is rotated from $-\mathbf{L}_1$ by the angle ϕ and which has a magnitude of r_2/L_2 times the length of L_1. Therefore, the *entire* first term in Eq. (12.9) represents the imbalance vector (relative to point O_1) of a mass of magnitude m_2 that is located at point C' shown in Fig. 12.12b. This mass moves with link L_1, and thus its motion must be compensated for.

The second term within the brackets in the third term in Eq. (12.9) represents a vector $\mathbf{r}_2''$ that is rotated from $\mathbf{L}_3$ by the angle ϕ and which has a magnitude of r_2/L_2 times the length of L_3. Therefore, the *entire* third term in Eq. (12.9) represents the imbalance vector (relative to point O_1) of a mass of magnitude m_2 that is located at point C'' shown in Fig. 12.12b. This mass moves with link L_3, and thus its motion must be compensated for.

The two masses at points C' and C'', then, can be considered to be attached to links L_1 and L_3, respectively, and thus if we balance these links considered as rotors which include their own masses *and* these hypothetically attached equivalent masses, the total imbalance vector of the linkage will not vary. That is, the total center of gravity will not move as the links move.

It will be noted that the triangles O_1AC' and BO_2C'' are geometrically similar to triangle BAC. Therefore, Procedure 12.4 provides static balancing of a four-bar linkage.

12.6   Static Balancing of Open Kinematic Chains

An *open kinematic chain* is an assemblage of links whose kinematics can be represented by a chain of vectors placed tail to head and in which the vectors do *not* form any loops. Examples of such open chains include articulated-arm robots (if the driving actuators are neglected), drafting machines, and articulated-arm desk lamps such as shown in Fig. 12.13. Perhaps the most famous examples of extensive open kinematic chains are the hanging so-called mobiles that were created and popularized by the sculptor Alexander Calder.

Open kinematic chains can be quite easily balanced by choosing first the link (body) that is furthest along the kinematic chain from ground.

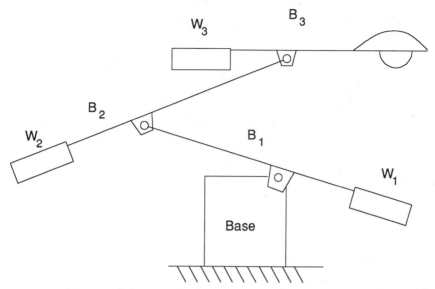

Figure 12.13   The static balancing of an articulated desk lamp.

In the case of the lamp in Fig. 12.13, this would be body B_3. Add a balance weight such as weight W_3 to that body so that the balanced center of gravity lies at the pivot that connects the body to the next body. Then, *considering the total mass of this first body with its balance weight to be concentrated at the pivot point of the first body,* add a balance weight to the next body in the chain (B_2 in this case) so that the resulting total center of gravity of this next body plus its balance weight and plus the mass of the first balanced body lies at the pivot point connecting this second body to the third body. Repeat this process progressively toward ground until all bodies have been balanced.

12.7   The Use of Springs for Static Balancing

The rotors that were discussed in the foregoing sections were balanced by adding appropriately placed balance masses to them. It also was shown in the derivation of Procedure 12.4 that the balancing of a four-bar linkage can be considered to be equivalent to the balancing of rotors. The addition of a balance weight to each of these systems applies a torque that varies in a desired manner as the rotor or links are moved. This torque is produced by the constant force vector (constant in magnitude and direction) that gravity exerts on the balance weight, acting at the varying moment arm that the geometry of the system produces. These effects are indicated in Fig. 12.14a, where it

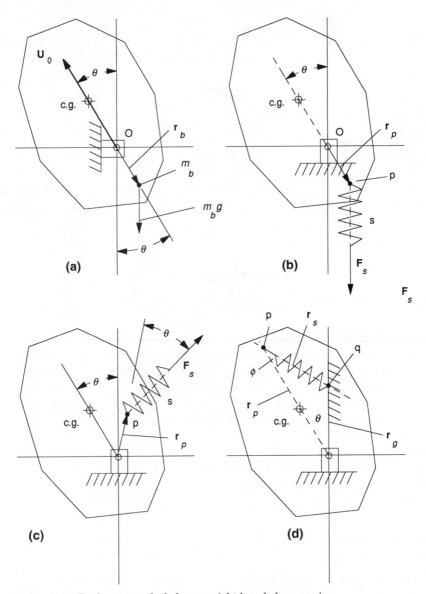

Figure 12.14 Replacement of a balance weight by a balance spring.

is seen that the balance weight m_b is placed in a position such that it is opposite to the imbalance vector $\mathbf{U}_o$ of the unbalanced rotor. It can be seen that the torque that this balance mass exerts on the rotor is given by

$$T_b = -m_b g r_b \sin \theta \qquad (12.10)$$

Figure 12.14*b* shows the same rotor but with the balance weight replaced by an extension spring *s* that is attached to the body at point *p*. The torque exerted by the spring is produced by the spring tension force vector $\mathbf{F}_s$ that acts at the varying moment arm produced by the geometry. In the positions shown in this figure, this torque is given by

$$T_s = -F_s \sin \theta \qquad (12.11)$$

It can be seen from both Fig. 12.14*a* and *b* and from Eq. (12.11) that *if the spring force were constant in magnitude and direction,* the spring could be substituted for the balance weight. (Note that a compression spring could be used to *press* downward rather than the extension spring, which, as shown, is *pulling* downward.)

An appreciable advantage to using a spring rather than a balance weight lies in the freedom of location and attachment of the spring. For example, the torque exerted on the rotor by the spring in the position shown in Fig. 12.14*c* is also seen to be given by Eq. (12.11).

If the lower end of the spring in Fig. 12.14*b* were attached to ground sufficiently far from the point at which it is attached to the rotor, and if the spring constant were made sufficiently low (small force change per unit change in stretch), the spring force magnitude and direction could be made to approach constant values. Providing such spring characteristics often can require a lot of space, so compromises are usually made. Often the changing spring force magnitude can be combined with a suitably arranged change in direction to give an adequate approximation to the desired balance torque. An example of such a combination is shown in Fig. 12.14*d*.

Figure 12.14*d* shows a frequently used balance spring arrangement. The center of gravity of the rotor is located as shown. In the position shown, the line connecting the center of gravity with the pivot is inclined from the vertical by an angle θ. A tension spring is attached to the rotor at point *p* and to ground at point *q*. It can be seen that the torque exerted on the rotor by the spring is given by

$$T_s = -F_s r_p \sin \phi \qquad (12.12)$$

However, from the law of sines,

$$\sin \phi = (r_g/r_s) \sin \theta \qquad (12.13)$$

Equations (12.12) and (12.13) combine to give

$$T_s = -F_s r_p (r_g/r_s) \sin \theta \qquad (12.14)$$

This torque T_s could provide the balance that Eq. (12.10) indicates to be desired if the factor $F_s r_p(r_g/r_s)$ were equal to $m_b g r_b$, which is constant for a given balance weight. The lengths r_p and r_g are constant, but for most extension coil springs, r_s varies as a function of F_s in the manner indicated in Fig. 12.15. When no tension is applied to these springs, each coil is in contact with its adjacent coil(s), and the resulting spring length r_{s_0} is known as the *free length*. If gradually increasing tension is applied to the spring, the spring length r_s will remain at this value r_{s_0} until the applied force reaches a value F_{s_0}, known as the *initial tension*. A further increase in tension pulls the coils apart and causes an increase in spring length r_s as indicated by the upward-slanting solid line. This stretched condition is the condition in which springs are used for balancing. As can be seen in Fig. 12.15, for tension values greater than this initial tension, the spring tension is related to spring length by

$$F_s = k(r_s - r_{s_0}) + F_{s_0} \qquad (12.15)$$

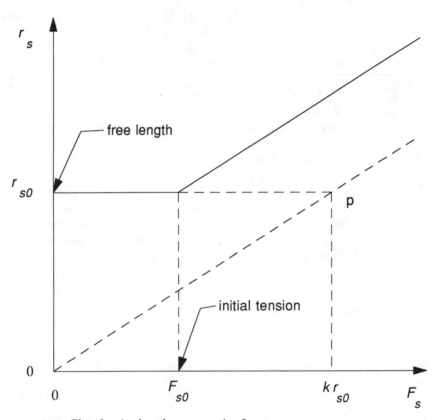

Figure 12.15 Plot of spring length versus spring force.

where k is the spring constant. Substituting Eq. (12.15) into Eq. (12.14) and suitably grouping terms will give

$$T_s = -[kr_g r_p + (F_{s_0} - kr_{s_0})r_p r_g/r_s] \sin \theta \qquad (12.16)$$

The first term in the brackets is constant, and all factors in the second term are constant except r_s. The second term therefore will vary as the spring length changes. If, however, $F_{s_0} = kr_{s_0}$, this second term vanishes, and by making $kr_g r_p = m_b gr_b$, Eq. (12.16) becomes identical to Eq. (12.10), and the spring would provide exactly the same balancing torque as would a balance weight.

It is possible to wind springs in which the initial tension $F_{s_0} = kr_{s_0}$. Such springs are known as *zero-length springs*, and they usually cost more than stock springs. Therefore, a compromise choice of a stock spring is often made. In Fig. 12.15 a dashed diagonal line is drawn from the origin representing the relationship $F_s = kr_s$. A horizontal dashed line is drawn for $r_s = r_{s_0}$, and this line intersects the diagonal line at point p. At this point, $F_s = kr_{s_0}$, as can be seen on the horizontal axis. Stock springs are usually wound so that $F_{s_0} < kr_{s_0}$, i.e., so that the solid curve is to the left and above the diagonal dashed curve. The second term in brackets in Eq. (12.16) therefore opposes the first term, thereby detracting from the spring's balancing torque. When stock springs are used, k, r_g, and r_p are usually chosen such that $kr_g r_p > m_b gr_b$ so that the spring provides excessive balancing torque when r_s is large (i.e., when θ is large), and therefore the second term in brackets in Eq. (12.16) is small. Then, as θ becomes smaller, r_s becomes smaller, the second term becomes larger, and spring balancing torque becomes equal to the desired value and then eventually becomes slightly inadequate. A compromise design results, but often such a compromise is satisfactory for the range of system motion of interest. With such a compromise, if the rotor is released, it will seek an equilibrium position that is between that at which $\theta = 0$ and the extreme of the designed range of motion. This equilibrium position is that at which the spring provides a torque that exactly balances the rotor.

Figure 12.16 schematically shows a rotor that is being balanced by use of a *compression* spring rather than an extension spring. In a manner similar to that used to derive Eq. (12.14), an equation for the torque provided by this spring system can be derived. The result is

$$T_s = -F_s r_p (r_g/r_s) \sin \theta \qquad (12.17)$$

which is identical to Eq. (12.14). Again, all factors on the right-hand side of this equation are constant except F_s and r_s. In the case of

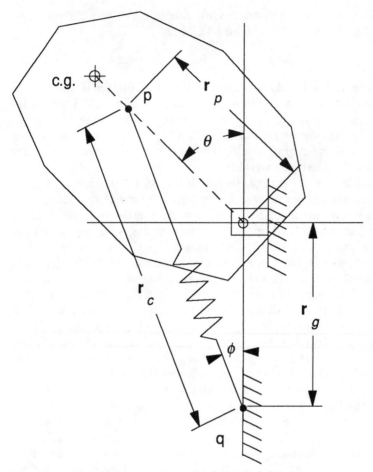

Figure 12.16   Use of a compression spring for static balancing.

extension springs, F_s increases as r_s increases, so the effects of their variations on the factor F_s/r_s tend to oppose each other. In the case of compression springs, however, F_s increases as r_s decreases, so the effects of their variations on the factor F_s/r_s tend to increase the variability of that factor. Even so, satisfactory compromises can be made in the design to provide adequate balancing with compression springs in many cases.

The use of compression springs involves problems of properly supporting their ends and of preventing buckling of the spring, particularly when large variations of r_s are to be accommodated. These problems can be avoided by using *gas springs*. A gas spring consists of a sealed pneumatic cylinder and piston assembly that contains a gas (usually nitrogen) under high pressure. As the unit is subjected to compression, tending to shorten it, the gas is further compressed and the compression force increases, usually by 30 to 40 percent as the length is decreased from its maximum value to its minimum value. These springs are specified in terms of their base length, their stroke, and their listed force. The *base length* is the total length of the assembly between its mounting pivots when no force is applied. The *stroke* is the maximum variation in total length possible with the given assembly. The *listed force* is that compressive force which is required to cause the assembly to start to shorten. (It is analogous to the initial tension in the case of extension springs.)

Figure 12.17 is a schematic drawing of a mechanism for balancing the rear hatch on an automobile. The hatch H is pivoted to the car body at point O. A gas spring s is pivoted to the hatch at point p and to the car body at point q. The hatch weighs 27 lb, and its center of gravity is 20 in from point O. The lengths shown are $r_p = 19$ in, $r_g = 6$ in, and the base length of the gas spring $r_{s_0} = 20$ in. The length r_g is inclined 41.6° from the vertical. In order to provide criteria for choosing an appropriate gas spring for this application, a plot will be made of spring force F_s required for exact balance versus spring length r_s.

Numbers for a plot of required F_s versus r_s were obtained by using dynamic simulation software called Working Model® (by Knowledge Revolution). In this software the gas spring was assumed to be a "length" actuator, and its length was chosen to have a sequence of values from the base length to a minimum length of 14 in. The software was made to output values of resulting actuator force. These values of force are plotted versus the values of actuator length as the solid curve

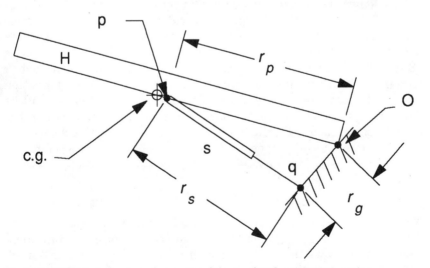

Figure 12.17 Static balancing of an automobile rear hatch using a gas spring.

REQUIRED HATCH SPRING FORCE (lb)

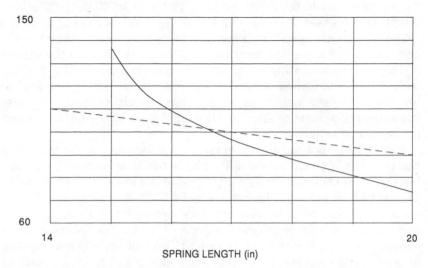

SPRING LENGTH (in)

Figure 12.18 Required spring force and available spring force versus spring length for static balancing of an automobile rear hatch.

in Fig. 12.18. Because this curve represents an increasing required force with decreasing spring length, an appropriate compression spring could be used to provide close to exact balancing over much of the range of motion.

In Fig. 12.18 a dashed straight line is drawn representing the force versus length for a gas spring that is commercially available from stock. It is seen that this spring provides more force than necessary when the hatch is fully open (spring length equals 20 in) and less force than needed when the hatch is fully closed (spring length equals 13 in). These actually are desirable features because, with such a spring, the hatch would stay fully open when placed in that position and would stay fully closed when placed in that position. It will be found that with the gas spring used in this example application, the hand forces needed to lift the hatch from the closed position and to pull it down from the fully open position are modest, and therefore the spring choice is satisfactory.

Comments and suggestions

Notice that if difficulty were encountered in finding a suitable spring for any of the preceding balancing examples, the mechanism dimensions r_g and r_p and the angle at which r_g is oriented could be varied. Also, the point at which the spring is attached to the rotor (point p)

need not lie on the straight line connecting the center of gravity of the rotor with the rotor pivot point O. By trying various values for these parameters, various spring requirements could be generated.

The use of curves of required force (or torque) versus deflection such as shown in Fig. 12.18 is applicable to the design of spring balancing systems involving not only compression springs but also extension springs, torsion springs, and springs with nonlinear relationships between force (or torque) and deflection. The data from which Fig. 12.18 was plotted were obtained by using dynamic simulation software. These data could have been obtained by direct calculation using moment arms and the geometry of the mechanism. Such calculations can become tedious, even if a CAD system is used. These data also could have been obtained by using kinematic analysis software and noting that the ratio of spring force to hatch weight at any position is equal to the ratio of the vertical velocity of the center of gravity of the hatch to the rate of change of spring length at that position. This last technique uses the principles of mechanical advantage, which were described in Sec. 3.9.

The foregoing examples considered springs in which the change in force was proportional to the change in length of the spring. It is also possible to use torsion springs over limited ranges of rotor rotation. Also, at added cost, springs that have special relationships of force versus deflection are available. Often, however, the use of kinematics (geometry) as done in the foregoing examples can eliminate the need for such special springs.

Notice that in Fig. 12.16 if the spring were an extension spring instead of the compression spring shown, the spring would contribute to *im*balance. Such an arrangement could be added to one of the links in Example 12.3 to cause the oven door to stay closed when it is in the fully closed position and to stay open when placed in the fully open position. This is but one of many possibilities for combining kinematics of spring mounting with the force characteristics of the spring itself to provide suitable balance or imbalance.

Forces, Torques, and Dynamic Balancing

13.1 Introduction

In this chapter procedures are presented for performing analyses to calculate dynamic forces and torques in mechanisms and for balancing those mechanisms. These procedures are similar to those presented in Chap. 12. However, knowledge of Newton's second law, which relates force and torque on a rigid body to the resulting translational and angular accelerations of that body, is important for understanding and/or deriving the procedures in this chapter. Therefore, that law is reviewed, and d'Alembert's principle is used to convert Newton's equations into a form in which they can be used in *quasi-static* analyses. Because these analyses involve both kinetics (Newton's law) and statics, they often are referred to as *kinetostatic analyses*.

Motion is an important part of the operation of most mechanisms. If a mechanism is not dynamically balanced, motion produces varying forces and couples that tend to shake the base to which the mechanism is attached. In Chap. 12 it was shown that static balance of a mechanism can be achieved by arranging the mass distribution of each of its parts such that no matter how that mechanism is driven through its range of motion, its total center of mass does not move. It will be shown in this chapter that when such a *statically balanced* mechanism is driven at any speed, it will transmit no net *shaking force* to the base on which it is mounted. However, if such a statically balanced mechanism is not also *dynamically balanced,* it will generally transmit a *shaking couple* to the base on which it is mounted. This chapter presents procedures for *dynamically balancing* rotors and four-bar linkages, thereby eliminating both the shaking forces and shaking couples.

Even when a mechanism has been dynamically balanced, it is still possible that in order to drive it with a constant input angular velocity the required driving torque will fluctuate appreciably with time. *Input torque smoothing* procedures presented in this chapter provide means for determining a distribution of mass in the mechanism such that the mechanism can be driven at a constant input angular velocity with a torque *that does not fluctuate.*

13.2   A Brief Review of Some Dynamics

Newton's second law

According to Newton's second law, if a body is acted on by a force, the time rate of change of the body's momentum is equal to that force (if appropriate units of mass, force, and distance are used in the computation). In dealing with mechanism kinematics, we can separate translational motions and rotational motions, so Newton's laws will be discussed first for pure translational motion.

Translational acceleration and forces

It is shown in texts on dynamics that if a force $\mathbf{F}$ is applied to a rigid body such that the line of action of that force passes through the center of mass (or center of gravity) of the body, the rate of change of the translational momentum vector of that body will be equal to the force applied. The translational momentum of a body is given by the product $m\mathbf{v}$, i.e., by its mass m times it velocity vector $\mathbf{v}$. Then, because we will deal with bodies whose masses do not change, we may write that the force $\mathbf{F}$ applied is equal to the time rate of change of $m\mathbf{v}$ or

$$\mathbf{F} = \frac{d(m\mathbf{v})}{dt} = m\left(\frac{d\mathbf{v}}{dt}\right) = m\mathbf{a}$$

where $\mathbf{a}$ is the acceleration vector of the body's center of gravity. This is the vector form of the familiar formula statement of Newton's second law ($F = ma$). If the quantity $m\mathbf{a}$ is subtracted from both sides of this equation, the resulting equation is

$$\mathbf{F} - m\mathbf{a} = 0 \qquad (13.1)$$

This equation looks like a static equilibrium equation in which the vector sum of two forces is zero. The second term, $-m\mathbf{a}$, when considered as a force, is called a *d'Alembert force* or often an *inertial reaction force.* It is seen that this force is directed in the direction opposite to the acceleration vector and has a magnitude equal to the mass of the body times the acceleration of the body. If the acceleration of a body is known, the d'Alembert force $-m\mathbf{a}$ can be considered to be a quasi-static force, and Eq. (13.1) can be used for a quasi-static analysis.

Rotational acceleration and torques

For purposes of studying rotational motion, consider a point mass m_i that is connected to a pivot O by a massless rod of length r_i, as shown in Fig. 13.1. If the rod is moving in such a manner as to produce on the mass a component $\mathbf{a}_i$ of acceleration perpendicular to the rod as shown, that acceleration force will produce a d'Alembert or inertial reaction force $-m_i\mathbf{a}_i$ as shown. The acceleration $\mathbf{a}_i$ is a tangential acceleration component because, as discussed in Chap. 2, it is directed perpendicular to the radius from the pivot to the mass. The magnitude of the tangential acceleration (see Chap. 2) is equal to the angular acceleration α of the rod times the radius r_i; that is, $a_t = r_i\alpha$. The magnitude of the d'Alembert force is therefore $m_i a_t = m_i r_i \alpha$. This force acts at a moment arm of length r_i from the pivot, so it will produce a d'Alembert torque $\tau_t = -r_i(m_i a_t) = -m_i r_i^2 \alpha$ about the pivot O. As drawn, α is counterclockwise, so $\mathbf{a}_i$ is upward and to the left. The d'Alembert force is therefore downward and to the right, and τ_i is a clockwise d'Alembert

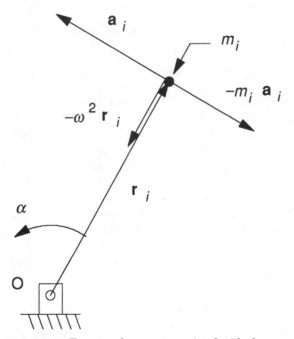

Figure 13.1 Forces and torques associated with the acceleration of a point mass.

torque. That is, the d'Alembert torque is in the opposite direction from the angular acceleration.

Real bodies can be considered to consist of collections of many such infinitesimal point masses, each of mass m_i, so the total d'Alembert or inertial reaction torque T_r that would be produced by a real body that was accelerating angularly at a rate α about pivot O would be the sum of the d'Alembert torques produced by all the infinitesimal masses or

$$T_r = -\sum_i \alpha m_i r_i^2 = -\alpha \sum_i m_i r_i^2$$

The moment of inertia of the body about point O is defined as

$$I = \sum_i m_i r_i^2$$

Therefore, $T_r = -I\alpha$.

It frequently is convenient to choose the center of mass (the center of gravity) as a reference point for computing torques and moments of inertia. When this is done, the resulting moment of inertia is denoted as I_0, and the last equation becomes

$$T_r = -I_0\alpha$$

For equilibrium, the d'Alembert torque T_r must be balanced by an externally applied torque T (which will produce α), so

$$T + T_r = T - I_0\alpha = 0 \tag{13.3}$$

where $I_0\alpha$ is the d'Alembert or inertial reaction torque. Equation (13.3) applies to angular acceleration and corresponds to Eq. (13.1), which applies to translational acceleration. Whereas Eq. (13.1) states that the d'Alembert force is equal and opposite to the rate of change $m\mathbf{a}$ of *translational* momentum $m\mathbf{v}$, Eq. (13.3) states that the d'Alembert torque is equal and opposite to rate of change $I\alpha$ of *angular* momentum $I\omega$ (also called *moment of momentum*).

It is shown in dynamics texts that the moment of inertia I' of a body about any point at a distance d from the center of gravity of the body is given by $I' = I_0 + md^2$, where m is the total mass of the body. This relationship is the result of the *parallel-axis theorem* for moments of inertia.

Combined translational and rotational acceleration

Figure 13.2a indicates a body that is experiencing a translational acceleration $\mathbf{a}$ of its center of gravity and a clockwise angular accelera-

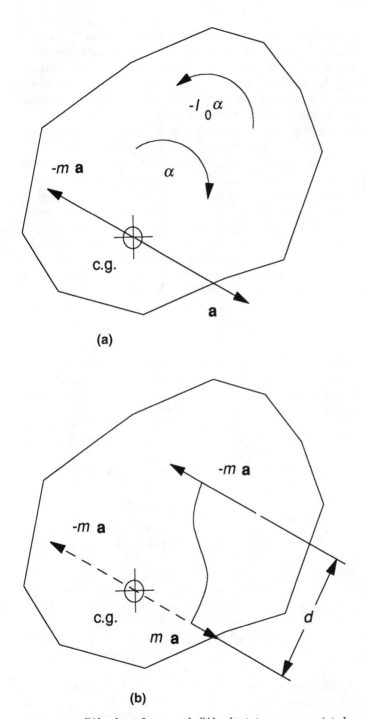

Figure 13.2 d'Alembert forces and d'Alembert torques associated with an accelerating rigid body.

tion α. In accordance with the foregoing discussion and Eqs. (13.1) and (13.3), this body can be considered to be subjected to a d'Alembert force $-m\mathbf{a}$ applied at its center of gravity as shown and to a counterclockwise d'Alembert torque $-I_0\alpha$ as shown. In Sec. 12.2 it was shown that a combination of a force applied at some point in a body plus a couple applied to that body is equivalent to an equal and parallel force applied at some other point in that body without a couple. Then the d'Alembert force $-m\mathbf{a}$ and the d'Alembert torque $-I_0\alpha$ can be replaced by the equivalent force $-m\mathbf{a}$ whose line of action is at some distance d from the center of gravity, as shown in Fig. 13.2b. In this figure it can be seen that the force $-m\mathbf{a}$ whose line of action is a distance d from the center of gravity produces a counterclockwise torque when viewed in conjunction with the dashed vector $m\mathbf{a}$ that is applied at the center of gravity. That dashed vector $m\mathbf{a}$ cancels the original d'Alembert force $-m\mathbf{a}$ that was shown in Fig. 13.2a. The displacement distance d must be such that the couple of magnitude $T = (d)(-ma)$ in Fig. 13.2b is equal to the couple $-I_0\alpha$ in Fig. 13.2a. That is, $-dma = -I_0\alpha$, which gives

$$d = \frac{I_0\alpha}{ma} \qquad (13.4)$$

Thus it is seen that for a body that is experiencing both translational and rotational accelerations, the effects of these two accelerations can be considered to be a single d'Alembert force $-m\mathbf{a}$ whose line of action is displaced a distance d from the body's center of gravity, where d is given by Eq. (13.4). In subsequent procedures and derivations this force will be balanced by other forces applied to the body to provide quasi-static equilibrium.

13.3 Dynamic Balancing of Bodies in Pure Rotation (Rotors)

As stated in Sec. 12.3, the motion of a body that is pivoted to ground consists of only pure rotation, and such a body will be referred to as a *rotor*. If such a body is not properly balanced dynamically, it will, when rotated at speed, produce forces and couples that will tend to shake the base to which it is pivoted. An example of such a problem is the severe shaking that occurs in the spin-dry operation of a washing machine when the load of clothes in the machine is greatly unbalanced. Another example is the vibration that an unbalanced automobile wheel imparts to the automobile when it is driven at high speed.

These shaking effects can consist of shaking forces and shaking couples. If the mass of the body is concentrated primarily in one plane that is perpendicular to the axis of rotation, the body will be called a *thin rotor,* and its imbalance will produce only shaking forces if the body is rotating at constant speed. If the mass of the rotating body is distributed appreciably *along the direction of the axis of rotation,* the body will be called an *extended rotor,* and its imbalance can produce both shaking forces and shaking couples, even if it is rotating at constant speed.

Dynamic balancing of thin rotors

A thin, *statically* balanced rotor (see Procedure 12.1), when rotated at constant angular velocity, will produce no shaking forces or torques. That is, it also will be *dynamically* balanced. This is shown in the following discussion.

Figure 13.3 shows a thin rotor that is rotating about the shaft at a constant angular velocity ω and which contains parts that have their centers of gravity located at points P_1, P_2, and P_3. These points are all

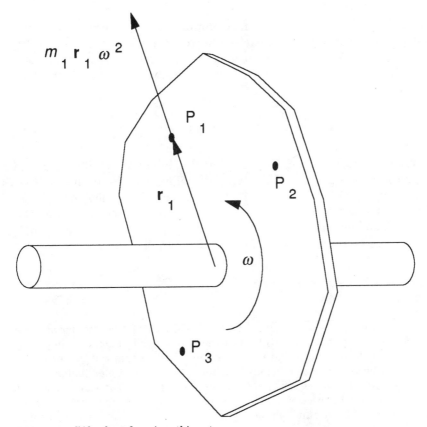

Figure 13.3 d'Alembert force in a thin rotor.

in the same plane perpendicular to the axis of rotation, and the masses of the parts do not extend far from this plane, so this is a thin rotor. The part whose center of gravity is at P_1 has a mass of m_1, and it is at a distance from the pivot represented by the vector $\mathbf{r}_1$. As discussed in Chap. 2 and as shown in Fig. 13.3, such a mass will have a centripetal acceleration component $\mathbf{a}_1 = -\mathbf{r}_1\omega^2$ directed radially inward, which produces a d'Alembert force $m_1\mathbf{r}_1\omega^2$ directed radially outward from the mass as shown. Each of the other parts of the rotor will produce a similar d'Alembert force that is directed radially outward from it. Therefore, the total d'Alembert force that will be produced by all parts of the rotor is the sum of all these forces, and this sum plus the force $\mathbf{F}_p$ applied to the rotor by the pivot bearings must be zero if the body is to be in equilibrium. The equilibrium condition is stated by Eq. (12.1), which for this particular case becomes

$$\mathbf{F}_p - \sum_{i=1}^{3} m_i\mathbf{a}_i = \mathbf{F}_p - \sum_{i=1}^{3} - (m_i\mathbf{r}_i\omega^2) = \mathbf{F}_p + \omega^2\sum_{i=1}^{3} m_i\mathbf{r}_i = 0 \qquad (13.5)$$

The rightmost summation in this equation is seen to be the sum of the imbalance vectors of the parts of the rotor as they were defined in Sec. 12.3. If a balance mass m_b is added to this rotor, the d'Alembert force on that mass will be $m_b\mathbf{r}_b\omega^2$ directed radially outward from it, so Eq. (13.5) becomes

$$\mathbf{F}_p + \omega^2 m_b\mathbf{r}_b + \omega^2\sum_{i=1}^{3} m_i\mathbf{r}_i = \mathbf{F}_p + \omega^2\left(\mathbf{U}_b + \sum_{i=1}^{3} m_i\mathbf{r}_i\right) = 0 \qquad (13.6)$$

where it is recalled from Sec. 12.3 that the imbalance vector $\mathbf{U}_b$ for the balance mass is given by $\mathbf{U}_b = m_b\mathbf{r}_b$. By comparing the term in parentheses in Eq. (13.6) with Eq. (12.1), it is seen that when the rotor is completely statically balanced, the term in parentheses is zero and the total d'Alembert force vanishes. The rotor is therefore dynamically balanced, and no pivot force $\mathbf{F}_p$ is required for equilibrium.

Dynamic balance of rotors having angular acceleration

If a rotor is driven in such a manner that it experiences angular accelerations, then that rotor will (even though statically balanced) impose a shaking torque on the base to which the driving actuator is mounted and to which the rotor is pivoted. That torque can be canceled (balanced) by providing a *compensating rotor* that is pivoted to that same base and whose angular acceleration is always in a direction opposite to that of the original rotor and whose angular acceleration is always proportional to that of the original rotor. This required proportionality is such that $I_c\alpha_c = -I_r\alpha_r$, where I_c is the moment of inertia of the compensating rotor,

α_c is the angular acceleration of the compensating rotor, I_r is the moment of inertia of the original rotor, and α_r is the angular acceleration of the original rotor.

The foregoing shaking torque and the corresponding compensation effect can be shown by reference to Fig. 13.4. Figure 13.4a depicts a

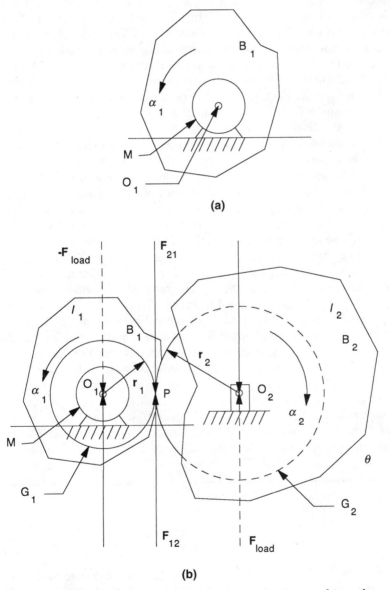

(a)

(b)

Figure 13.4 Dynamic balancing of a rotor that is experiencing angular acceleration (use of a compensating rotor).

statically balanced body that is pivoted at point O_1. This body is rigidly attached to the rotating portion of the motor M by means of a shaft, and together they constitute a rotor B_1. The total moment of inertia of this rotor is I_1. The magnetic forces in the motor apply a torque T_M to the rotor, thereby producing an angular acceleration of the rotor of α_1, where $T_M = I_1\alpha_1$. As shown, the motor magnetic forces are applying a counterclockwise torque to the rotor, so the angular acceleration of the rotor is counterclockwise. Newton's third law tells us that for every action there is an equal and opposite reaction. In this case, this means that because the motor magnetics are applying a counterclockwise torque T_m to the rotor, the motor magnetics are applying a clockwise torque $-T_M$ to the motor frame and thus to the base to which the motor frame is mounted. Angularly accelerating the rotor therefore produces a shaking torque $-T_m = -I_1\alpha_1$ on the base.

In this illustrative case the rotor is considered to be driven by the magnetic forces in an electric motor. This rotor could have been driven by a cam and follower system, by a hydraulic or pneumatic actuator, by a linkage, or by any other type of mechanism. The same shaking torque would be transmitted to the base as a result of rotor angular acceleration regardless of the mechanism used to drive the rotor as long as the driving mechanism is attached to the base. Of course, the driving mechanism itself also might produce *additional* shaking forces and torques. These shaking forces and torques are discussed in Secs. 13.4 and 13.5.

Figure 13.4*b* shows the same rotor and motor, but now a gear G_1 of radius r_1 has been firmly attached to the rotor B_1. Consider this gear to be part of the rotor, so the total moment of inertia of the rotor plus the gear will be considered to be I_1. A second gear G_2 of radius r_2 is firmly attached to a second rotor B_2 that has a moment of inertia I_2 about point O_2 at which it is pivoted to the base. Consider this second gear to be part of the second rotor, so the total moment of inertia of the rotor plus the gear is I_2. The two gears are in mesh, so consequently, $\alpha_2 r_2 = -\alpha_1 r_1$.

At point P where the two gears mesh, gear G_1 on rotor B_1 applies an upward force $\mathbf{F}_{12}$ to gear G_2 on rotor B_2, and the gear on rotor B_2 applies an equal and opposite (downward) force $\mathbf{F}_{21}$ to the gear on rotor B_1. The force $\mathbf{F}_{12}$ produces a clockwise (negative) torque of magnitude $F_{12}r_2$ about pivot point O_2. Then $F_{12}r_2 = -I_2\alpha_2$ because α_2 is negative (clockwise). Then

$$F_{12} = \frac{-I_2\alpha_2}{r_2} \tag{13.7}$$

To be in translational equilibrium, body B_2 must be acted on by an equal and opposite downward force from the base at O_2. The reaction (shaking) force that the body B_2 applies to the base will therefore be an equal and opposite upward force, shown dashed and labeled $\mathbf{F}_{\text{load}}$. Body B_1 is acted on by a downward force $\mathbf{F}_{21}$ at the gear mesh point P, so to be in translational equilibrium, the base must apply an equal and opposite upward force to body B_1 at point O_1. The reaction (shaking) force that the body B_1 applies to the base therefore will be an equal and opposite downward force, shown dashed at point O_1. Because $\mathbf{F}_{21} = -\mathbf{F}_{12}$, the dashed shaking forces at points O_1 and O_2 will each have a magnitude of F_{12}, and these two dashed d'Alembert forces, being the only *forces* applied to the base, constitute a counterclockwise shaking couple:

$$T_c = F_{12}(r_1 + r_2)$$

Substituting F_{12} from Eq. (13.7) gives

$$T_c = \frac{-I_2\alpha_2(r_1 + r_2)}{r_2}$$

or, because $\alpha_2 r_2 = -\alpha_1 r_1$,

$$T_c = \frac{I_2\alpha_1 r_1(r_1 + r_2)}{(r_2)^2} \tag{13.8}$$

Summing the torques that are applied to body B_1,

$$T_M - r_1 F_{21} - I_1\alpha_1 = 0 \quad \text{or} \quad T_M = r_1 F_{21} + I_1\alpha_1 \tag{13.9}$$

The magnitude of $\mathbf{F}_{21}$ is equal to the magnitude of $\mathbf{F}_{12}$, so, from Eq. (12.7),

$$F_{21} = \frac{-I_2\alpha_2}{r_2}$$

Substituting this into Eq. (13.9) gives

$$T_M = \frac{-I_2\alpha_2 r_1}{r_2} + I_1\alpha_1$$

However, as pointed out previously, $\alpha_2 r_2 = -\alpha_1 r_1$, so

$$T_M = I_2\alpha_1\left(\frac{r_1}{r_2}\right)^2 + I_1\alpha_1 \tag{13.10}$$

T_M is the torque that the magnetic forces in the motor apply to body B_1, so those magnetic forces also will apply a reaction torque $-T_M$

to the motor frame and thus to the base. Then the total shaking moment that the system shown in Fig. 13.4b will apply to the base is $T_{\text{shake}} = -T_M + T_c$. Substituting from Eqs. (13.8) and (13.10) gives

$$T_{\text{shake}} = -I_2 \alpha_1 \left(\frac{r_1}{r_2}\right)^2 - I_1 \alpha_1 + \frac{I_2 \alpha_1 r_1 (r_1 + r_2)}{(r_2)^2}$$

Collecting terms gives

$$T_{\text{shake}} = \alpha_1 \left[I_2 \left(\frac{r_1}{r_2}\right) - I_1 \right] \tag{13.11}$$

If we make $I_2 = I_1(r_2/r_1)$, the factor in brackets in Eq. (13.11) becomes zero and the shaking torque vanishes. Remembering that $\alpha_2 r_2 = -\alpha_1 r_1$, this condition for causing the shaking torque to vanish can be written $I_2 \alpha_2 = -I_1 \alpha_1$. Then we can repeat the statement made at the start of this subsection:

> To cancel the shaking couple (torque) that is applied by an angularly accelerating rotor to the base from which that rotor is being driven and to which it is pivoted, provide a compensating rotor that is pivoted to the same base and is driven from the same base in a direction opposite to the driving of the original rotor and the *absolute value* of whose product $I\alpha$ is always equal to that of the original rotor.

Dynamic balancing of an extended rotor

Figure 13.5 shows an example rotor that consists of two bodies that are attached to and spaced along a common shaft that is rotating at an angular velocity ω. The mass of body B_1 is m_1, and its center of gravity is at a vector distance $\mathbf{r}_1$ directly above the axis of rotation. The mass of body B_2 is m_2, and its center of gravity is at a vector distance $\mathbf{r}_2$ directly below the axis of rotation. This rotor is assumed to be statically balanced, so the imbalance vectors for the two masses are equal in magnitude and opposite in direction. That is,

$$\mathbf{U}_1 = m_1 \mathbf{r}_1 = -\mathbf{U}_2 = -m_2 \mathbf{r}_2 \tag{13.12}$$

Because of the angular velocity of the rotor, the center of gravity of each of these masses will experience a centripetal acceleration that will produce a radially outward d'Alembert force. The d'Alembert force on body B_1 will be $\omega^2 m_1 \mathbf{r}_1$ as shown, and the d'Alembert force on body B_2 will be $\omega^2 m_2 \mathbf{r}_2$ as shown. The static balance condition stated as Eq. (13.12) implies that these two d'Alembert forces are equal and opposite, so there will be no net resulting shaking force transmitted to the base. However, because the lines of action of these forces are not coincident, they produce a shaking couple that for the conditions of Fig. 13.5 will be a

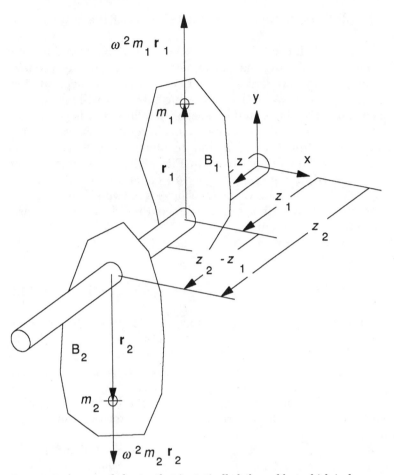

Figure 13.5 An extended rotor that is statically balanced but which is dynamically unbalanced.

torque about an axis parallel to the x axis shown. For an observer placed along the positive x axis and looking toward the origin, this torque would appear counterclockwise, i.e., in a direction that would tend to carry the positive y axis toward the positive z axis. A torque in such a direction will be referred to as a *positive torque*.[1] Its magnitude will be

$$T = \omega^2 m_1 r_1 (z_2 - z_1) \qquad (13.13)$$

---

[1]This is in accordance with the "right-hand rule" for rotations, which states that rotation that is in a direction from the positive x axis toward the positive y axis is positive rotation about the z axis, rotation that is in a direction from the positive y axis toward the positive z axis is positive rotation about the x axis, and rotation that is in a direction from the positive z axis toward the positive x axis is positive rotation about the y axis.

because the distance between the lines of action of the two forces is $(z_2 - z_1)$.

To dynamically balance this rotor, a balancing couple would have to be provided. Remembering that a couple can be considered to consist of two equal and opposite forces whose lines of action are separated by some distance, it can be seen that balancing this rotor will require two masses, each of which is located in a separate plane that is perpendicular to the rotation axis. Dynamic balancing of rotors is therefore often referred to as *two-plane balancing,* and the planes are called *correction planes.* Although the example rotor in Fig. 13.5 was assumed to be statically balanced, in general, a rotor will be both statically and dynamically unbalanced. The two balance masses in two-plane balancing can be used to provide both types of balance. Procedure 13.2 will provide such complete balancing.

As background for Procedure 13.2, note that Eq. (13.13) can be written

$$T = \omega^2 m_1 r_1 (z_2 - z_1) = \omega^2 m_2 r_2 z_2 - \omega^2 m_1 r_1 z_1 = T_2 - T_1 \qquad (13.14)$$

where T_1 and T_2 are the d'Alembert torques that would be produced about the x axis in Fig. 13.5 by the masses m_1 and m_2, respectively. This total torque could be balanced by a balance mass m_b in a plane that is at a distance z_b from the xy plane shown. Such a balance mass would produce a torque $-\omega^2 m_b r_b z_b$ about the x axis if $\mathbf{r}_b$ were vertically upward, thereby opposing the original imbalance torque T. If appropriate values are chosen for m_b, r_b, and z_b, the net torque would be reduced to zero.

Then an analogous computation could be performed about an axis parallel to the x axis but in the plane of the mass m_b. The mass m_b would produce no torque about this axis, and a second balance mass could be placed in the original xy plane to balance the imbalance torque about this axis in the plane of m_b. Then, because there would be no torque about each of two noncoincident axes, there could be no net force and no net couple, and the rotor would be both statically and dynamically balanced.

Note that the factor ω^2 appears in each d'Alembert torque, and therefore it can be canceled out from all the terms in the computations for the balance masses. The quantity of interest would then be $m\mathbf{r}z = \mathbf{U}z$ for each mass, where z is the distance of that mass from one of the correction planes. The dynamic balancing process then becomes for *each* correction plane very much like the static balancing Procedure 12.1 except that each imbalance vector $\mathbf{U}_i$ is multiplied by the corresponding z_i.

The d'Alembert torque produced about the x axis by each point mass in the yz plane is seen from the preceding discussion to be proportional to that mass times the distance of the mass from the origin, measured parallel to the axis of rotation, times the distance of the mass from the origin, measured parallel to a third orthogonal axis (such as y). In the case discussed above, this product is $m_i z_i y_i$. The total torque about the x axis is the sum of the torques produced by all the masses and is therefore proportional to the sum of all those products. That is,

$$T_x = \omega^2 \sum_i m_i z_i y_i \tag{13.15}$$

In mechanics and dynamics texts, the summation in this equation is known as a *product of inertia* of the collection of masses relative to the coordinate axis system, and it is denoted as I_{zy} or I_{yz}, so

$$T_x = \omega^2 I_{yz} \tag{13.16}$$

where ω_z is the angular velocity of the group of masses about z. This product of inertia I_{yz} and the other products of inertia I_{xy} and I_{xz} are defined not only for collections of discrete masses but also for continuously distributed masses such as variously shaped bodies. Analogous expression can be written for torques about other axes. This will be discussed further in Examples 13.1 and 13.2.

Procedure 13.1: Dynamic Balancing of Extended Rotors   This procedure consists of choosing the two correction planes in which to put the balancing masses, determining the imbalance moments about axes in the first of these planes, computing the magnitude and location of a balance mass needed in the second plane, and then determining the imbalance moments about axes in the second of these planes and computing the magnitude and location of a balance mass needed in the first plane.

1. Choose the locations of the two correction planes to be used. These planes will be perpendicular to the axis of rotation, and the distance between these planes will be referred to as L. In order to minimize the balance mass required to eliminate the shaking couple, these planes usually will be chosen to be as far from each other as is convenient. Otherwise, they can be placed at any desired locations.

2. Choose one of the correction planes and call it plane A, and call the other correction plane B. Establish a coordinate system *with its origin in this plane A* and with its z axis aligned along the axis of rotation of the rotor. This will place the x and y axes in correction plane A. It usually is most convenient to orient the coordinate system such that the positive z axis points toward the majority of the imbalance masses. (An example rotor consisting of three imbalance masses and correction planes A and B is shown in Fig. 13.6.)

3. Determine the distance z_i from the origin to each mass m_i. Note that the sign (positive or negative) of each z_i will depend on whether the corresponding mass is in the positive or negative direction from the origin along the z axis. Examples of such distances are indicated in Fig. 13.6.

4. Draw an end view of the rotor showing the x and y axes and the locations of the center of gravity of each imbalance mass in the rotor. Figure 13.7a is such

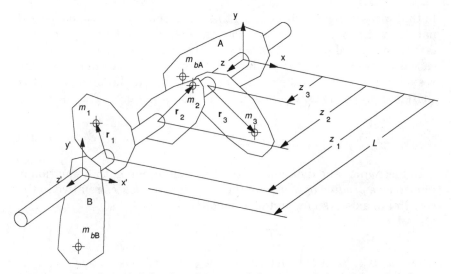

Figure 13.6 The dynamic balancing of an extended rotor consisting of three imbalance masses.

a drawing, and it corresponds to the rotor in Fig. 13.6. A vector from the axis to each mass m_i will be its radius vector $\mathbf{r}_i$ at an angle θ_i from the x axis and with a magnitude r_i, as shown in Fig. 13.7a.

5. Choose one of the imbalance masses and number it mass 1. On a transparency laid over the drawing of step 4 or on a separate CAD layer, draw a product of inertia vector $\mathbf{U}_1 z_1 = m_1 \mathbf{r}_1 z_1$ for that mass. This vector will have its tail at the origin, will lie on top of $\mathbf{r}_1$ for that mass, and will have a magnitude $m_1 r_1 z_1$ drawn to a convenient scale. Figure 13.7b is such a drawing.

6. Choose another one of the imbalance masses and number it mass i, where i is a number that is equal to 1 plus the number of the previous mass chosen. On the transparency laid over the drawing of step 4 or on the separate CAD layer, draw from the tip of the previous vector a product of inertia vector $\mathbf{U}_i z_i = m_i \mathbf{r}_i z_i$, where the subscripts i signify that the values are those associated with this mass i. This vector will have its tail at the tip of the previous vector, will have a magnitude $m_i r_i z_i$, and will be drawn to the same scale as used in step 5.

7. Repeat step 6 until vectors have been drawn tail to head for all the imbalance masses. Figure 13.7b shows such a drawing, including the product of inertia vectors for the three example masses in Fig. 13.6.

8. From the tip of the last vector drawn, draw a vector to the origin. This vector is $\mathbf{U}_b L = m_b \mathbf{r}_b L$ for the balance mass in plane B. The distance factor L will be positive if the plane B is in a positive direction along the z axis from the origin in plane A; otherwise, it is negative. Divide the magnitude of this vector $\mathbf{U}_b L$ by the distance L between the correction planes. If L is negative, the direction of the resulting vector is *opposite* to the direction of the vector before the division. The result is a vector $\mathbf{U}_b$ that points in the direction

of the location of the balance weight in correction plane B and whose magnitude is equal to the mass m_b times the radius r_b from the axis of the rotor to this balance mass. A value for either m_b or r_b can be chosen, and the other can be calculated from the magnitude of the vector as scaled from the drawing.

9. Choose the correction plane B and establish a coordinate system *with its origin in this plane* and with its z axis aligned along the axis of rotation of the rotor. This will place the x and y axes in this correction plane B. The *directions* of these axes can be the same as those established in step 2, or they can have x and z directions reversed so that the z_i and L values have more convenient signs.

10. Repeat steps 3, 5, 6, and 7.

11. From the tip of the last vector drawn, draw a vector to the origin. This vector is $\mathbf{U}_b L = m_b \mathbf{r}_b L$ for the balance mass in plane A. The distance L will be negative if the plane A is in a negative direction along the z axis from the origin in

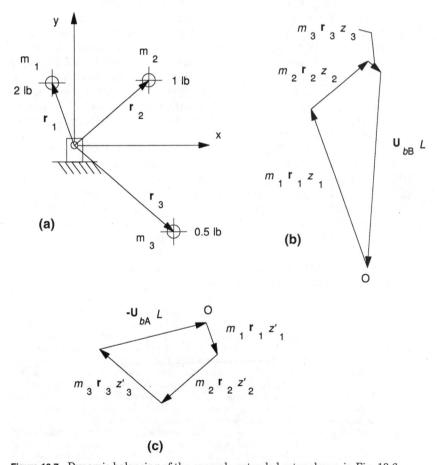

Figure 13.7   Dynamic balancing of the example extended rotor shown in Fig. 13.6.

plane B; otherwise, it is positive. Divide the magnitude of this vector $\mathbf{U}_b L$ by the distance L between the correction planes. If L is negative, the direction of the resulting vector is opposite to the direction of the vector before the division. The result is a vector $\mathbf{U}_b$ that points in the direction of the location of the balance weight in correction plane A and whose magnitude is equal to the mass m_b times the radius r_b from the axis of the rotor to this balance mass. A value for either m_b or r_b can be chosen, and the other can be calculated from the magnitude of the vector as scaled from the drawing.

12. If the rotor is to be subjected to angular acceleration about its axis of rotation, remember that a compensating rotor may be required, as discussed in a preceding subsection on dynamically balanced rotors with angular acceleration.

This procedure will be illustrated by the following example.

Example 13.1: Dynamic Balancing of an Extended Rotor Figure 13.6
shows a rotor that consists of three imbalance masses m_1, m_2, and m_3 whose centers of gravity are at the tips of the vectors $\mathbf{r}_1$, $\mathbf{r}_2$, and $\mathbf{r}_3$, respectively. The weights of these masses are, respectively, 2, 1, and 0.5 lb. Figure 13.7a is an end view of the locations of the masses (in accordance with step 4 of Procedure 13.1), and it can be seen that this figure is identical to Fig. 12.4a. However, the centers of gravity of the masses in Fig. 12.4a were all in the same plane, which was perpendicular to the rotation axis, whereas in this example the masses are distributed along the axis of rotation, as shown in Fig. 13.6. These masses are equally spaced 1 in apart along the rotation axis. The angles that the radius vectors $\mathbf{r}_1$, $\mathbf{r}_2$, and $\mathbf{r}_3$ make with the x axis are 110°, 40°, and $-40°$, respectively, just as in the case of Fig. 12.4a.

For purposes of balancing, a correction plane A is established arbitrarily at 0.5 in from the plane of mass m_3, and a correction plane B is established arbitrarily 0.5 in from mass m_1 as shown in Fig. 13.6 (step 1 of Procedure 13.1). Although balance masses are shown in planes A and B, the weights and locations of these masses are yet to be determined. Then, in accordance with step 3, the distances from plane A to the masses m_1, m_2, and m_3 are $z_1 = 2.5$ in, $z_2 = 1.5$ in, and $z_3 = 0.5$ in, respectively, and the separation distance between plane A and plane B is $L = 3.0$ in.

Then, in preparation for steps 5, 6, and 7, the products of inertia of the masses are computed[2] as

---

[2]Each of these quantities is a product of inertia of a mass referred to a coordinate system that consists of the z axis, an axis parallel to the $\mathbf{r}$ vector for that particular mass, and a third axis perpendicular to those two axes. These products of inertia are therefore referred to three different axis systems. However, all these systems share a common axis (the z axis). Therefore, each $\mathbf{r}$ vector, and thus each magnitude r, can easily be resolved into components in the x and y axes in a common axis system such as the xyz system in Fig. 13.6. Therefore, each of these products of inertia also can be resolved into components in that common system. These components can be added just like vector components to give components of the total product of inertia, so these products of inertia will be represented by "product of inertia vectors." Each product of inertia vector will be parallel to the corresponding $\mathbf{r}_i$ vector and will have a corresponding magnitude of $m_i r_i z_i$. These vectors therefore can be added and subtracted vectorially without resolution to give the total effects that we seek.

$$m_1 r_1 z_1 = (2)(1)(2.5) = 5.0 \text{ lb in}^2$$

$$m_2 r_2 z_2 = (1)(1.5)(1.5) = 2.25 \text{ lb in}^2 \qquad (13.17)$$

and
$$m_3 r_3 z_3 = (0.5)(2)(0.5) = 0.5 \text{ lb in}^2$$

For step 5, Fig. 13.7b shows a product of inertia vector $\mathbf{I}_{rz_1} = m_1\mathbf{r}_1 z_1$ that represents the product of inertia of mass m_1. This vector has a magnitude of 5 lb in^2, as calculated above, and is parallel to $\mathbf{r}_1$ in Fig. 13.7a. It is drawn from the origin O at a scale of 2 lb in^2 per inch, so it is 2.5 in long.

For steps 6 and 7, the vectors $\mathbf{I}_{rz_2} = m_2\mathbf{r}_2 z_2$ and $\mathbf{I}_{rz_3} = m_3\mathbf{r}_3 z_3$ were calculated and plotted in the same manner but connected tail to head as shown in Fig. 13.7b.

In step 8 the vector polygon in Fig. 13.7b is closed by the vector $\mathbf{I}_{bB} = \mathbf{U}_{bB}L$, which is the product of inertia of the required balance mass in plane B. That balance mass will be located in plane B in a direction from the rotation axis that is the same as the direction of vector $\mathbf{U}_{bB}L$. The distance of the mass from that axis is computed from the magnitude of $\mathbf{U}_{bB}L$, which is $U_{bB}L = m_b r_b L$. The vector $\mathbf{U}_{bB}L$ as plotted in Fig. 13.7b is 2.92 in long and at an angle of 266° from the x axis, so because the drawing scale is 2 lb in^2 per inch, $U_{bB}L = m_b r_b L = 5.84$ lb in^2. Then, because $L = 3.0$ in, $m_b r_b = 5.84/3 = 1.95$ lb in. A balance weight of 1.0 lb is therefore placed in plane B at 1.95 in from the axis at an angle of 266° from the xz plane. This balance mass m_{bB} is shown in plane B in Fig. 13.6.

Next, plane B is chosen (step 9), and an axis system $x'y'z'$ is placed in it with its axes parallel to those which were used in the previous computations, as shown in Fig. 13.6. The products of inertia of the three masses are calculated, but now the z distances are measured from the B plane. These axes in the B plane are displaced by a distance L in the positive z direction from those in the A plane. Therefore, lengths $z_i - L$ are used instead of lengths z_i, and these lengths will be negative. The computations are

$$m_1 r_1(z_1 - L) = (2)(1)(-0.5) = -1.0 \text{ lb in}^2$$

$$m_2 r_2(z_2 - L) = (1)(1.5)(-1.5) = -2.25 \text{ lb in}^2 \qquad (13.18)$$

and
$$m_3 r_3(z_3 - L) = (0.5)(2)(-2.5) = -2.5 \text{ lb in}^2$$

The product of inertia vectors that result from these computations are plotted in Fig. 13.7c. The vector $\mathbf{I}_{bA} = -\mathbf{U}_{bA}L$ that closes the polygon is 1.69 in long and is at an angle of 13° from the x axis. Then, because the drawing scale is 2 lb in^2 per inch, the required product of inertia of the balance weight in plane A is 3.38 lb in^2. However, because the factor by which this quantity must be divided is $-L = -3.0$, the imbalance vector for this weight has a magnitude of 1.13 lb in but in a direction 180° from the direction of $\mathbf{I}_{bA} = -\mathbf{U}_{bA}L$. A weight of 0.56 lb in plane A at a distance of 2.0 in from the axis and at an angle of 193° will provide the required balance. Such a balance weight is shown in plane A in Fig. 13.6.

It will be noted that the balance weight in plane A is located angularly in a direction such that it tends to more nearly oppose the d'Alembert force effects of masses m_3 and m_2 than of mass m_1. This can be seen to be because d'Alembert forces produced by masses m_3 and m_2 have larger moment arms relative to axes in plane B than does that of mass m_1, and the d'Alembert force produced by mass in plane A must balance these moments. Similarly, and for the same reasons, the balance weight in plane B is located angularly in a direction such that it tends to more nearly oppose the d'Alembert force effects of masses m_1 and m_2 than of mass m_3.

Example 13.1a: Computed Dynamic Balancing of an Extended Rotor
The results found in Example 13.1 by graphical means also can be obtained by direct numerical computation. To do so, the product of inertia vector magnitudes computed in Eqs. (13.17) and (13.18) are resolved into x and y components and added numerically to give the components of the required balance weight locations. For these computations for the balance mass in plane B, the following table may be constructed using the results in Eqs. (13.17):

i = mass no.	$I_{rzi} = m_i r_i z_i$	θ_i	$I_{rzi} \cos \theta_i$	$I_{rzi} \sin \theta_i$
1	5.0 lb in²	110°	−1.710 lb in²	4.698 lb in²
2	2.25 lb in²	40°	1.724 lb in²	1.446 lb in²
3	0.5 lb in²	−40°	0.383 lb in²	−0.321 lb in²
Totals			0.397 lb in²	5.823 lb in²

The total in the fourth column is the product of inertia that must be *counteracted* by the product of inertia $I_{bxz} = m_b x_b L$ of the balance mass in the B plane, and thus that balance product of inertia must be the negative of that total. In the foregoing expression, m_b is the mass of that balance weight, x_b is the x distance from the rotation axis to that weight, and $L = 3.0$ in is the distance separating the correction planes A and B. Therefore,

$$m_b x_b (3.0) = -0.397 \text{ lb in}^2 \quad \text{or} \quad m_b x_b = -0.132 \text{ lb in} \qquad (13.19)$$

The total in the fifth column is the product of inertia that must be *counteracted* by the product of inertia $I_{byz} = m_b y_b L$ of the balance mass in the B plane, where y_b is the y distance from the rotation axis to that weight. Therefore,

$$m_b y_b (3.0) = -5.823 \text{ lb in}^2 \quad \text{or} \quad m_b y_b = -1.941 \text{ lb in} \qquad (13.20)$$

The quantities computed in Eqs. (13.19) and (13.20) are x and y components, respectively, of the imbalance vector $\mathbf{U}_{bB}$ of the required balance mass in the B plane. This vector therefore has a magnitude that is the square root of the sum of the squares of the components or $U_{bB} = [(-0.312)^2 + (-1.941)^2]^{0.5} = 1.946$ lb in. The angle of this vector is found from the fact that the tangent of the angle is equal to the y component [from Eq. (13.20)] divided by the x component [from Eq. (13.19)]. Therefore,

$$\tan \theta_{bB} = \frac{-1.941}{-0.132}$$

so $\theta_{bB} = 266°$. These are the same results as were obtained by the graphical method.

For the computations for the balance mass in plane A, a similar procedure is followed. The products of inertia are computed relative to axes in plane B. Because these axes are displaced in the positive z direction from those in the A plane, the lengths $z_i - L$ are used instead of lengths z_i, and these lengths will be negative. The following table may then be constructed using the results in Eq. (13.18):

i = mass no.	$I_{rzi} = m_i r_i (z_i - L)$	θ_i	$I_{rzi} \cos \theta_i$	$I_{rzi} \sin \theta_i$
1	-1.0 lb in^2	110°	0.342 lb in^2	-0.940 lb in^2
2	-2.25 lb in^2	40°	-1.724 lb in^2	-1.446 lb in^2
3	-2.5 lb in^2	$-40°$	-1.915 lb in^2	1.607 lb in^2
Totals			-3.297 lb in^2	-0.779 lb in^2

The total in the fourth column is the product of inertia that must be *counteracted* by the product of inertia $I_{bxz} = m_b x_b (-L)$ of the balance mass in the A plane, and thus this balance product of inertia must be the negative of that total. In the foregoing expression, m_b is the mass of that balance weight, x_b is the x distance from the rotation axis to that weight, and $L = 3.0$ in is the distance separating the correction planes A and B. The factor $-L$ is used in this computation because the A plane is at a *negative* distance from the B plane when measured along the z' axis. Therefore,

$$m_b x_b (-3.0) = 3.297 \text{ lb in}^2 \quad \text{or} \quad m_b x_b = -1.099 \text{ lb in} \quad (13.21)$$

The total in the fifth column is the product of inertia that must be *counteracted* by the product of inertia $I_{byz} = m_b y_b (-L)$ of the balance mass in the A plane, where y_b is the y distance from the rotation axis to that weight. Therefore,

$$m_b y_b (-3.0) = 0.779 \text{ lb in}^2 \quad \text{or} \quad m_b y_b = -0.260 \text{ lb in} \quad (13.22)$$

The quantities computed in Eqs. (13.21) and (13.22) are x and y components, respectively, of the imbalance vector $\mathbf{U}_{bA}$ of the required balance mass in the A plane. This vector therefore has a magnitude that is the square root of the sum of the squares of the components or $U_{bA} = [(-1.099)^2 + (-0.260)^2]^{0.5} = 1.129$ lb in. The angle of this vector is found from the fact that the tangent of the angle is equal to the y component [from Eq. (13.22)] divided by the x component [from Eq. (13.21)]. Therefore,

$$\tan \theta_{bA} = \frac{-0.260}{-1.099}$$

so $\theta_{bB} = 193°$. These are the same results as were obtained by the graphical method.

Example 13.2: Dynamic Balancing of a Rotor Having Continuously Distributed Masses In Examples 13.1 and 13.1a, each of the imbalance and balance masses was represented by a point mass located at the center of gravity of the corresponding actual mass. For rotor-balancing procedures, such a representation for a mass is valid if (1) the mass is distributed symmetrically relative to a plane perpendicular to the axis of rotation or (2) the mass does not have such symmetry but has product of inertia referred to its center of gravity of $I_{xz} = I_{yz} = 0$, where z is an axis parallel to the axis of rotation. This latter case corresponds to a *principal axis* of the mass being parallel to the axis of rotation. There are frequent occasions in which these conditions do not prevail but in which the rotor must still be dynamically balanced. Such a case is illustrated in Fig. 13.8, which shows top and front views of a scanning mirror assembly.

In Fig. 13.8, mass m_1 is a mirror that is mounted at 45° to the axis of rotation of a scanning mechanism that is spun about the horizontal axis shown. The mirror and the prismatic block to which it is mounted are the only portions of the scanner that are not symmetrical about the z axis.

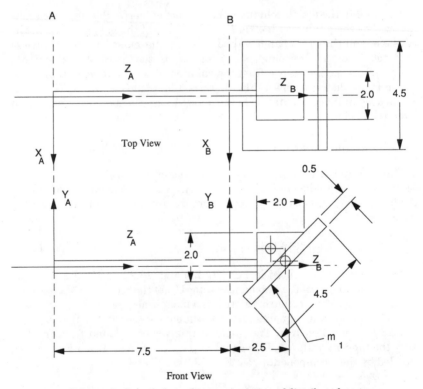

Front View

Figure 13.8 A dynamically unbalanced rotor consisting of distributed masses.

Therefore, they are the only components that produce d'Alembert imbalance forces and torques.

The dimensions in centimeters of the mirror and block are as given in the figure. It has been decided that the balance weights will be placed in a plane A at a distance of 10 cm from the center of the mirror face and in a plane B at a distance of 2.5 cm from the center of the mirror face as shown. This establishes axes X_A, Y_A, and Z_A in plane A and axes X_B, Y_B, and Z_B in plane B. The mirror has a mass m_1 of 26 g, and its center of gravity is 0.177 cm to the left of and 0.177 cm above the center of the mirror face. The block has a mass m_2 of 11.2 g, and its center of gravity is 0.687 cm to the left of and 0.687 cm above the center of the mirror face.

The balancing procedure is very much like previous procedures, consisting of computing the products of inertia of the imbalance masses, first referred to axes in the A plane and then referred to axes in the B plane. However, these products of inertia are not simply the product of the mass times the y distance times the z distance from the origin of the axis system in the A or B plane. The parallel-axis theorem for products of inertia must be used.

The *parallel-axis theorem for products of inertia* states that if I_{yzG} is a product of inertia of a body referred to a body axis system with its origin at the body's center of gravity, the corresponding product of inertia I_{yz} of that body referred to a displaced axis system whose axes are parallel to those of the body axis system is given by

$$I_{yz} = my_0z_0 + I_{yzG} \qquad (13.23)$$

where m is the mass of the body and y_0 and z_0 are the y and z distances, respectively, from the displaced origin to the center of gravity of the body measured in the displaced axis system.

The my_0z_0 term in Eq. (13.23) is the same as the product of inertia terms that have been used in Procedure 13.3 and Examples 13.1 and 13.1a. Therefore, it can be seen that the computations are the same as described in Procedure 13.2, except that the product of inertia of each body *about its center of gravity* must be added in order to obtain the total product of inertia of the rotor as referred to each correction plane.

Products of inertias of bodies referred to their centers of gravity can be computed automatically by some CAD systems and by some dynamic simulation systems. This is particularly true if the body is readily composed of so-called primitives in the software library. If the products of inertia must be computed manually, procedures for doing so are presented in texts and handbooks on mechanics, classical mechanics, and dynamics. Such references give formulas for simple shapes and describe integration methods and coordinate transformation methods.

The products of inertia for the mirror and for the block were computed using methods in a book by Den Hartog.[3] The results are for the mirror

---

[3]Den Hartog, J. P., *Mechanics,* Dover Publications, New York, 1961, Chap. XII.

(which is mass 1), $I_{yzG_1} = 21.7$ g cm², and for the block (which is mass 2), I_{yzG_2} = 1.24 g cm². Then, because the centers of gravity of all masses are in the yz plane and the masses are symmetrical about the yz plane, the balance masses will be in the yz plane and there is no need to perform a graphical construction. We therefore skip directly to step 5 of Procedure 13.2 and perform the subsequent steps by direct computation. The resulting equations, which are analogous to Eqs. (13.17), are then as follows:
Referred to the axes in the A plane:

$$m_1 y_1 z_1 + I_{yzG_1} = (26)(0.177)(10 - 0.177) + 21.7 = 45.20 + 21.7 = 66.90 \text{ g cm}^2$$

$$m_2 y_2 z_2 + I_{yzG_2} = (11.2)(0.687)(10 - 0.687) + 1.24 = 71.66 + 1.24 = 72.90 \text{ g cm}^2$$

The total product of inertia that must be *balanced* by the product of inertia I_{yzB} of a mass m_B in plane B is the sum of these two numbers, or 66.90 + 72.90 = 139.80, so

$$I_{yzB} = -139.80 \text{ g cm}^2$$

Because m_B is 7.5 cm from the A plane, the imbalance magnitude U_B of that balance mass is $U_B = I_{yzB}/7.5 = (-139.80)/(7.5) = -18.64$ g cm. This imbalance can be provided by a mass in plane B of 9.32 g placed 2 cm downward from the axis of rotation.

Next, referred to the axes in the B plane, the equations, which are analogous to Eqs. (13.18), are

$$m_1 y_1 z_1 + I_{yzG_1} = (26)(0.177)(2.5 - 0.177) + 21.7 = 10.69 + 21.7 = 32.39 \text{ g cm}^2$$

$$m_2 y_2 z_2 + I_{yzG_2} = (11.2)(0.687)(2.5 - 0.687) + 1.24 = 13.95 + 1.24 = 15.19 \text{ g cm}^2$$

The total product of inertia that must be *balanced* by the product of inertia I_{yzA} of a mass m_A in plane A is the sum of these two numbers, or 32.39 + 15.19 = 47.58, so

$$I_{yzA} = -47.58 \text{ g cm}^2$$

Because m_A is -7.5 cm from the B plane (it is in the negative z direction from the origin of the axes in plane B), the imbalance magnitude U_A of that balance mass is $U_A = I_{yzA}/(-7.5) = (-47.58)/(-7.5) = +6.34$ g cm. This imbalance can be provided by a mass in plane A of 6.34 g placed 1.0 cm upward from the axis of rotation.

Comments and suggestions

In Example 13.2 the centers of gravity of all the masses were in a single plane containing the axis of rotation of the rotor. In addition, there were no products of inertia I_{xz}'s that involve axes out of that plane. As a consequence, the net d'Alembert shaking force produced by these masses lay in that plane (acting at the center of gravity of the rotor), and the d'Alembert shaking couple or torque produced acted about an axis perpendicular to that plane. As shown in Sec. 12.2, such a combination of a force and couple is equivalent to a single force, equal in magnitude and direction to the original force but applied at some displaced point. Therefore, the dynamic imbalance could have been balanced by a single balance mass appropriately placed. Unfortunately, in the case of Example 13.2, that single balance mass would be to the right of the mirror and below the z axis, so it and/or its support structure could interfere with the optical path of the device unless carefully designed. In other situations, such a single balance mass might easily provide a satisfactory solution.

Again, as in static balancing, the appropriate *removal* of mass also can be used for dynamic balancing.

13.4   Dynamic Forces and Moments in a Mechanism

Section 12.3 described procedures for determining the forces and torques that will exist at various locations in a mechanism when forces and torques are applied statically to that mechanism. If a mechanism consists of parts that have appreciable mass, motion of that mechanism can produce appreciable inertial forces and torques on the parts of the mechanism and on the joints between those parts. These forces and torques are, of course, the results of the translational and angular accelerations of the masses involved. In Sec. 13.2 it was shown that accelerations can be considered to produce d'Alembert or inertial reaction forces and torques. Such d'Alembert forces and torques can be calculated from the known accelerations and masses and moments of inertia. Then these forces and torques can be used in *kinetostatic analysis* procedures that are identical to the static analysis procedures of Sec. 12.3.

Procedure 13.2 and Example 13.3: Finding Dynamic Forces in a Four-Bar Linkage   This procedure for kinetostatic analysis of a linkage will be illustrated by analyzing the four-bar linkage shown in Fig. 13.9a. The link lengths are $L_1 = 1.0$ ft, $L_2 = 3.0$ ft, $L_3 = 2.5$ ft, and $L_4 = 3.5$ ft. The scale in the drawing is 1.0 ft per inch. The weight of the coupler link is 100 lb, so its mass is $100/32.2 = 3.11$ slugs. The center of gravity of the coupler is 2.0 ft from pivot A and 1.25 ft from pivot B in the directions shown in the figure. The moment of inertia of the coupler about that center of gravity is 3.11 slugs ft^2. For this example, the inertial effects of the input crank L_1 and the rocker L_3 will not be considered because their effects can simply be added to those of the coupler, as was shown in Procedure 12.3 and Example 12.2.

The linkage is driven through link L_1, and the angle of that link at the instant used in this example is $\theta_1 = 60°$. The input angular velocity is

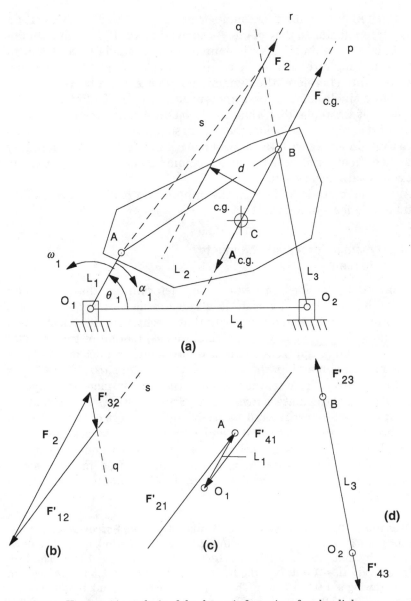

Figure 13.9 Kinetostatic analysis of the dynamic forces in a four-bar linkage.

$\omega_1 = 3.0$ rad/s (i.e., counterclockwise), and the input angular acceleration is $\alpha_1 = -6.0$ rad/s² (i.e., clockwise). The position, velocity, and acceleration analyses of this linkage for these conditions give the angular acceleration of the coupler as $\alpha_2 = 3.713$ rad/s² (i.e., counterclockwise) and the acceleration of the center of gravity of the coupler as $\mathbf{A}_{c.g.} = 4.382$ ft/s² at an angle of 241.37° (measured from the positive x axis, which is directed horizontally and to the right). This acceleration is shown as a vector downward and to the left from the center of gravity in Fig. 13.9.

1. Calculate the d'Alembert (inertial reaction) *torque* on each individual link, considering it to be independent of all other links. This is computed from the relation $T = -I\alpha$, where T is the d'Alembert torque, I is the moment of inertia of the link about its center of gravity, and α is the angular acceleration of the link. For the coupler in this example, $T = -(3.11)(3.713) = -11.55$ ft · lb. Note that because the angular acceleration is counterclockwise, the d'Alembert torque is clockwise.

2. Calculate the d'Alembert (inertial reaction) *force* resulting from translational acceleration of the center of gravity of each individual link, considering it to be independent of all other links. This is computed from the relation $F = -ma$, where F is the d'Alembert force, m is the mass of the link and a is the translational acceleration of the center of gravity of the link. This force is a vector that acts at the center of gravity and is directed in the opposite direction from the acceleration vector. For the coupler in this example, $F = (3.11)(4.382) = 13.63$ lb. Note that because the acceleration is downward and to the left, the d'Alembert force is upward and to the right from the center of gravity, as shown as $\mathbf{F}_{c.g.}$ in Fig. 13.9. Just by *pure accident,* this force vector passes through the pivot at point B *in this example.* This is of no significance.

3. For each link, combine the torque and force from steps 1 and 2 to obtain a single resulting force and its line of action. This consists of displacing the force computed in step 2 by a distance $d = T/F_{c.g.}$ in a direction such that the displaced force will produce a torque about the center of gravity in the same direction as the torque computed in step 1. The displaced force vector *replaces* the vector that was through the center of gravity. For the coupler in this example, $d = (-11.55)/(13.63) = -0.847$ ft. Readers can either interpret the negative sign to indicate that the displacement is to the left of the original vector when looking in the direction in which that vector is pointing, or they can simply choose the displacement direction that gives the correct direction of torque about the center of gravity. In this example, the displaced force vector is shown in Fig. 13.9 as $\mathbf{F}_2$ at distance d from the force vector through the center of gravity.

4. The resulting force vectors computed in step 3 can now be treated in the same manner as static forces for the remainder of the analysis. That is, the analysis becomes a *kinetostatic* analysis. The joint forces and torques that result from the inertial effects of each link in turn will be computed to provide partial answers, just as in Procedure 12.3 and Example 12.2.

From this point onward, perform the steps that follow step 2 in Procedure 12.3.

For this example, the direction of force $\mathbf{F}_{32}$ that is applied to link L_2 at point B by link L_3 is parallel to link L_3, so line q is drawn in Fig. 13.9a. The line of action of $\mathbf{F}_2$ (line r) intersects line q as shown. The line of action (line s) of the force $\mathbf{F}_{12}$ that is applied to coupler L_2 at point A by crank L_1 must intersect $\mathbf{F}_2$ and $\mathbf{F}_{32}$ at the same intersection point, so line s is drawn as shown (steps 4 and 5 of Procedure 12.3). This determines the *directions* of $\mathbf{F}_{32}$ and $\mathbf{F}_{12}$.

In Fig. 13.9b the vector $\mathbf{F}_2$ is drawn to a scale of 5 lb per inch (and parallel to $\mathbf{F}_2$ in Fig. 13.9a) with a line q through its tip and a line s through its tail where lines q and s are parallel, respectively, to line q and s in Fig. 13.9a. The first partial answers, vectors $\mathbf{F}'_{32}$ and $\mathbf{F}'_{12}$ are then drawn, respectively, from the tip and tail of $\mathbf{F}_2$ to the intersection of lines q and s.

Measuring these vectors and applying the chosen scale factor of 5 lb in gives magnitudes of $F_{32} = 3$ lb and $F_{12} = 11.5$ lb.

In Fig. 13.9c link L_1 is drawn, and the force $\mathbf{F}'_{21}$ is drawn as applied to that link by link L_2 at point A. (Note that $\mathbf{F}'_{12}$ is equal and opposite to $\mathbf{F}'_{21}$.) For link L_1 to be in equilibrium, an equal and opposite force $\mathbf{F}'_{41}$ must be applied to it by ground (L_4) at point O_1 as shown. The two forces $\mathbf{F}'_{21}$ and $\mathbf{F}'_{41}$ are not colinear, so they produce a couple or torque. The separation between the lines of action of these two forces is measured (scaled) from the drawing as 0.140 ft. The magnitude of the torque produced is therefore $F'_{21}(0.140) = (11.5)(0.140) = 1.6$ ft $\cdot$ lb in a clockwise direction. This torque must be balanced by a counterclockwise drive torque at point O_1 of $T'_{41} = 1.6$ ft $\cdot$ lb.

Link L_3 is drawn in Fig. 13.9d, and the force $\mathbf{F}'_{23}$ is drawn as applied to that link by link L_2 at point B. (Note that $\mathbf{F}'_{23}$ is equal and opposite to $\mathbf{F}'_{32}$.) For link L_3 to be in equilibrium, an equal and opposite force $\mathbf{F}'_{43}$ must be applied to it by ground (L_4) at point O_2 as shown. The forces $\mathbf{F}'_{43}$ and $\mathbf{F}'_{23}$ are colinear, so no torque is generated.

At this point all the first partial answers for the joint forces and torques have been computed. The remaining partial answers and the totals are computed just as in Procedure 12.3 and Example 12.2.

Alternatively, once the steps 1, 2, and 3 of Procedure 13.3 have been completed, the answers could have been computed simultaneously by the method described following Example 12.2.

13.5 Dynamic Balancing of Mechanisms

General considerations

It is generally very difficult or impossible to both statically and dynamically balance any but the simplest mechanism *without adding gears and/or cams and additional compensating rotors*. In cases where such complete balancing is important, consideration should first be given *in the concept phase of the synthesis* to the nature and arrangement of the functions that are to be performed by the mechanism. Such considerations should include attempts to *minimize the masses* of components that must have appreciable motion. The considerations also should include attempts to devise *symmetry* in the ways in which the functions are to be performed. While no specific methods of attack are given, the following two examples illustrate two ways in which complete balancing was accomplished. A formal procedure for dynamically balancing four-bar linkages is presented in the next section.

Example 13.4: Balancing a Swing-Die Mechanism   It often is desirable to perform printing, punching, perforating, laminating, or other intermittent operations on products that are continuously passing a particular station in a manufacturing process. If the continuous motion of the product is not to be severely disrupted, the mechanism that performs the operation must match the velocity of the product at least approximately and at least for a brief time while performing the operation. In response to this requirement, a mechanism such as that shown schematically in Fig. 13.10 has been devised.

Figure 13.10 schematically shows a punch-and-die set that consists of a punch P and a die D that are used to punch holes in a web of material M that is continuously moving from left to right. To do this, the protrusion on the bottom of the punch is lowered until it extends slightly into the closely matching cavity of the die, thereby shearing through the web of material. Because the punch must be very accurately aligned with the die, the base plate B_2 to which the die is attached is fitted with rigid posts C that slide in guides G that are rigidly attached to the base plate B_1 to which the punch P is attached.

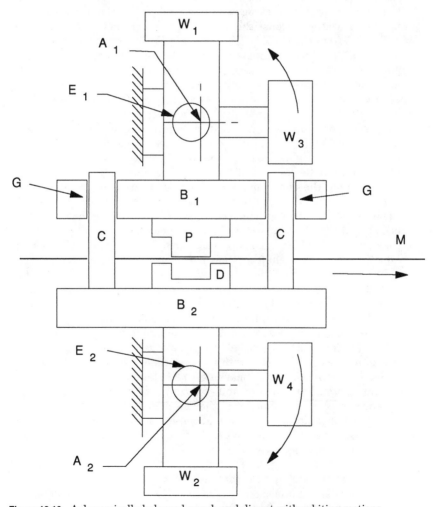

Figure 13.10 A dynamically balanced punch-and-die set with orbiting motions.

While the punch is penetrating the web, the punch and die set must move rightward with the web material. In the mechanism shown, the punch assembly PB_1G is caused to move along a circular orbital path in a counterclockwise direction while the die assembly is caused to move along a circular orbital path in a clockwise direction. When the punch is at the bottom of its path, the die is at the top of its path, the punch is extended into the die cavity, and they are both traveling rightward. These motions are produced by the continuous, synchronous, opposite rotations of eccentrics E_1 and E_2. Eccentric E_1 is attached to a shaft that rotates counterclockwise about the axis A_1 in a bearing that is fastened to ground. The bearing that surrounds the eccentric itself is attached to the base plate B_1 so that the plate is caused to move as the eccentric rotates. A similar arrangement is provided for the die set by the eccentric E_2, axis A_2, and plate B_2.

The guides G prevent rocking rotation of the plates B_1 and B_2 relative to each other. However, because the center of gravity of the punch assembly is at a distance from the center of the eccentric E_1, the orbital motion of that assembly will produce d'Alembert torques about the eccentric that will attempt to rock the assembly clockwise and counterclockwise. Similarly, the orbital motion of the die assembly will attempt to rock it back and forth. These rocking d'Alembert torques can impose large forces on the guides G because the guides are quite limited in their vertical length. To eliminate these d'Alembert torques and their associated guide loads, balancing weights W_1 and W_2 are added to the punch assembly and die assembly, respectively. These weights cause the centers of gravity of their respective assemblies to lie at the centers of the respective eccentrics. Acceleration of these assemblies therefore can produce no d'Alembert torques about the eccentrics.

The upper shaft then was balanced as a rotor, considering the mass of the punch assembly and its balance weight W_1 to be concentrated at the center of the eccentric E_1. This balancing was accomplished by adding a balance weight W_3 to the upper shaft as shown. The lower shaft was similarly balanced by adding balance weight W_4. Care was taken in the placement of all four of the balance weights to ensure that the total mechanism was dynamically balanced as well as being statically balanced.

A machine such as shown in Fig. 13.10 has been built using a punch-and-die set and base plates weighing about 70 lb and with eccentric offsets of about 0.5 in. This machine was balanced as discussed above and has run at speeds of in excess of 450 rpm with no perceptible vibration. The ease and completeness with which this machine was balanced result from the fact that all its motions are circular and at constant angular velocity, and thus balancing can be rotor balancing. Other mechanisms have been built and are available to perform the same function but which used some straight-line motions. Readers will find that such straight-line motions are generally more difficult to balance.

Example 13.5: Balancing a Reciprocating-Saw Mechanism Figure 13.11 schematically depicts a mechanism for providing reciprocating motion to a saw. The mechanism is driven by a constant-speed shaft located at S. Attached to this shaft is an arm L that carries a pivot P_1. A spur gear G_1 is

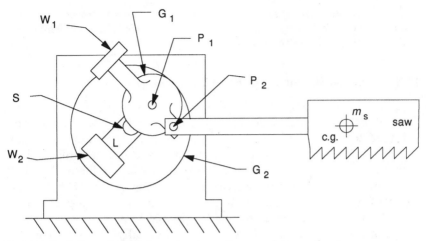

Figure 13.11 A dynamically balanced reciprocating-saw drive mechanism.

pivoted to arm L at pivot P_1. The pitch radius of gear G_1 is equal to the distance SP_1. Spur gear G_1 rolls around the inside of internal gear G_2, which is concentric with shaft S and which is fixed to ground. The pitch radius of G_2 is twice that of G_1, so any point on the pitch circle of G_1 will trace a 2:1 hypocycloid inside G_2. It can be shown that this hypocycloid will be a straight line lying along a diameter of G_2. The arm connected to the saw is pivoted to gear G_1 at a point P_2 that is on the pitch circle of G_1 and which therefore reciprocates along the horizontal diameter of G_2.

The motion of the saw is pure horizontal translation. The d'Alembert shaking force generated by the motion of the saw mass m_s will be a sinusoidally reversing horizontal force with a line of action through points S and P_2. This is exactly the same as the d'Alembert force that would be produced by an identical mass attached to G_1 at point P_2. For balancing purposes, then, we may consider the saw mass to be concentrated at point P_2 on G_1. Then balance G_1 as a rotor about point P_1 by adding a balance weight W_1 to G_1, so that the total center of gravity of W_1, G_1, and m_s (where m_s is considered to be at P_2) will be at point P_1.

Once the foregoing balancing has been done, the center of gravity of the saw plus gear G_1 plus the balance weight W_1 can be considered to be concentrated at point P_1 and to travel at constant speed along a circular path of radius SP_1 centered at point S. Then a balance weight W_2 can be added to arm L as shown, such as to cause the total center of gravity of arm L plus the concentrated weight at point P_1 plus the weight W_2 to lie at point S. That is, arm L is balanced as a rotor considering the weights of the saw, the gear G_1, and the balance weight W_1 to be concentrated at P_1.

Some cleverness is required in placing the weights W_1 and W_2 in a direction normal to the plane of the drawing to ensure that shaking couples are eliminated.

As in the preceding example, the fact that the moving masses can be considered to be moving along circular paths relative to parts of the mechanism simplifies the balancing task. Motions that differ from circular or sinusoidal motions are usually more difficult to balance. A general procedure for balancing four-bar linkages in which motions generally are *not* circular or sinusoidal is given in the next section.

13.6   Dynamic Balancing of Four-Bar Linkages

In general, an unbalanced four-bar linkage will produce a shaking (d'Alembert) force and a shaking (d'Alembert) torque or couple that will tend to shake the base to which the linkage is mounted. Because static balancing causes the total center of gravity of the linkage to remain fixed, such balancing eliminates the net shaking force. The shaking torque will persist because the angular momentum of the linkage will continue to vary as the individual linkage parts move, producing a total varying d'Alembert torque. A linkage that is an exact mirror image of the linkage (each linkage being statically balanced) could be mounted to the same base, and because the angular momentums of the links in this compensating linkage would exactly oppose those of the first, the net shaking torque would be zero. Such an arrangement is shown in Fig. 13.12a.

In Fig. 13.12a the original, statically balanced linkage is shown as $L_1L_2L_3L_4$ with static balance weights at m_{b_1} and m_{b_3}. The linkage is driven by the rotation of link L_1. The mirror-image compensating linkage is shown as $L_1'L_2'L_3'L_4$, including static balance weights. The compensating linkage would be driven by rotation of link L_1', whose angular position and velocity would be made to be the negatives of those of L_1. This can be accomplished by gearing the two drives together or by separate synchronized drives. Notice that because of the symmetrical placement of these two linkages, any vertical shaking forces would cancel each other, even if the individual linkages were not statically balanced. The same is not true for horizontal shaking forces, however.

Notice that because neither of these two linkages, when statically balanced, produces a shaking force, the compensating linkage may be moved about and oriented arbitrarily, as long as the motions of both linkages remain *parallel to* the same plane (the picture plane). The compensating linkage also can be enlarged or shrunk (without changing its proportions) as long as the masses and moments of inertia of its links are *carefully scaled*[4] to give the same (but opposite) d'Alembert

---

[4]The dimensions of the enlarged or shrunk compensating linkage can be obtained by multiplying the dimensions of the original linkage by a scale factor c_s. The moments of inertia of the compensating linkage links must be equal to those of the original. However, the mass of each of the compensating links must each be equal to $\frac{1}{c_s}$ times that of the corresponding original link.

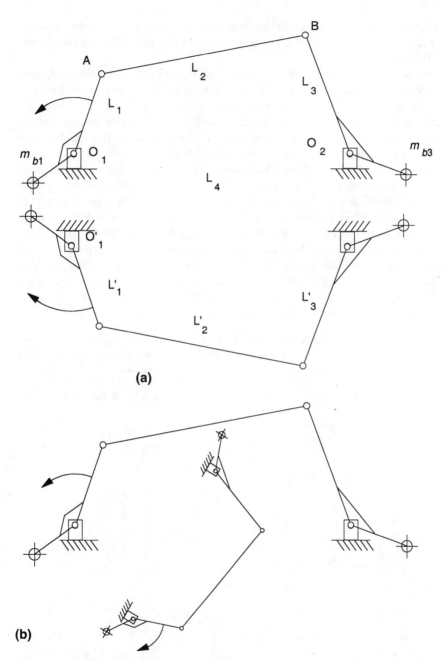

Figure 13.12 Dynamic balancing of a four-bar linkage by using a compensating four-bar linkage.

forces and torques as the corresponding links in the first linkage. Figure 13.12b shows such a possible position and size for the image linkage.

The shaking torque also can be eliminated *without* adding a compensating linkage by suitably redistributing the masses of the links and by adding one or two compensating rotors such as those described in the Sec. 13.3 subsection entitled "Dynamic balance of rotors having angular acceleration." Dynamic balancing of four-bar linkages in such a manner has many features that are similar to those of static balancing. However, whereas static balancing of such linkages can always be accomplished without modifying the coupler mass distribution, this generally is not true for this type of dynamic balancing. This dynamic balancing therefore starts with modification of the coupler. Once the coupler has been modified, the resulting masses that are equivalent to the coupler mass are considered to be parts of the links that are pivoted to ground. Then each of the links is dynamically balanced as a rotor using Procedure 13.1 (or Procedure 12.4 if it is a thin rotor). This four-bar linkage dynamic balancing procedure is as follows.

Procedure 13.3: Dynamic Balancing of Four-Bar Linkages   This procedure starts with consideration of the location of the center of gravity of the coupler and with consideration of the moment of inertia of that coupler.

1. On a drawing of the coupler (such as shown in Fig. 13.13), establish an axis system with its origin at one of the coupler pivots and in which the positive x axis

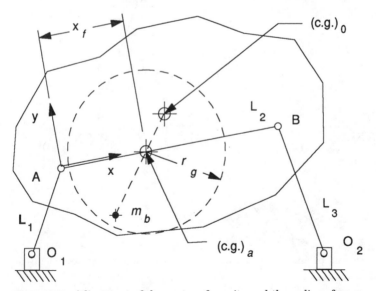

Figure 13.13 Adjustment of the center of gravity and the radius of gyration of the coupler.

points toward the other pivot. Draw the y axis perpendicular to the x axis. In this axis system, determine the x and y coordinates x_0 and y_0, respectively, of the center of gravity of the coupler.

2. Add a balance weight of mass m_b to the coupler such that the total center of gravity of the coupler plus this weight lies on the x axis. To do this, make $m_b y_b = -m_0 y_0$, where m_b is the mass of the balance weight, y_b is the y coordinate of the center of gravity of the balance weight, m_0 is the original mass of the coupler, and y_0 is the y coordinate of the original center of gravity of the coupler. Determine the coordinates of the resulting total center of gravity. As shown in Fig. 13.13, the resulting center of gravity of the balanced coupler will lie at the intersection $(c.g.)_a$ of the line connecting the coupler's pivots and the line from the original center of gravity to the center of gravity of the balance weight.

3. Determine the moment of inertia of the balanced coupler that results from step 2 about its center of gravity. If the coupler has been drawn and balanced on a CAD system, it may be possible to determine this moment of inertia automatically. Otherwise, it may be computed from formulas for simple shapes and with the use of the parallel-axis theorem for moment of inertia, which was discussed previously.

4. It is now necessary to add mass to or remove mass from the coupler in such a manner that its final center of gravity remains somewhere on the line connecting its pivots and also such that Eq. (13.23) is satisfied.

$$x_f(L_2 - x_f) = \left(r_g\right)_f^2 \tag{13.23}$$

where, by definition, $\left(r_g\right)_f^2 = I_f/m_f$ and x_f is the x coordinate of the final center of gravity of the coupler, L_2 is the distance between the coupler pivots, $\left(r_g\right)_f$ is the final (i.e., balanced) radius of gyration of the coupler, I_f is the final moment of inertia of the coupler about its center of gravity, and m_f is the final mass of the coupler.

Step 4 can be performed by trial-and-error adjustment of any of a large number of mass and dimension parameters. There are, however, a few guidelines that can be used to simplify this process. First, it will generally be most convenient if the added or removed masses do not move the total coupler center of gravity and therefore do not change the values on the left-hand side of Eq. (13.23). This can be accomplished by adding or removing masses whose centers of gravity coincide with the total center of gravity. The objective of adding or removing these masses is then simply to increase or decrease the radius of gyration $\left(r_g\right)_f$. Second, if a circle of radius equal to the coupler radius of gyration is drawn with its center at the coupler center of gravity, removing mass from anywhere inside that circle or adding mass anywhere outside that circle will increase the radius of gyration. Such a circle is shown in Fig. 13.13. Conversely, adding mass inside that circle or removing mass from outside that circle will decrease the radius of gyration. Third, the further the added or removed mass is from the center of gravity, the greater will be its effect on the radius of gyration.

Note that the balance mass m_b also can be changed in magnitude m_b and location to vary the radius of gyration as long as in so doing the total center of gravity remains on line AB.

Some CAD systems and dynamic simulation systems have what are referred to as *sensitivity and/or optimization study or search features.* The guidelines in the preceding paragraph can be used to set up rules that such study or search features can be used to automatically search for mass distributions that will satisfy Eq. (13.23).

5. Once the coupler mass distribution has been adjusted by step 4, the coupler mass *effects* can be replaced by an equivalent point mass at point A plus an equivalent point mass at point B. These two masses, taken together, will have the same total mass, the same center of gravity, and the same moment of inertia as the coupler plus all its mass adjustments from steps 2 and 4. Compute these equivalent masses from

$$m_A = \frac{m_f(L_2 - x_{0f})}{L_2} \tag{13.24}$$

$$m_B = \frac{m_f(x_{0f})}{L_2} \tag{13.25}$$

where m_A is the equivalent mass at point A, m_B is the equivalent mass at point B, m_f is the total mass of the adjusted coupler, L_2 is the distance between the coupler pivots, and x_{0f} is the x coordinate of the center of gravity of the adjusted coupler mass distribution.

6. Consider the equivalent mass m_A to be attached at point A on the link that is pivoted to the coupler at point A and also to ground. Consider the equivalent mass m_B to be attached at point B on the link that is pivoted to the coupler at point B and also to ground. Then balance each of these links dynamically as a rotor using Procedure 13.1 (or Procedure 12.4 if it is a thin rotor).

7. At least one of the two links that are pivoted to ground will experience varying angular velocity. Provide each of these links that experiences such varying angular velocity with a compensating rotor in accordance with the subsection entitled "Dynamically balanced rotors with angular acceleration." The moments of inertia used in computing parameters for these compensating rotors will be the moments of inertia of the links that are pivoted to ground after they have been dynamically balanced in step 6 and which are *considered* to include the *hypothetical* equivalent masses at points A and B.

Derivation of Procedure 13.3

In this derivation it will be shown that for generally encountered distributions of mass in the coupler of a four-bar linkage, dynamic balancing of that linkage is very complicated. It is then shown that if that coupler mass distribution is modified to meet certain conditions, the linkage can be dynamically balanced as though considered to be two rotors.

In Sec. 13.2 it was pointed out that a body that is experiencing both translational acceleration and rotational acceleration can be considered to produce both a d'Alembert force and a d'Alembert torque or couple. The line of action of the force will pass through the center of

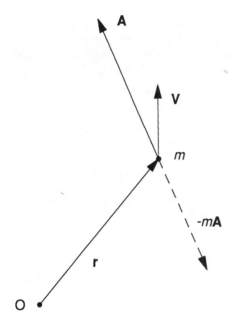

Figure 13.14 The moment of the momentum of a point mass referred to a chosen reference point.

gravity of the body and will be associated with that translational acceleration. The d'Alembert torque (or couple) will be associated with that rotational acceleration. Each link in a moving linkage will therefore produce such a d'Alembert force and such a d'Alembert torque. These *forces* will add to produce a total d'Alembert force with its line of action through the total center of gravity of the linkage. This total force will be proportional to the total mass of the linkage times the acceleration of the total center of gravity of the linkage. Static balancing ensures that the total center of gravity of the linkage will remain stationary, so such balancing eliminates the total d'Alembert shaking force.

To investigate the remaining d'Alembert torque, consider Newton's second law as applied to rotational motion. As indicated in Sec. 13.2, this can be expressed as: the *torque* applied to a rigid body is equal to the rate of change of the body's moment of momentum (which is also called the *angular momentum of the body*). To see how this can be applied to a four-bar linkage, refer to Fig. 13.14.[5]

---

[5]We will be dealing with planar motion, so all lengths, displacements, velocities, accelerations, and forces will be parallel to the reference plane and can be treated as vectors in that plane. All angles, angular velocities, angular momentums, angular accelerations, and torques will be measured about axes that are perpendicular to the reference plane. Strictly speaking, these angular quantities are vector quantities that are perpendicular to the plane, but in this text they frequently will be considered to be positive or negative scalar quantities.

In Fig. 13.14 a point mass of magnitude m is located relative to an origin O as indicated by the position vector $\mathbf{r}$. The velocity of this mass is given by $\mathbf{V} = \dot{\mathbf{r}}$, and the acceleration is given by $\mathbf{A} = \ddot{\mathbf{r}}$. The force that must be applied to cause the acceleration of the mass is a vector:

$$\mathbf{F}_{app} = m\mathbf{A} = m\ddot{\mathbf{r}} \tag{13.26}$$

Such a force would represent an applied moment or torque $\mathbf{T}_{app}$ about the chosen origin given by the vector cross-product

$$\mathbf{T}_{app} = \mathbf{r} \times \mathbf{F}_{app} = \mathbf{r} \times m\ddot{\mathbf{r}} = \mathbf{r} \times m\mathbf{A} \tag{13.27}$$

This result also can be obtained by using the concept of moment of momentum. The momentum of the point mass is the vector $m\dot{\mathbf{r}}$, and by analogy with the force vector, that momentum vector can be considered to have a moment of

$$\mathbf{H} = \mathbf{r} \times m\dot{\mathbf{r}} = \mathbf{r} \times m\mathbf{V} \tag{13.28}$$

relative to the chosen origin.

This moment $\mathbf{H}$ is defined as the *moment of momentum* (also called *angular momentum*) of the point mass relative to that chosen origin. If, then, we recall that according to Newton's law, the applied torque is equal to the rate of change of angular momentum, then

$$\mathbf{T}_{app} = \left(\frac{d\mathbf{H}}{dt}\right) = \left(\frac{d(\mathbf{r} \times m\dot{\mathbf{r}})}{dt}\right) = \mathbf{r} \times m\ddot{\mathbf{r}} + \dot{\mathbf{r}} \times m\dot{\mathbf{r}} \tag{13.29}$$

or, because the last cross-product in this equation is zero, $\mathbf{T}_{app} = \mathbf{r} \times m\ddot{\mathbf{r}}$, just as in Eq. (13.27). Equation (13.29) can be rewritten as

$$\mathbf{T}_{app} - \left(\frac{d\mathbf{H}}{dt}\right) = 0$$

where $-(d\mathbf{H}/dt)$ is the d'Alembert torque associated with the acceleration of the point mass.

For a system consisting of a number of masses, the total applied torque and the total d'Alembert torque are the sums of the corresponding torques associated with the individual masses, all referred to the same chosen origin, so we may write

$$\sum_i \mathbf{T}_i - \sum_i \frac{d\mathbf{H}_i}{dt} = 0 \tag{13.30}$$

where $\mathbf{T}_i$ is the torque applied to the ith mass and $\mathbf{H}_i$ is the moment of momentum of the ith mass, all referred to the same chosen origin. The second term in Eq. (13.30) is seen to be the sum of the d'Alembert torques resulting from the rates of change of the moments of momentum of all the point masses.

The total moment of momentum of a system of masses is the sum of the moments of momentum of the individual masses or

$$\mathbf{H} = \sum_i \mathbf{H}_i \qquad (13.31)$$

so $\dot{\mathbf{H}} = \sum_i \dot{\mathbf{H}}_i$ = negative of the sum of the d'Alembert torques (13.32)

Equation (13.32) shows that if the rate of change of the total moment of momentum $\mathbf{H}$ of a system of masses is known, all referred to the same chosen origin, the total d'Alembert or shaking torque experienced by the system can be found. Therefore, note the following characteristics of that total angular momentum:

First, it can be shown that if $\mathbf{H}_{c.g.}$ is the angular momentum of the system as calculated about its center of gravity, then

$$\mathbf{H} = \mathbf{H}_{c.g.} + \mathbf{R}_{c.g.} \times m\mathbf{V}_{c.g.} \qquad (13.33)$$

where $\mathbf{H}$ is the angular momentum about some chosen reference axis, $\mathbf{R}_{c.g.}$ is the vector from that axis to the instantaneous center of gravity of the system, m is the total mass of the system, and $\mathbf{V}_{c.g.}$ is the velocity of the center of gravity of the system. Static balancing of a four-bar linkage causes the total center of gravity of that linkage to remain fixed, so $\mathbf{V}_{c.g.}$ for a statically balanced linkage will be zero. Then the total angular momentum calculated *relative to any axis* will be equal to that which would be calculated about its center of gravity. The total shaking torque for a statically balanced linkage therefore can be computed from knowledge of the rate of change of that total angular momentum as calculated about any axis.

Second, the total angular momentum is the linear sum of the angular momentums of the individual masses. This was stated as Eq. (13.31). The two links in a four-bar linkage that are pivoted to ground are rotors, and as shown in Sec. 13.3, static balancing of them would cause their centers of gravity to be stationary, and the use of compensating rotors would remove any of their remaining angular momentum. The effects of these links therefore can be considered separately from the coupler effects. Attention therefore will be directed to the angular momentum of the coupler alone. The effects of the coupler angular momentum and the effects of its dynamic balancing can be superposed on those of the two links that are pivoted to ground.

Figure 13.15 shows a four-bar linkage with a point mass of m_2 at point c on the coupler and with the balance masses m_{b_1} and m_{b_3} attached to links L_1 and L_3, respectively. These balance masses are chosen such that they provide static balance of only the point mass at c (see Sec. 12.5). The masses of links L_1 and L_3 are not shown or considered in the following derivation because, as indicated above, their effects can be considered separately. Also, the coupler mass initially will be assumed to consist of a single point mass. Later, this will be extended to consideration of a coupler consisting of more than one point mass. The total angular momentum about the total center of gravity of the parts shown is equal to the total angular momentum about any axis such as the pivot O_1 because static balancing has eliminated any velocity of the total center of gravity. Then we may write the total angular momentum as the sum of the individual angular momentums of the three masses relative to pivot O_1 as

$$\mathbf{H}_t = \mathbf{H}_c + \mathbf{H}_{b_1} + \mathbf{H}_{b_3} \qquad (13.33)$$

where $\quad \mathbf{H}_c = \mathbf{r}_c \times m_c \dot{\mathbf{r}}_c$

$$\mathbf{H}_{b_1} = \mathbf{r}_{b_1} \times m_{b_1} \dot{\mathbf{r}}_{b_1}$$

$$\mathbf{H}_{b_3} = (\mathbf{L}_4 + \mathbf{r}_{b_3}) \times m_{b_3} \dot{\mathbf{r}}_{b_3}$$

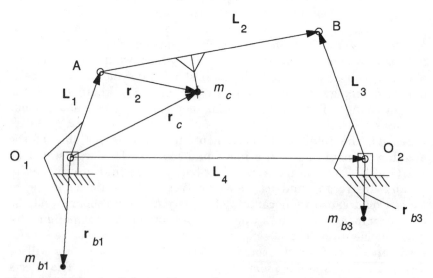

Figure 13.15 A four-bar linkage with a coupler mass consisting of a point mass.

The expression for $\mathbf{H}_c$ can be expanded by noting that

$$\mathbf{r}_c = \mathbf{L}_1 + \mathbf{r}_2 \qquad (13.34)$$

and also that (written in complex variable form)

$$\mathbf{r}_2 = \mathbf{L}_2 \frac{r_2}{L_2} e^{j\phi} \qquad (13.35)$$

Further, the vector loop closure equation for this linkage is $\mathbf{L}_1 + \mathbf{L}_2 - \mathbf{L}_3 - \mathbf{L}_4 = 0$, which can be solved for $\mathbf{L}_2$ to give

$$\mathbf{L}_2 = -\mathbf{L}_1 + \mathbf{L}_3 + \mathbf{L}_4 \qquad (13.36)$$

Substituting Eqs. (13.35) and (13.36) into Eq. (13.34) gives

$$\mathbf{r}_c = \mathbf{L}_1\left(1 - \frac{r_2}{L_2}e^{j\phi}\right) + \mathbf{L}_3\frac{r_2}{L_2}e^{j\phi} + \mathbf{L}_4\frac{r_2}{L_2}e^{j\phi} \qquad (13.37)$$

This equation may be differentiated with respect to time, noting that the link lengths, r_2, and ϕ are constant, to give

$$\dot{\mathbf{r}}_c = \dot{\mathbf{L}}_1\left(1 - \frac{r_2}{L_2}e^{j\phi}\right) + \dot{\mathbf{L}}_3\frac{r_2}{L_2}e^{j\phi} \qquad (13.38)$$

Then, substituting Eqs. (13.37) and (13.38) into the expression for $\mathbf{H}_c$ in Eq. (13.33) gives

$$\mathbf{H}_t = \mathbf{A} + \mathbf{B} + \mathbf{C} + \mathbf{D} + \mathbf{E} + \mathbf{F}$$

where $\quad \mathbf{A} = \mathbf{L}_1\left(1 - \frac{r_2}{L_2}e^{j\phi}\right) \times m_c\dot{\mathbf{L}}_1\left(1 - \frac{r_2}{L_2}e^{j\phi}\right) + \mathbf{r}_{b_1} \times m_{b_1}\dot{\mathbf{r}}_{b_1}$

$\mathbf{B} = \left(\mathbf{L}_4 + \mathbf{L}_3\frac{r_2}{L_2}e^{j\phi}\right) \times m_c\dot{\mathbf{L}}_3\left(\frac{r_2}{L_2}e^{j\phi}\right) + (\mathbf{L}_4 + \mathbf{r}_{b_3}) \times m_{b_3}\dot{\mathbf{r}}_{b_3}$

$\mathbf{C} = \mathbf{L}_1\left(1 - \frac{r_2}{L_2}e^{j\phi}\right) \times m_c\dot{\mathbf{L}}_3\frac{r_2}{L_2}e^{j\phi}$

$\qquad\qquad\qquad\qquad\qquad\qquad\qquad\qquad\qquad (13.39)$

$\mathbf{D} = \mathbf{L}_3\frac{r_2}{L_2}e^{j\phi} \times m_c\dot{\mathbf{L}}_1\left(1 - \frac{r_2}{L_2}e^{j\phi}\right)$

$\mathbf{E} = \mathbf{L}_4\frac{r_2}{L_2}e^{j\phi} \times m_c\dot{\mathbf{L}}_1\left(1 - \frac{r_2}{L_2}e^{j\phi}\right)$

$\mathbf{F} = \left(\mathbf{L}_4\frac{r_2}{L_2}e^{j\phi} - \mathbf{L}_4\right) \times m_c\dot{\mathbf{L}}_3\frac{r_2}{L_2}e^{j\phi}$

Notice that an extra $\mathbf{L}_4$ has been added artificially in the first parentheses in the expression for $\mathbf{B}$ and that an extra $-\mathbf{L}_4$ has been added artificially in the parentheses in the expression for $\mathbf{F}$. It can be seen that the effects of these two additions cancel each other. They were made only for purposes of the following discussion.

As mentioned previously, elimination of shaking torques requires the elimination of variations in the total angular momentum of the linkage. Careful examination of the expression for $\mathbf{A}$ in Eq. (13.39) shows that it represents the moment of momentum of an equivalent point mass m_c at point p_1 in Fig. 13.16 plus the moment of momentum of a mass m_{b_1} in the position shown in that figure and where both those masses are attached to link L_1. The static balancing discussions in Chap. 12 show that if the coupler is statically balanced by masses on the other two moving links, the mass m_{b_1} balances the mass m_c at point p_1, so the center of gravity of the pair is at the pivot O_1. The moment of momentum component $\mathbf{A}$ therefore could be canceled by the use of a compensating rotor. There would then be no variation in the component of the total angular momentum that is represented by term A plus that compensating rotor.

A careful examination of the expression for $\mathbf{B}$ in Eq. (13.39) shows that it represents the moment of momentum of an equivalent point mass m_c at point p_3 in Fig. 13.16 plus the moment of momentum of a mass m_{b_3} in the position shown in that figure and where both of those masses are attached to link L_3. As in the case of the moment of momentum component $\mathbf{A}$, this moment of momentum component $\mathbf{B}$ also can be canceled by a compensating rotor. There would then be no vari-

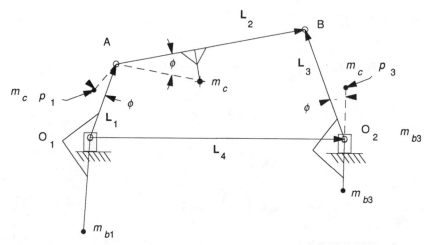

Figure 13.16 A four-bar linkage showing a coupler point mass, its equivalent point masses on the other moving links, and the associated static balancing masses.

ation in this component of the total angular momentum plus the angular momentum of the compensating rotor.

The expressions for $\mathbf{C}$, $\mathbf{D}$, $\mathbf{E}$, and $\mathbf{F}$ are also vector cross-products of momentum arm vectors times translational momentum vectors. However, the velocity involved in each translational momentum is not equal to the rate of change of the moment arm in the expression. For example, in the expression for $\mathbf{C}$, the translational momentum of the mass is given by the mass m_c times the motion of a vector attached to link L_3, but the moment arm is a vector attached to the vector L_1, and vectors $\mathbf{L}_1$ and $\mathbf{L}_3$ do not move in unison. Therefore, the equivalent point mass could not be considered connected to the body represented by the moment arm. Although these terms are canceled by the previously described mirror-image compensating linkages, the situations represented individually by terms $\mathbf{C}$, $\mathbf{D}$, $\mathbf{E}$, and $\mathbf{F}$ are not physically realizable, and variations in these angular momentum components cannot be compensated for with simple combinations of compensating rotors.

However, note that if $r_2 = 0$, all terms except $\mathbf{A}$ and $\mathbf{B}$ vanish. In Figs. 13.15 and 13.16 it can be seen that if $r_2 = 0$, point p_3 would coincide with point O_3, and the balance arm $\mathbf{r}_{b_3}$ would be zero, so $\dot{\mathbf{r}}_{b_3}$ would also be zero and the term $\mathbf{B}$ would vanish. Then $\mathbf{H}$ can be dynamically balanced (canceled) by means of a compensating rotor. This condition would correspond to an equivalent coupler point mass m_c being located at point A in Fig. 13.16. Note also, by similar reasoning, that if $r_2 = L_2$ and $\phi = 0$, all terms except $\mathbf{B}$ vanish, and $\mathbf{H}$ can be dynamically balanced (canceled) by means of a compensating rotor. This condition would correspond to an equivalent coupler point mass m_c being located at point B in Fig. 13.16. Thus, if the coupler mass distribution were to consist of or be equivalent to a point mass placed on the coupler at point A *and* another placed on the coupler at point B, the effects of the varying angular momentum of that coupler could be eliminated by using *two* compensating rotors.

Consider now the inertial properties of such a coupler whose mass distribution is dynamically equivalent to two point masses, one at point A and the other at point B. Figure 13.17a shows a body consisting of two point masses m_A and m_B at points A and B, respectively, connected by a massless rod of length L_2. The center of gravity of this coupler will be located at a distance x_f from point A such that

$$m_A x_f = m_B(L_2 - x_f) \quad \text{so} \quad m_B = \frac{m_A x_f}{L_2 - x_f} \qquad (13.40)$$

The total mass of this coupler is

$$m_2 = m_A + m_B \qquad (13.41)$$

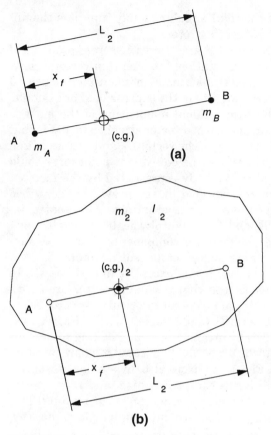

Figure 13.17 A mass distribution that is dynamically equivalent to a point mass at point A plus a second point mass at point B.

The moment of inertia of this coupler about its center of gravity is

$$I_2 = m_A x_f^2 + m_B (L_2 - x_f)^2 \qquad (13.42)$$

From Eqs. (13.40) and (13.41), the total mass m_2 of the coupler is

$$m_2 = m_A \left(\frac{L_2}{L_2 - x_f} \right) \qquad (13.43)$$

From Eqs. (13.40), (13.42), and (13.43) (and some collection of terms),

$$I_2 = m_2 x_f (L_2 - x_f) \qquad (13.44)$$

The radius of gyration of a body is that radius measured from the center of gravity at which all its mass could be considered to be concentrated without changing its moment of inertia. Therefore, if k_2 is the radius of gyration of the coupler,

$$I_2 = m_2 x_f(L_2 - x_f) = m_2 k_2{}^2 \qquad (13.45)$$

from which we conclude that $k_2{}^2 = x_f(L_2 - x_f)$.

Whereas Fig. 13.17a shows a coupler that consists of two point masses, Fig. 13.17b shows a coupler that consists of a distributed mass. Pivot points A and B are located on this distributed mass at a distance L_2 apart, just as similar points are in Fig. 13.17a. The center of gravity of the distributed mass is located on the line from A to B at a distance x_f from point A, just as was the case in Fig. 13.17a. The total mass of the coupler in Fig. 13.17b is the same as that in Fig. 13.17a, i.e., m_2. If the radius of gyration k_2 of the body in Fig. 13.17b satisfies the conditions of Eq. (13.45), then the moment of inertia of this body will equal that of the coupler in Fig. 13.17a. Then, from the standpoint of inertial effects in planar motion and for purposes of static and dynamic balancing, these two coupler configurations are identical because their masses are equal, their centers of gravity are located identically, and their moments of inertia are equal.

From the foregoing it is concluded that if a coupler's mass is distributed such that its center of gravity lies on a straight line connecting its two pivots, and if Eq. (13.45) is satisfied, then that coupler's mass can be considered to be equivalent to two point masses, one at each pivot point. The equivalent point masses at points A and B will be

$$m_A = m_2 \left(\frac{L_2 - x_f}{L_2} \right)$$

and

$$m_B = m_2 \left(\frac{x_f}{L_2} \right)$$

Then the linkage can be dynamically balanced by using one or two compensating rotors as in Procedure 13.3.

13.7 Input Torque Smoothing

In the discussion of compensating rotors in Sec. 13.3 it was shown that such a compensating rotor can be used to cancel the shaking torque produced by a pure rotor that is subjected to angular acceleration. That is, by making the ratio of the moment of inertia of the compensating rotor to the moment of inertia of the original rotor equal to the gear ratio r_2/r_1, the shaking torque T_{shake} in Eq. (13.11) is reduced to zero. However, it should be noted that when this is done, the *driving* motor torque given by Eq. (13.10) becomes $T_M = (1 + r_1/r_2)I_1\alpha_1$, which is $(1 + r_1/r_2)$ times as great as that required without the compensating rotor.

The shaking forces and torques produced in mechanisms such as four-bar linkages are more complicated than those produced by simple rotors. However, in Secs. 12.5 and 13.6 it was shown that the static and dynamic balancing of four-bar linkages can be reduced to the equivalent of the balancing of rotors. Also, in most cases of interest, the driving actuator (such as a motor) is mounted to the same base as is the mechanism, just as was the case in the discussion in Sec. 13.3. As a consequence, even though a mechanism has been balanced so that its shaking forces and torques have been reduced to zero, the torque (or force) required to drive that mechanism will not be reduced. Indeed, because balancing masses in addition to the original mechanism masses must be accelerated, the fluctuations in the required drive torques will be increased—often by a large amount. In many cases these fluctuations are excessive even without balancing.

Although the fluctuations in the torque required to drive a mechanism do not increase the average power that must be supplied to drive it, these fluctuations can tax the ability of a control system to provide the desired drive characteristics. Such characteristics can include features such as a desired constant velocity and/or desired synchronism with other mechanisms or with other functions of a machine. In addition, such fluctuations can appreciably increase the electrical energy losses in electric drive motors, thereby causing excessive heating and/or a requirement for a larger motor.

The power input to a mechanism by a rotary actuator such as an electric motor is equal to the input torque times the angular velocity of that motor. In the mechanism, that power is absorbed (1) by friction that converts mechanical energy to heat, (2) by the lifting of weights, (3) by the increasing of kinetic energy of the mechanism masses, and (4) by the output work (such as cutting, bending, printing, etc.) done by the mechanism. Friction can be minimized by the use of good bearings. The lifting of weights can be balanced by static balancing. The varying kinetic energy in a mechanism will absorb input power when such kinetic energy is increased and will feed power back to the drive when that kinetic energy decreases. The output work is a necessary productive use of the input power.

The kinetic energy possessed by an individual part of the mechanism is equal to the mass of that part times the square of the translational velocity of the part plus the moment of inertia of that part times the square of the angular velocity of the part. Of course, *the first consideration in minimizing the drive torque or force fluctuations is to minimize the mass of any mechanism parts that experience appreciable variations in their velocities.* Because the squares of the velocities of a part determine the kinetic energy, the directions of those velocities are not important with regard to input torque considerations—only their magnitudes are important. For this reason, although the angular velocity of a compensating rotor is opposite to that of the part it is compensating, and therefore the motion of the compensating rotor cancels the angular momentum of the compensated part, the kinetic energy of the compensating rotor actually *adds* to the kinetic energy of the original part.

If the magnitude of the angular and/or translational velocity of one part can be made to decrease while that of those of another part are increasing, the decrease in kinetic energy in the first part will tend to cancel all or part of the increase in kinetic energy in the second part. Such a technique can be used to smooth out fluctuations in the input drive torque by tending to keep the total kinetic energy in the mechanism constant. In Example 3.5 the suggestion was made that if a second slider of mass equal to the mass of the first were added to the system, and if that second slider were to be driven with such timing that its motion were 90° out of phase with that of the first, the fluctuations in the input torque could be nearly canceled. It can be seen in Fig. 3.21 that such would be the case. Readers should examine the way in which the velocities of these two sliders would vary and note that as the kinetic energy of one increases, the kinetic energy of the other would decrease by an equal amount. The effects of the piston masses in a V-configuration crank-and-piston compressor or pump are the same as in this foregoing example. That is, in such a compressor or pump, the input torque fluctuations due to inertial effects are canceled. The work done in compressing or pumping the fluid, however, is another matter.

As noted in Example 3.5, the closer the motion of the slider is to being sinusoidal, the more effectively the torque can be smoothed. That is, *sinusoidally varying velocities produce kinetic energy variations that are most easily canceled.* Conversely, motions that differ greatly from sinusoidal can produce input drive force or torque variations that are difficult to minimize or eliminate. Examples of such difficult-to-smooth motions are those produced by cam-driven mechanisms and linkages having complicated motions.

Another method for producing a smooth, relatively unvarying required input torque is to *design the driven mechanism in such a manner that its total kinetic energy inherently remains constant during operation.* This approach is best applied to the concept stages of the design, and Examples 13.4 and 13.5 illustrate mechanisms in which the total kinetic energy is constant. In Example 13.4 notice that all rotating parts rotate at constant angular velocity, so their rotational kinetic energies are constant (see Fig. 13.10). The centers of gravity of the nonrotating parts travel at constant velocities along circular paths, so their translational kinetic energies are constant. In Example 13.5 the kinetic energy of the saw is the same as though its mass were concentrated at point P_2 on gear G_1 (see Fig. 13.11). Considering it to be concentrated in this manner, it is seen that all parts are rotating at constant angular velocity. It is also seen that because the total center of gravity of parts G_1, W_1, and W_s is located at point P_1, their total center of gravity travels at constant translational velocity along a circular path about point S. The total kinetic energy of this entire mechanism is constant during operation. The only fluctuations that will remain in the required drive torque will be those due to the variations in the sawing force produced by the cutting action of the saw. These necessary saw force variations will be extremely difficult to compensate for.

Index

ABOUT THE AUTHOR

Homer D. Eckhardt, P.E., received his B.S. and M.S. from MIT. He is a consulting engineer specializing in the design, analysis, synthesis, and debugging of automatic manufacturing machinery. He has more than 40 years of experience applying kinematics to the design of dynamic systems at such companies as Polaroid, Rockwell International, and R.C.A. Aerospace Systems. He has also taught courses on kinematics and machine design at Tufts University and the Worcester Polytechnic Institute. Mr. Eckhardt resides in Lincoln, Massachusetts.